LATTICE GAS METHODS:

THEORY, APPLICATIONS, AND HARDWARE

edited by

Gary D. DOOLEN

A Bradford Book
The MIT Press
Cambridge, Massachusetts
London, England

First MIT Press edition, 1991

Reprinted from *Physica D,* Volume 47, Numbers 1–2, 1991. The MIT Press has exclusive license to sell this English-language book edition throughout the world.

Printed and bound in the Netherlands.

Library of Congress Cataloging-in-Publication Data

Lattice gas methods : theory, applications, and hardware / edited by Gary D. Doolen. — 1st MIT Press ed.
p. cm. — (Special issue of physica D)
Papers presented at a NATO advanced research workshop held in Los Alamos, N.M. on Sept. 6–9, 1989.
"A Bradford book."
"Reprinted from Physica D, volume 47, numbers 1–2, 1991"—T.p. verso.
Includes bibliographical references and index.
ISBN 0-262-54063-0
1. Differential equations, Partial—Congresses. 2. Lattice gas—Congresses. I. Doolen, Gary D. II. Series.
QA374.L343 1991
515'.353—dc20 91-10010
CIP

PREFACE

Since 1986, the reality of solving partial differential equations using logical, lattice gas methods has been demonstrated for a wide variety of equations, including hydrodynamic, magnetohydrodynamic, wave, Burgers, Poisson, radiation transport, and reaction–diffusion equations. Several computer designers have created hardware which exploits the fact that lattice gas methods are completely parallel, producing significant gains in speed. The possibility now exists to create a Cray-size computer which implements lattice gas algorithms at speeds of many millions of times faster than existing computers. To study the many options for hardware and for algorithms, a NATO Advanced Research Workshop was held in Los Alamos, New Mexico on September 6–9, 1989. The title of this workshop was "Lattice Gas Methods for Partial Differential Equations: Theory, Applications, and Hardware."

The workshop was mainly sponsored by the NATO Special Programme Panel on Chaos, Order and Patterns: Aspects of Nonlinearity. Additional support was provided by DARPA grant DPP8850 and by the Earth and Environmental Sciences Division and the Center for Nonlinear Studies at Los Alamos National Laboratory. Conference management was skillfully and graciously provided by Marian Martinez of the CNLS.

Gary Doolen,
Center for Nonlinear Studies
Los Alamos, NM 87545 USA

Physica D 47 (1991) viii–ix
North-Holland

CONTENTS

CHAPTER 1

OVERVIEWS AND FRONTIERS

Physica D 47 (1991) 3–8
North-Holland

STATISTICAL MECHANICS AND HYDRODYNAMICS OF LATTICE GAS AUTOMATA: AN OVERVIEW

Jean Pierre BOON
Physique Non-Linéaire et Mécanique Statistique, Université Libre de Bruxelles, Campus Plaine, C.P. 231, Bruxelles, Belgium

Received 10 November 1989

Some of the issues raised by recent work on lattice gas automata are reviewed.

1. Microscopic simulation of fluid dynamics

When considering the microscopic simulation of fluid dynamical phenomena, one is most logically led to construct a molecular model of the system to be simulated. Such a description would involve the molecular characteristics of the fluid, that is the specification of an interaction potential. This is the basis of the *molecular dynamics* approach, which over more than thirty years has produced extremely valuable results and previously unavailable information mostly for equilibrium and transport properties [1]. In recent years the method has been extended to the investigation of non-equilibrium phenomena; with the restriction to *microhydrodynamical* simulations imposed by the limited number of particles used in the computation (10^2 to 10^5) [2], these results have set lower limits to the validity of the description of fluid dynamics in terms of continuous media. The difficulty raised by the simulation of macroscopic (large scale) phenomena stems from the ratio of the hydrodynamic scales to the molecular length and time scales set as the basic quantities in the computations. As a most typical example consider the Reynolds number for a system of N particles enclosed in a box of size L,

$$\mathrm{Re} = L_{\mathrm{H}} V \nu^{-1} \tag{1}$$

V is the flow velocity, ν the kinematic viscosity, and L_{H} the characteristic hydrodynamic length: $L_{\mathrm{H}} = \alpha L$, where α is a constant ($\alpha \le 1$) and $L \sim l N^{1/D}$ with l the mean free path and D the space-dimension; $\nu = \beta l v$, where β is a constant of order one and v is the root-mean-squared velocity which is roughly the sound speed c_{s}. Hence

$$\mathrm{Re} = \mathrm{const.}\ N^{1/D} M, \tag{2}$$

where $M = V c_{\mathrm{s}}^{-1}$ is the Mach number and the constant is $\mathscr{O}(1)$. So for a system with 10^6 particles, maximum Reynolds number values would not exceed $\sim 10^3$ for $D = 2$ and $\sim 10^2$ for $D = 3$, at $M \le 1$. This evaluation holds for an incompressible fluid in the low-density limit. Thus a crucial issue in microscopic simulations of hydrodynamic flows is the number of particles that can be implemented.

In the last five years, *lattice gas automata* (LGA) methods have been developed which provide an alternative to molecular dynamics for microscopic simulations of macroscopic fluid dynamical phenomena [3]. The lattice gas is constructed as an oversimplified fictitious microworld operated as a cellular automaton. A physical picture could be provided by viewing the lattice gas as a collection of zero-dimensional particles confined to one-

Table 1
Simulation of the physics of fluid systems with cellular automata

Cellular automata		Physics of fluid systems
propagation: single velocity and multispeed models	⟶	lattice gas thermodynamics
collision rules and conservation laws	⟶	dynamical behavior and transport properties
particle properties: mass, species (color), charge	⟶	multispecies, multiphase and reactive systems
lattice geometry and symmetries	⟶	2D and 3D hydrodynamics

dimensional displacements on the links of a lattice embedded in a 1-, 2-, 3-, 4-D space, and undergoing collisions when they meet at a lattice node. Operational efficiency of the lattice gas automaton follows from (i) space, time, and velocity discretization, (ii) exclusion principle, (iii) displacement synchronization, and (iv) simplicity of collision rules. Updating the automaton universe is performed by two-step sequences: propagation and collision. Collision rules are set according to conservation laws (particle number, momentum, and energy) and conserved quantities are related to the lattice symmetries. In this sense, the lattice gas is best defined as an *automaton universe with symmetry group and updating rules yielding an appropriate set of conserved quantities*. Thus the lattice gas automaton should be viewed as a model system *per se*, and neither as an oversimplified version of molecular dynamics nor as a mere cellular automaton.

Considering the basic properties defining the lattice gas leads quite logically to some of the important issues that arise in simulating the physics of fluid systems with cellular automata; see table 1. Obviously neither are these properties nor these topics unrelated. They enter the broader framework of statistical mechanics and fluid dynamics of lattice gases at essentially three levels as schematically shown in table 2. Some of these issues are reviewed in the present introductory article and, along with further topics, are discussed in detail in the present volume.

Table 2

Microdynamics ⟶	Mesodynamics ⟶	Macrodynamics
microdynamical equations	lattice Boltzmann equation	LG Navier–Stokes equation

microdynamical equations ⟶ CA and LB simulations
lattice Boltzmann equation ⟶ CA and LB simulations
CA and LB simulations ⟶ statistical mechanics
CA and LB simulations ⟶ lattice Boltzmann equation ⟶ statistical mechanics
CA and LB simulations ⟶ fluid dynamics
LG Navier–Stokes equation ⟶ fluid dynamics

statistical mechanics

H-theorem
correlation functions
thermodynamic properties (e.g. c_s)
transport properties (η, D_s)
Boltzmann and
non-Boltzmann effects, . . .

fluid dynamics

athermal/compressible fluids
high Re: 2D, 3D turbulence
low Re: boundary layers
multispecies systems
multiphase fluids
reactive systems
3D complex flows
(viscous fingering, capillarity, . . .)

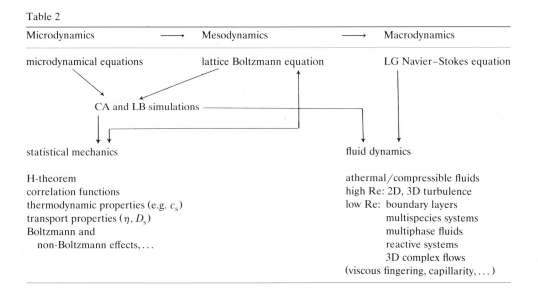

2. Statistical mechanical aspects

The long-time behavior of the Green–Kubo integrands defining the transport properties [4] and the implications of their non-exponential decay have been an important issue in non-equilibrium statistical mechanics [6]. Indeed the persistence of long-time correlations is in contradiction of the Boltzmann–Enskog theory, and the existence of vortex back-flow at the molecular scale appears as a manifestation of the violation of the molecular chaos assumption. Stimulated about twenty years ago largely by molecular dynamics simulation results for the velocity autocorrelation function [6], the now conventionally called *long-time-tail* problem met a recent regain of interest through the computational capabilities offered by LGA methods for simulating systems with large number of particles. One method is to compute the time-integrated autocorrelation function, i.e. to measure the corresponding transport coefficient. For the momentum current autocorrelation function, this technique amounts to perform a "LGA viscometer experiment" [7]. The kinematic viscosity ν was so measured by 2D Poiseuille channel flow with an FHP gas as a function of the size L of the system for various densities ($d = 0.2$–0.5); the simulations can be viewed as the analog of a set of scattering experiments performed at different wavenumbers. The results show the long wavelength divergence of the 2D transport coefficient in the linear form: $\nu = F(\ln L)$ with $\mathrm{d}\nu/\mathrm{d}(\ln L) = f(d)$.

The other method is by direct measurement. For the velocity autocorrelation function, one proceeds by tracking a tagged particle in the lattice gas. A particular feature of the FHP-IV model (using maximum colored momentum transfer in the collision rules [8]) is that it produces at moderate densities ($d \sim 0.3$) a negative part in the velocity autocorrelation function, i.e. momentum re-correlation typical of a cage effect at sufficiently high densities [9]. The long-time-tail effect was detected by the presence of a zero-frequency centered peak in the spectral density corresponding to the long-time power law divergence in the velocity autocorrelation function ($\sim t^{-1}$, in 2D) [10]. The t^{-1} behavior was also clearly evidenced in the FHP-III and FCHC gases (with post collisional random redistribution [8]) where highly accurate averaging procedure allowed quantitative analysis with mode coupling theory [11].

Kinetic theory aspects are discussed in a companion overview paper [12].

3. The Reynolds number

Turbulence remains one of the last poorly understood problems of classical physics. In particular one of the basic questions is the 2D to 3D symmetry breaking transition in open flows. This transition and the ensuing evolution towards turbulence generally take place at rather high Reynolds numbers. Consequently, if LGA methods are to be useful in the study of turbulent flows, high Reynolds number values must be reached, which turns out to be a highly nontrivial problem from the point of view of practical implementation. The LGA Reynolds number is given by [13]

$$\mathrm{Re} = L_{\mathrm{H}} V/\nu_{\mathrm{R}}, \quad \nu_{\mathrm{R}} = \nu(d)/g(d), \tag{3}$$

where the kinematic viscosity is rescaled with the Galilean correction factor, which arises because of the discrete nature of the lattice gas. By rewriting the Reynolds number as

$$\mathrm{Re} = M L_{\mathrm{H}} R_*(d), \quad R_* = c_{\mathrm{s}}/\nu_{\mathrm{R}}(d)$$

it is clearly of interest for flow simulations to operate under conditions that maximize the Reynolds coefficient R_*, i.e. to set up collision tables which minimize ν and maximize $g(d)$. Considerable effort is being invested in such a program [14, 15] and the temporary status of the art is schematically shown in table 3.

Table 3

LGA model	R_*^{max} theory	R_*^{max} simulation
FHP-IV	2.65	
FHP + 3 rest particles	4.56	
FCHC	7.57	
FCHC + 3 rest particles	17.22	7.13
with semi-detailed balance violation	40 (99)	13.5

The figures for R_*^{max} are obtained within the Boltzmann approximation. The values 7.13 and 13.5 are the so far best implemented values. For instance, the remarkable simulations demonstrating the tridimensionalization of the wake behind a flat disk [16] were performed at $R_* = 7.11$ yielding a maximum Re value of 191. It should be stressed that contact between LGA simulations and laboratory experiments has been very scarce and quantitative comparison should be extremely valuable not only for the validation of the LGA method but also for an analytical understanding of the transitional regimes in the evolution towards turbulence. In this respect recent work investigating quantitatively open flow turbulence in connection with dynamical systems theory [17, 18] should be profitably considered as useful guidelines for LGA simulations at moderate [19] and high Reynolds numbers.

4. Multispecies / multiphase systems

The "colored automata" technique, which assigns a species tag (a color) to the particles, has proved useful for a variety of problems, mostly 2D low-Re complex flows. There are two alternatives for colored LGA, where by adding an extra bit to specify the site state (site denotes link and center at a lattice node) one has [9, 20–22]:

00	empty site,	white hole
01	red particle (species A),	white particle
10	blue particle (species B),	black particle
11	–or new species particle,	black hole

The simplest application concerns binary systems with mutual diffusion. In this case, since particles are identical except for their color specification, such systems are well suited for measuring the diffusion coefficient D_s [19]. Now by optimizing colored momentum transfer and color grouping in the collision rules (FHP-IV model), D_s can be minimized and the model can be used for color tracing and for flow visualization when the characteristic hydrodynamic time is short compared to the diffusion time. When the two components (A and B) are implemented with specific collision rules so as to exhibit different physical properties (e.g. d and/or ν) interesting dynamical phenomena can be investigated such as the Rayleigh–Taylor instability [20], the Saffman–Taylor instability and viscous fingering in porous media [23].

Immiscible fluids are realized by defining local color field and local color flux which requires non-local (i.e. nearest neighbor) information and correspondingly designing collision rules which minimize flux against field [21]; this is a procedure which mimics "attractive forces" between like color particles and "repulsive forces" between particles with different colors. This technique was used in simulations showing that an initially homogeneous state destabilizes to produce phase separation and that an initial bilayer structure, with the upper layer denser than the lower one, reverses configuration via a Rayleigh–Taylor instability [24]. Capillary flow of immiscible fluids in an idealized porous medium was demonstrated by the same technique [21]. Interfaces have also been modeled by a connected set of nodes forming a "boundary" that deforms according to the direction of net momentum across the boundary. Surface tension can so be varied by tuning the threshold of the momentum balance required to modify the link connecting set [25]. A recent model for LGA surface tension [22] proceeds, as mentioned above, by color assignment not only to the particle but also to the holes (empty sites); the virtue of the model is that no non-local information is required. The color of the hole is set by majority rule of the post-collisional state. Thus the vectors of the next pre-collisional states contain information about color distribution and surface tension is achieved by selecting those collisions which scatter particles predominantly along the direction of the momentum vector corresponding to their color. Simulations have shown the efficiency of the model for capillary pressure measurements in a wetting channel [22]. In this respect it should be stressed that while LGA have now proved to perform quite efficiently for complex flow simulations, effort should be invested into quantitative measurements and subsequently to comparative analysis with experimental data.

Reactive systems constitute a particular class of multispecies and multiphase systems. Reactive recombination has been used for species separation to simulate the Kelvin–Helmholtz instability [20] and the coupling of chemistry with hydrodynamics in the diffusion flame [26]. Recently a more systematic theoretical approach has been developed as an automaton model for reaction–diffusion systems [27]; the LGA simulations have shown that the model exhibits nucleation and pattern formation in agreement with the corresponding phenomenological PDEs. As a final comment it should be emphasized that, in such reactive systems as well as in hydrodynamics, there exists a class of LGA capable of matching the macroscopic description given by a set of PDEs; further explorations along this paradigm should provide a useful classification of the minimal basics required not only to produce the correct phenomenology but also to obtain a deeper understanding of their fundamental significance.

References

[1] J.P. Boon and S. Yip, Molecular Hydrodynamics (McGraw-Hill, New York, 1980).

[2] M. Mareschal, ed., Microscopic Simulations of Complex Flows (Plenum Press, New York, 1990).

[3] G.D. Doolen, ed., Lattice Gas Methods for Partial Differential Equations (Addison-Wesley, Reading, MA, 1990).

[4] M.H. Ernst, Mode coupling theory and tails in CA fluids, Physica D 47 (1991) 198–211, these Proceedings.

[5] Y. Pomeau and P. Résibois, Phys. Rep. 19 (1975) 63.

[6] B.J. Alder and T.E. Wainright, Phys. Rev. A 1 (1970) 18.

[7] L.P. Kadanoff, G.R. McNamara and G. Zanetti, Phys. Rev. A 40 (1989) 4527–4541.

[8] D. d'Humières, P. Lallemand, J.P. Boon, D. Dab and A. Noullez, in: Chaos and Complexity, eds. R. Livi, S. Ruffo, S. Ciliberto and M. Buiatti (World Scientific, Singapore, 1988) pp. 278–301.

[9] J.P. Boon and A. Noullez, in: Discrete Kinetic Theory, Lattice Gas Dynamics, and Foundations of Hydrodynamics, ed. R. Monaco (World Scientific, Singapore, 1989) pp. 399–407.

[10] A. Noullez and J.P. Boon, Long time correlations in a 2D lattice gas, Physica D 47 (1991) 212–215, these Proceedings.

[11] D. Frenkel and M.H. Ernst, Phys. Rev. Lett. 63 (1989) 2165;
M.A. van der Hoef and D. Frenkel, Tagged particle diffusion in 3D lattice gas cellular automata, Physica D 47 (1991) 191–197, these Proceedings.

[12] X.P. Kong and E.G.D. Cohen, A kinetic theorist's look at lattice gas cellular automata, Physica D 47 (1991) 9–18, these Proceedings.

[13] U. Frisch, D. d'Humières, B. Hasslacher, P. Lallemand, Y. Pomeau and J.P. Rivet, Complex Systems 1 (1987) 648.

[14] B. Dubrulle, U. Frisch, M. Hénon and J.P. Rivet, Low-viscosity lattice gases, Physica D 47 (1991) 27–29, these Proceedings;

[15] J.A. Somers and P.C. Rem, in: Cellular Automata and Modeling of Complex Physical Systems, eds. P. Manneville, N. Boccara, G.Y. Vichniac and R. Bidaux (Springer, Berlin, 1989) pp. 161–177.

[16] J.P. Rivet, Compt. Rend. Acad. Sci. Paris II 305 (1987) 751.

[17] M. Bonetti and J.P. Boon, Phys. Rev. A 40 (1989) 3322.

[18] W.I. Goldburg, P. Tong and H.K. Pak, Physica D 38 (1989) 128.

[19] J.P. Rivet, M. Hénon, U. Frisch and D. d'Humières, Europhys. Lett. 7 (1988) 231.

[20] P. Clavin, P. Lallemand, Y. Pomeau and G. Searby, J. Fluid Mech. 188 (1988) 437.

[21] R. Santos, D. Rothman and B. Boghosian, Lattice gas studies of immiscible two-phase flow in inhomogeneously wet 2D porous media, presented at the Workshop on Lattice Gas Methods for PDE's, Los Alamos, September 1989.

[22] J.A. Somers and P.C. Rem, Analysis of surface tension in two-phase lattice gases, Physica D 47 (1991) 39–46, these Proceedings.

[23] M. Bonetti, A. Noullez and J.P. Boon, in: Discrete Kinetic Theory, Lattice Gas Dynamics, and Foundations of Hydrodynamics, ed. R. Monaco (World Scientific, Singapore, 1989) pp. 394–398.

[24] A. Gunstensen and D. Rothman, A lattice-gas model for three immiscible fluids, Physica D 47 (1991) 47–52, these Proceedings;
A. Gunstensen and D. Rothman, A Galilean-invariant immiscible lattice gas, Physica D 47 (1991) 53–63, these Proceedings.
[25] F. Hayot, Viscous fingering in a lattice gas, Physica D 47 (1991) 64–71, these Proceedings.
[26] V. Zehnlé and G. Searby, J. Phys. (Paris) 50 (1989) 1083.
[27] A. Lawniczak, D. Dab, R. Kapral and J.P. Boon, Reactive lattice gas automata, Physica D 47 (1991) 132–158, these Proceedings.

Physica D 47 (1991) 9–18
North-Holland

A KINETIC THEORIST'S LOOK AT LATTICE GAS CELLULAR AUTOMATA

X.P. KONG and E.G.D. COHEN
The Rockefeller University, New York, NY 10021, USA

Received 7 December 1989

The diffusion behavior of a number of Lorentz lattice gases is studied in its dependence on collision rules and its similarity to corresponding Lorentz gases that are continuous in space, but have the same discrete velocity space. The difference in behavior resulting from probabilistic or deterministic collision rules is discussed.

1. Introduction

In this paper, we want to discuss aspects of two basic questions that involve all lattice gas cellular automata: (i) What is the relation between the behavior of a continuum fluid that the lattice gas is supposed to mimic and a lattice gas? (ii) What is the influence of the collision rules that determine the motion of the particles of the lattice gas and the macroscopic behavior of the lattice gas?

In order to focus on these two questions in the simplest context, we will discuss the Lorentz [1] lattice gas, rather than the general lattice gas. In the Lorentz lattice gas one particle moves – or many particles move independently of each other – through randomly placed scatterers on a lattice. In this paper, we will restrict ourselves to a square lattice. Before we discuss this lattice gas, we will first discuss the continuum case, which the lattice gas under consideration represents: the Ehrenfest [2] wind-tree model (cf. fig. 1b). The physical process studied in Lorentz (lattice) gases is that of particle diffusion through stationary scatterers rather than that of a flow of mutually interacting particles.

We briefly summarize the main formulae for a normal diffusion process. If $P(\boldsymbol{r},t)$ is the probability density to find the diffusing particle at the position $\boldsymbol{r}$ at t, when it started at the origin at $t=0$, then $P(\boldsymbol{r},t)$ satisfies the diffusion equation:

$$\frac{\partial P(\boldsymbol{r},t)}{\partial t} = D\,\nabla^2 P(\boldsymbol{r},t), \tag{1}$$

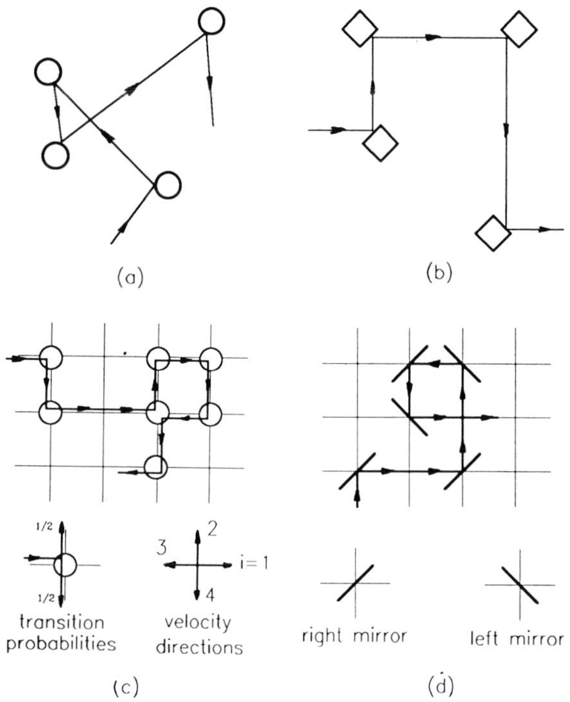

Fig. 1. Particle trajectories in various Lorentz models: (a) circular Lorentz model; (b) Ehrenfest wind-tree model; (c) probabilistic square lattice Lorentz model with transition probability and velocity directions shown (for clarity, the trajectory of the particle has been drawn slightly away from the lattice); (d) mirror model with both kinds of mirrors shown.

with, in two dimensions, the Gaussian solution:

$$P(\boldsymbol{r},t) = \frac{1}{4\pi Dt}\exp\left(-\frac{r^2}{4Dt}\right), \tag{2}$$

0167-2789/91/$03.50 © 1991 – Elsevier Science Publishers B.V. (North-Holland)

where D is the diffusion coefficient. The mean square displacement of the particle is then

$$\Delta(t) \equiv \langle \boldsymbol{r}^2(t) \rangle = \int \mathrm{d}\boldsymbol{r}\, \boldsymbol{r}^2 P(\boldsymbol{r},t) = 4Dt, \tag{3}$$

or

$$\langle x^2(t) \rangle = 2Dt, \tag{4}$$

where the average is over an appropriate ensemble of scatterers. For a Gaussian distribution, all odd moments and the cumulants, which can all be expressed in terms of the second moment, vanish. In particular, the kurtosis K, given by

$$K = \frac{\langle x^4 \rangle - 3\langle x^2 \rangle^2}{\langle x^2 \rangle^2}, \tag{5}$$

vanishes. Thus for a Gaussian diffusion process (class I) the only quantity to compute is the diffusion coefficient in its dependence on the density n of the scatterers, i.e., the number of scatterers per unit area: $D = D(n)$. This is a problem of kinetic theory.

2. The Ehrenfest wind-tree model

In the wind-tree model, a (wind) particle moves through and is elastically scattered by an array of randomly placed diamonds (trees) with corresponding diagonals parallel to each other. If a particle starts out in one of four velocity directions along the $\pm x$ axis or the $\pm y$ axis, it maintains these four discrete velocity directions for all times. In this model, the speed of a particle $v = |v_1| = |v_2| = |v_3| = |v_4|$ does not change and the only relevant conserved quantity is the particle number (or mass), since the momentum is not conserved and the particle energy is trivially connected with the particle mass.

Ehrenfest introduced this model to elucidate the nature of the Stoszzahl Ansatz or the assumption of molecular chaos in the derivation of the Boltzmann equation. Later, Hauge and one of us [3] used the model to clarify the nature of the divergences that occur in a density expansion of the transport coefficients of a moderately dense gas. Two cases were studied: the trees are "hard" for each other, so that they cannot overlap (non-overlapping (NOV) case) or the trees are "soft" for each other and can overlap (overlapping (OV) case). The results, as far as they are relevant for this paper, are the following:

NOV-case, normal diffusion (class I): The mean square displacement of the wind particle, $\Delta(t)$, behaves for long times $t \gg t_{\mathrm{mfp}}$, the mean free time between successive collisions, which is $1/2avn$ for the wind-tree model, as

$$\Delta(t) = 4Dt, \tag{6}$$

where the diffusion coefficient D is given by

$$D^{\mathrm{wt}} = D_{\mathrm{B}}^{\mathrm{wt}} + \mathcal{O}(n^0), \tag{7}$$

with the Boltzmann (low density) approximation, $D_{\mathrm{B}}^{\mathrm{wt}}$, to the diffusion coefficient D for both NOV and OV trees given by

$$D_{\mathrm{B}}^{\mathrm{wt}} = \frac{v}{4na} \xrightarrow{a=v=1} \frac{1}{4n}. \tag{8}$$

Here a is half of the length of the diagonal of a tree, which we will set equal to one in the following. The computation of the correction to D_{B}, indicated in (7), was the main purpose of ref. [3]. The contributions to this correction term due to ring, orbiting and retracing events are sketched in fig. 2 (for more precise definitions of those events, see ref. [3]).

OV-case: Here one has to distinguish two subcases:

(A) Abnormal diffusion (class III) for $n < n_c$ [3,4]:

$$\Delta(t) \sim t^{1-4n/3+\mathcal{O}(n^2)}, \tag{9}$$

so that $\Delta(t)$ grows slower than t. This is due to the large contribution of the retracing events that prevent the wind particle from spreading sufficiently fast ($\sim t$) to obtain normal diffusion [3].

(B) No diffusion (class IV) for $n > n_c$:

$$\Delta(t) < \text{constant}. \tag{10}$$

In this case a particle finds itself always trapped in a cage of overlapping trees (cf. fig. 2e). We have used n_c for the critical density [5,6], where the chance of trapping becomes finite. We should point out that in the Boltzmann approximation the diffusion coefficient for this case is again given

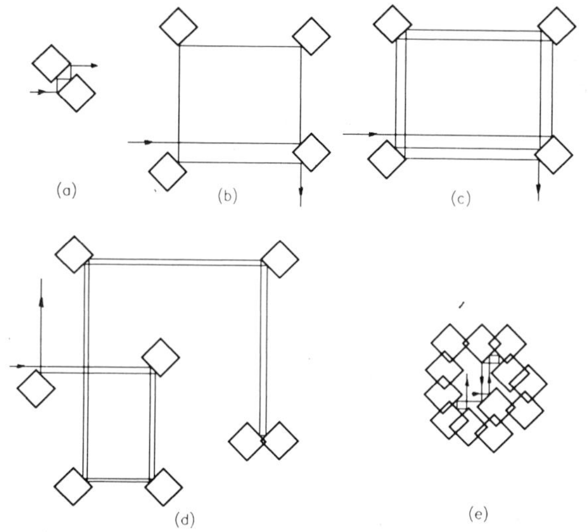

Fig. 2. Four kinds of events which contribute to the corrections to the Boltzmann approximation of the diffusion coefficient and a trapping configuration (cage) in the Ehrenfest wind-tree model: (a) repeated collisions between two nearby scatterers; (b) ring events; (c) orbiting events; (d) retracing events; (e) cage in the OV case.

by (8), since in this low-density approximation higher density effects, like tree cages as in fig. 2e, that involve four scatterers at least, are not taken into account.

Wood and Lado [7] have confirmed the normal diffusion behavior of the NOV wind-tree model and the abnormal diffusion behavior of the OV wind-tree model (cf. fig. 3) for $n < n_c$. The slope of $\Delta(t)/t$ agrees well for low density with that given by (9). Preliminary calculations by Wood [8] also confirmed the absence of diffusion at high tree concentration at a critical density of trees $n_c \approx 1$.

3. Lorentz lattice gases

The Lorentz lattice gases we consider here are all obtained by restricting the motion of the wind particle to the bonds of a square lattice. The particle moves in discrete unit time steps from one site to a neighboring site in the direction of its velocity. A number of these lattice sites have been randomly occupied by stationary scatterers, with which the particle collides according to certain collision rules. We note that the only difference between the Ehrenfest wind-tree model and the

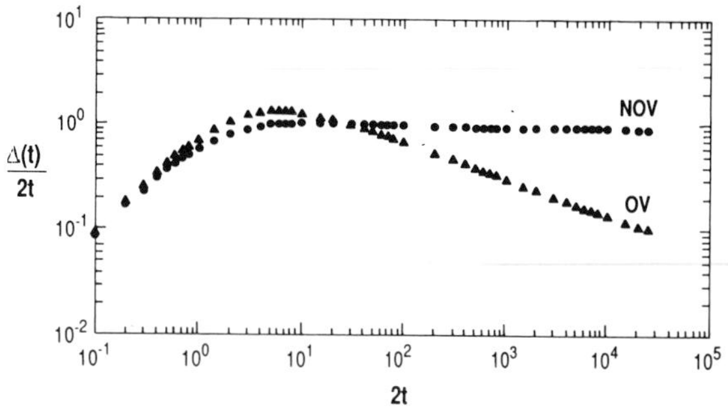

Fig. 3. Normal and abnormal diffusion in the NOV and OV Ehrenfest wind-tree model respectively, with 8192 scatterers (from ref. [7]).

Lorentz lattice gas is that space is continuous in the first case while discrete in the second case; the velocities have the same four discrete values in both cases. Thus we can study the influence of space discretization alone on the behavior of the system, while usually, space and velocities are discretized together.

As to the choice of the scattering rules that correspond to the continuum model, one could consider a probabilistic model (fig. 1c) or a deterministic model (fig. 1d). In the probabilistic model, there is a probability 1/2 that the particle scatters to the right or the left of its original direction. In a deterministic model, a scatterer reflects the particle, as if it was a right or left mirror (cf. fig. 1d), so that choosing an equal number of right and left scatterers would assure on the average an equal number of right and left deflections of the wind particle.

Probabilistic Lorentz gas models have been considered extensively by Binder, Ernst and van Velzen [9–12] and the deterministic mirror model was introduced by Ruijgrok and Cohen [13]. In the next section we will consider the probabilistic model.

4. Probabilistic model

If the lattice distance is denoted by a, one finds empirically that for all concentrations of scatterers the diffusion appears to be normal and the diffusion coefficient to be given to a very high degree of approximation by its low concentration (Boltzmann) value:

$$D^{\rm lg}(c) \approx D_{\rm B}^{\rm lg}(c)$$

$$= \frac{av}{2c} - \frac{av}{4} \xrightarrow{a=v=1} \frac{1}{2c} - \frac{1}{4}. \tag{11}$$

Here $c = N/M$ is the fraction of the M lattice sites occupied by the N scatterers. The term $-1/4$ on the right-hand side of (11) is due to the discreteness of the lattice [14]. Taking the continuum limit in (11), one gets

$$D^{\rm lg}(c) \xrightarrow{a\to 0,\, na\to {\rm const.}} D_{\rm B}^{\rm wt}(c), \tag{12}$$

where $D_{\rm B}^{\rm wt}$ is given by (8). In obtaining (8) one has to remember that the $2a$ in (8) corresponds to a in (12), being the relevant cross sections for the wind-tree and the lattice gas, respectively.

We remark that while $D_{\rm B}^{\rm lg}$ appears to be valid for all c [11,12], $D_{\rm B}^{\rm wt}$ is only applicable at low densities. Using the correction term of $\mathcal{O}(1)$ on the right-hand side of (7) as given by Hauge and Cohen [3] for the NOV case, one finds that $D^{\rm lg}(c)$ can only be mapped onto $D_{\rm B}^{\rm wt}(c)$ for $c < 0.1$.

5. Mirror model

In the mirror model, one introduces three types of occupation variables (cf. fig. 1d):

$n_i(\boldsymbol{r}, t) = 1$ or 0, if a particle with velocity $\boldsymbol{e}_i$ ($i = 1, 2, 3, 4$),
$m_{\rm R}(\boldsymbol{r}, t) = 1$ or 0, if a right mirror,
$m_{\rm L}(\boldsymbol{r}, t) = 1$ or 0, if a left mirror

is or is not at lattice site $\boldsymbol{r}$ at time t, respectively. Here $\boldsymbol{e}_i$ is the unit velocity along direction i.

The microscopic equations of motion for these occupation variables are

$$n_i(\boldsymbol{r} + \boldsymbol{e}_i, t+1) = (1 - m_{\rm L} - m_{\rm R})n_i + m_{\rm R}n_{i+1} + m_{\rm L}n_{i-1}, \qquad i = 1, 3, \tag{13}$$

$$n_i(\boldsymbol{r} + \boldsymbol{e}_i, t+1) = (1 - m_{\rm L} - m_{\rm R})n_i + m_{\rm R}n_{i-1} + m_{\rm L}n_{i+1}, \qquad i = 2, 4. \tag{14}$$

On the right-hand sides of (13) and (14) all variables are taken at $\boldsymbol{r}$ and t. The 1 in the first terms on the right-hand sides represents the free streaming term, when no scatterer is present at $\boldsymbol{r}, t$, while the other bilinear terms represent losses (negative signs) and gains (positive signs) due to collisions with scatterers (mirrors)

The Boltzmann approximation is obtained from (13) and (14) in two steps: (i) one introduces an average occupation number:

$$f_i(\boldsymbol{r}, t) = \langle n_i(\boldsymbol{r}, t)\rangle, \qquad i = 1, 2, 3, 4, \tag{15}$$

where the average is over different initial random configurations of the scatterers; (ii) one assumes "molecular chaos" by approximating:

$$\langle m_{\rm R}n_i\rangle \to \langle m_{\rm R}\rangle\langle n_i\rangle = c_{\rm R}f_i, \tag{16}$$

$$\langle m_{\rm L}n_i\rangle \to \langle m_{\rm L}\rangle\langle n_i\rangle = c_{\rm L}f_i, \tag{17}$$

where $c_{\rm R} = N_{\rm R}/M$ and $c_{\rm L} = N_{\rm L}/M$ ($c = c_{\rm L} + c_{\rm R}$) are the fractions of lattice sites occupied with right and left mirrors, if $N_{\rm R}$ and $N_{\rm L}$ are the number of right and left mirrors, respectively. The "molecular chaos" assumption implies that there is no correlation between the presence of a particle with velocity $\boldsymbol{e}_i$ and that of a right or a left mirror at any site $\boldsymbol{r}$ at any t.

Using (13)–(17), one obtains the linear Boltzmann equation for the mirror model:

$$f_i(\boldsymbol{r} + \boldsymbol{e}_i, t+1) = f_i(\boldsymbol{r}, t) + \sum_{j=1}^{4} T_{ij} f_j(\boldsymbol{r}, t), \tag{18}$$

where T_{ij} are the elements of a symmetric collision matrix T:

$$T = \begin{pmatrix} -c & c_{\rm R} & 0 & c_{\rm L} \\ c_{\rm R} & -c & c_{\rm L} & 0 \\ 0 & c_{\rm L} & -c & c_{\rm R} \\ c_{\rm L} & 0 & c_{\rm R} & -c \end{pmatrix}. \tag{19}$$

The Boltzmann approximation of the diffusion coefficient can be obtained from (18) by using either the Chapman–Enskog method [15] or the method used in ref. [10]. For $c_{\rm L} \neq c_{\rm R}$, one has a diffusion tensor with xx or yy component:

$$D_{\rm B}^{xx} = D_{\rm B}^{yy} = \frac{c}{8c_{\rm L}c_{\rm R}} - \frac{1}{4}, \tag{20}$$

which for $c_{\rm L} = c_{\rm R} = c/2$ reduces to

$$D_{\rm B}^{\rm mirror}(c) = \frac{1}{2c} - \frac{1}{4}, \tag{21}$$

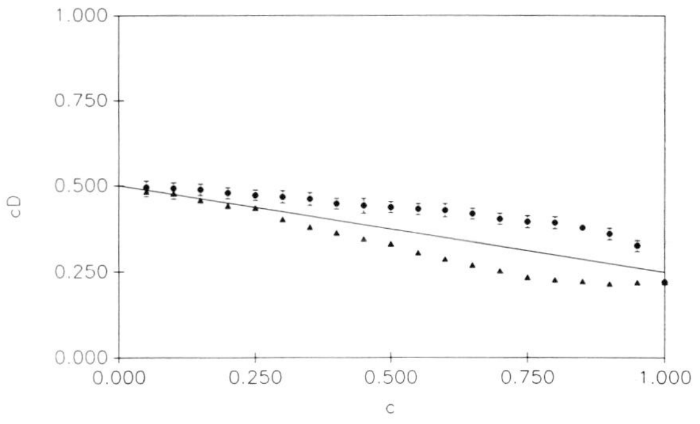

Fig. 4. Concentration dependence of $cD(c)$ for the mirror model (filled circles) and the GO model (filled triangles) at $t = 4000$ mean free times. The drawn line is the Boltzmann approximation for both. The error bars for the GO model are smaller than the symbols.

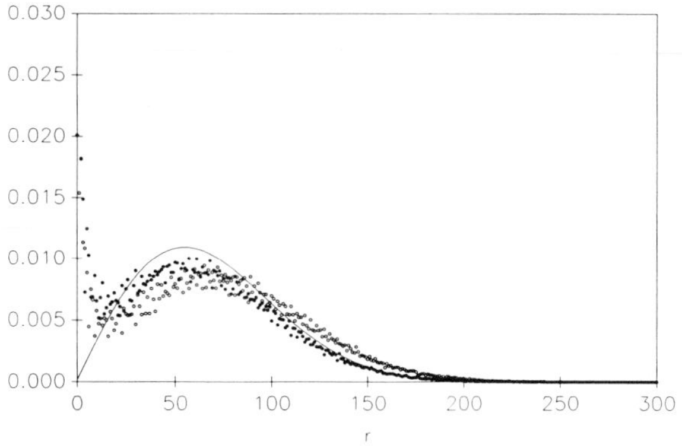

Fig. 6. Density distribution functions, $\hat{P}$, of the mirror model (open circles) and GO model (filled circles) for $c = 0.5$ at $t = 2048$. The drawn curve is for a Gaussian using the measured diffusion coefficient of the mirror model.

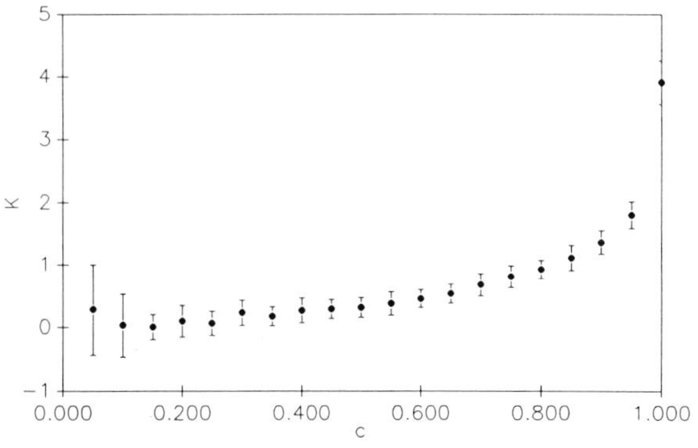

Fig. 5. Concentration dependence of the kurtosis, K, of the mirror model at $t = 4000$ mean free times.

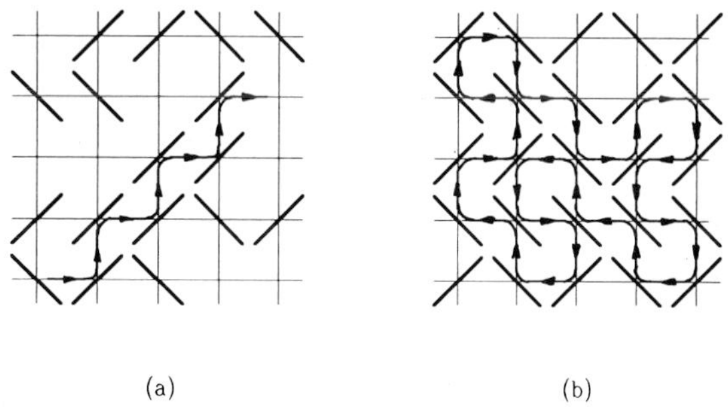

Fig. 7. Typical zigzag motions (a) and closed orbits (b) of the mirror model. For clarity, the trajectories are drawn curved when a collision occurs.

the same value that was found for the probabilistic model discussed in the previous section (cf. (11)). Thus in the continuum limit the Boltzmann diffusion coefficients reduce for both models to that of the Ehrenfest wind-tree model, so that it seems that we can consider both models as a lattice version of the wind-tree model.

6. Simulation results for the mirror model

To check the diffusion behavior of the mirror model we carried out simulations on a MicroVax II for a 1024×1024 square lattice, with periodic boundary conditions for the scatterers. About 2600 particles placed randomly on the lattice were followed on the infinite checkerboard to determine their mean square displacement and other properties, discussed below, over typically several thousand mean free times, here defined as $1/c$.

Since the mean square displacement proved to be proportional to the time, eq. (3) allows us to determine a diffusion coefficient D. For $c_R = c_L = c/2$, cD is plotted in fig. 4 as a function of c. The positive deviations of D from the Boltzmann value D_B can amount to about 17%. They can be related to patches of parallel mirrors, leading to fast zigzag motions through the scatterers (cf. fig. 7a). In order to check that the diffusion process is normal (class I), we computed the kurtosis, given by (5) and plotted in fig. 5. The non-vanishing of the kurtosis prompted us to determine the density distribution function $P(\boldsymbol{r}, t)$. In fig. 6, we plot

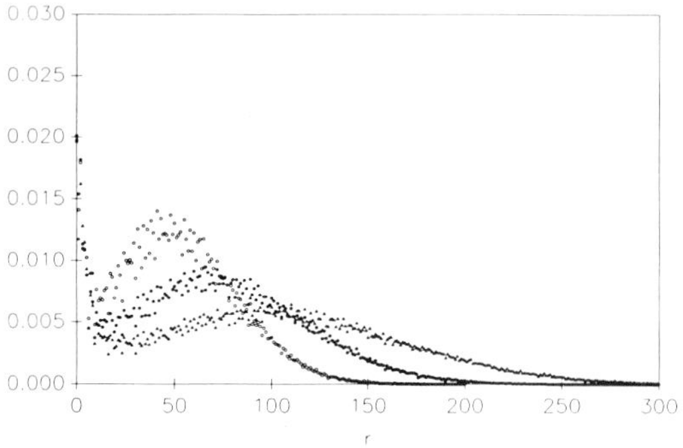

Fig. 8. Density distributions, $\hat{P}$, of the mirror model for $c = 0.5$ at $t = 1024$ (open circles), 2048 (filled circles), 4096 (filled triangles).

a typical $\hat{P}(r,t) = 2\pi r P(r,t)$ as a function of r for $t = 2048$ and $c = 0.5$. $\hat{P}(r,t)$ was determined for each r by counting the number of particles on the lattice sites in a ring between a distance r and $r+1$ from the origin. The dots represent the results of $\hat{P}$ for the lattice gas, while the drawn line gives $\hat{P}$ if P is a Gaussian, using the measured diffusion coefficient. The jumping up and down of the dots as a function of r is due to the fact that half of the lattice sites can be reached at even time steps, while the other half is reached at odd time steps. This makes the number of particles in each ring vary discontinuously from ring to ring. The sharp peak of $\hat{P}$ near the origin is due to the large chance for a particle to find itself in a closed periodic orbit near the origin (cf. fig. 7b). The excess of $\hat{P}$ for large r compared to what a Gaussian P would give, is due to zigzag trajectories in patches of parallel mirrors (cf. fig. 7a). Fig. 8 shows how $\hat{P}$ evolves in the course of time, leading asymptotically to a monotonically decreasing curve with a maximum at the origin $r = 0$. This is in sharp contrast to the asymptotic behavior of $\hat{P}$ for a Gaussian P which leads to a vanishingly small $\hat{P}$ along the r axis. We remark that the behavior of $\Delta(t)$, K and $\hat{P}(r,t)$ for the mirror model with $c_R \neq c_L$ is qualitatively similar to that for $c_R = c_L$ [14].

Thus we have here a diffusion process where the mean square displacement is proportional to the time and yet the distribution function is non-Gaussian. We will call this anomalous diffusion (class II), to distinguish it from the abnormal diffusion (class III) of the OV wind-tree model, where $\Delta(t)$ is not proportional to the time.

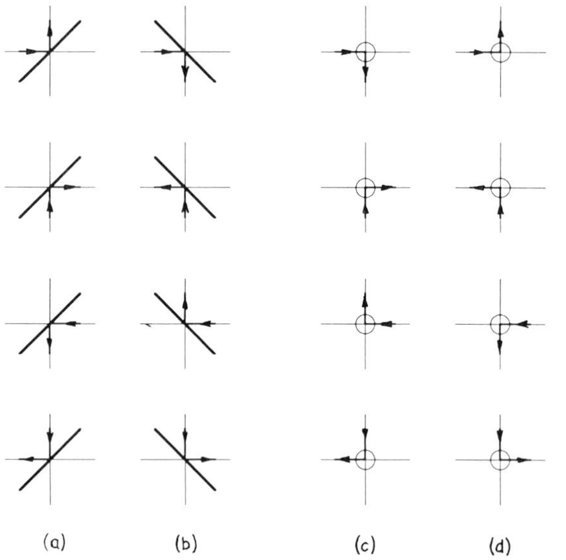

Fig. 9. Comparison of the scattering rules of the mirror model and the GO model: (a) right mirrors; (b) left mirrors; (c) right GO scatterers; (d) left GO scatterers.

Comparing this deterministic model with the probabilistic model, we notice that the Boltzmann approximations are the same for these two models. However, the deterministic nature of the collision rules makes the actual behavior of the mirror model very different from that of the corresponding probabilistic model.

7. Gunn–Ortuño model

We now make what appears to be a slight change in the collision rules of the mirror model in order to study the sensitivity of the diffusion process in a Lorentz lattice gas to the imposed collision rules of the particles and the scatterers. To that end we consider a special case of a class of Lorentz lattice gas models introduced by Gunn and Ortuño [16], where a scatterer scatters a particle either always to the right or always to the left. The collisions in this (GO) model and the mirror model are compared in fig. 9. While in the GO model the velocity of the particle after collision is always turned over $+\pi/2$ (right scatterer) or $-\pi/2$ (left scatterer), a (right or left) mirror scatters for 50% of the cases to the right and for 50% of the cases to the left, depending on the way the particle is coming in.

Introducing the same occupation variables as before for the mirror model, except that m_R and m_L now refer to right and left scatterers, respectively, the microscopic equations of motion for the GO model are:

$$n_i(\boldsymbol{r}+\boldsymbol{e}_i, t+1) = (1 - m_L - m_R)n_i + m_R n_{i+1} + m_L n_{i-1}, \qquad i = 1,2,3,4, \tag{22}$$

where the right-hand side is taken at $\boldsymbol{r}, t$. In the same way as before, a linear Boltzmann equation is obtained for the average particle occupation numbers $f_i(\boldsymbol{r}, t)$, with now a cyclic collision matrix:

$$T = \begin{pmatrix} -c & c_R & 0 & c_L \\ c_L & -c & c_R & 0 \\ 0 & c_L & -c & c_R \\ c_R & 0 & c_L & -c \end{pmatrix}. \tag{23}$$

This leads to a Boltzmann diffusion tensor with

$$D_B^{xx} = D_B^{yy} = \frac{c}{4(c_R^2 + c_L^2)} - \frac{1}{4}, \tag{24}$$

which, for $c_L = c_R$, reduces to

$$D_B^{GO}(c) = \frac{1}{2c} - \frac{1}{4}, \tag{25}$$

the same value as for the mirror and probabilistic models.

We did the same kind of simulations for this model as for the mirror model and found for $c_L = c_R = c/2$ very simular results: $\Delta(t) \sim t, cD^{GO}(c)$ is not too different from cD_B^{GO} with maximum negative deviations of about 17% (cf. fig. 4) and $\hat{P}(r,t)$ is non-Gaussian with a sharp peak at the origin (cf. fig. 6). We note that the peak of $\hat{P}(r,t)$ near $r = 0$ is more pronounced for the GO model than for the mirror model, while the depletion with respect to the Gaussian for intermediate r is less, as is the excess for large r.

However, for $c_R \neq c_L$ the GO model behaves very differently from the mirror model in that while the latter has anomalous diffusion (class II) for all c, the former has a dynamical phase

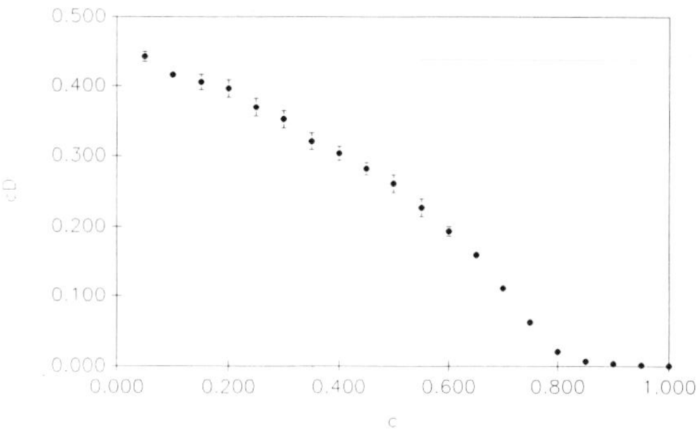

Fig. 10. Concentration dependence of $cD(c)$ of GO model for $c_R/c_L = 2$ at $t = 4000$.

transition from anomalous (class II) to no diffusion (class IV). A typical behavior of the diffusion coefficient as a function of concentration for $c_R/c_L = 2$ at $t = 4000$ is given in fig. 10. Clearly at $c \approx 0.8$, the diffusion coefficient vanishes and a non-diffusion regime starts. A phase diagram is sketched in fig. 11. The phase boundaries between class II and class IV behavior are approximately straight lines. These lines can be inferred as follows. In the case that there are only right or left scatterers (i.e. on the axes of fig. 11), a particle is always trapped if it starts inside a site percolation cluster of the scatterers, which obtains for $c = 0.593\ldots$, the threshold for site percolation on a square lattice [17]. Furthermore, since the point $c_L = c_R = 1/2$, where the GO model is identical to the mirror model with the same concentration, corresponds to another percolation threshold – the bond percolation threshold of two sublattices [14,16] – it must also be on the phase boundary [14]. Exploratory calculations show that straight lines between these points appear to give a fairly good rendition of the phase boundary. We remark that the point $c_R = c_L = 1/2$ belongs to the class II region [14].

8. A model with mixed deterministic and probabilistic scatterers

The behavior of all the lattice gases discussed so far differs qualitatively from that of the OV wind-tree model. A model that incorporates the analogue of the retracing events, which are at the

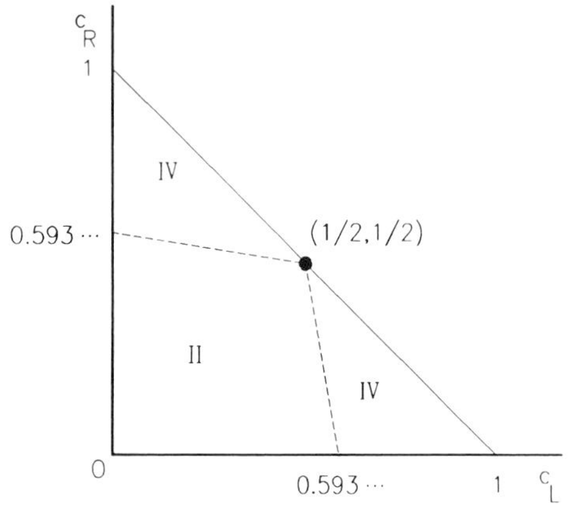

Fig. 11. Phase diagram of the GO model.

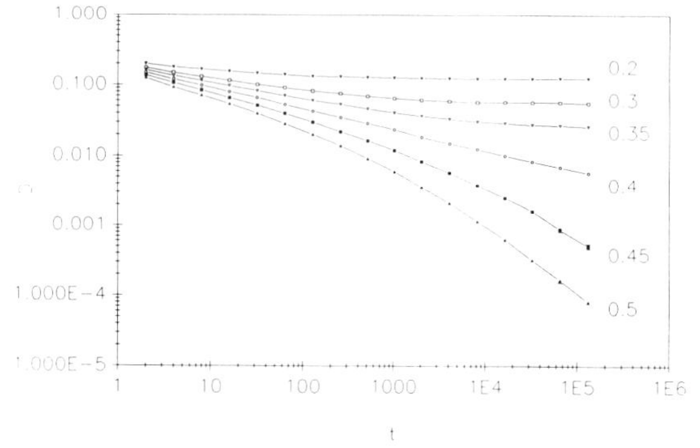

Fig. 12. Log–log plot of the time dependence of the diffusion coefficient, D, of the mixed scatterer model with $c = 1$. The numerals are the fractions of the back-scatterers.

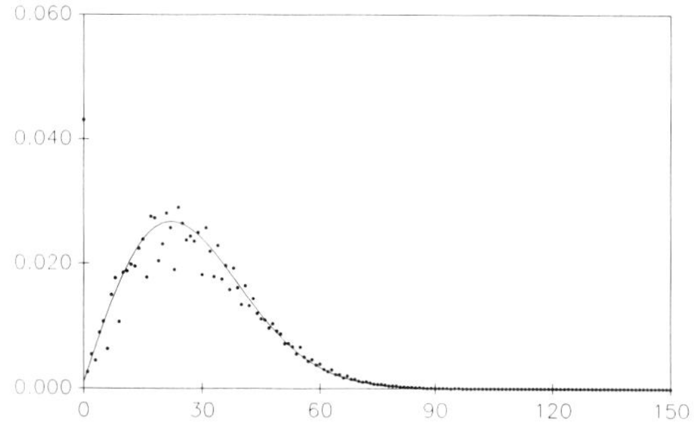

Fig. 13. Density distribution function, $\hat{P}$, of the mixed scatterer model for $c = 1, c_B = 0.2$. The drawn line is for a Gaussian using the measured diffusion coefficient.

root of the abnormal diffusion (class III) in the OV wind-tree model, has been introduced by Binder [18]. We consider here a simple case of this model with two types of scatterers: those that have the same probabilistic scattering law as considered in section 2 (scattering to the right or the left with probability 1/2 each), and those that have a deterministic scattering law, reflecting the particle back in its original direction at collision, i.e., reversing its velocity. These "back-scattering" scatterers are the lattice gas analogues of those tree configurations in the OV wind tree model that make the wind particle retrace its path (cf. fig. 2d). There are two variables in this model: the total concentration c of all the scatterers and the concentration of the back scatterers, c_B. One can convince oneself that if the moving particle starts from inside a site percolation cluster of the back-scatterers, it cannot get out of the cluster, i.e., it is trapped. Hence, like in section 7, for $c_B > 0.593\ldots$, the threshold for site percolation on a square lattice, one has class IV behavior. In fig. 12, the results for $c = 1$ and several values of c_B for $t = 2^{17}$ are plotted. We see that for $c_B \approx 0.4$, the model has class III behavior and a dynamical phase transition from class IV to class III occurs for a back-scatterer concentration, c_B, much lower than the site percolation threshold. The data also show that there is yet another dynamical phase transition, from class III (cf. figs. 12, 13) to class II behavior, for $c_B \approx 0.35$. Fig. 13 shows the density distribution function for $c_B = 0.2$. The peak at $r = 0$ is due to periodic orbits of period 2 between nearest neighbor back scatterers, which are the only possible periodic orbits for this model. Noting that normal diffusion (class I) will occur for $c_B \to 0$, this mixed scatterer model with $c = 1$ exhibits all four classes of diffusion behavior.

9. Discussion

(A) We first discuss the two questions raised at the beginning of this paper.

1. Although all lattice gas models considered here are qualitatively the same in the Boltzmann approximation and then also correspond to the NOV and OV wind-tree models, the actual behavior of all these models is very different. In particular, four different sets of collision rules in the lattice gas case give rise to four different types of diffusion. The comparison between the behavior of

Table 1
Comparison of the diffusion behavior of the continuous wind-tree model and discrete lattice gas models

	Model	$\Delta(t)$	$P(\boldsymbol{r},t)$	Concentration
continuum	NOV wind-tree	$\sim t$	Gaussian	all n
	OV wind-tree	$\sim t^{1-\alpha(n)}$	?	$n < n_c$
	OV wind-tree	$<$ const.	–	$n > n_c$
discrete	probabilistic	$\sim t$	Gaussian	all c
	mirror	$\sim t$	non-Gaussian	all c
	GO	$\sim t$	non-Gaussian	$c_R = c_L$
				$c_R \neq c_L$, small c
	GO	$<$ const.	–	$c_R \neq c_L$, high c
	mixed	$\sim t$	non-Gaussian	$c = 1; c_B \leq 0.35$
	mixed	$\sim t^{1-\alpha}$	non-Gaussian	$c = 1; c_B \approx 0.4$
	mixed	$<$ const.	–	$c = 1; c_B \geq 0.45$

Table 2
Diffusion classes

Class	$\Delta(t)$	$P(\boldsymbol{r},t)$	Diffusion	Model
I	$\sim t$	Gaussian	normal	NOV wind-tree
				probabilistic lattice gas
II	$\sim t$	non-Gaussian	anomalous	mirror lattice gas
				GO lattice gas for $c_R = c_L$ or $c_R \neq c_L$ with c small
				mixed lattice gas with $c = 1; c_B \leq 0.35$
III	$\sim t^{1-\alpha}$	non-Gaussian	abnormal	OV wind-tree for $n < n_c$
				mixed lattice gas with $c = 1; c \approx 0.4$
IV	$<$ const.	–	non	OV wind-tree for $n > n_c$
				GO lattice gas with $c_R \neq c_L$; high c
				mixed lattice gas with $c = 1$; $c_B \geq 0.45$

the continuous wind-tree model and the discrete lattice gas models discussed here is summarized in table 1.

2. A comparison between the behavior of the probabilistic and the deterministic Lorentz gas models leads to the following:

(a) While a return of a particle to its original position can occur in both models, such a return is only temporary in the probabilistic case, while permanent in the deterministic case. This is because in the probabilistic case there is, at each collision, a finite chance for the particle to jump out of its orbit. Therefore, the phase space in the probabilistic case is "metrically transitive", while it is "metrically intransitive" in the deterministic case. In fact, the permanently closed orbits in the deterministic case can be considered as constants of the motion that subdivide the phase space into non-connected regions.

(b) While the probabilistic models all appear to exhibit normal diffusion (class I), the deterministic models show a great variety of diffusion behavior, ranging at least from class II to class IV.

(B) We finally discuss a number of open questions.

1. What equation does the $P(\boldsymbol{r},t)$ for a deterministic Lorentz lattice gas obey? This equation, describing non-Gaussian diffusion, would incorporate the hidden constraints imposed by the existence of closed orbits in the system. As such, it might be a paradigm for equations that describe diffusion in other constrained geometries, like in polymer systems [19].

2. Are there "universality classes" of Lorentz lattice gases for similar dynamical behavior and, if so, would at least some of these universality classes contain continuum models? In the latter case, Lorentz lattice gases could be considered

as simplified models of continuum Lorentz gases. Table 2 presents a first attempt to group various models into classes that exhibit similar diffusion behavior. It seems likely that these (universality) classes will be replaced by others, once a better understanding of diffusion in Lorentz gases as well as corresponding continuum systems has been obtained.

3. One could wonder what the connection is between the behavior of lattice gas cellular automata in general (and Lorentz lattice gas automata in particular) under certain collision rules and that of more abstract cellular automata under certain "game" rules, which need not obey any conservation laws. In both cases one is interested in the time evolution of a discrete system. It might be interesting to see under what conditions the two sets of rules correspond

Acknowledgements

This work was performed in part under Department of Energy (DOE) grant No. DE-FG02-88ER13847.

References

[1] H.A. Lorentz, Proc. Roy. Acad. Amst. 7 (1905) 438, 585, 684.
[2] P. Ehrenfest, Collected Scientific Papers (North-Holland, Amsterdam, 1959) p. 229.
[3] E.H. Hauge and E.G.D. Cohen, J. Math. Phys. 10 (1969) 397.
[4] H. van Beijeren and E.H. Hauge, Phys. Lett. A 39 (1972) 397.
[5] E.H. Hauge, in: Transport Phenomena, eds. G. Kirczenow and J. Marro (Springer, Berlin, 1974) p. 337.
[6] S.W. Haan and R. Zwanzig, J. Phys. A 10 (1977) 1547.
[7] W.W. Wood and F. Lado, J. Comp. Phys. 7 (1971) 528.
[8] W.W. Wood, unpublished.
[9] P.M. Binder, Complex System 1 (1987) 559.
[10] M.H. Ernst and P.M. Binder, J. Stat. Phys. 51 (1988) 981.
[11] M.H. Ernst, G.A. van Velzen and P.M. Binder, Phys. Rev. A 39 (1989) 4327.
[12] M.H. Ernst and G.A. van Velzen, J. Phys. A 22 (1989) 4611.
[13] Th.W. Ruijgrok and E.G.D. Cohen, Phys. Lett. A 133 (1988) 415.
[14] X.P. Kong and E.G.D. Cohen, Phys. Rev. B 40 (1989) 4838.
[15] E.G.D. Cohen, in: Théories Cinétiques Classiques et Relativistes, Proceedings of the Colloques Internationaux du Centre National de la Recherche Scientifique, 1974, ed. M.G. Pichon (Centre National de la Recherche Scientifique, Paris, 1975) p. 269.
[16] J.M.F. Gunn and M. Ortuño, J. Phys. A 18 (1985) L1035.
[17] D. Stauffer, Introduction to Percolation Theory (Taylor and Francis, London, 1985).
[18] P.M. Binder, Complex Systems 3 (1989) 1.
[19] X.P. Kong and E.G.D. Cohen, J. Stat. Phys., to be published.

Physica D 47 (1991) 19–23
North-Holland

SPONTANEOUS CURVATURE IN A CLASS OF LATTICE GAS FIELD THEORIES ⋆

Brosl HASSLACHER
Department of Physics, University of California/San Diego, La Jolla, CA 92093, USA
and Theoretical Division and Center for Nonlinear Studies, Los Alamos National Laboratory, Los Alamos, NM 87545, USA

Received 15 February 1990

We describe a class of cellular automata having a natural lattice gas interpretation which also develop non-perturbative curvature singularities. Variations of these models could be useful in describing curvature transitions in crystals, membranes and superconducting materials.

The lattice gas model for fluid systems [1] originated by considering hyper-discrete descriptions of a field theory which had a natural underlying molecular dynamics. Starting with a discrete Euclidean space and time and the simplest discrete dynamics, only a few local scattering rules are necessary to recover macroscopic fluid equations. The Navier–Stokes equations are a perturbation theoretic result, derived from the Boltzmann equation. This situation is peculiar to fluid systems, which contain no essential constraints beyond number, momentum and if one wishes, energy conservation. Even in the fluid context, it is not clear how to evade the Fermi statistics restriction, which makes lattice gas models so simple, or to show why simulations of lattice gas fluids should lead to early and robust thermodynamics in the absence of ergodic behavior for finite sample systems [2].

One would like to understand how to construct a lattice gas system for dynamics with more sophisticated constraints - for example various gauge theories - the classic one being Maxwell's equations and their consistent coupling to matter. An example is plasma systems. For gauge theories, no natural microscopic physical model exists, so there is no perturbative approach from a master equation with which to develop field equations.

It would be valuable to have some simple examples of totally discrete models containing a simple gauge symmetry and see if the final field equations emerge in a non-perturbative way. We can make the problem even simpler by asking for a true cellular automaton model which develops a lattice gas picture as the effective dynamics of a system with gauge symmetry. The simplest gauge model is discrete two-dimensional Euclidean quantum gravity, where gauge symmetry is a relabeling of points on the underlying lattice. The essential feature of any model of gravitation is the development of spontaneous curvature which is eventually interpreted as mass. Spontaneous curvature is a non-perturbative effect.

The simplest models along these lines are in 1+1 dimensions and the theatre is Minkowski rather than Euclidean space. Several models of this kind have been proposed [3], but recently a particularly rich one was put forward by 't Hooft [4], which demonstrates how lattice gas rules emerge in the absence of an underlying microscopic dynamics. In this model curvature arises as a non-perturbative instability. Recently, we have found that there is a tower of such models all related to soluble statistical mechanical systems at criticality in two dimensions [5]. This already means that this construction of hyper-discrete gauge models is a non-trivial undertaking.

To see in detail how lattice gas models can arise

⋆ Supported in part by DOE grant KC-04-01-030.

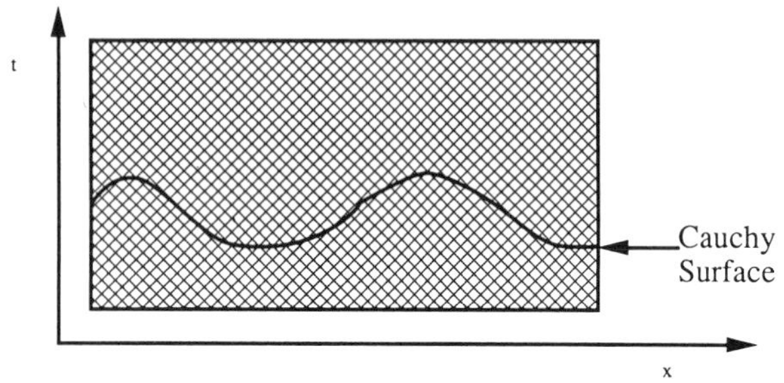

Fig. 1.

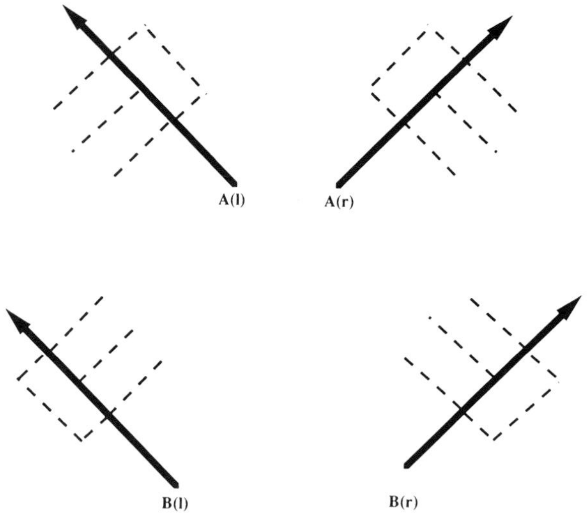

Fig. 2.

from thinking about consistent kinetics in discrete space–time, we will review 't Hooft's model. Here, it is only meant as an illustration of how to construct such models.

If there is no microscopic dynamics on which to build a cellular automaton model, one can create one by going to a space with more structure than a flat Euclidean one. We begin in a 1+1 dimensional space–time with Minkowski metric and the associated Lorentz transformations among frames. If we make this space–time totally discrete and consider only Lorentz transformations which have boosts differing discretely by powers of 2, we can write down a very simple space–time cellular automaton, which is local and deterministic.

Take a single 1+1 dimensional space time, and make infinite copies of it. Display them on a plane as a collection of light-cones connected along null defect lines. This will be the fundamental cellular automaton lattice and our kinetic background (fig. 1). Such a collection of frames, to be consistent, must transform among one another by Lorentz transformations including boosts. This is the analog to gluing together one-particle distribution functions in a picture of an abstract fluid.

Let us try and introduce local, causal, Lorentz invariant dynamics on this picture. We allow only four types of Lorentz boosts, corresponding to a boost angle of absolute value $\theta_0 = \cosh^{-1}(5/4)$, agreeing with our picture of light cones glued together in this very simple way. For Lorentz transformations in light-cone coordinates (u, v), this is a Lorentz transformation of the form $(u, v) \to (2u, v/2)$ and the angles can have both positive and negative signs. Introducing Lorentz transformations causes patching lines between transformations. These resemble defects in the dislocation dynamics of crystals, in this case a space–time crystal. Introduce simple dynamics into this kinetic description by making these defect lines observables and giving the analog of scattering rules for colliding defects. In this way, we go from a cellular automaton picture of space–time points to a lattice gas picture of dislocations. If the boost angle is negative the defect is labeled A_l or A_r according to in which direction it propagates; if the boost angle is positive the defect is labeled B_l or B_r, similarly. See fig. 2.

To develop the model further, we will need the analog of propagating particles. To do that, consider a Cauchy surface C of initial data passing through the lattice (fig. 1). The generic situation is that it intersects the lattice in a sequence of left and right lattice links and lattice defects. As the surface evolves in time, this intersection pattern becomes quite complex, but we can decompose it into simple elements. If we reinterpret the lattice links and defects as mathematical particles, we get a collection of right and left moving particles. We must fix their scattering rules or interactions. But these already have a large number of constraints on them if we insist on a deterministic particle gas with P and CT invariance. The last requirement is not ad hoc. It is necessary if the continuum limit of this model is to make sense as a quantum field theory.

Labeling lattice link particles by O_l and O_r, and

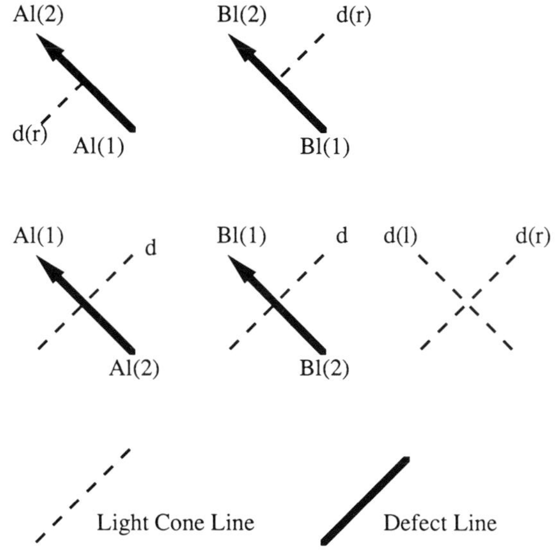

Fig. 3.

employing a subscript 1 or 2 to label defect particles before and after interaction with link particles, we have the interaction rules shown in fig. 3. Interpreting O as its own antiparticle and the pairs (A_1, B_2), (A_2, B_1) as antiparticles, these rules are P and CT invariant.

Parity invariance forces symmetric initial states to scatter to symmetric final states. Thus the interaction rule in this sector must be a map

$$\{A_1A_1, A_2A_2, B_2B_2\} \rightarrow \{A_1A_1, B_1B_1, B_2B_2\},$$

where the left and right labels are to be understood. The only fine point here is that the triple vertex rule implies that B_1B_1 (A_2A_2) does not appear in the set of in (out) states since B_1 (A_2) particles decay spontaneously (are only created by interactions with O particles). There are thus six possible reversible symmetric rules; only four of them, however, are CT invariant.

Finally we consider interaction rule for asymmetric states which must be a map

$$\{A_1A_2, A_1B_2, A_2B_2\} \rightarrow \{A_1B_1, A_1B_2, B_1B_2\},$$

where each state listed also has a parity transform. We will take only some of these rules, since they break down into separate consistent classes and we take 't Hooft's choice. Our final set of collision rules for defect particles is

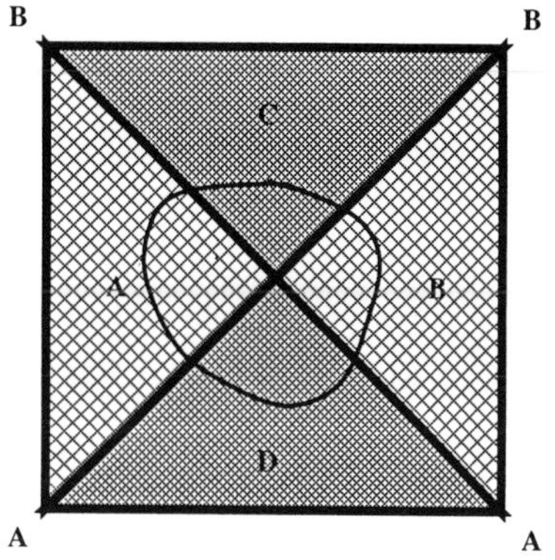

Fig. 4.

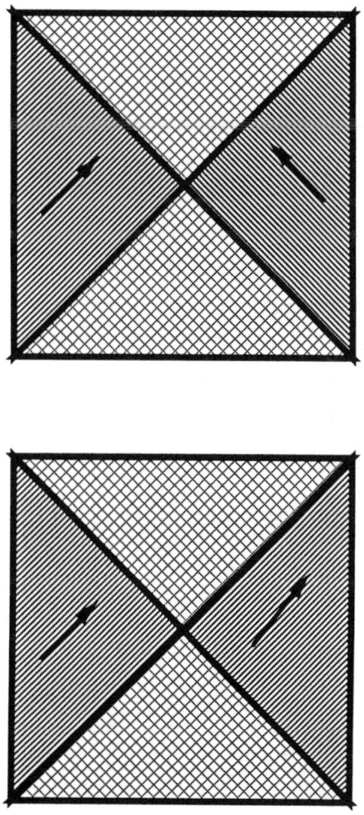

Fig. 5. Schematic of two locally flat regions.

$$\begin{array}{ll} A_{r2}A_{l2} \rightarrow A_{l1}A_{r1} & B_{r2}B_{l2} \rightarrow B_{l2}B_{r1} \\ A_{r1}A_{l1} \rightarrow B_{l2}B_{r2} & A_{r1}A_{l1} \rightarrow B_{l1}B_{r2} \\ A_{r2}B_{l2} \rightarrow A_{l1}B_{r1} & B_{r2}A_{l2} \rightarrow B_{l1}A_{r1} \end{array}$$

What makes this class of models interesting is that any set of consistent rules forces a locally flat space–time (see fig. 5 to have a global nonzero curvature. To see this, consider a closed path around a defect collision (fig. 4). Moving around such a path in a counterclockwise direction, there is a boost as each defect is crossed. Since ingoing and outgoing defects of the same type contribute boosts of opposite sign, at each interaction where particles of type A and B are separately conserved there is no net Lorentz transformation connected with the closed path. When there is a loss of n A particles and a gain of n B particles, however, there is a total boost of $-2n\theta_0$; if there is a gain of n A particles and a loss of n B particles the total boost is $2n\theta_0$. Since the total boost around a closed path measures the curvature, we find that there can be positive, negative, or no curvature associated with a given defect collision.

Since 't Hooft wants to construct a model of black holes, there is no need to consider other than negative or zero boosts around intersections. So with his rules we can at most get models with negative curvature.

That there must be such curvature in the model is guaranteed by the triple vertex graphs of fig. 3. One sees that it implies there must exist intersections of the form fig. 4. It is clear from the diagram that there is no global Lorentz transformation that connects regions A, B of this graph with regions C, D since the unit area of space–time patches is not preserved. This implies that this diagram violates the flatness condition and corresponds to a non-perturbative conical singularity of negative curvature.

In order not to allow this curvature effect to evaporate, we must postulate that the amount of information in a unit cell of space–time be finite. Otherwise, the curvature could leak away by a sequence of 2^n blocking transformations taken to the limit point. One simply inserts more and more coordinate lines between the ones one already has in the obvious way so as to convert the 3-vertex to a 4-vertex. It is not possible to do this if we must stop the grid refinement at some finite level n.

To summarize this simple but elegant model: we have a deterministic, local cellular automaton model of Minkowski space–time which has the following properties: local Lorentz invariance, PCT invariance, local flatness and finite information density. These conditions allow us to reinterpert lattice defect lines or coordinate patch lines as particles in a lattice gas construction. The scattering rules for defects force non-perturbative negative curvature.

Because the present model was designed as a model for black holes, the rule set was chosen to have only zero or negative curvature. But examining all of the rule sets compatible with our assumptions and looking at the global circuit around a vertex argument given above, it is clear that rules inducing positive zero and negative curvature are also allowed. This fact allows one to construct a richer and in a sense more natural set of space–times. In a physical theory, one expects both signs of curvature to exist with equal probability.

It is possible to show that this class of spontaneous curvature models corresponds to solutions to the Yang–Baxter functional equations of statistical mechanics as well as a unitarity and cross channel unitarity condition [5]. This class of models then has both a knot theory interpretation as a dynamics realizing Kauffman-like exchange identities directly [6] and as exactly soluble two dimensional statistical mechanical models at criticality [7]. Both of those view points provide additional information on non-perturbative aspects of their dynamics.

Besides the intrinsic interest as a class of models that evolve lattice gas behavior with spontaneous curvature, variants of these curvature models can be used to describe the Kosterlitz–Thouless crumpling phase transition in certain types of crystals and membranes that support a non-zero shear modulus. They may be particularly good at modeling the crinkled phase of undulating hexatic membranes and analogous effects in superconducting materials [8].

References

[1] U. Frisch, B. Hasslacher and Y. Pomeau, Lattice gas automata for the Navier–Stokes equation, Phys. Rev. Lett. 56 (1986) 1505–1508.

[2] G. Doolen, ed., Lattice Gas Methods for Partial Differential Equations, Santa Fe Institute Studies in the Sciences of Complexity (Addison–Wesley, New York, 1990).

[3] C. Bennet, Logical reversibility of computation, IBM J. Res. Devel. 6 (1973);

N. Margolis, Physics-like models of computation, Physica D 10 (1984) 81–95;
T. Toffoli, Cellular automata mechanics, University of Michigan Computer and Communication Sciences Dept. Technical Report 208 (1977).
[4] G. 't Hooft, A two-dimensional model with discrete general coordinate-invariance, Duke University preprint (1989); Equivalence relations between deterministic and quantum mechanical systems, J. Stat. Phys. 53 (1988) 323–344.
[5] B. Hasslacher and D. Meyer, Department of Physics, University of California/San Diego, La Jolla, CA 92093, preprint (March, 1990).
[6] L.H. Kauffman, State models for link polynomials, University of Illinois at Chicago preprint (1989); State models for knot polynomials, University of Illinois at Chicago preprint (1989); Spin networks and knot polynomials, Int. J. Mod. Phys. A5 (1990) 93–115; Knots, abstract tensors and the Yang–Baxter equation, University of Illinois at Chicago preprint (1989); On knots, Ann. Math. Stud. 115 (Princeton Univ. Press, Princeton, 1987); State models and the Jones polynomial, Topology 26 (1987) 395–407.
[7] Y. Akutsu and M. Wadati, Knots, braids and exactly solvable models in statistical mechanics, Commun. Math. Phys. 117 (1988) 243–259; Knot invariants and the critical statistical systems, J. Phys. Soc. Japan 56 (1987) 839–842.
[8] D. Nelson, Theory of the crumpling transition, in: Statistical Mechanics of Membranes and Surfaces, eds. D.R. Nelson, T. Piran and S. Weinberg (World Scientific, Singapore, 1989).

CHAPTER 2

VISCOSITY AND HYDRODYNAMIC MODES

Physica D 47 (1991) 27–29
North-Holland

LOW-VISCOSITY LATTICE GASES

B. DUBRULLE[a, b], U. FRISCH[c], M. HÉNON[c] and J.-P. RIVET[c, d]
[a] *Observatoire Midi-Pyrénées, 14 avenue E. Belin, 31400 Toulouse, France*
[b] *École Normale Supérieure, 45 rue d'Ulm, 75005 Paris, France*
[c] *C.N.R.S., Observatoire de Nice, B.P. 139, 06003 Nice Cedex, France*
[d] *Laboratoire de Physique Statistique, 24 rue Lhomond, 75231 Paris Cedex 05, France*

Received 29 December 1989

New three-dimensional lattice gas models with very low (and possibly negative) viscosities are studied theoretically and tested in numerical implementations.

We have studied a class of lattice gas models which are variants of the "FCHC model" [1–3]. The aim is to achieve the highest possible Reynolds coefficient (inverse non-dimensionalized viscosity) for efficient simulations of the three-dimensional incompressible Navier–Stokes equations. The models include an arbitrary number of rest particles (as in ref. [4]) and, in addition, violations of semi-detailed balance. Specifically, the correspondence between collision input and output states is chosen deterministic but not one-to-one, thereby allowing non-definite-positive viscosities. Alternative strategies for obtaining negative viscosities are known [5].

Within the framework of the Boltzmann approximation exact expressions are obtained for the Reynolds coefficients. The minimization of the viscosity is done by solving a Hitchcock-type optimization problem for the fine-tuning of the collision rules.

Fig. 1 shows the dependence of the Boltzmann-based kinematic viscosity on the (reduced) density d for various values of the number of rest particles. When the number of rest particles exceeds 1, it is seen that there is a range of densities at which the viscosity takes negative values.

Various optimal models with up to three rest particles (26 bits) per node have been implemented on a CRAY-2 and their true transport coefficients have been measured with good accuracy.

Fairly large discrepancies with Boltzmann values are observed when semi-detailed balance is violated; in particular, no negative viscosity is obtained. This is exemplified by fig. 2, which compares Boltzmann-based and true quantities (such as the viscosity) for the model FCHC-7, which has three rest particles and violations of semi-detailed balance consistent with duality-invariance, that is particle–hole symmetry. (Dropping the latter constraint as in ref. [6] turned out not to improve performances.)

In spite of these discrepancies, model FCHC-7, the best obtained so far, has a Reynolds coefficient of 13.5, twice that of the best previously implemented model and thus is about 16 times more efficient computationally in three-dimensional simulations.

Further improvements seem possible because some of the discrepancies with Boltzmann-based values can be traced back to the fact that our implementation of the FCHC models is only two nodes wide in the fourth dimension. (We recall that FCHC models reside on a four-dimensional lattice.) Models with very high Reynolds coefficients can probably be used for sub-grid-scale modelling of turbulent flows.

A detailed version of this paper may be found in ref. [7].

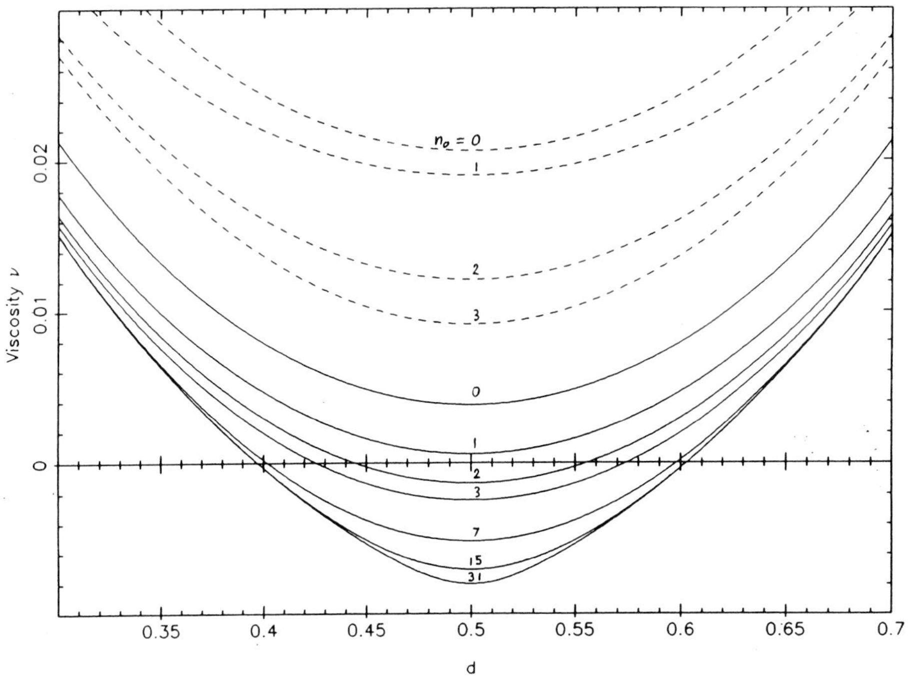

Fig. 1. Boltzmann-based kinematic viscosity ν as a function of d. The full lines represent the case with no semi-detailed balance with, from top to bottom, $n_0 = 0, 1, 2, 3, 7, 15, 31$ rest particles. For comparison, the dashed lines represent the viscosity in the case of semi-detailed balance [4], for 0, 1, 2, and 3 rest particles. (From ref. [7].)

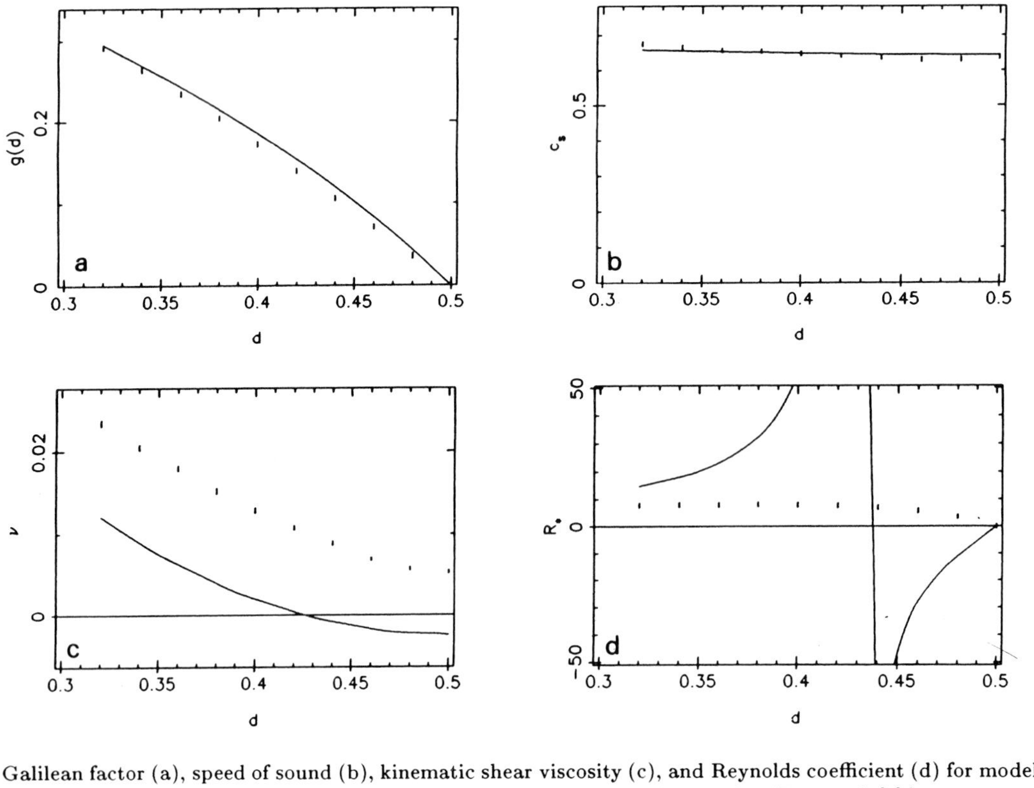

Fig. 2. Galilean factor (a), speed of sound (b), kinematic shear viscosity (c), and Reynolds coefficient (d) for model FCHC-7. Solid curves: theoretical Boltzmann values. Error bars : numerical simulation results. (From ref. [7].)

Acknowledgements

We thank D. d'Humières for useful discussions. CRAY-2 resources have been provided by the Centre de Calcul Vectoriel pour la Recherche. This work has been supported by DRET (88/1450), the EEC (SC1-0212-C) and Los Alamos (9-L38-3634R). One of us (B.D.) received partial support from an Amelia Earhart fellowship of the Zonta foundation.

References

[1] D. d'Humières, P. Lallemand and U. Frisch, Lattice gas models for 3-D hydrodynamics, Europhys. Lett. 2 (1986) 291–297.

[2] U. Frisch, D. d'Humières, B. Hasslacher, P. Lallemand, Y. Pomeau and J.-P. Rivet, Lattice gas hydrodynamics in two and three dimensions, Complex Systems 1 (1987) 649–707; reprinted in: Lecture Notes on Turbulence, eds. I.R. Herring and J.C. McWilliams (World Scientific, Singapore, 1989) pp. 297–371.

[3] J.-P. Rivet, M. Hénon, U. Frisch and D. d'Humières, Simulating fully three-dimensional external flow by lattice gas methods, Europhys. Lett. 7 (1988) 231–236.

[4] M. Hénon, Optimization of collision rules in the FCHC lattice gas, and addition of rest particles, in: Discrete Kinetic Theory, Lattice Gas Dynamics and Foundations of Hydrodynamics, Torino, September 20-24, 1988, ed. R. Monaco (World Scientific, Singapore, 1989) pp. 146–159.

[5] D.D. Rothman, Negative-viscosity lattice gases, J. Stat. Phys. 56 (1989) 517–524.

[6] J.A. Somers and P.C. Rem, The construction of efficient collision tables for fluid flow computations with cellular automata, in: Cellular automata and the Modelling of Complex Systems, Les Houches (Springer, Berlin, 1989).

[7] B. Dubrulle, U. Frisch, M. Hénon and J.-P. Rivet, Low viscosity lattice gases, J. Stat. Phys. 59 (1990) 1187–1226.

Physica D 47 (1991) 30–35
North-Holland

COUNTING HYDRODYNAMIC MODES IN LATTICE GAS AUTOMATA MODELS

Gianluigi ZANETTI
Program in Applied and Computational Mathematics, Princeton University, Princeton, NJ 08544, USA

Received 15 February 1990

A Monte Carlo scheme for the search of extensive conserved quantities in lattice gas automata models is described. It is based on an approximation to the microscopic dynamics and it amounts to estimating the dimension of the eigenspace with eigenvalue 1 of a linear operator related to the lattice gas automata model evolution operator linearized around equilibrium distributions. The applicability of this technique is limited to models with collision rules satisfying semi-detailed balance.

1. Introduction

In this short note I describe a Monte Carlo technique that could be useful in the exploration of the hydrodynamic properties of some lattice gas automata (LGA) models.

The best known example of this class of models was introduced by Frisch, Hasslacher and Pomeau [1] as an alternative approach to the numerical solution of the two-dimensional incompressible Navier–Stokes equation. It can be described as a gas of particles that inhabit the sites of a regular lattice, are allowed to hop from a lattice site to its nearest neighbors, and can experience a primitive form of collision at the lattice sites. The "collisions" between particles are implemented as a set of deterministic collision rules that redistribute the particles at each site between the possible hopping directions ("velocities"). The rules are constructed to conserve the total number of particles and the total linear momentum present at each site. These conservation laws reappear in the macroscopic dynamics (described by coarse grained conserved densities, e.g., momentum density, obtained by averaging their macroscopic equivalents over subregions of the lattice) as differential conservation laws. It is usually argued that, when the underlying regular lattice has been properly chosen (i.e., an hexagonal lattice for the two-dimensional LGA) the form of the hydrodynamic equations appropriate for the LGA is very similar to that found for simple fluids [2].

It has been recently shown that the physics of the simple model discussed above is actually more complicated than the one of simple fluids. Together with the hydrodynamics modes corresponding to the conservations laws enforced microscopically (e.g., two sound waves and a shear wave) there are three spurious conserved densities [3, 4] (staggered momentum modes) that are a peculiarity of the discrete character of the model. Since these spurious densities are non-linearly coupled to the first three, the lattice gas automata can produce flows that are not solutions of the Navier–Stokes equation [4].

The nature of these three spurious densities is rather trivial, they are due to the interplay between the microscopic conservation laws enforced at the lattice sites and the discrete dynamics, and they amount to a splitting of the lattice into weakly coupled sublattices. However, they were discovered in a somewhat fortuitous way [5] about two years after ref. [1] was published.

I will describe now a simple algorithm that, once the microscopic model has been defined, should be helpful in searching for spurious conserved quantities.

The algorithm is based on:

(A) The transformation of the microscopic LGA model to a simplified model whose evolution is given by a linear operator $\widetilde{DL}$ [6], which approximates the LGA evolution operator linearized around an equilibrium distribution.

(B) A Monte Carlo search for eigenvectors of $\widetilde{DL}$ with eigenvalue 1. The number of linearly independent vectors found is a lower bound to the number of hydrodynamic modes of the model.

This scheme can be easily applied to the models described above and to many of its modifications (see review articles in these Proceedings). The result is what expected: it finds the three quantities conserved by construction, and the three spurious ones.

As a first nontrivial application of this scheme we can test whether the LGA model can be modified so that there are no spurious conserved quantities. The spurious quantities of ref. [4] can be easily destroyed by extending the number of hops possible to the particles (and modifying accordingly the set of collision rules[#1] thus providing a coupling between the previously independent sublattices. It is not a priori obvious, however, that extending the set of particles "velocities" will not introduce other (different) spurious quantities. Surprisingly enough[#2] this does not happen: applying the procedure described above to the model with extended set of velocities gives only three conserved densities, i.e., the expected particle number and two components of the linear momentum.

It appears that this simple modification is enough to free, within the approximation used, the hydrodynamic behavior of the LGA from spurious conserved quantities.

The rest of this note is organized as follows. In the next section I explain the algorithm in detail on a simple example. In the last section I describe some of the limitations of this technique.

2. Boring details

I will describe the technique on an example: the original FHP model of ref. [1]. In this model the particles have momenta chosen from the vectors

$$\boldsymbol{C}_a = \left(\cos\left[\tfrac{1}{3}\pi(a-1)\right], \sin\left[\tfrac{1}{3}\pi(a-1)\right]\right),\quad a = 1,\dots,6. \tag{1}$$

The macroscopic densities corresponding to the number of particles and momentum conservation are, respectively,

$$n(\boldsymbol{r},t) = \sum_{a=1,\dots,6} f_a(\boldsymbol{r},t),$$

and

$$\boldsymbol{g}(\boldsymbol{r},t) = \sum_{a=1,\dots,6} \boldsymbol{C}_a f_a(\boldsymbol{r},t),$$

where $f_a(\boldsymbol{r},t)$ is the Boolean population field that indicates the presence (1) or absence (0) of a particle moving with momentum $\boldsymbol{C}_a$ at site $\boldsymbol{r}$ and time step t. The time evolution of the particle populations can then be written, assuming that the particles first hop in the direction of their velocities and then are subject to collisions, as

$$f_a(\boldsymbol{r},t+1) = f_a(\boldsymbol{r}-\boldsymbol{C}_a,t) + T_a(\{f_b(\boldsymbol{r}-\boldsymbol{C}_b,t)\}_{b=1,\dots,6}), \tag{2}$$

where T_a is the microscopic collision operator,

[#1] The collision rules I used are the usual FHP rules between particles of the same speed plus all the possible collisions between three particles with velocities such that their total momentum is equal to two.

[#2] Thus providing contrary evidence to the analogy proposed in ref. [4] between the spurious conserved quantities in LGA and the fermion-doubling problem in lattice gauge theory.

that is the Boolean algebra expression that corresponds to the chosen set of collision rules. See ref. [7] for the explicit definition of T_a.

2.1. Simplified model

I would like to reduce the search for extensive conserved quantities to finding eigenvectors of eigenvalue one of an appropriate matrix.

The first step is to transform eq. (2) in the corresponding Boltzmann equation by understanding $f_a(\boldsymbol{r}, t)$ as a continuous variable between zero and one and replacing the Boolean operations in T_a with the appropriate arithmetic operations. From now on I will use $f_a(\boldsymbol{r}, t)$ to indicate the continuous variables. We will consider a rhombic region, Ω, of the lattice bounded by periodic boundary conditions. As a short hand notation I will use capitalized Latin letters to indicate a "probability distribution" of the Boltzmann model, e.g., $X(t) = \{f_a(\boldsymbol{r}, t)\}_{a=1,\dots,6;\, \boldsymbol{r}\in\Omega}$, and T, $X(t=1) = T[X(t)]$, to indicate the evolution operator.

We can now randomly generate an initial condition, $X(0)$, i.e., we randomly assign $0 < f_a(\boldsymbol{r}, t) < 1$, where $\boldsymbol{r}$ runs on all sites of Ω and a on all the possible directions. Then we iterate eq. (2) forward in time.

Hénon [2] has shown that for lattice gas automata models that satisfy the property of "semi-detailed" balance,[#3] as, for instance, the original FHP model, it is possible to prove that the quantity $\Theta(t)$, defined by

$$\Theta(t) = \sum_{\boldsymbol{r}\in\Omega} \sum_{a=1,\dots,6} \{f_a(\boldsymbol{r}, t) \log[f_a(\boldsymbol{r}, t)] + [1 - f_a(\boldsymbol{r}, t)] \log[1 - f_a(\boldsymbol{r}, t)]\}, \quad (3)$$

is non-increasing, i.e., $\Theta(t+1) \le \Theta(t)$.

[#3] That is, the microscopic collision are designed so that all the possible configurations of a lattice site with the same value of the microscopically conserved quantities, e.g., number of particles and momentum, have the same equilibrium probability.

The system we are considering is finite and thus $X(t)$ will converge, in a finite time t^* [#4], to a probability distribution, $X^*(t) = X(t)$, $t \ge t^*$, such that $\Theta(t) = \Theta(t^*)$ for $t \ge t^*$. If we assume that $f_a^*(\boldsymbol{r}, t)$ is homogeneous, $f_a^*(\boldsymbol{r}, t) = f_a^*(t)$, then it is easy to see that it will also be time-independent, $f_a^*(t) = f_a^*$. Moreover, f_a^* will have the expected Fermi–Dirac form [2]:

$$f_a^* = \frac{1}{1 + \exp(\alpha - \boldsymbol{\beta} \cdot \boldsymbol{C}_a)}, \quad (4)$$

where $\alpha, \boldsymbol{\beta}$ are the intensive parameters conjugated, respectively, to the number and momentum densities.

There could also be $X^*(t)$ that are not homogeneous and time-independent, but rather periodic in both time and space (this, in fact, corresponds to the presence of the spurious conserved quantities mentioned above). In the latter case $X^*(t)$ will not be a fixed point of the evolution operator T, but rather of $L = T^q$, where q is the period of $X^*(t)$, $X^*(t+q) = X^*(t)$. Let us now assume that X^* depend only on the presence of microscopic conservation laws and not on the details of the collision operator T_a. Then X^* should also be a fixed point of the streaming operator S^q, where $S\{f_a(\boldsymbol{r})\} = \{f_a(\boldsymbol{r} - \boldsymbol{C}_a)\}$, and thus

$$f_a^*(\boldsymbol{r}) = f_a^*(\boldsymbol{r} - q\boldsymbol{C}_a). \quad (5)$$

If we apply what we just discussed to the FHP model we immediately find that $X(t)$ converges to a fixed point, i.e., $q = 1$, if both sides of the rhombic region Ω have an odd number of sites, and to a cycle of period two, $q = 2$, if at least one of the sides has an even number of sites. This apparently odd behavior is a direct consequence of the presence of the staggered momentum modes. See below.

[#4] Actually $X(t)$ it is exponentially convergent to X^*. On the other hand, since the calculations are done in finite precision, after a time t^*, $X(t)$, $t > t^*$ will not be distinguishable from $X^*(t)$. The time t^* should be of the order of the hydrodynamic relaxation time and thus scale as the square of the size of Ω.

We can now linearize L around X^* and call the linearized evolution operator DL. This can be done exactly if we have the analytic expressions for the collision operator, eq. (2), but I will argue that it is not necessary.

In correspondence to each independent extensive conserved quantity, e.g., the total linear momentum, the total number of particles, the alleged spurious conserved quantities ..., there is a conjugated intensive parameter and the latter should parameterize the resulting equilibrium distribution as α and $\boldsymbol{\beta}$ parameterize f_a^* in eq. (4). Thus the "equilibrium probability distribution" X^* will be parameterized by the intensive parameters $\alpha, \boldsymbol{\beta}, \ldots$, i.e., $X^* = X^*(\alpha, \boldsymbol{\beta}, \ldots)$, $X^*(\alpha, \boldsymbol{\beta}, \ldots) = L[X^*(\alpha, \boldsymbol{\beta}, \ldots)]$. Therefore the vectors obtained by taking the derivative of X^* with respect to one of the intensive parameters will be eigenvectors with eigenvalue 1 of DL, the Jacobian matrix of the evolution operator L. Conversely to each eigenvector of eigenvalue 1 is associated an additive conserved quantity. In fact, let $|1\rangle, |2\rangle, \ldots, |m\rangle$, $|i\rangle = \{\phi_a^i(\boldsymbol{r})\}_{a,\boldsymbol{r}}$ be right eigenvectors of DL with eigenvalue 1. Eq. (5) implies that $|i\rangle$ should also be an eigenvector of the streaming operator S^q thus $\{\phi_a^i(\boldsymbol{r})\} = \{\phi_a^i(\boldsymbol{r} - q\boldsymbol{C}_a)\}$. Now let $\langle i|$, $i = 1, \ldots, m$ be the corresponding left eigenvectors of DL, normalized so that $\langle i|j\rangle = \delta_{i,j}$. By the same reasoning used above, $\langle i|$ should be a left eigenvector of S^q. But S is unitary and thus $\langle i|$ is proportional to the transpose of $|i\rangle$. Consider now a small-amplitude, slowly varying perturbation of X^*, $X^* + \epsilon|\Psi(t)\rangle$ with $|\Psi(t)\rangle$ defined as $\{\Psi_a(\boldsymbol{r}, t)\}_{a,\boldsymbol{r}}$. Then, to order ϵ, $|\Psi(t=1)\rangle = (DL)|\Psi(t)\rangle$, but also

$$|\Psi(t+1)\rangle = S^q|\Psi(t)\rangle + [(DL) - S^q]\,|\Psi(t)\rangle. \tag{6}$$

Thus

$$\langle i|\Psi(t+1)\rangle = \langle i|S^q\Psi(t)\rangle, \tag{7}$$

which can be rewritten, after a trivial simplification, as

$$\sum_{a,\boldsymbol{r}} \phi_a(\boldsymbol{r})\,\Psi_a(\boldsymbol{r}, t+q) = \sum_{a,\boldsymbol{r}} \phi_a(\boldsymbol{r})\,\Psi_a(\boldsymbol{r}, t). \tag{8}$$

Hence $\sum_{a,\boldsymbol{r}} \phi_a(\boldsymbol{r})\,\Psi_a(\boldsymbol{r}, t)$ is, to order ϵ, globally conserved. One can also note that $f_a(\boldsymbol{R}, t+q)$ depends only on the $f_a(\boldsymbol{R} + \boldsymbol{r}, t)$ with $\boldsymbol{r} \in \Lambda_q$,

$$\Lambda_q = \left\{\boldsymbol{r} = \sum_{a=1,\ldots,6} m_a \boldsymbol{C}_a, m_a \geq 0; \sum_a m_a = q\right\}, \tag{9}$$

with the m_a integers. Define now $\langle i(\boldsymbol{R})| = \{\phi_a^i(\boldsymbol{r})\,\chi^{\Lambda_q}(\boldsymbol{r} - \boldsymbol{R})\}_{a,\boldsymbol{r}}$ where $\chi^{\Lambda_q}(\boldsymbol{r}) = 1$ if $\boldsymbol{r} \in \Lambda_q$ and 0 otherwise, and the densities

$$\begin{aligned}\rho^i(\boldsymbol{R}, t) &= \langle i(\boldsymbol{R})|\Psi\rangle \\ &= \sum_{a, \boldsymbol{r} \in \Lambda_q} \phi_a^i(\boldsymbol{r} + \boldsymbol{R})\,\Psi_a(\boldsymbol{r} + \boldsymbol{R}, t).\end{aligned} \tag{10}$$

We can now project eq. (6) on $\langle i(\boldsymbol{R})|$, to obtain

$$\begin{aligned}&\{\rho^i(\boldsymbol{R}, t+q) - \rho^i(\boldsymbol{R}, t)\} \\ &\quad - \sum_{a, \boldsymbol{r} \in \Lambda_q} \phi_a^i(\boldsymbol{R} + \boldsymbol{r}) \\ &\quad \times \{\Psi_a(\boldsymbol{R} + \boldsymbol{r} - q\boldsymbol{C}_a, t) - \Psi_a(\boldsymbol{R} + \boldsymbol{r}, t)\} = 0.\end{aligned} \tag{11}$$

Assuming that Ψ changes slowly on distances of the order of the size of Λ_q, eq. (11) can be rewritten as the differential conservation law

$$\begin{aligned}&\partial_t \rho^i(\boldsymbol{R}, t) \\ &\quad + \boldsymbol{\nabla} \cdot \left(\sum_{a, \boldsymbol{r} \in \Lambda_q} \phi_a^i(\boldsymbol{R} + \boldsymbol{r})\,\boldsymbol{C}_a \Psi_a(\boldsymbol{R} + \boldsymbol{r}, t)\right) = 0.\end{aligned} \tag{12}$$

In correspondence to each extensive conserved quantity there will be an eigenvector of DL with eigenvalue 1. Thus d, the dimension of the eigenspace of DL corresponding to the eigenvalue 1, is the number of extensive quantities preserved by the dynamics.

In the proceeding discussion I have used the assumption that X^* does not depend on the

Table 1

Conserved quantity	Corresponding vector
number of particles	$\{1,1,1,1,1,1\}$
x momentum	$\{1,\frac{1}{2},-\frac{1}{2},-1,-\frac{1}{2},\frac{1}{2}\}$
y momentum	$\{0,1,1,0,-1,-1\}$

details of the microscopic collision rules but rather on the existence of microscopic conservation laws. If this is the case, as it should be if the microscopic conservation laws satisfy the property of semi-detailed balance, then the details of the linearized operator DL should be irrelevant too, and we can simplify the task by searching for eigenvectors with eigenvalue 1 of a simplified matrix $\widetilde{DL}$ defined as $\widetilde{DL}=((LT)S)^q$ where (LT) is a block diagonal matrix, each block of which is a $n_v \times n_v$ matrix B, where n_v is the number of microscopic velocity vectors, e.g., $n_v=6$ for the FHP. The matrix B is chosen so that the vectors corresponding to the microscopic conservation laws are eigenvectors of B with eigenvalue 1 and $0=Bv$ for any vector v orthogonal to the conservation laws vectors. For instance for the FHP model, see table 1.

2.2. *Search for eigenvectors*

We can now search for an orthonormal base $\{dY_i\}$ of the eigenspace of $\widetilde{DL}$ corresponding to eigenvalue 1, by using the following, almost obvious, procedure.

(0) Decide how many trial vectors we will consider, put this number in the variable TRIES. Set $d=0$, i.e., no eigenvector with eigenvalue 1 found thus far.

(1) If TRIES is equal to zero, report d (the number of vectors in $\{dY_i\}$), and stop; otherwise decrement TRIES by one.

(2) Randomly generate a vector dX. Eliminate the part of dX which is linearly dependent on the $\{dY_i\}$ found thus far. Rename the result dX.

(3) Repeatly apply DL on dX until it does not change any more.

(4) If the result of the previous operation is linearly dependent from the $\{dY_i\}$ known thus far go back to (1).

(5) Put the result in the set $\{dY_i\}$ and orthonormalize it again.

(6) Go back to (1).

The procedure defined above is numerically well natured because, since the system is finite, there is a clear separation of time scales, i.e., the multiplicity of the eigenvalue 1 of $\widetilde{DL}$ is (hopefully) small and all the other eigenvalues have modulo smaller than $1-\eta$, where η scales as the inverse of the squared size of the system[#5]

Applying this scheme to the FHP model (or to any of the derived models with a single speed) gives $d=3$ when both sides of Ω have an odd number of sites, $d=6$ when both sides have an even number of sites and $d=4$ otherwise. The six eigenvectors in the even–even case are easily recognized to correspond to the number of particles, the two components of the linear momentum and the three staggered momentum modes. I remind the reader that the staggered momentum densities can be written as

$$h_a(\boldsymbol{r},t)=(-1)^t(-1)^{\boldsymbol{B}_a\cdot\boldsymbol{r}}\boldsymbol{C}_a^{\perp}\cdot\boldsymbol{g}(\boldsymbol{r},t) \tag{13}$$

where $\boldsymbol{C}_a^{\perp}$ is obtained by rotating $\boldsymbol{C}_a$ by $\pi/4$ counterclockwise, $\boldsymbol{B}_a$ is the reciprocal space vector perpendicular to $\boldsymbol{C}_a$, i.e., $\boldsymbol{B}_a=(2/\sqrt{3})\boldsymbol{C}_a^{\perp}$, and $a=1,2,3$. The density h_a in equation (13) corresponds to the eigenvector $\{\phi_b^{h_a}(\boldsymbol{r})_{b,\boldsymbol{r}}=\{(-1)^{\boldsymbol{B}_a\cdot\boldsymbol{r}}\boldsymbol{C}_a^{\perp}\cdot\boldsymbol{C}_b\}_{b,\boldsymbol{r}}$, and it explains the period for $X^*(t)$ observed in step (B) of the procedure. It also explains the dependence of d on the "eveness" of the sides of Ω. In fact, the periodic boundary conditions imposed on the sides of Ω that are not parallel to $\boldsymbol{C}_a$ are effectively antiperiodic b.c. for h_a if the number of sites of the side is odd [8].

[#5] I am withholding some obvious technical results, e.g., deciding when a vector is (numerically) zero.

3. Conclusions

In the previous section, q was defined as the period of $X^*(t)$, the asymptotic solution the model Boltzmann equation. It is, however, clear that the direct use of the model Boltzmann equation is not necessary and one can directly search for the eigenvectors with eigenvalue 1 of $\widetilde{DL}$ and see what happens when q is changed. In this form the algorithm can be applied to models with a large number of velocities such as the currently proposed 3D schemes [2].

The scheme I described is basically a check on the "geometry" of a LGA model. I believe that this should be enough for models whose microscopic collision rules satisfy semi-detailed balance.[#6]

[#6] This assumption was already implicit in the decision of substituting the original Boolean model with a Boltzmann equation description. The latter is a closed equation for the one-particle probability distribution and it ignores completely multi-particles correlations. Thus it is not an accurate description of the original Boolean gas. However, if the microscopic collision rules satisfy semidetailed balance, the grand canonical probability distribution for the Boolean version of the model is expected to factorize in a product of single particles distribution functions with the latter being the one found as the equilibrium solution of the related Boltzmann equation. (We are actually dealing with the microcanonical equilibrium distribution here, but this should only introduce corrections of order (volume of $\Omega)^{-1}$.)

Acknowledgements

I would like to thank Dwight Barkley and Dan Rothman for useful discussions. This research was supported in part by NSF contract No. DMS-8906292 and DARPA contract No. N00014-86-K-0759.

References

[1] U. Frisch, B. Hasslacher and Y. Pomeau, Phys. Rev. Lett. 56 (1986) 1505.
[2] U. Frisch, D. d'Humières, B. Hasslacher, P. Lallemand, Y. Pomeau and J. Rivet, Complex Systems 1 (1987) 649.
[3] D. d'Humières, Y.H. Qian and P. Lallemand, Invariants in lattice gas models, in: Discrete Kinematic Theory, Lattice Gas Dynamics, and Foundations of Hydrodynamics, ed. R. Monaco (World Scientific, Singapore, 1989) pp. 102–113.
[4] G. Zanetti, Phys. Rev. A 40 (1989) 1539.
[5] G. McNamara, unpublished.
[6] F. Higuera, S. Succi and R. Benzi, Lattice gas dynamics with enhanced collisions IBM ECSEC preprint (1989).
[7] D. d'Humières and P. Lallemand, Complex Systems 1 (1987) 599.
[8] L. Kadanoff, G. McNamara and G. Zanetti, Complex Systems 1 (1987) 791.

CHAPTER 3

MULTIPHASE AND POROUS MEDIA

Physica D 47 (1991) 39–46
North-Holland

ANALYSIS OF SURFACE TENSION IN TWO-PHASE LATTICE GASES

J.A. SOMERS and P.C. REM
Koninklijke / Shell-Laboratorium, Amsterdam (Shell Research B.V.), Postbus 3003, 1003 AA Amsterdam, The Netherlands

Received 24 January 1990

One of the promising fields of application for lattice gas methods is two-phase flow simulation. An important issue in the design of two-phase lattice gas models is managing complexity. Continuing the basic principles of single-phase lattice gases, the two-phase collision operator should be kept as simple as possible, for the sake of both easy implementation and theoretical validation. At the same time it is required that parameters such as surface tension, diffusion and viscosity obtain physically relevant values. In this paper we present a two-dimensional two-phase model, based on a strictly local 16-bit collision operator, which incorporates a respectable surface tension relatively independent of the density.

1. Introduction

A two-dimensional collision operator which accomplishes surface tension was first defined by Rothman et al. [1]. At the workshop in Turin, September 1988, we introduced an alternative two-phase approach in three dimensions, in which both the holes and the particles carry information about the colour of the fluid [2]. The algorithm uses a strictly local collision operator, thus obeying the basic principles of a cellular automaton [3, 4] and reducing the complexity of the model. Recently, we have refined this approach and applied it to the hexagonal lattice gas with up to three stationary particles.

From the simulations with this two-phase lattice gas model we have learned that time correlations play a very important role. The Boltzmann approximation is by no means valid in a two-phase simulation, unless multiple lattices are combined in an ensemble-like system. For small ensembles (say up to five lattices) the surface tension increases appreciably with the ensemble size. Also the isotropy of the collision rules can be recovered much better macroscopically, when the ensemble size is increased.

The experiments which are presented in this paper have all been run on a 36-node Transputer network. All features of the implementation (i.e. initialization, propagation, collision, boundary conditions, post processing) have been parallelized, such that the overall performance of the algorithm scales with the size of the machine it runs on [5]. Currently, the algorithm runs at 750 000 node updates per second.

In the next section we will introduce the model and indicate how its collision tables can be constructed. The third section shows how Laplace's law is derived in the limit of a thin interface. It is shown that the surface tension results from a second-order perturbation of the equilibrium distribution of the particles, and that it is correlated with the mean free path in the single-phase fluids. The fourth section addresses the issue of isotropy. The perturbation of the equilibrium distribution at the interface will be studied in more detail, showing that it decays at a sub-lattice scale in the Boltzmann limit.

2. A two-phase hexagonal lattice gas

A two-phase hexagonal lattice gas simulates a two-phase flow by tracing explicitly a bulk of coloured particles, moving with unit speed along the edges of a hexagonal lattice. Particles may collide at the nodes of the lattice conserving mass of each kind and total momentum locally. Macroscopic phenomena, such as surface tension and diffusion, depend on the detailed form of the collision rules.

We will define a strictly local collision operator of a hexagonal lattice gas with two kinds of particles, coloured red and green, respectively. The velocities along the edges of the lattice are labeled $\boldsymbol{c}_i$, $0 \le i < n$. Two-dimensional models without stationary particles have $n = 6$. The state of a single node in the lattice is represented by a pair of sequences (s, f), $s = (s_i\colon\ 0 \le i < n)$, $f = (f_i\colon\ 0 \le i < n)$. The binary variable s_i indicates whether a particle ($s_i = 1$) or a hole ($s_i = 0$) is moving along edge $\boldsymbol{c}_i$. The binary variable f_i represents one unit of colour information on edge $\boldsymbol{c}_i$. Red particles are represented by $s_i = 1$, $f_i = 1$, green particles by $s_i = 1$, $f_i = 0$. Not only the particles but also the holes carry colour information. The collision rules are tuned such that the colour information moving along $\boldsymbol{c}_i$ is positively correlated with the colour of the bulk in direction $-\boldsymbol{c}_i$. This correlation is quite natural for the colour of the particles, and will be introduced artificially for the colour of the holes.

Given a pre-collision state (s, f) the collision operator produces an after-collision state (s', f'), thereby conserving mass of each kind, and total momentum

$$\sum_i f_i' s_i' = \sum_i f_i s_i,$$
$$\sum_i (1 - f_i') s_i' = \sum_i (1 - f_i) s_i,$$
$$\sum_i s_i' \boldsymbol{c}_i = \sum_i s_i \boldsymbol{c}_i. \tag{1}$$

Complete separation of the two fluids can only be maintained if the collision rules are capable of separating the two colours on a microscopic level. Therefore some information about the shape of the interface locally between the two fluids should be available from the discrete state of each single node. In ref. [1] the local estimate of the normal of the interface is called the local colour field, and it was derived from the colours of the particles at neighbouring nodes. In our case we exploit the positive correlation of f_i with the colour of the bulk in direction $-\boldsymbol{c}_i$. The local colour current $\boldsymbol{F}$ is defined strictly in terms of the state of a single node without distinguishing between particles and holes.

$$\boldsymbol{F} = \sum_i \boldsymbol{c}_i f_i. \tag{2}$$

Just to get started, we will assume that this local colour current $\boldsymbol{F}$ is a sufficiently accurate estimate of the normal of the interface locally. Hence we choose the collision rules such that the red and the green particles are scattered as close as possible respectively antiparallel and parallel to the local colour current, subject to the conservation laws (1). This is achieved if the collision operator maximizes

$$\sum_i (1 - 2f_i') s_i' (\boldsymbol{c}_i \cdot \boldsymbol{F}) \tag{3}$$

for each individual collision, which involves both kinds of particles.

Neither the above criterion nor the conservation laws specify what colour information will be carried by the holes in the after-collision state. Note that the propagation step conserves the colour current. So if the local colour current $\boldsymbol{F}$ in the pre-collision state should be positively correlated with the normal of the interface, the same should hold for the (similarly defined) local colour current $\boldsymbol{F}'$ in the after-collision state. So we will colour the holes in the after-collision state (s', f'), such that the scalar product $\boldsymbol{F}' \cdot \boldsymbol{F}$ is maximized. More specifically, a secondary criterion for the collision operator will maximize

$$\sum_i (1 - 2f_i')(s_i' - 1)(\boldsymbol{c}_i \cdot \boldsymbol{F}). \tag{4}$$

If multiple optimal collisions are found maximizing the criteria (3) and (4), one is picked arbitrarily. We have found experimentally, however, that a significantly higher surface tension is obtained when after-collision states with many stationary particles are preferred at the interface.

The above collision strategy implies that, in fact, the source of the colour current is the colour of the particles at the interface combined with the expanding nature of the gas. The holes only try to conserve the colour current through the collision, and propagate it into the bulk at both sides of the interface. However, the holes should not propagate the colour current too far away from the interface, as two independent interfaces at respectable distance should not "feel" each other.

The decay of the local colour current is dealt with in the collision rules for those states in which all particles have the same colour. The scattering of the particles in a single-phase collision is tuned to optimize the shear viscosity only. The colour information f' in the after-collision state is identical to the colour information f in the pre-collision state, except where a hole of different colour in the pre-collision state is replaced by a particle in the after-collision state. So the distance that, say, a red hole can penetrate into the green bulk of particles is related to the mean free path of the single-phase collision rules.

A complete time step of the two-phase hexagonal lattice gas, with one stationary particle, is shown in fig. 1. The actual interface where particles of both kinds coexist is only a few cells wide and very mobile due to the Brownian motion. However, the area with a positive colour current is much wider, ensuring that individual particles cannot occasionally penetrate the bulk of the opposite colour. Note that the width of the area with a positive colour current does not influence the single-phase bulk properties, as the scattering of particles due to the single-phase collision rules does not depend on the local colour current.

It is evident that the above two-phase hexagonal lattice gas can be implemented with a 16-bit look-up table (involving three stationary particles). Isotropy of the collision operator can easily be established by permuting the pre-collision states

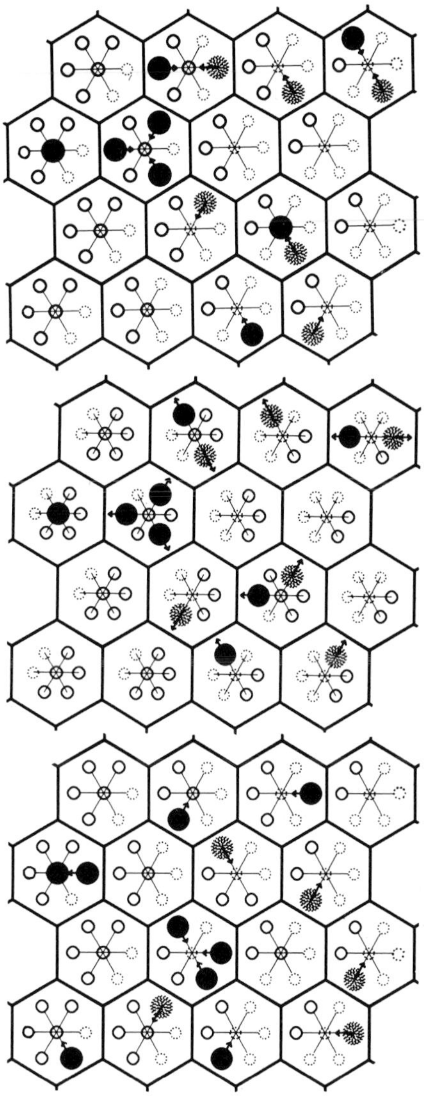

Fig. 1. Pre-collision state (a), after-collision state (b) and after-propagation state (c) of a complete time step in a two-phase hexagonal lattice gas with one stationary particle. Filled dots indicate particles, open circles indicate holes (absence of particles). Both the particles and the holes carry a colour (solid or dotted).

at run time with a random choice from the lattice symmetries and applying the inverse permutation to the after-collision state from the table. We would like to emphasize that it is very important to recover isotropy exactly at the microscopic level, as even the slightest deficiencies will become apparent in the macroscopic shapes of interfaces.

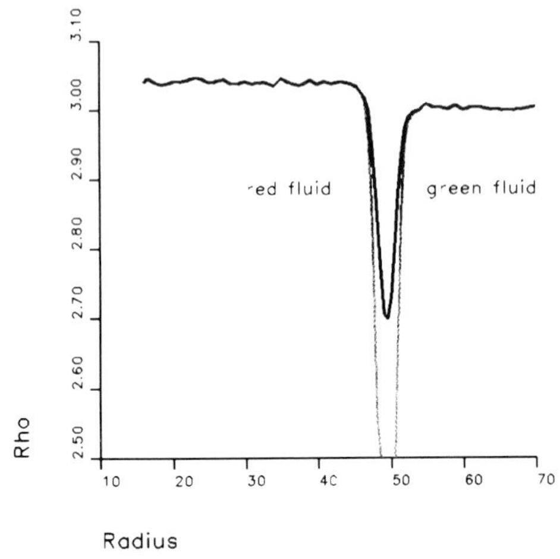

Fig. 2. The density within a bubble is plotted against the distance to the origin of the bubble. The density drop at the interface is related to the surface tension. The experiment was run with the model with three stationary particles. Tangential averages were taken over 100 time steps for 30 lattices in the ensemble.

3. Laplace's law

The design of the two-phase collision tables in the previous section is based on the intuitive idea about separation at a microscopic level. From experiments we have learned that the proposed two-phase hexagonal lattice gas is indeed capable of separating the two fluids spontaneously. Simulations of a red bubble with radius R in a green environment have shown a pressure difference between the red fluid within the bubble and its environment (see fig. 2). This pressure difference appears to satisfy the law of Laplace

$$\Delta P = \frac{\sigma}{R} \tag{5}$$

with σ indicating the surface tension at the interface. Next, we will explain how Laplace's law can be derived from the dynamics of the model, in the limit where the Boltzmann approximation is valid.

Let $N_i^{\mathrm{R}}(\boldsymbol{r})$ and $N_i^{\mathrm{G}}(\boldsymbol{r})$ denote the equilibrium distribution of, respectively, the red and green particles moving along edge $\boldsymbol{c}_i$ at position $\boldsymbol{r}$, relative to the centre of a red bubble with radius R. Arguments of symmetry, the finite lattice structure and a zero flow velocity imply that the equilibrium distribution at position $\boldsymbol{r}$ will depend on the densities of the fluids in the following way:

$$\begin{aligned} N_i^{\mathrm{R}} &= d^{\mathrm{R}} + \xi^{\mathrm{R}}\left[(\boldsymbol{c}_i\cdot\boldsymbol{e}_r)^2 - \tfrac{1}{2}\right] \\ &\quad + \zeta^{\mathrm{R}}\left[(\boldsymbol{c}_i\cdot\boldsymbol{e}_r)^3 - \tfrac{3}{4}(\boldsymbol{c}_i\cdot\boldsymbol{e}_r)\right], \\ N_i^{\mathrm{G}} &= d^{\mathrm{G}} + \xi^{\mathrm{G}}\left[(\boldsymbol{c}_i\cdot\boldsymbol{e}_r)^2 - \tfrac{1}{2}\right] \\ &\quad + \zeta^{\mathrm{G}}\left[(\boldsymbol{c}_i\cdot\boldsymbol{e}_r)^3 - \tfrac{3}{4}(\boldsymbol{c}_i\cdot\boldsymbol{e}_r)\right]. \end{aligned} \tag{6}$$

The quantities d^{R} and $d^{\mathrm{G}}, d^{\mathrm{R}} + d^{\mathrm{G}} = d$, are the usual relative densities, i.e. the mean particle occurrence of moving particles in the single-phase, zero-velocity limit. They relate to the macroscopic densities ρ^{R} and ρ^{G} depending on the number of stationary particles involved. ($\rho = 6d$, $\rho = 7d$ or $\rho = 7d + 2d^2/[d^2 + (1-d)^2]$ for respectively zero, one or three stationary particles. See also ref. [6].) The vector $\boldsymbol{e}_r$ is the radial unit vector. The coefficients $\xi^{\mathrm{R,G}}$ and $\zeta^{\mathrm{R,G}}$ are unknown, but they will depend on the two-phase collision rules. Note that these coefficients will both be zero far from the interface.

As usual, the time evolution of the $N_i^{\mathrm{R,G}}$ is specified by the Boltzmann transport equation, reflecting the dynamics of a lattice gas:

$$\begin{aligned} \partial_t N_i^{\mathrm{R}} + \boldsymbol{c}_i\cdot\nabla N_i^{\mathrm{R}} &= \Omega_i^{\mathrm{R}}, \\ \partial_t N_i^{\mathrm{G}} + \boldsymbol{c}_i\cdot\nabla N_i^{\mathrm{G}} &= \Omega_i^{\mathrm{G}}. \end{aligned} \tag{7}$$

Ω_i^{R} and Ω_i^{G} are the contributions of the collision operator on respectively N_i^{R} and N_i^{G}. In case of a steady state experiment the $\partial_t N_i^{\mathrm{R,G}}$ are both equal to zero, hence total momentum conservation yields

$$\begin{aligned} \boldsymbol{0} &= \left(\sum_i \boldsymbol{c}_i\Omega_i^{\mathrm{R}}\right) + \left(\sum_i \boldsymbol{c}_i\Omega_i^{\mathrm{G}}\right) \\ &= \left(\nabla\cdot\sum_i \boldsymbol{c}_i\boldsymbol{c}_i N_i^{\mathrm{R}}\right) + \left(\nabla\cdot\sum_i \boldsymbol{c}_i\boldsymbol{c}_i N_i^{\mathrm{G}}\right) \\ &= (\nabla\cdot\Pi^{\mathrm{R}}) + (\nabla\cdot\Pi^{\mathrm{G}}). \end{aligned} \tag{8}$$

Explicit formulae for the $\Pi^{\mathrm{R,G}}$ can be obtained from eq. (6) and the symmetries of the hexagonal lattice:

$$\Pi^{R,G}_{\alpha\beta} = \left(3d^{R,G} - \tfrac{3}{4}\xi^{R,G}\right)\delta_{\alpha\beta} + \tfrac{3}{2}\xi^{R,G} e_{r\alpha} e_{r\beta}. \quad (9)$$

In this equation α and β are coordinate indices in Cartesian space, and $\delta_{\alpha\beta}$ is the usual Kronecker delta. But before we substitute the Π^R and Π^G into eq. (8), we will first simplify the exercise by transforming (9) into a (r, ϕ) polar coordinate system:

$$\Pi^{R,G}_{r\phi} = \begin{pmatrix} 3d^{R,G} + \tfrac{3}{4}\xi^{R,G} & 0 \\ 0 & 3d^{R,G} - \tfrac{3}{4}\xi^{R,G} \end{pmatrix}. \quad (10)$$

Here we have assumed that the shape of the bubble is perfect, i.e. the $d^{R,G}$ and $\xi^{R,G}$ fields are cylinder-symmetric, varying only with r. Later we will show that this assumption is not valid a priori.

After substituting the radial components of the divergence of Π^R and Π^G, i.e. $(\nabla \cdot \Pi)_r = \partial_r \Pi_{rr} + (\Pi_{rr} - \Pi_{\phi\phi})/r$, one finds for the radial component of eq. (8)

$$0 = \partial_r\left[3d + \tfrac{3}{4}(\xi^G + \xi^R)\right] + \frac{3}{2r}(\xi^R + \xi^G). \quad (11)$$

This differential equation can be integrated analytically. The integration constant follows from the boundary condition at the origin $d(0) = d^R(0)$:

$$d(r) = d^R(0) - \tfrac{1}{6}F(r) - \frac{1}{3}\int_0^r \frac{F(r')}{r'}\,\mathrm{d}r' \quad (12)$$

with

$$F(r) = \tfrac{3}{2}\left[\xi^R(r) + \xi^G(r)\right].$$

The higher-order perturbations from the homogeneous particle distribution are generated at the interface where the two fluids meet, and will decay exponentially. So the $\xi^{R,G}$ field and thus $F(r)$ will find its maximum value exactly at the interface but drop to zero very quickly at some distance from the interface. Let us neglect the value of $F(r)$ outside the interval $(R - \Delta, R + \Delta)$, and assume Δ to be small compared to the radius R of the bubble. Then we can derive Laplace's law from eq. (12) using the boundary condition at infinity, i.e. $d(\infty) = d^G(\infty)$, and the relation between the pressure p and the relative density d in the zero velocity limit, $p = 3d$:

$$p^R(0) - p^G(\infty) = \frac{1}{R}\int_{R-\Delta}^{R+\Delta} F(r)\,\mathrm{d}r. \quad (13)$$

Thus the surface tension of the lattice gas appears to depend on the thickness of the interface and the coefficients $\xi^{R,G}$. We have not yet found a way to derive these coefficients directly from the collision rules, though some insight into their nature has been acquired experimentally.

The Boltzmann limit can be approximated in an actual lattice gas experiment by running multiple lattices simultaneously in order to achieve an ensemble-like behaviour. Convergence of the whole system to one single solution is guaranteed by mixing among the lattices in the ensemble during the propagation step. Indeed we have found that a two-phase lattice gas ensemble simulation can show totally different phenomena compared to an ordinary lattice gas run (with just one member of the ensemble). Even a few (say five) lattices are sufficient to undo the most disturbing influences of time correlations.

In fig. 2 experimental values of d^R, d^G and d are plotted against the distance r from the centre of the bubble, such that the form of eq. (12) can be verified. The data were obtained by taking averages in time and over the ensemble (100 time steps, 30 lattices) for each distance. The model in this experiment involved three stationary particles. Semi-detailed balance was satisfied for the single-phase collision. The plot shows a very steep pressure gradient near the interface, corresponding to a density decay of about a factor of 10 per site. A similar plot for just one single lattice in the ensemble would show a much smoother pressure gradient and a lower level within the bubble.

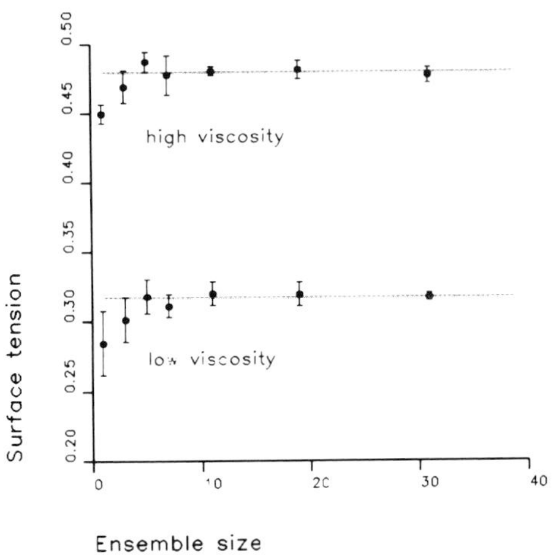

Fig. 3. Surface tension in a bubble experiment with radius 50. The measured values increase with the size of the ensemble. Convergence is observed when at least five lattices are run. Three stationary particles were included.

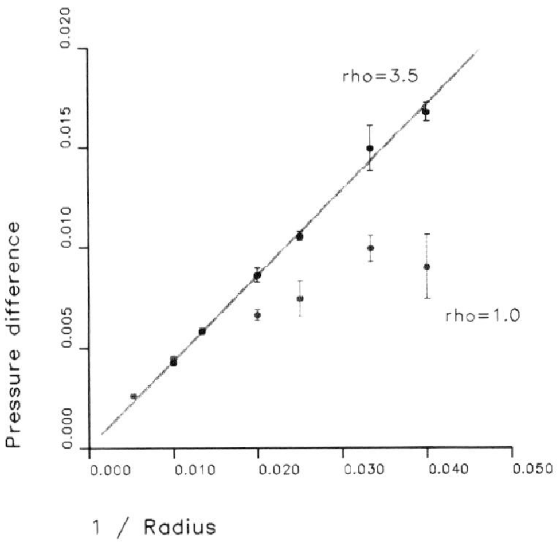

Fig. 4. According to Laplace's law, the pressure difference between a bubble and its environment is proportional to the inverse of its radius. Significant deviations are found for small bubbles at a low density when the width of the interface is of the same order as the radius of the bubble. A model with three stationary particles was used with one single lattice in the ensemble.

In fig. 3 we have plotted the surface tension as a function of the ensemble size. Small ensembles exhibit a notably lower surface tension, but fortunately the behaviour of the lattice gas converges very soon. An ensemble of five lattices seems to be reasonable enough to approximate the Boltzmann limit.

The pressure difference between a bubble and its environment has been measured for bubbles with various radii at various densities. Fig. 4 shows the results as a function of $1/R$. The experiments were run with only one lattice in the ensemble, averaging over 5000 time steps. Significant deviations from Laplace's law are found for small bubbles at a low density. In the latter experiments the width of the interface was relatively large compared to the radius R of the bubble. Fig. 5 shows the measured surface tension as a function of the relative density, again from experiments with only one lattice in the ensemble. The highest surface tension is obtained when both single phase fluids exhibit a high viscosity. Apparently the decay of the $F(r)$ field depends on the mean free path in the single-phase. A long mean free path (high viscosity)

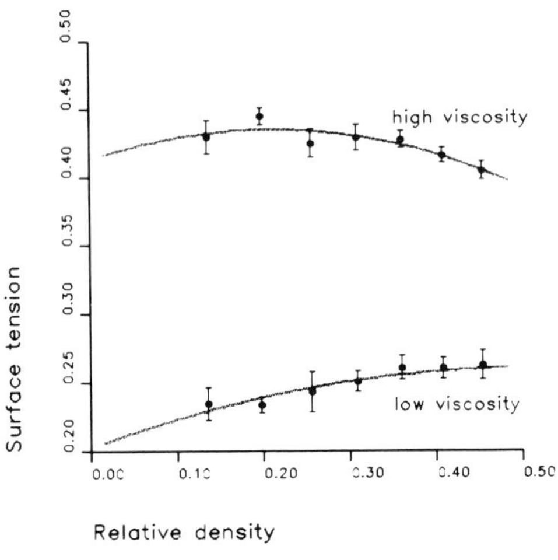

Fig. 5. Surface tension of a high-viscosity and a low-viscosity fluid measured at various densities on the basis of the model with three stationary particles.

enlarges the outcome of the integral in eq. (13). This difference in surface tension between highly viscous fluids and low-viscosity fluids seems to persist as well in the Boltzmann limit, even though the area with a non-zero $F(r)$ becomes very small (see fig. 3).

4. Isotropy of the surface tension

In order to obtain better insight into the mechanisms which derive the behaviour of a two-phase lattice gas near the interface, we will transform the $N_i^{\mathrm{R,G}}$ of eq. (6) linearly into a carefully chosen vector space $\langle \boldsymbol{v}_i \rangle$, $0 \le i < n$ (see also ref. [7]). For the sake of simplicity, we will just consider the model without stationary particles. The vector space consists of the five degrees of freedom that appear in the Navier–Stokes equations plus a uniquely remaining sixth, which does not show any direct hydrodynamic relevance at the macroscopic level but comes into play when the anisotropy of higher-order terms is studied. The six base vectors $\boldsymbol{v}_i$ are defined in the following way:

$$\begin{aligned}
\boldsymbol{v}_0 &= (1,1,1,1,1,1),\\
\boldsymbol{v}_1 &= (1,\tfrac{1}{2},-\tfrac{1}{2},-1,-\tfrac{1}{2},\tfrac{1}{2}) = \boldsymbol{c}_{ix},\\
\boldsymbol{v}_2 &= (0,\tfrac{1}{2}\sqrt{3},\tfrac{1}{2}\sqrt{3},0,-\tfrac{1}{2}\sqrt{3},-\tfrac{1}{2}\sqrt{3}) = \boldsymbol{c}_{iy},\\
\boldsymbol{v}_3 &= (1,-\tfrac{1}{2},-\tfrac{1}{2},1,-\tfrac{1}{2},-\tfrac{1}{2}) = \boldsymbol{c}_{ix}^2 - \boldsymbol{c}_{iy}^2,\\
\boldsymbol{v}_4 &= (0,\tfrac{1}{2}\sqrt{3},-\tfrac{1}{2}\sqrt{3},0,\tfrac{1}{2}\sqrt{3},-\tfrac{1}{2}\sqrt{3}) = 2\boldsymbol{c}_{ix}\boldsymbol{c}_{iy},\\
\boldsymbol{v}_5 &= (1,-1,1,-1,1,-1) = \boldsymbol{c}_{ix}^3 - 3\boldsymbol{c}_{ix}\boldsymbol{c}_{iy}^2. \qquad (14)
\end{aligned}$$

The equilibrium distribution of the red and the green particles with respect to the vector space $\langle \boldsymbol{v}_i \rangle$ will be denoted by E_i^{R} and E_i^{G}, respectively. Hence the $N_i^{\mathrm{R,G}}$ are related to the $E_i^{\mathrm{R,G}}$ by

$$N_i^{\mathrm{R,G}} = \left(\sum_j E_j^{\mathrm{R,G}} \boldsymbol{v}_j \right)_i . \qquad (15)$$

Obviously, trivial equalities are satisfied for the densities $\rho^{\mathrm{R,G}}$, and the spatial velocity components U_x and U_y in terms of $E_i^{\mathrm{R,G}}$:

$$E_0^{\mathrm{R,G}} = \tfrac{1}{6}\rho^{\mathrm{R,G}},$$

$$E_1^{\mathrm{R}} + E_1^{\mathrm{G}} = \tfrac{1}{3}\rho U_x, \quad E_2^{\mathrm{R}} + E_2^{\mathrm{G}} = \tfrac{1}{3}\rho U_y. \qquad (16)$$

Now it is interesting to consider the equilibrium distribution in the case that a macroscopic bubble with radius R is assumed. By substituting $\boldsymbol{e}_r = (\cos\phi, \sin\phi)$ the following relations emerge for the coefficients $\xi^{\mathrm{R,G}}$ and $\zeta^{\mathrm{R,G}}$ in eq. (6):

$$\begin{aligned}
E_3^{\mathrm{R,G}}(r) &= \tfrac{1}{2}\xi(r)^{\mathrm{R,G}}\cos(2\phi),\\
E_4^{\mathrm{R,G}}(r) &= \tfrac{1}{2}\xi(r)^{\mathrm{R,G}}\sin(2\phi),\\
E_5^{\mathrm{R,G}}(r) &= \tfrac{1}{4}\zeta(r)^{\mathrm{R,G}}\cos(3\phi). \qquad (17)
\end{aligned}$$

These equations show the required functional form of $E_3^{\mathrm{R,G}}$ and $E_4^{\mathrm{R,G}}$, such that the $\xi^{\mathrm{R,G}}$ coefficients will not depend on ϕ (which is the case for an isotropic bubble). One might think of obtaining quantitative values for $E_3^{\mathrm{R,G}}$, $E_4^{\mathrm{R,G}}$ and $E_5^{\mathrm{R,G}}$ directly from the collision rules in a way similar to Hénon's linear analysis to obtain a theoretical value for the viscosity ν. However, such an exercise would not work, as is clear from our finding (see previous section) that these fields decay by about a factor of 10 per lattice unit in the Boltzmann limit. Hence the second-order perturbations of the particle distribution at the interface evolve at a sub-lattice scale, and a linear analysis does not converge!

We have been able to obtain experimental values for the $E_3^{\mathrm{R,G}}$, $E_4^{\mathrm{R,G}}$ and $E_5^{\mathrm{R,G}}$ fields. These experimental values provide an alternative way to calculate the surface tension. By eliminating the $\xi^{\mathrm{R,G}}$ in eqs. (13), (17) and integrating over ϕ, an explicit formula for the surface tension is obtained which can be evaluated using experimental values of the fields $E_3^{\mathrm{R,G}}$ and $E_4^{\mathrm{R,G}}$ only:

$$\sigma = \frac{3}{2\pi}\int_0^{2\pi}\int_{R-\Delta}^{R+\Delta}\left(\frac{(E_3^{\mathrm{R}} + E_3^{\mathrm{G}})\cos 2\phi}{r} + \frac{(E_4^{\mathrm{R}} + E_4^{\mathrm{G}})\sin 2\phi}{r}\right)\mathrm{d}r\,\mathrm{d}\phi. \qquad (18)$$

For the model with three stationary particles the surface tension, computed this way, agrees astonishing well (within 1%) with the surface tension as it was obtained by measuring the pressure difference between the bubble and its environment. (The experiments involved 20 lattices in an ensemble.) This indicates that the assumptions which had to be made to derive the final relations (13) and (18) are realistic.

No agreement at all is found for the model without stationary particles. The experiments reveal a truly hexagonal bubble with each side of the hexagon perpendicular to an edge of the lattice. Obviously, the assumption of the bubble being circular does not hold here and significant parts of the theory are not valid anymore. New expressions for the $E_i^{R,G}$ fields can easily be obtained by substituting an anisotropic $\boldsymbol{e}_r$ in eq. (6). An anisotropic generalization of eq. (13), however, is not straightforward.

Note, that for the two-dimensional models the local colour current $\boldsymbol{F}$, which was defined in eq. (2), always points into one out of thirteen directions, i.e. zero or along $(\cos(i\pi/12), \sin(i\pi/12))$, $0 \le i < 12$. The hexagonal shape of a bubble indicates that the separation of the fluids along the directions $(\cos(i\pi/6), \sin(i\pi/6))$ is biased wrongly with the separation along $(\cos[(2i+1)\pi/12], \sin[(2i+1)\pi/12])$. It happens to be a lucky hit that this problem does not emerge in the two-dimensional model with three stationary particles. We believe that the tuning of the collision rules, in order to recover a macroscopically isotropic surface tension will become an issue again, when three-dimensional models are constructed.

5. Conclusions

The lattice gas automata technique is known to excel in tackling complexity through simplicity. Back in ref. [1] it was shown how moving interfaces in two-phase flows of any complexity can be modelled explicitly within the rules of a relatively simple cellular automaton. In this paper we have simplified the dynamics of such a two-phase lattice gas even further. This way we have improved the efficiency of simulation runs. This step also gave us the opportunity to obtain some theoretical basis for the model and a better understanding of the mechanisms that govern its behaviour.

A three-dimensional generalization of the proposed model seems feasible, and indeed would bring in a competitive edge for the lattice gas automata as compared to classical numerical techniques. However, considering the effort it takes to implement even merely single-phase three-dimensional lattice gas, we are convinced that still more simplicity is required to make a two-phase version run efficiently.

References

[1] D.H. Rothman and J.M. Keller, Immiscible cellular-automaton fluids, J. Stat. Phys. 52 (1988) 1119.

[2] P.C. Rem and J.A. Somers, Cellular automata algorithms on a Transputer network, in: Discrete Kinematic Theory, Lattice Gas Dynamics, and Foundations of Hydrodynamics, ed. R. Monaco (World Scientific, Singapore, 1989) pp. 268–275.

[3] U. Frisch, B. Hasslacher and Y. Pomeau, Lattice gas automata, Phys. Rev. Lett. 56 (1986) 1505–1508.

[4] U. Frisch, D. d'Humières, B. Hasslacher, P. Lallemand, Y. Pomeau and J.-P. Rivet, Lattice gas hydrodynamics in two and three dimensions, Complex Systems 1 (1987) 649–707.

[5] J.A. Somers and P.C. Rem, A parallel cellular automata implementation on a Transputer network for the simulation of small scale fluid flow experiments, in: Proceedings of Shell Conference on Parallel Computing, ed. G.A. van Zee, Lecture Notes on Computer Science, No. 384 (Springer, Berlin, 1989).

[6] D. d'Humières and P. Lallemand, Numerical simulations of hydrodynamics with lattice gas automata in two dimensions, Complex Systems 1 (1987) 599–632.

[7] U. Frisch, Relation between the lattice Boltzmann equation and the Navier–Stokes equations, Physica D 47 (1991) 231–232, these Proceedings.

Physica D 47 (1991) 47–52
North-Holland

A LATTICE-GAS MODEL FOR THREE IMMISCIBLE FLUIDS

Andrew K. GUNSTENSEN and Daniel H. ROTHMAN
Department of Earth, Atmospheric, and Planetary Sciences, Massachusetts Institute of Technology, Cambridge, MA 02139, USA

Received 20 January 1990

Lattice-gas methods have recently proven very useful for the study of immiscible mixtures of two fluids, with applications ranging from two-phase flow in porous media to spinodal decomposition of binary fluids. Whereas the original one-phase lattice gas models the fluid as a collection of identical particles, in the immiscible two-phase lattice gas the particles are colored red or blue and the collisions between particles are chosen to achieve surface tension.

We introduce a new lattice-gas model which extends the two-phase immiscible lattice gas to the simulation of a mixture of three immiscible fluids, i.e., red, green and blue. This extension achieves more than the obvious generalization: immiscible mixtures of three fluids yield phenomena that can be qualitatively different from analogous phenomena observed with two fluids. To demonstrate this point, we show simulations of phase separation of three immiscible fluids and three-phase flow in porous media.

1. Introduction

Lattice gases are a new class of numerical methods for the solution of the Navier–Stokes equations. In the original, one-phase model proposed by Frisch, Hasslacher and Pomeau (FHP) [1], the fluid is represented by particles of unit mass and momentum moving on a triangular lattice. When the particles reach a node on the lattice they undergo a collision which redistributes the particles among the available directions while conserving the total mass and total momentum present at the node. By taking large-scale averages from the lattice, the full Navier–Stokes equations can be recovered in an asymptotic limit [1, 2].

Lattice gases offer particularly intriguing opportunities for the study of binary fluid mixtures, both miscible [3–5] and immiscible [6, 7, 14]. In the case of immiscible lattice gases (ILG), the particles are colored red or blue and the collision rules are chosen to send particles towards neighboring sites containing other particles of the same color, achieving surface tension.

In this paper, we introduce a new lattice-gas model for the simulation of three immiscible fluids. This extension achieves more than a simple generalization of the previous two-fluid ILG, because mixtures of three fluids can yield qualitatively different phenomena than those observed in binary fluids. For example, due to the presence of three competing surface tensions, bubbles are not necessarily round. Also, for flow in the presence of solid boundaries, the same fluid can now be wetting relative to one fluid and non-wetting relative to the other.

The extension of the two-phase ILG to the simulation of three immiscible fluids is straightforward. The particles are now colored red, blue or green and the collisions are chosen to yield three surface tension coefficients, which are not necessarily equal.

We demonstrate the utility of the new model with two sets of simulations. In the first example, we show the spontaneous separation of three fluids for two different combinations of surface tensions. For the first combination of surface tension coefficients, the three-phase contact

0167-2789/91/$03.50

points (contact lines in three dimensions) are stable and the fluid separates into polygonal bubbles. For the second set of coefficients, the three-phase contact points are unstable and the fluid separates into two big bubbles separated by a thin layer of the third fluid.

In the second example, we demonstrate an application of the new model to the simulation of three-phase flow in porous media. Our extension thus allows us to simulate the simultaneous flow of, say, oil, water, and gas in sedimentary rocks. We compare our three-phase simulation to an analogous two-phase simulation, and observe that the third acts to block the flow of the invading fluid.

2. The three-phase model

We begin by reviewing ILG model of ref. [6]. The three-phase model then follows as a simple extension.

In the ILG model the particles are colored either red or blue and surface tension is obtained by the correct choice of collision rules. Let the presence or absence of a red particle with velocity $\boldsymbol{c}_i$ be indicated by the Boolean variable r_i and a blue particle by b_i. Only one of r_i or b_i can be equal to 1 since there is a maximum of one particle per node with a given velocity. In addition to the six allowed moving particles, up to one rest particle per site, with unit mass and zero velocity, is added to the model.

We define the local color "flux"

$$\boldsymbol{q}(s(\boldsymbol{x})) = \sum_i \boldsymbol{c}[r_i(\boldsymbol{x}) - b_i(\boldsymbol{x})], \tag{1}$$

where $s(\boldsymbol{x})$ is a number representing the state of the particles at the site at location $\boldsymbol{x}$ and is composed of the individual r_i and b_i. The local color "field" (essentially the local color gradient) is defined by

$$\boldsymbol{f}(\boldsymbol{x}) = \sum_i \boldsymbol{c}_i \sum_j [r_j(\boldsymbol{x}+\boldsymbol{c}_i) - b_j(\boldsymbol{x}+\boldsymbol{c}_i)]. \tag{2}$$

The "work" done by the flux against the field is then

$$W(s) = -\boldsymbol{f}\cdot\boldsymbol{q}(s). \tag{3}$$

The collision rules are then chosen to minimize W, while conserving the number of each color of particle and the total momentum at a site. In symbols, the collision results in the state s that satisfies

$$\min_{s\in\mathscr{S}(s_{\text{in}})} W(s), \tag{4}$$

where $\mathscr{S}(s_{\text{in}})$ is the set of all allowed output states for input state s_{in}, or all states with the same number of each color of particle and the same total momentum as s_{in}. These rules give the same Navier–Stokes equations in the homogeneous region as in the original FHP lattice gas, but with the addition of surface tension at interfaces [6].

In the three-phase model, the particles are colored either red, green or blue and the Boolean variables r_i, g_i and b_i are used to indicate the presence of a particle of the corresponding color at a site. Again, there is a maximum of one particle per site with a given velocity. The color flux is now defined separately for each of the three colors; thus

$$\boldsymbol{q}_{\text{r}}(\boldsymbol{x}) = \sum_i \boldsymbol{c}_i r_i(\boldsymbol{x}),$$
$$\boldsymbol{q}_{\text{g}}(\boldsymbol{x}) = \sum_i \boldsymbol{c}_i g_i(\boldsymbol{x}),$$
$$\boldsymbol{q}_{\text{b}}(\boldsymbol{x}) = \sum_i \boldsymbol{c}_i b_i(\boldsymbol{x}) \tag{5}$$

instead of the one, signed flux used in the two-phase model. Similarly, the three local color gradients or fields are defined to be

$$\boldsymbol{f}_{\text{r}}(\boldsymbol{x}) = \sum_i \boldsymbol{c}_i \sum_j r_j(\boldsymbol{x}+\boldsymbol{c}_i),$$
$$\boldsymbol{f}_{\text{g}}(\boldsymbol{x}) = \sum_i \boldsymbol{c}_i \sum_j g_j(\boldsymbol{x}+\boldsymbol{c}_i),$$
$$\boldsymbol{f}_{\text{b}}(\boldsymbol{x}) = \sum_i \boldsymbol{c}_i \sum_j b_j(\boldsymbol{x}+\boldsymbol{c}_i). \tag{6}$$

The work done by the flux against the field is now

$$W(s) = -(\alpha \boldsymbol{f}_{\mathrm{r}} \cdot \boldsymbol{q}_{\mathrm{r}} + \beta \boldsymbol{f}_{\mathrm{g}} \cdot \boldsymbol{q}_{\mathrm{g}} + \gamma \boldsymbol{f}_{\mathrm{b}} \cdot \boldsymbol{q}_{\mathrm{b}}), \tag{7}$$

where α, β, and γ are weighting factors which allow the surface tension between any two of the three pairs of fluids to be varied relative to the other surface tensions. The collision optimization may once again be written by eq. (4), where $\mathscr{S}(s_{\mathrm{in}})$ now includes the additional constraint that the number of green particles must be conserved.

The relative strengths of the surface tensions can be varied by changing the values of α, β and γ. For example, if $\alpha = \beta = 1$ and $\gamma = 0.5$ then the surface tension σ_{rg}, between red and green would be greater than the other two surface tensions σ_{rb} and σ_{bg}.

This algorithm was implemented on a MIPS M/120 computer and achieved a speed of approximately 30 000 site updates per second, which is about 40% of the speed of the two-phase algorithm.

3. Surface tension measurements

The presence of surface tension in the three-phase immiscible lattice gas was confirmed using an improved version of the bubble test used in ref. [6]. In two dimensions, Laplace's law of surface tension is

$$p_{\mathrm{inner}} - p_{\mathrm{outer}} = \frac{\sigma}{R}, \tag{8}$$

where p_{inner} is the pressure inside the bubble, p_{outer} is the pressure outside the bubble, σ is the surface tension coefficient and R is the radius of the bubble.

The numerical experiment is initialized with a bubble of one color in a sea of another color. The system is allowed to evolve for 500 time steps to attain equilibrium and then the average pressure difference between the inside and outside of the bubble is measured over the next 1500 time steps. The pressure inside the bubble is measured by computing the colorblind pressure at all lattice sites inside a circle of radius $0.7R$, centered on the bubble. Similarly, the external pressure is obtained by measuring the colorblind pressure at all lattice sites greater than $1.3R$ from the center of the bubble. This method of measurement eliminates any biases in the measured pressure resulting from the presence of a small number of particles of one color inside a region of another color due to the small, but finite, diffusivity of one color into the other. The experiment was repeated for a range of bubble radii from $R = 16$ to $R = 60$. By plotting the pressure difference against the inverse radius the surface tension can be found from the slope of the resulting line.

The surface tension coefficients corresponding to two different sets of collision weights were measured and the results are shown in fig. 1. The points on the curve are the average of six independent tests with error bars of ± 1 standard

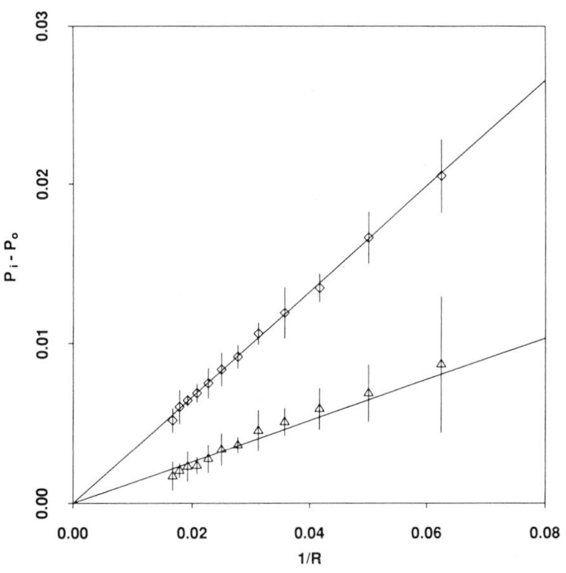

Fig. 1. The results of a series of bubble tests. The points shown represent the average of six independent numerical experiments with error bars of ± 1 standard deviation. The lines are the best fitting straight lines through the origin. The surface tension is equal to the slope of the curve. The diamonds correspond to values of $\alpha = \gamma = 1.0$ and $\sigma = 0.33 \pm 0.03$. The triangles correspond to values of $\alpha = 1.0$ and $\gamma = -0.5$ and $\sigma = 0.13 \pm 0.03$.

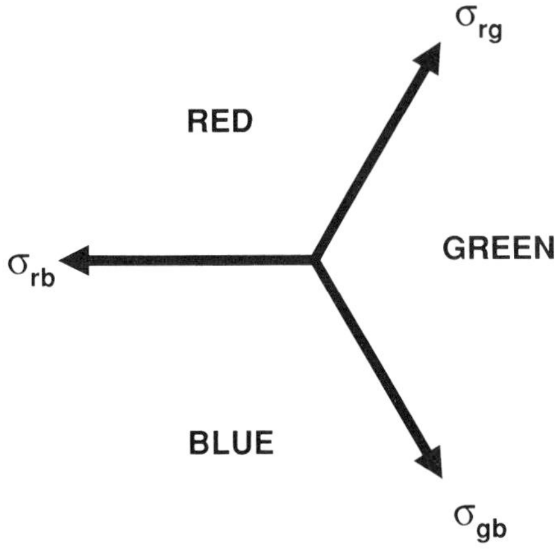

Fig. 2. The forces exerted by the surface tension on the three-phase contact point, where all three fluids meet. If one of the surface tensions is greater than the sum of the other two then no stable contact point can exist.

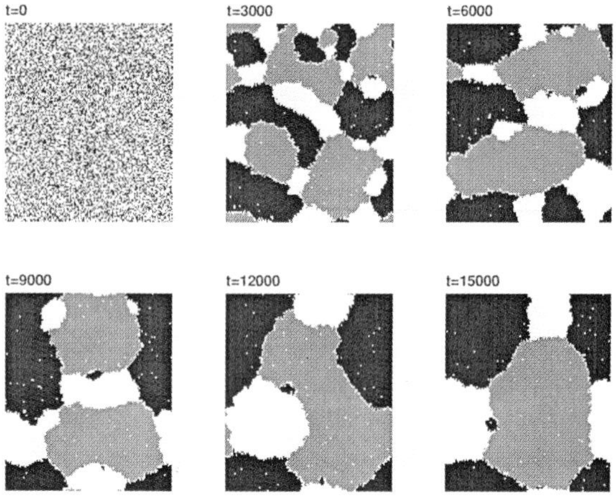

Fig. 3. Phase separation for equal surface tensions. The blue fluid is shown as black, the green as grey and the red as white. The fluid separates into polygonal bubbles.

deviation and the solid lines are the best fitting straight lines through the origin. The linearity of the two sets of experimental points indicates that the model has the correct dependence of the pressure difference on radius. The points shown by the diamonds correspond to values of $\alpha = \beta = \gamma = 1$ and give a surface tension of $\sigma_{rg} = \sigma_{rb} = \sigma_{gb} = 0.33 \pm 0.03$. The points shown by the triangles correspond to values of $\alpha = \beta = 1$, $\gamma = -0.5$ and give a surface tension of $\sigma_{rb} = \sigma_{gb} = 0.13 \pm 0.03$ and $\sigma_{rg} = 0.33 \pm 0.03$.

4. Phase separation

The ability of one pair of fluids to have a surface tension greater than the sum of the other two surface tensions is an important feature of the model. At a three-phase contact point the surface tensions exert forces on the contact point as shown in fig. 2. If the largest surface tension is greater than the sum of the other two, then the forces cannot balance and the contact point will not be stable. If stable contact points can exist then the three fluids will separate into polygonal blobs but if no stable contact points can exist then a thin layer of one fluid will insert itself between any boundary between the two fluids with the strong surface tension. These differences in the nature of the contact points give rise to large changes in the overall dynamics.

An example of the spontaneous separation of three initially mixed fluids with three equal surface tension coefficients is shown in fig. 3. The black represents the blue fluid, the grey the green fluid and the white the red fluid. As expected the fluid separates into polygonal blobs. Another

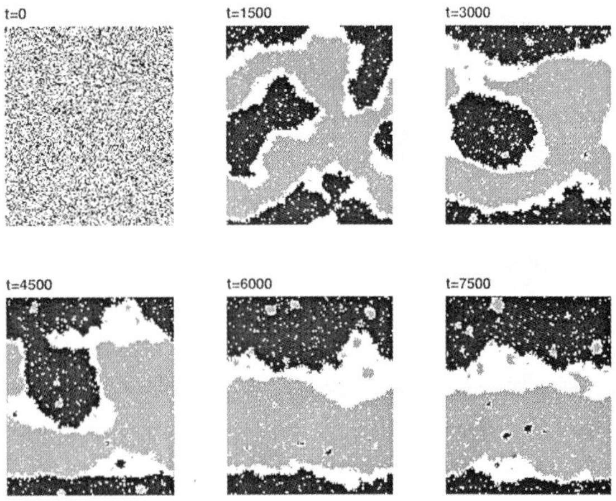

Fig. 4. Phase separation for unequal surface tensions. The blue fluid is shown as black, the green as grey and the red as white. Since no stable contact points can exist, the fluid separates into bubbles of blue and green separated by a thin layer of red.

phase separation, this time with one large and two small surface tensions, is shown in fig. 4. For this case the red fluid inserts itself between the other two fluids due to the instability of the contact points.

5. Three-phase flow in porous media

An important example of multiphase flow occurs in porous media [8, 9]. The utility of lattice-gas methods for the simulation of one- and two-phase flow in microscopic models of porous media has already been demonstrated [10–12]; here we extend these applications to the flow of three immiscible phases.

In the following examples, the wetting properties of the solid boundaries are set such that the blue fluid wets the solid relative to both the red and green fluids and the red fluid wets the solid boundaries relative to the green fluid. This is similar to the flow of oil (red), water (blue), and gas (green) in typical sedimentary rocks.

The porous medium used in our numerical experiments is similar to that used by Lenormand et al. in their experimental studies [13]. The medium consists of random sized squares of solid material located on a regular square grid. A 180 by 180 lattice was used for the simulation with the blocks placed on a 5 by 5 macroscopic grid. The minimum channel width is set to 8 lattice units wide, which, by previous tests, is wide enough to avoid serious discrepancies from the presumed continuum limit. The variation in size between the largest and smallest blocks spans a factor of three. The surface tensions were set such that $\sigma_{bg} = 0.33$ and $\sigma_{rb} = \sigma_{rg} = 0.13$; thus three-phase contact points are unstable.

For the first run, the lattice was initialized with a mixture of red and green fluids and the blue fluid injected at the top. A constant pressure difference was applied between the top and bottom of the lattice by adding a fixed quantity of momentum per time step to the system along the top of the lattice. Fig. 5 shows the evolution of

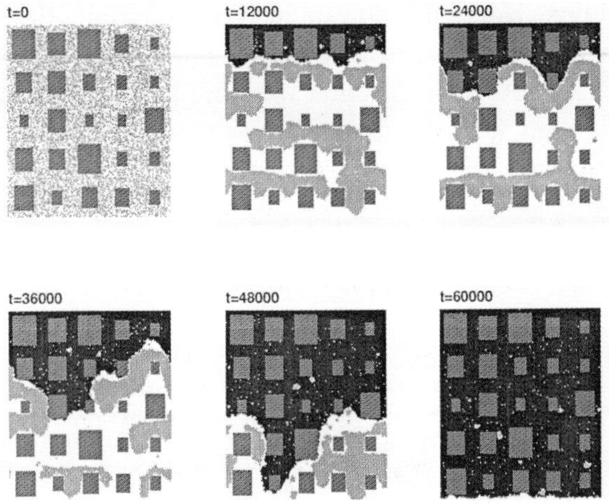

Fig. 5. Imbibition of an invading fluid into a medium initially containing two resident fluids. The color scheme is the same as for the phase separation pictures with the solid blocks shown as dark grey.

this experiment. The color scheme used here is similar to the one used to show the phase separation: red is white, blue is black and green is grey. In addition, the blocks are shown in another shade of grey.

It is interesting to compare the result of fig. 5 with a similar experiment; but with only two

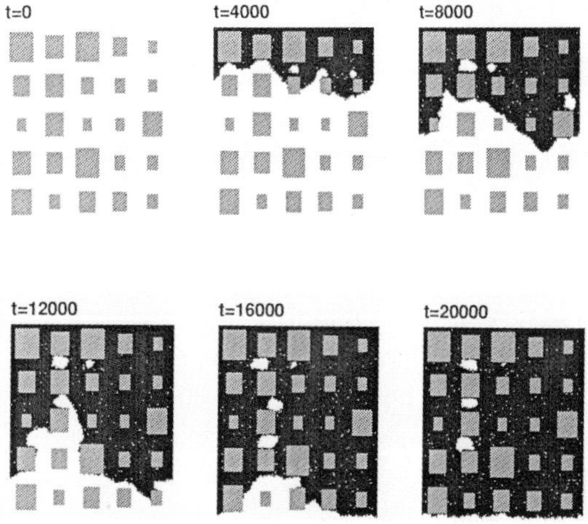

Fig. 6. Imbibition of an invading fluid into a medium initially containing only one resident fluid (note the difference in time scales from fig. 5). The color scheme is the same as for the phase separation pictures with the solid blocks shown as dark grey.

fluids. Here, the porous medium was initialized with only one resident phase, red. From fig. 6, one sees that the flow of the invading fluid through the porous medium is now much faster than the flow in the previous simulation. The third phase acts to block the pores to both the passage of the invading fluid and escape of the other resident fluid. It is not yet clear how much of this effect is due to the two-dimensional nature of the simulation.

6. Conclusions

We have introduced a generalization of the two-fluid ILG for the simulation of three immiscible fluids. Examples of phase separation of three-component fluid mixtures with both stable and unstable three-phase contact points have been illustrated, in addition to simulations of multiphase flow through a microscopic model of a porous medium. In the latter case we observed that the presence of the third fluid in the porous medium acts to block the flow of the invading fluid. In both of these examples, the presence of the third fluid yields new phenomena not seen in the analogous binary system.

Acknowledgements

We thank F. Kalaydjian and R. Lenormand for interesting discussions. This work was supported by NSF Grant EAR-8817027 and the sponsors of the MIT Porous Flow Project.

References

[1] U. Frisch, B. Hasslacher and Y. Pomeau, Phys. Rev. Lett. 56 (1986) 1505.
[2] U. Frisch, D. d'Humières, B. Hasslacher, P. Lallemand, Y. Pomeau and J. Rivet, Complex Systems 1 (1987) 649.
[3] C. Burges and S. Zaleski, Complex Systems 1 (1987) 31.
[4] D. d'Humières, P. Lallemand and G. Searby, Complex Systems 1 (1987) 633.
[5] P. Clavin, P. Lallemand, Y. Pomeau and G. Searby, J. Fluid Mech. 188 (1988) 437.
[6] D.H. Rothman and J. Keller, J. Stat. Phys. 52 (1988) 1119.
[7] D.H. Rothman and S. Zaleski, J. Phys. (Paris) 50 (1989) 2161.
[8] J. Bear, Dynamics of fluids in porous media, (Dover Publications, New York, 1972).
[9] A.E. Scheidegger, The physics of flow through porous media (Macmillan Company, New York, 1960).
[10] S. Succi, E. Foti and F. Higuera, Europhys. Lett. 10 (1989) 433.
[11] D.H. Rothman, J. Geophys. Res. 95 (1990) 8663.
[12] D.H. Rothman, Geophysics 53 (1988) 509.
[13] R. Lenormand, C. Zarcone and A. Sarr, J. Fluid Mech. 135 (1983) 337.
[14] C. Appert and S. Zaleski, Phys. Rev. Lett. 64 (1990) 1.

Physica D 47 (1991) 53–63
North-Holland

A GALILEAN-INVARIANT IMMISCIBLE LATTICE GAS

Andrew K. GUNSTENSEN and Daniel H. ROTHMAN
Department of Earth, Atmospheric, and Planetary Sciences, Massachusetts Institute of Technology, Cambridge, MA 02139, USA

Received 20 January 1990

Recently, lattice-gas methods have been introduced as a technique for the simulation of one- and two-phase fluid flow. These methods model the fluid as a collection of particles which propagate on a regular lattice and undergo collisions at the nodes of the lattice. In an asymptotic limit, lattice gases simulate the Navier–Stokes equations. However, these models suffer from a lack of Galilean invariance. An important physical manifestation of the lack of invariance is that the fluid vorticity advects with a velocity different from the velocity of the fluid.

We introduce a new, Galilean-invariant, model for simulating immiscible two-phase flow. Unlike previous Galilean-invariant models, the collisions in this new model satisfy semi-detailed balance, which is achieved by the inclusion of a large number of rest particles with zero velocity. Since adding many rest particles is not computationally tractable, the presence of a large number of such particles is simulated by weighting the outcome of the collisions by a factor related to the frequency with which the collisions would have occurred if the rest particles had been explicitly included in the model. We demonstrate that, in the new model, the vorticity advects at the same velocity as the fluid. We also show that the model obeys Laplace's formula for surface tension and demonstrate an application of the new model to the Rayleigh–Taylor instability. Growth rates as a function of wavenumber computed in the early stages of the instability compare well to theoretical predictions.

1. Introduction

Simulating the motion of interfaces between two immiscible fluids is one of the outstanding problems in computational fluid dynamics [1]. Although traditional methods such as finite difference and finite-element algorithms can efficiently handle simple cases in which the interface has a simple form, more complicated situations involving complex flows and complex interfacial geometries can be difficult to model.

Recently, Rothman and Keller introduced a new method for modeling immiscible two-phase flow [2]. Their method is an example of a new class of methods known as lattice-gas automata, introduced by Frisch, Hasslacher and Pomeau (FHP) in 1986 to simulate the incompressible Navier–Stokes equations [3]. Lattice-gas automata model the fluid as particles of unit mass and unit velocity moving on a regular lattice. When particles meet at a node of the lattice they undergo a collision which redistributes the particles among the various lattice directions while conserving the number of particles and the total momentum at each node. Rothman and Keller extended the FHP lattice gas to model two immiscible phases by coloring the particles, either red or blue, and by altering the collision rules to achieve surface tension between the two fluids.

Frisch et al. [3] showed that when large-scale averages in time and space are taken from the lattice, the lattice gas models the incompressible Navier–Stokes equations. Empirical results show that the immiscible lattice gas (ILG), in addition to obeying the incompressible Navier–Stokes equations in homogeneous regions, obeys Laplace's law for surface tension at fluid interfaces [2].

However, both the FHP and ILG models suffer from a lack of Galilean invariance. The loss of Galilean invariance manifests itself in a variety of ways, the most serious of which is that vorticity in the fluid advects with a velocity different from the velocity of the fluid. While this problem can be easily corrected in a one-phase fluid by a simple rescaling of the fluid velocity, the rescaling will not work for two-phase flow since the velocity of the fluid interfaces is constrained by the conservation of particles to be equal to the average particle velocity and is therefore not subject to the same rescaling. The effect of the lack of Galilean invariance can be seen by running a numerical experiment simulating the motion of a shear-wave in the lattice-gas fluid [4, 5].

Here we show how Galilean invariance can be restored in the two-phase model by adding a large number of rest particles with zero velocity. A specific Galilean-invariant variation of the ILG model with 18 rest particles is constructed. Since explicitly coding the additional particles into the model would be computationally intractable, the presence of a large number of rest particles is simulated by weighting the outcome of the collisions by a factor derived from the frequency with which the outcome would have occurred if all 18 rest particles had been included in the model. A similar shear wave test confirms that the new model is indeed Galilean invariant.

The lack of Galilean invariance, although of little significance in one-phase flow, becomes crucially important for two-phase flow at non-negligible Reynolds numbers, where inertial effects are significant and cannot be corrected with a simple rescaling of the velocity. To illustrate the utility of our new Galilean-invariant model, we include here a simulation of a Rayleigh–Taylor instability. Growth rates measured in the early stages of the instability compare closely to theoretical predictions.

2. Galilean invariance

The generic lattice gas models the fluid as a collection of identical particles of unit mass and unit velocity which move on a regular lattice.

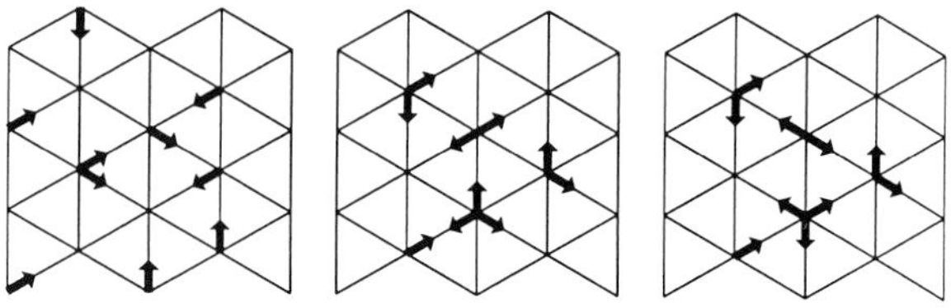

Fig. 1. The evolution of a small part of the FHP lattice gas is shown in the three panels. The first panel shows the gas before movement or collision, the second after movement and the third after the collision step.

When the particles meet at a node of the lattice they can undergo a collision which redistributes the particles among the lattice directions while conserving the total number and total momentum of the particles at the node. In two dimensions, the simplest lattice that yields isotropic fluid dynamics is triangular. A small portion of a triangular lattice and the evolution of the gas over one time step is shown in fig. 1. Here, and elsewhere, we restrict our discussion to two dimensions although generalizations to three dimensions are conceptually straightforward.

Frisch et al. [3, 6], showed that the equations obeyed by the 2D lattice gas are

$$\nabla \cdot \boldsymbol{u} = 0 \tag{1}$$

and

$$\partial_t \boldsymbol{u} + g(\rho)\, \boldsymbol{u} \cdot \nabla \boldsymbol{u} = -\nabla p + \nu(\rho)\, \nabla^2 \boldsymbol{u}, \tag{2}$$

where

$$g(\rho) = \frac{3-\rho}{6-\rho}, \tag{3}$$

ρ is the average particle density, $\boldsymbol{u}$ the average particle velocity, p the pressure, and ν the viscosity. The first of these equations is an expression for conservation of mass and the second an expression for the conservation of momentum similar to the usual Navier–Stokes equation except for the $g(\rho)$ factor preceding the inertial term. This extra factor is the result of the discretization of the particle velocities, all of which are of unit magnitude, and of the lattice, which has only six possible directions. The presence of this factor causes the FHP model to lack Galilean invariance.

An important physical manifestation of the lack of Galilean invariance in the model is the advection of vorticity with a velocity different from the fluid velocity. This can be seen by taking the curl of eq. (2), recognizing the vorticity $\boldsymbol{\omega} = \nabla \times \boldsymbol{u}$, using eq. (1), and noting that the curl of the divergence of the pressure vanishes to obtain

$$\frac{\partial \boldsymbol{\omega}}{\partial t} + g(\rho)\, \boldsymbol{u} \cdot \nabla \boldsymbol{\omega} = g(\rho)\, \boldsymbol{\omega} \cdot \nabla \boldsymbol{u} + \nu \nabla^2 \boldsymbol{\omega}. \qquad (4)$$

This equation is the same as in the usual case, save for the presence of the $g(\rho)$ factor. Noting that the $g(\rho)$ factor always appears with a $\boldsymbol{u}$, Galilean invariance may be restored to the FHP model by a simple rescaling

$$\boldsymbol{u}' = g(\rho)\, \boldsymbol{u}, \qquad \omega' = \omega, \qquad (5)$$

which changes eq. (4) to

$$\frac{\partial \boldsymbol{\omega}'}{\partial t} + \boldsymbol{u}' \cdot \nabla \boldsymbol{\omega}' = \boldsymbol{\omega}' \cdot \nabla \boldsymbol{u}' + \nu \nabla^2 \boldsymbol{\omega}. \qquad (6)$$

This equation is the usual expression that describes the motion of vorticity. Thus, both the fluid and the vorticity now advect at the same velocity, $\boldsymbol{u}'$.

The rescaling restores Galilean invariance to the FHP model because, in the original lattice gas, the vorticity moves on the lattice at a velocity $\boldsymbol{u}_{\text{vort}} = g(\rho)\, \boldsymbol{u}$, which differs from the average particle velocity $\boldsymbol{u}$ by the factor $g(\rho)$. In order for the vorticity to advect at the same velocity as the fluid, one must reinterpret the meaning of the velocity of the particles on the lattice, scaling it by $g(\rho)$. After doing the rescaling, the advection of the vorticity and the fluid occurs at the same velocity.

3. Immiscible lattice gas

The immiscible lattice gas of ref. [2] is a variation of the original FHP model. One rest particle, with a zero velocity, was added to the model and the particles were colored either red or blue. The particles move in the same fashion as the original particles in the FHP model. However, the particle collisions are now chosen such that surface tension emerges from the model. This is done by maximizing the flux of color in the direction of the color gradient, which can achieve a negative diffusivity [7]. Essentially, the collisions send particles of one color towards neighboring sites containing particles of the same color while conserving the total number of particles, the number of each color, and the total momentum at each site.

In addition to obeying the same Navier–Stokes law in the homogeneous regions as does the original FHP model, the ILG simulates the dynamics of surface tension for both fluids at rest (Laplace's law of surface tension) [2] and fluids in motion [8].

However, since the ILG model obeys the same equations as the original FHP model, it suffers from the same lack of Galilean invariance. This lack of Galilean invariance can be easily illustrated by a simple numerical experiment [4]. The lattice is initialized with a shear wave,

$$v_x = V_x,$$
$$v_y = V_y \cos(k_x x), \qquad (7)$$

which has vorticity

$$\omega = \nabla \times \boldsymbol{u} = k_x V_y \sin(k_x x). \qquad (8)$$

This lattice is allowed to evolve over time and the velocity of the vorticity is measured by computing the phase of the k_x component of the y-velocity, as a function of time. The slope of this curve should be equal to $g(\rho) v_x k_x$ or $v_{\text{vort}} k_x$, where

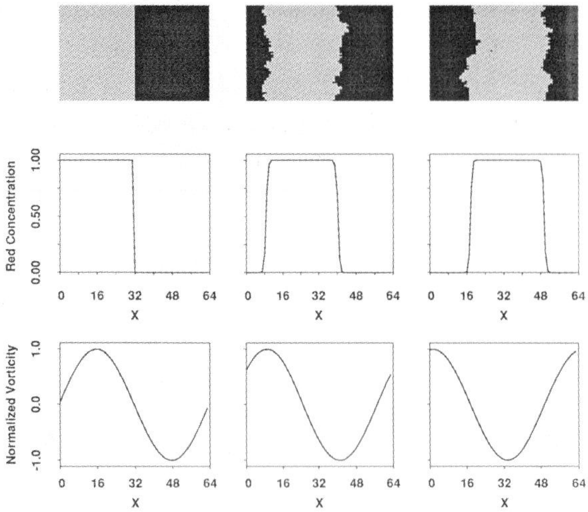

Fig. 2. The results of a shear-test with the ILG model. The top three plots show the fluid, the middle three the red concentration, and the bottom three the vorticity, which is moving in the opposite direction to the fluid.

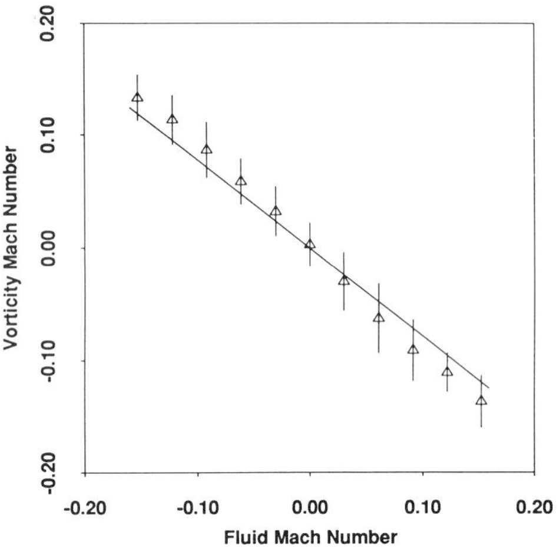

Fig. 3. Plot of vorticity velocity against the fluid velocity for the ILG model. Both velocities have been normalized to a Mach number. The measured points are shown as triangles with error bars of ± 2 standard deviations and the predicted slope $v_{\mathrm{vort}} = g(\rho)\, v_{\mathrm{fluid}}$ by the solid line.

v_{vort} is the velocity with which the vorticity is advected.

The test was run on a 64×64 lattice and k_x was chosen so that there was one wavelength of the velocity variation present in the lattice. All the particles with a positive ω were colored red and the particles with a negative ω were colored blue. The test was run at an average density of 4.9 particles per site, which is the density where the maximum value of surface tension was found in ref. [2]. At this density $g(\rho)$ is negative, and the vorticity should move in the direction opposite to the motion of the fluid.

A sample run of the experiment is shown in fig. 2. The top row of plots shows the evolution of the fluids over time. The black represents the blue fluid, and the gray the red fluid. The next row of plots shows the normalized red concentration as a function of the horizontal coordinate x. The third set of plots shows the k_x component of the vorticity. The lack of Galilean invariance in the ILG model is clearly observed by noting that the vorticity moves in the opposite direction to the motion of the red (gray) concentration. Fig. 3 summarizes the results of a number of tests with a graph of the velocity of the vorticity as a function of the fluid velocity. The velocities have been non-dimensionalized by converting them into Mach numbers $M = u/c_s$, where c_s is the speed of sound [3]. The triangles are the measured points, representing the mean of 10 independent simulations with error bars ± 2 standard deviations, and the solid line is the theoretical curve for $v_{\mathrm{vort}} = g(\rho)\, v_{\mathrm{fluid}}$. There appears to be a systematic error in the slope of the theoretical curve, since the measured vorticity velocity is more negative than is predicted. The origin of this discrepancy is unknown; however, it vanishes in the Galilean-invariant model discussed below.

As noted in section 1, a simple rescaling of the fluid velocity will not correct the two-phase model since the advection of the vorticity and the motion of the interfaces must occur at the same velocity in a Galilean-invariant model. Because the velocity of the interface is fixed by the conservation of particles to be equal to the average velocity of the particles, the necessary solution to this problem is to find a model in which $g(\rho) = 1$.

The lack of invariance is not important in studies of low Reynolds number flow since the flow is dominated by viscous forces and the relative effects of the inertial term are insignificant. However, the lack of Galilean invariance does become important in the study of flows at significant Reynolds numbers, where the inertial term plays an important role in the dynamics.

4. The origin of the $g(\rho)$ factor

For completeness, it is useful to summarize the origin of the $g(\rho)$ factor in the lattice-gas equations. Physically, it is a manifestation of the discreteness of particle velocities and of the lattice. However, it is instructive to review its theoretical origin.

The derivation of the lattice-gas equations is performed as follows [3, 6]. One considers a probabilistic description of the lattice gas, and works with N_i, the probability that there is a particle with velocity $\boldsymbol{c}_i$ at a site. The time evolution of the gas can be described by two equations which describe the conservation of mass and momentum in the collisions:

$$\sum_i N_i(t+1, \boldsymbol{r}+\boldsymbol{c}_i) = \sum_i N_i(t, \boldsymbol{r}),$$

$$\sum_i \boldsymbol{c}_i N_i(t+1, \boldsymbol{r}+\boldsymbol{c}_i) = \sum_i \boldsymbol{c}_i N_i(t, \boldsymbol{r}), \tag{9}$$

where $\boldsymbol{c}_i$ is the velocity associated with the ith lattice direction, and $\boldsymbol{r}$ is the location of the site on the lattice, and the sums are taken over the 6 lattice directions and the rest particles.

If the only conserved quantities in the collisions are the total mass and the total momentum at a site, and the collisions are reversible (more precisely, "semi-detailed balance" holds [6]), then the distribution of N_i can be shown to obey a Fermi–Dirac distribution,

$$N_i = \frac{1}{1+\exp(h+\boldsymbol{q}\cdot\boldsymbol{u})}, \tag{10}$$

where h and $\boldsymbol{q}$ are a real number and vector respectively, corresponding to the conservation of mass and momentum, and $\boldsymbol{u}$ is the average particle velocity. That the particles obey a Fermi–Dirac distribution is not surprising since an exclusion principle is present with only one particle of a particular velocity being allowed at a given lattice site.

Density and velocity for the lattice gas are defined to be

$$\rho = \sum_i N_i,$$

$$\rho\boldsymbol{u} = \sum_i \boldsymbol{c}_i N_i. \tag{11}$$

The Lagrange multipliers h and $\boldsymbol{q}$ are expanded in terms of $\boldsymbol{u}$, and a Taylor expansion of eq. (10) is performed for low u. The expansion of h and $\boldsymbol{q}$ is substituted into the definitions of ρ and $\boldsymbol{u}$, and h and $\boldsymbol{q}$ are then found in terms of ρ and $\boldsymbol{u}$. One then finds that the population densities are equal to

$$N_i = \frac{\rho}{6}\left(1 + \frac{2\rho}{c^2} u_\alpha c_{i\alpha} + \frac{4\rho}{c^4} g(\rho)\left(c_{i\alpha}c_{i\beta} - \tfrac{1}{2}c^2\delta_{\alpha\beta}\right)u_\alpha u_\beta\right), \tag{12}$$

where Greek subscripts refer to the components of a vector and

$$g(\rho) = \frac{3-\rho}{6-\rho}. \tag{13}$$

Eq. (12) is then substituted into the propagation equations, and after doing some asymptotics, one obtains the final equations that the lattice gas obeys at the large scale, eqs. (1) and (2).

Similar equations can be derived for a balanced-collision model with a maximum of $M = M_r + 6$ possible particles per site, where M_r is the maximum number of rest particles per site. The presence of balanced collisions means that the reduced density $d = \rho/M$ is the same for moving and rest particles. The derivation can be done in terms of the moving particles, save that $g(\rho)$ is changed by a factor to account for the rest particles, so that

$$g(\rho) = \frac{M}{6}\left(\frac{\frac{1}{2}M - \rho}{M - \rho}\right), \tag{14}$$

a result expressed differently in ref. [4]. The new factor in $g(\rho)$ arises because the derivation is done in terms of moving particles but the fluid velocity is defined as the average velocity of all particles, including rest particles.

Essentially, the FHP lattice gas is not Galilean invariant because of all the particles move with unit speed. From eq. (14) we see that the introduction of zero-velocity particles provides enough control over the average particle speed to enable the model to be Galilean invariant.

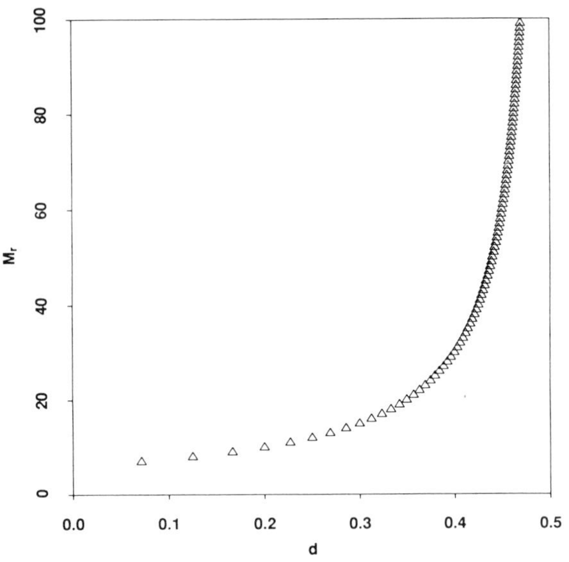

Fig. 4. Graph showing the possible combinations of reduced density $d = \rho/M$ and number of rest particles M_r, which give $g(\rho) = 1$.

5. A Galilean-invariant model

There have been a number of attempts to correct the lack of Galilean invariance in the FHP model by changing the model slightly to make $g(\rho) = 1$ [4, 5]. Generally, these attempts all add one or more rest particles and unbalance the collisions in the model to produce a higher density of rest particles than moving particles, thus resulting in $g(\rho) = 1$ for a particular density. However, a theoretical disadvantage of this solution is that the property of semi-detailed balance is lost for a subset of the collisions. The derivation of the large-scale equations as performed in ref. [6] does not apply, but the Navier–Stokes equations may still be recovered [9]. However, the derivation is more complicated and the $g(\rho)$ factor becomes dependent on the particular collisions used in the model.

An alternative solution to the lack of invariance is suggested by eq. (14): one simply seeks integers $M > 6$ that yield $g(\rho) = 1$. The possible combinations of M_r and reduced density, $d = \rho/M$, which yield $g(\rho) = 1$ are shown in fig. 4. For example, a reduced density $d = \frac{1}{3}$, and number of rest particles $M_r = 18$, gives $g(\rho) = 1$. A reduced density of $d = \frac{1}{3}$ was chosen as a compromise between the number of rest particles needed, which affects the amount of computer memory needed to implement the model, and the surface tension, which is expected to increase as the density increases [2].

The new model was coded in a similar fashion to the ILG model. The ILG model is implemented by explicitly storing every particle as two bits of data, one to indicate the presence of a particle and one bit to store the color of the particle. The movement step is a simple propagation of the moving particles one lattice unit in the direction of their velocity. The collision step consists of computing a local color "field" at each lattice site from the color density of the six nearest neighbors, and then using this information with the site information to compute an outcome index. This index is then used to locate the outcome of the collision from a table where the output of all possible collisions has been precomputed and stored. Surface tension is obtained by choosing the collision outcomes to optimally send particles of one color towards neighboring sites containing other particles of the same color.

However, explicitly including 18 rest particles in a model would be computationally intractable. A better approach is to note that the rest particles are all identical and cannot be distinguished from each other. Thus, one only needs to keep track of the number of rest particles of each color present at a site. This information is combined with the state of the six moving particles to make up a reduced configuration. When computing the collision tables, only the reduced configurations, each of which represents a large number of lattice configurations, need to be considered. For a state with n_c rest particles ($n_c \le M_r$) including n_{redc} red rest particles, this number of equivalent configurations, or multiplicity, is given by

$$\mathcal{M}(s) = \frac{18!}{(18-n_c)!(n_c - n_{redc})!\,n_{redc}!} \,. \qquad (15)$$

Thus the storage requirements for the collision tables for a set of lattice configurations which have a reduced configuration s can be reduced by a factor of $\mathcal{M}(s)$. This reduces the total number of different configurations from $3^{24} \approx 2.8 \times 10^{11}$ to a much more tractable 138 510.

Despite the simplicity of the model, some care is needed when choosing the outcome of a collision to ensure that the model has balanced collisions. To account for the multiplicity of actual configurations represented by one reduced computer configuration, the result of a collision is selected according to the probability of that output configuration existing if all 18 rest particles had been explicitly included in the model. The computations of the probability are done by finding the total number of possible output configurations, and then using this number to normalize the result of eq. (15) to the probability

$$P(s_{input} \rightarrow s_{output}) = \frac{\mathcal{M}(s_{output})}{\Sigma_{s'}\mathcal{M}(s')} \,, \qquad (16)$$

where s' is the set of all the possible output states after a collision with s_{input} as the input state. The result of a collision is then chosen by randomly selecting an output state from the set of possible output states according to the probability of that particular output state existing.

6. Results

6.1. Shear wave tests

In order to confirm that the new model is indeed Galilean invariant, the same shear wave test described earlier was run on the new model. Fig. 5 shows the results of the same test as fig. 2 but with the Galilean-invariant model. The vorticity now advects with the same velocity as does the fluid. Fig. 6 shows the results of a number of tests, plotting the vorticity velocity against the fluid velocity. The measured points, shown as triangles, represent the mean of 5 independent simulations with error bars of ± 2 standard deviations. The theoretical line for $g(\rho) = 1$ is shown as the solid line. The measured points and the theory agree very well, confirming that the new model is Galilean invariant.

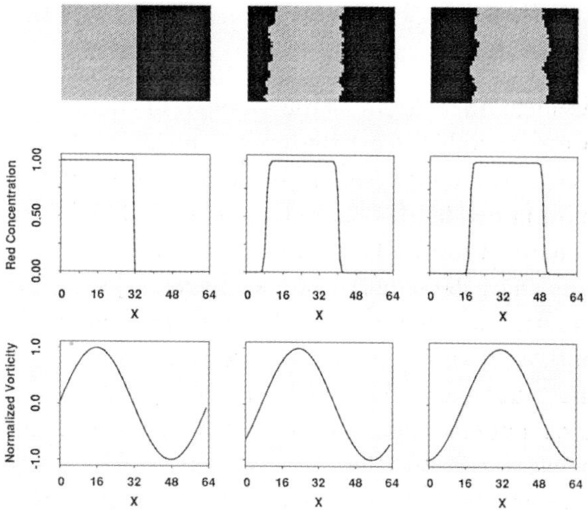

Fig. 5. The results of a shear-wave test with the Galilean invariant model. The top three pictures show the fluid. The red fluid is light gray and the blue fluid is black. The middle three plots show the red concentration as a function of horizontal position x. The bottom three plots show the vorticity, which is moving at the same velocity as the fluid.

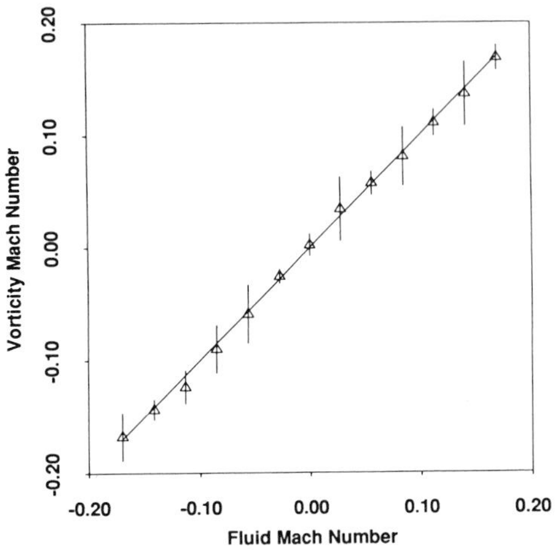

Fig. 6. Plot of vorticity velocity against the fluid velocity for the Galilean-invariant model. Both velocities have been normalized to a Mach number. The measured points are shown as triangles with error bars of ± 2 standard deviation, and the predicted slope $v_{\text{vort}} = v_{\text{fluid}}$ by the solid line.

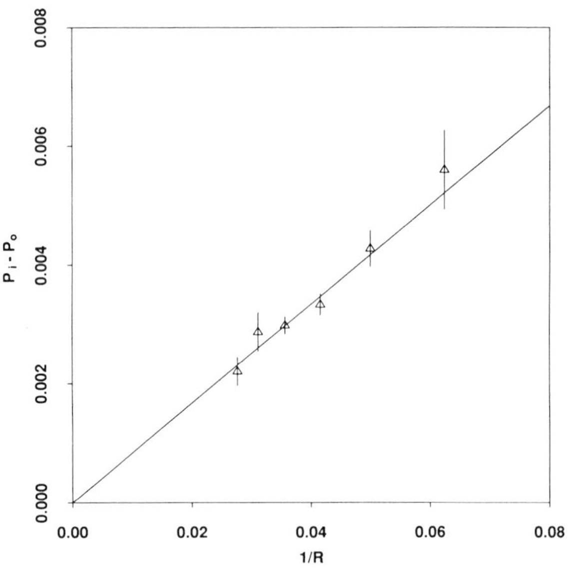

Fig. 7. Numerical verification of Laplace's law for surface tension. The triangles are the results of numerical experiments and the line is the best fitting line through the origin. The surface tension is equal to the slope of the line.

6.2. Laplace's law

Laplace's law for surface tension in two dimensions,

$$P_{\text{inside}} - P_{\text{outside}} = \frac{\sigma}{R}, \tag{17}$$

predicts that for a static bubble of fluid, the pressure difference between the inside and outside of the bubble should be equal to the surface tension divided by the radius of the bubble. Our numerical test of Laplace's law was an improved version of the bubble test used in ref. [2] to verify the ILG's adherence to the law. The lattice is initialized with a bubble of one color in a sea of the other color. The lattice is allowed to evolve over time and the equilibrium pressure difference between the two liquids is measured. The pressure in the bubble is found by measuring the colorblind pressure in a bubble of radius $0.7R$, while the external pressure is obtained by measuring the colorblind pressure at all sites at a distance greater than $1.3R$ from the center of the bubble. By measuring the pressure differences in this manner, any possible biases due to the slight diffusion of one color into another are eliminated. Fig. 7 shows the results of a series of tests run with bubbles of varying radii, plotted with the best fitting straight line that intersects the origin. The triangles are the mean of four independent measurements with error bars of ± 1 standard deviation. The good linear fit of the results indicates that the model shows the correct dependence of the pressure difference on bubble radius. The value of the surface tension coefficient is equal to the slope of the line, giving $\sigma = 0.085 \pm 0.008$.

6.3. The Rayleigh–Taylor instability

A simulation of the Rayleigh–Taylor instability demonstrates the utility of the new model. The Rayleigh–Taylor instability is a gravitational instability which occurs at the interface between two fluids of different densities when the denser fluid lies above the less dense fluid. Examples of this instability in nature include the formation of salt domes and many mixing phenomena [11]. During the evolution of the instability, the two fluids will overturn in four characteristic phases [11]. First, in the linear regime, small-amplitude perturbations of the interface grow exponentially. The presence of surface tension between the two fluids acts to damp out perturbations above a critical wavenumber and also causes the existence of a maximally unstable wavenumber where the growth rate is the highest. In the next stage of the instability the perturbations grow nonlinearly into a number of fingers with the maximally unstable wavenumber becoming a dominant wavenumber of the interface perturbation. The bubbles then develop, via a Kelvin–Helmholtz instability, a characteristic mushroom shape during the third stage of growth as the fingers penetrate into regions of the other fluid. During this stage the smaller fingers are absorbed into the larger ones. The last stage of the instability occurs when the fingers and bubbles break up into smaller bubbles and mix chaotically. Eventually the two fluids separate with the denser fluid now lying below the less dense fluid.

To simulate the Rayleigh–Taylor instability, gravity was incorporated in the system by applying a force which accelerates the red particles upward and the blue particles downward [10]. Buoyancy forces are simulated in the Boussinesq limit, where the density differences change the equations of motion only through the inclusion of a gravitational force term.

If we define $\Delta q(x)$ as the change in colored momentum (red momentum minus blue momentum) due to the application of gravity at lattice location x, then the magnitude of the gravitational force density applied to the fluids is

$$G = \frac{2}{n\sqrt{3}} \sum_{x} \Delta \boldsymbol{q}(\boldsymbol{x}) \cdot \hat{\boldsymbol{y}}, \qquad (18)$$

where n is the number of lattice sites and $\hat{y}$ is the unit vector pointing up. By keeping G constant in a time-averaged sense it may be equated to the usual gravity term by

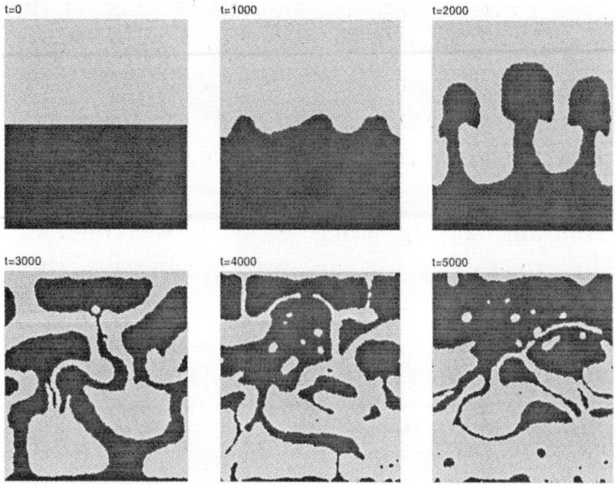

Fig. 8. An example of the evolution of a Rayleigh–Taylor instability over time. The lattice is 256 × 256; time is given in time steps.

$$\langle G \rangle = g(\rho_{\text{blue}} - \rho_{\text{red}}). \qquad (19)$$

Note, however, that in the Boussinesq limit the Atwood number $A = (\rho_{\text{blue}} - \rho_{\text{red}})/(\rho_{\text{blue}} + \rho_{\text{red}}) = 0$.

Fig. 8 shows the evolution of a simulation of the Rayleigh–Taylor instability over time. Initially the interface is flat, but random motion of the particles on the lattice disturb it. Perturbations of wavenumber less than the critical wavenumber grow exponentially while disturbances of greater wavenumber are damped. As the lattice evolves over time the initial perturbations grow into large fingers which develop via a Kelvin–Helmholtz instability into mushroom-shaped heads. These fingers break up into smaller bubbles and the two fluids eventually separate again.

In order to help confirm that the Galilean-invariant lattice gas correctly simulates the Rayleigh–Taylor instability, the growth rate as a function of wavenumber was measured in the early, linear states of the instability. (We have not yet performed any quantitative studies in the non-linear regime.) A 256 by 128 lattice was initialized with a flat interface. A small gravitational force was applied and the location of the interface measured for 2000 time steps. Seven independent numerical experiments were performed and the time evolution of the magnitude of each Fourier component of the interface location was averaged to obtain an average time evolution corresponding to each wavenumber. The growth rate for each wavenumber was then found from the slope of a graph of the logarithm of the magnitude of the Fourier component versus time. If any ambiguity in the slope was present then the slope nearer to $t = 0$ was chosen. Fig. 9 compares our empirical results with the theoretical prediction from linear instability theory [12], modified here to include the finite height of our "box". The essential results from linear stability theory are the occurrence of a dominant mode of instability (due to viscosity and surface tension) and a critical wavenumber k_c (due only to surface tension). The occurrence of a dominant wavenumber is evident not only from the dispersion curves in fig. 9, but also from the simulation itself in fig. 8. The critical wavenumber occurs because the surface tension damps out perturbations with high curvatures. The linear theory predicts that infinitesimal perturbations of wavenumber k will grow if k is less than the critical wavenumber

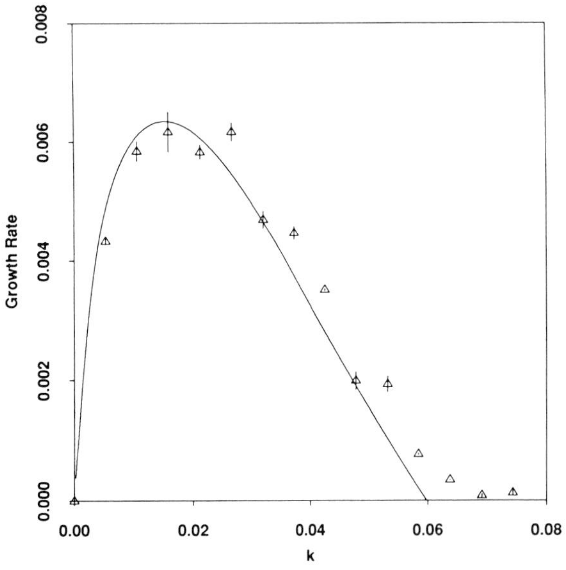

Fig. 9. The measured growth rates of the Rayleigh–Taylor interface as a function of wavenumber. The measured points are shown as triangles with error bars of ± 1 standard deviation. The prediction from linear theory is shown as the solid line.

$$k_c = \frac{1}{2\pi}\sqrt{\frac{G}{\sigma}}, \tag{20}$$

where G is the average amount of gravitational force applied per unit area and σ is the surface tension. For the experimental parameters chosen in the simulation the critical wavenumber is $k_c = 0.060$. Approximately the same wavenumber cutoff is seen in the measured growth rates. In general the growth rates at other wavenumbers also agree fairly well with the theoretical prediction. We attribute the scatter in the measured values to the delicate nature of the experiment.

The advantage of using the Galilean-invariant ILG to model the Rayleigh–Taylor instability is the ease with which it can simulate the late stages of the instability where the interfaces are extremely convoluted and the fingers and bubbles are merging and splitting. Conventional numerical methods appear to require more elaborate codes to track such complex interfacial motion [11, 13].

7. Conclusions

A Galilean-invariant model for immiscible two-phase flow has been created. Our new model uses a large number of rest particles and balanced collisions, unlike other Galilean-invariant lattice-gas models. This extension to the basic ILG model opens a wider range of multiphase flow problems to study by lattice-gas methods, since flows at significant Reynolds number can now be studied correctly.

An application of the new model to the Rayleigh–Taylor instability was demonstrated. The characteristic phases in the development of the instability were observed. Growth rates of the initial stages of the instability were measured and compared well with predictions from linear theory.

Acknowledgements

We thank Stephane Zaleski for his many interactions with us during the course of this work. This work was supported by NSF Grant EAR-8817027 and by the sponsors of the MIT Porous Flow Project.

References

[1] J.M. Hyman, Physica D 12 (1984) 396.
[2] D. Rothman and J. Keller, J. Stat. Phys. 52 (1988) 1119.
[3] U. Frisch, B. Hasslacher and Y. Pomeay, Phys. Rev. Lett. 56 (1986) 1505.
[4] D. d'Humières, P. Lallemand and G. Searby, Complex Systems 1 (1987) 633.
[5] G. Searby, V. Zehnlé and B. Denet, in: Discrete Kinetic Theory, Lattice-Gas Dynamics, and Foundations of Hydrodynamics, ed. R. Monaco (World Scientific, Singapore, 1989) p. 300.
[6] U. Frisch, D. d'Humiéres, B. Hasslacher, P. Lallemand, Y. Pomeau and J. Rivet, Complex Systems 1 (1987) 649.
[7] D. Rothman and S. Zaleski, J. Phys. 50 (1989) 2161.
[8] D.H. Rothman, in Discrete Kinetic Theory, Lattice-Gas Dynamics, and Foundations of Hydrodynamics, ed. R. Monaco (World Scientific, Singapore, 1989) p. 286.
[9] B. Dubrulle, Complex Systems 2 (1988) 577.
[10] C. Burges and S. Zaleski, Complex Systems 1 (1987) 31.
[11] D.H. Sharp, Physica D 12 (1984) 3.
[12] S. Chandrasekhar, Hydrodynamic and Hydromagnetic Stability (Oxford Univ. Press, Oxford, 1961).
[13] J. Glimm, R.O. McBryan, Menikoff and D.H. Sharp, SIAM J. Sci. Stat Comput. 7 (1986) 230.

Physica D 47 (1991) 64–71
North-Holland

FINGERING INSTABILITY IN A LATTICE GAS

F. HAYOT
Department of Physics, The Ohio State University, 174 W. 18th Avenue, Columbus, OH 43210, USA

Received 10 January 1990

I describe work done over the last two years concerning a Saffman–Taylor-type instability in lattice gas hydrodynamics. The emergence of typical macroscopic laws, such as Darcy's and Laplace's, from a microscopic lattice gas is shown. The successful modelling of a fingering instability between two immiscible fluids hinges on the development of an appropriate interface algorithm.

1. Introduction

This contribution summarizes the work contained in refs. [1–4]. The aim of the work described in these references was to develop a formalism for the study of Saffman–Taylor-type instabilities in lattice gas hydrodynamics. The aim presented a challenge, since it was not clear how a macroscopic phenomenon such as an interface with surface tension, and its deformations, would emerge from the microscopic dynamics of the hexagonal lattice gas.

The Saffman–Taylor instability [5] concerns the interface between two immiscible fluids, the less viscous one driving the more viscous one. The instability consists of a finger-like deformation of the interface, allowing the pushing fluid (water, for example) to penetrate into the other one (oil, for example). Typical experiments involve flow in Hele-Shaw cells, or in porous media. The width of the finger is of size comparable to the width of the cell (the only relevant length scale), or of size much larger than any pore in the medium. What unites the two sets of experiments is the underlying mathematical description, which is based on Darcy's law for fluid velocities together with their incompressibility.

It is therefore clear that two typical macroscopic laws – Darcy's for fluid velocity, and Laplace's for surface tension – must be implemented for a lattice gas. In section 2, the emergence of Darcy's law is described. In section 3, an interface is developed and its properties studied. The stage is thus set for section 4, where the fingering instability of the interface is shown. Section 5 contains a discussion of the lattice gas approach to the instability problem.

The model used throughout is the hexagonal lattice gas with two-particle, three-particle symmetric and asymmetric, and four-particle collisions. There are no stationary particles. Densities are normally of two particles per lattice site. All simulations are two-dimensional. Macroscopic quantities, such as fluid velocity, are obtained after averaging microscopic ones over regions of the hexagonal lattice.

While the work summarized in this contribution was advancing, a parallel but different approach was taken by Rothman [6], who introduced special microscopic laws leading to the separation of two fluids. The approach, though describing well flow through the pores of some medium (for which ours is inappropriate), appears incapable of dealing with the Saffman–Taylor instability itself.

2. Darcy's law

Darcy's law is a typical macroscopic law. It states that in a porous medium, or in Hele-Shaw cell flow, the velocity of the fluid is proportional to the gradient of pressure which drives it. The expression of the law is as follows:

$$\boldsymbol{u} = -\frac{K}{\mu}\boldsymbol{\nabla} p, \tag{1}$$

where K is defined as the medium permeability, and μ is the viscosity. For Hele-Shaw cell flow between two parallel plates, K is replaced by the square of the distance between the plates, divided by 12.

Since flow in our lattice gas model is two-dimensional, Darcy's law can only be obtained by mimicking the restrictions to the flow caused by pores. We therefore introduce into the flow a random distribution of fixed point scatterers, which absorb momentum by scattering any impinging gas particle back into its incoming direction. Without the presence of scatterers, the velocity profile is the usual parabolic one of Poiseuille. The action of scattering points is to flatten this profile (cf. fig. 1). The scatterers are treated as external agents which do not affect the dynamics of the flow. Their density must be sufficiently low so that the corresponding mean free path is much larger than the one corresponding to particle interactions. In our simulations the density of scatterers is therefore one tenth of the particle density. The flat profile can be described by introducing a damping term into the Navier–Stokes equation, which becomes [1]

$$\frac{\mu}{\rho}\frac{\mathrm{d}^2 u}{\mathrm{d}x^2} - \alpha u = \frac{1}{\rho}\frac{\mathrm{d}p}{\mathrm{d}y}, \tag{2}$$

where u is the fluid velocity in the y direction, and x is the transverse direction. Far from the boundaries where viscosity can be neglected, velocity is proportional to the gradient of pressure, given directly from (2) by

$$u = -\frac{1}{\rho\alpha}\frac{\mathrm{d}p}{\mathrm{d}y}. \tag{3}$$

The coefficient α can be calculated in the Boltzmann approximation, and it is found equal to twice the density of scatterers (see also eq. (16)). However, numerically, the coefficient of proportionality is determined to be closer to 1. We thus rewrite eq. (3) as [1]

$$u = -\frac{1}{\rho P_s}\frac{\mathrm{d}p}{\mathrm{d}y}, \tag{4}$$

where P_s represents the density of scatterers. This is Darcy's effective law for the lattice gas. What is noteworthy is that viscosity does not enter into eq. (4), while it appears in the empirical law of Darcy (cf. eq. (1)). It is replaced by the density of particles, while clearly the permeability is related to the inverse of the density of scatterers. This feature of the lattice gas entails that in the discussion of interface instability, the fingering instability which occurs does not correspond to the original Saffman–Taylor one. It is, however, of the same type, as will be shown in section 4.

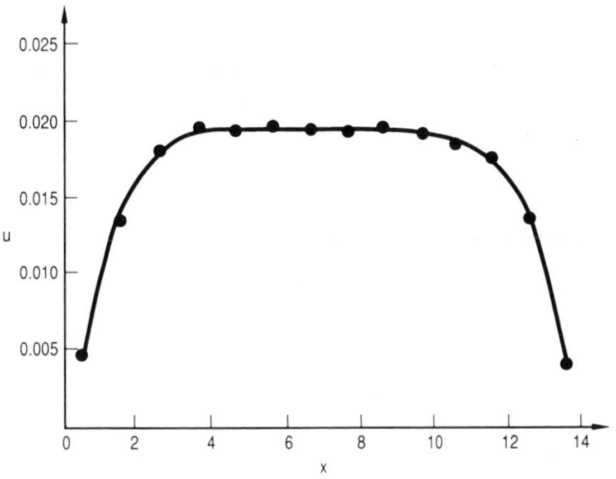

Fig. 1. Velocity profile of a pressure driven flow in the presence of scatterers. Lateral boundary conditions are no-slip.

3. Interface model and fluctuations

After the derivation of Darcy's law, the next step is the construction of an interface model and its study [2]. Our approach is semi-macroscopic. The interface is a continuous chain of links, lying along the hexagonal lattice, which separates two fluids. The interface deforms under impact of particles from either fluid. The algorithm essentially describes how deformations take place. The challenge is to devise one that shows the right physical properties.

We developed two algorithms. The first one is link based: deformation results from the net momentum perpendicular to a link. Particles either bounce back or reflect specularly from interface sites. The fluids only interact through deformation of the interface, which absorbs momentum. The second algorithm is site based. Particles can cross from one fluid into the other, while they push out the interface. When they penetrate into the opposite fluid, they change their identity to that of particles in the new fluid. Momentum is conserved on the interface, and particle number is on the average for each separate fluid.

While the two algorithms have similar static properties, the second one only is appropriate for a flow situation where one fluid pushes the other, as is the case in Hele-Shaw cell flow. The interaction between the two fluids in the first algorithm is too indirect to allow for an interface moving along with the fluids. The results of section 4 are therefore based on the second algorithm. First, though, we will describe static properties. The first requirement is to verify that our interface algorithm leads to the expected physical behavior. These are some of the numerical experiments we performed [2]:

(i) We took a flat interface between fluids at rest and deformed it. We then followed its restoration to flatness.

(ii) We followed the expansion (or contraction) of a square inclusion of one fluid inside the other into a bubble.

From the final state bubble we determined surface tension φ using Laplace's law,

$$p_1 - p_2 = \frac{\varphi}{R}\,. \tag{5}$$

In fig. 2 is shown the value of the pressure difference between the inside and outside of a bubble as a function of bubble radius. The coefficient of proportionality is constant and equal to surface tension. We are able to change the value of surface tension by changing the threshold value of momentum required to deform the interface between the two fluids. A simple, low-density, calculation of surface tension can be made [2]. If N_i and N_o designate respectively the number of inside and outside fluid links attached to the interface, then at equilibrium one should have

$$\rho_\mathrm{i} N_\mathrm{i} = \rho_\mathrm{o} N_\mathrm{o}\,,$$

where ρ_i and ρ_o are respectively inside and outside particle densities. Now, for a closed curve on a hexagonal lattice, one has the relation

$$N_\mathrm{o} - N_\mathrm{i} = 12.$$

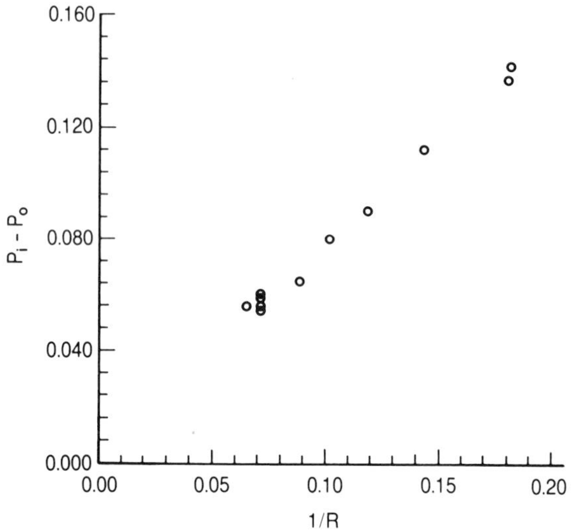

Fig. 2. Pressure difference between the inside and outside of a bubble as a function of the inverse bubble radius. The linear increase is an agreement with Laplace's law and the slope is equal to surface tension.

Writing that $N_\mathrm{i} = C2\pi R$, namely that the number of inside links is proportional to the bubble radius, and using the lattice gas relation between pressure and density ($p = \frac{1}{3}\rho$), one then gets from Laplace's law the following expression for surface tension:

$$\varphi = \frac{18}{\pi C}\rho_\mathrm{o}\,. \tag{6}$$

This relation works best for the stiffest interface, for which the preceding relation (for a measured $C = 2.12$) leads to $\varphi = 0.95$ for $\rho_\mathrm{o} = 1/3$ whereas the simulation gives $\varphi = 0.79$. The dependence on density is found to be less strong than the linear one indicated by the calculation. A more precise expression will be derived below.

4. Fluctuations

We wish to study the thermodynamics of the interface in the heat bath provided by the two fluids it separates [3]. (The interface model is the site-based one.) The interface fluctuates under the random impact of particles belonging to the fluids. Call x the direction along the interface, and L the width of the system in the x direction. The height of the one-dimensional interface is denoted by $h(x)$. The length of the interface itself along the lattice links is larger than L, and is denoted by $\mathcal{L}$. The probability $P(h(x))$ of an interface configuration is given by

$$P(h(x)) \sim \exp\left(\frac{\partial S}{\partial \mathcal{L}} \triangle \mathcal{L}\right), \tag{7}$$

where S is the entropy. $\triangle\mathcal{L}$ is the change of length relative to L and is equal to

$$\triangle\mathcal{L} = \frac{1}{2}\int_0^L \left(\frac{\mathrm{d}h}{\mathrm{d}x}\right)^2 \mathrm{d}x.$$

Setting $\partial S/\partial\mathcal{L} = -\alpha/kT$, $P(h(x))$ takes the form [3]

$$P(h(x)) \sim \exp\left(-\frac{\alpha}{kT}\triangle\mathcal{L}\right), \tag{8}$$

which implies that the interface behaves like a surface of energy proportional to its length, according to

$$E = \alpha\mathcal{L}.$$

Here α is the microscopic surface tension, which is distinct from the macroscopic one as defined by Laplace's law. The temperature T is an effective one, associated with random particle motions. The coefficient α in expression (8) is determined from measuring interface fluctuations, namely the quantity $\langle[h(x) - h(0)]^2\rangle$. We find the following typical one-dimensional, random walk behavior [3] (for $|x|$ sufficiently small), shown in fig. 3,

$$\langle[h(x) - h(0)]^2\rangle = \frac{kT}{\alpha}|x|. \tag{9}$$

The value of the coefficient in front of $|x|$ turns out to be 1.54 ± 0.13. We can do more and check that our interface satisfies a Langevin equation with noise of the type

$$\frac{\partial h}{\partial t} = D\frac{\partial^2 h}{\partial x^2} + \eta(x,t), \tag{10}$$

with noise correlations

$$\langle\eta(x,t)\,\eta(x',t')\rangle = \gamma\delta(x-x')\,\delta(t-t').$$

By averaging over an ensemble of systems with the same initial conditions we get rid of the noise

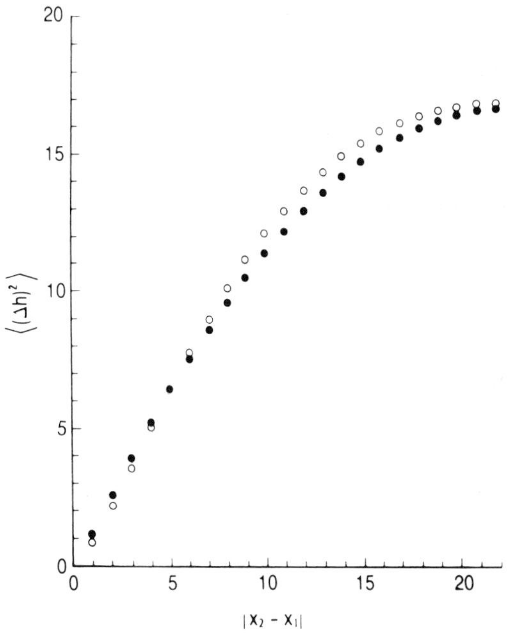

Fig. 3. Interface fluctuations as a function of distance. Calculated curve (·), numerical curve (○).

term, and obtain the diffusion equation. We compare the numerical evolution of the interface with the solution of equation, starting from the same identical condition. Both evolutions can be superimposed. There is also a fluctuation dissipation theorem which states that

$$\frac{kT}{\alpha} = \frac{\gamma}{2D}.$$

We verify this relationship numerically.

In view of the series of results reported in this section, we are convinced that our interface algorithm indeed describes a physical interface. We are then ready to deal with the problem of interface instability. Our results are the subject of the following section.

Before proceeding however, we wish to report an entropy calculation which is in agreement with our numerical results. The lattice gas is a Fermi system, since two particles cannot occupy the same link in the same time step. The number of states accessible to the bulk fluid is

$$\Gamma = \frac{N_S!}{N!(N-N_S)!}, \tag{11}$$

where N is the number of particles and N_S the number of velocity states. N_S is related to the area A occupied by the fluid and the number N_ℓ of interface links by

$$N_S = 6A - N_\ell. \tag{12}$$

From (11) the entropy S is derived as

$$\frac{S}{k} = -N_S\left[(1-f)\log(1-f) + f\log f\right],$$

where $f = N/N_S$ is the average particle density. In order to obtain Laplace's law and therefore an expression for surface tension, one needs to consider the situation of one bubble of say red fluid immersed in blue fluid, and maximize the entropy of this system relative to the bubble radius. This procedure leads to the following expression for surface tension [3]:

$$\varphi = (1-f)\left(2\log\frac{1}{1-f} - \frac{1}{4\pi}\frac{\partial S_{\text{surface}}}{\partial R}\right).$$

For small f and a stiff surface (small $\partial S_{\text{surface}}/\partial R$) one finds again that surface tension is proportional to density, as the previous simplified argument given above indicated.

5. Fingering instability [4]

The Saffman–Taylor instability is an interface instability between two immiscible, moving fluids. It occurs when the less viscous fluid is pushing the more viscous one. The instability takes the form of a finger of the pushing fluid penetrating into the pushed one. The classic example is water pushing oil through a porous medium. The basic laws are Darcy's

$$\boldsymbol{u} = -\frac{K}{\mu}\boldsymbol{\nabla}(p + \rho g y), \tag{13}$$

where u is the velocity field, μ the viscosity, K the permeability, g the acceleration due to gravity, and p the pressure, and that of fluid incompressibilty

$$\boldsymbol{\nabla}\cdot\boldsymbol{u} = 0.$$

These two laws combined lead to the fundamental equation for Hele-Shaw cell or porous medium flow, namely

$$\triangle p = 0. \tag{14}$$

The direction y is that of the flow, perpendicular to the interface separating the fluids. What drives the instability? There is a balance between the pressure gradient which causes deformation of the interface, and the restoring force due to surface tension. The former is of the form $\mu U/K$, where U is a characteristic velocity, the latter is of the form φ/W^2, where W is a characteristic width, that of the Hele-Shaw cell, or that relevant to a porous medium. The ratio of the two forces is the capillary number,

$$C_\text{a} = \frac{W^2}{K}\frac{\mu U}{\varphi}. \tag{15}$$

(We have assumed here that, as in the case of water pushing oil, one viscosity dominates.)

When surface tension is zero, the interface instability occurs at all wavelengths (for an infinite system), and the finger to channel width is undetermined. When surface tension is different from zero, however small, the finger width tends to one half the channel width from above for large capillary numbers. This is what is shown to occur analytically, numerically and experimentally, in the absence of effects due to the third dimension. What is also found, both numerically and experimentally, is that at very high capillary numbers the finger itself becomes unstable to tip splitting. I will discuss this issue further on.

6. Simulation results

The interface instability we will observe is of the Saffman–Taylor type. It is not the actual Saffman–Taylor one, because – as mentioned before – the coefficient in Darcy's law is particular to our algorithm. We derive Darcy's law in the following form (the implementation is somewhat different from the one reported in section 2):

$$\boldsymbol{u} = -\frac{1}{2\rho P^\text{s}}(\boldsymbol{\nabla}p - P^\text{f}e_y), \tag{16}$$

where P^s is the density of scatterers, and ρ the average number of particles per site. Here P^f represents a one-body particle force, which when acting adds one unit of momentum to the flow. The reason for its introduction is the flexibility it provides to reduce the magnitude of the pressure difference, and thus counteract compressibility effects inherent to the gas. Note from (16), that even if the density is space dependent, we still have

$$\nabla(\rho \boldsymbol{u}) = 0 \tag{17}$$

and therefore $\triangle p = 0$, which remains the basic equation of the phenomenon under study.

There are two problems with a lattice gas description of the fingering instability:

(i) The pressure gradient, which in this case drives the instability and cannot therefore be made very small, induces a density gradient. The latter leads to compressibility effects, which have prevented us from investigating questions of finger width as compared to channel width.

(ii) The lattice gas algorithm has intrinsic noise, because of the random motions of particles. The presence of this noise affects the stability of the finger itself.

Linear stability analysis leads to the following equation for the growth rate of a perturbation of amplitude A and wavevector q:

$$\frac{\mathrm{d}A}{\mathrm{d}t} = \frac{[(P_1^{\mathrm{f}} - P_2^{\mathrm{f}}) + 2U_\rho(P_2^{\mathrm{s}} - P_1^{\mathrm{s}})]q - \varphi q^3}{2\rho(P_1^{\mathrm{s}} + P_2^{\mathrm{s}})} A. \tag{18}$$

This equation has the same form as the corresponding equation derived by Saffman and Taylor, with K replaced by $1/2\rho P^{\mathrm{s}}$ and $-\rho g$ replaced by P^{f}. In particular the term linear in q depends on the velocity U of the interface. In the absence of the body force, instability occurs only if P_2^{s} is bigger than P_1^{s}, which is the analogue of the statement that instability occurs whenever the lower viscosity fluid drives the higher viscosity one.

In order to minimize density gradients we put $P_1^{\mathrm{s}} = P_2^{\mathrm{s}} = P^{\mathrm{s}}$, and $P_2^{\mathrm{f}} = 0$, $P_1^{\mathrm{f}} = P^{\mathrm{f}}$. In this case eq. (18) takes the form

$$\frac{\mathrm{d}A}{\mathrm{d}t} = \frac{P^{\mathrm{f}} q - \varphi q^3}{4\rho P^{\mathrm{s}}} A. \tag{19}$$

The critical wave vector at which the growth rate of a perturbation vanishes, is given by

$$q_{\mathrm{c}} = \left(\frac{P^{\mathrm{f}}}{\varphi}\right)^{1/2}. \tag{20}$$

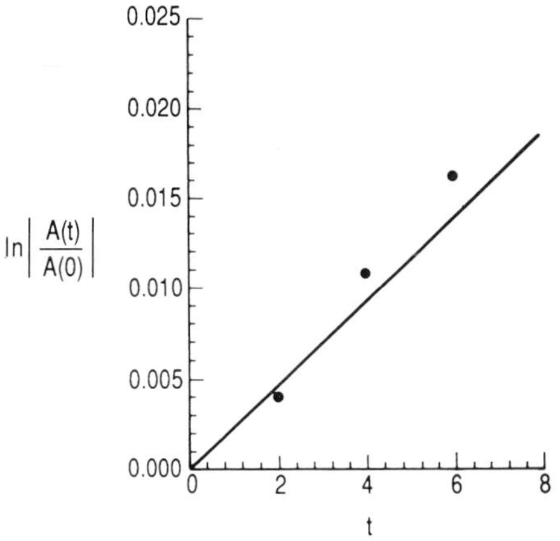

Fig. 4. Growth of a linear perturbation of amplitude A as a function of time. The linear growth of the logarithm is drawn according to eq. (19) as a straight line. Simulation results are indicated by the dots. The relevant values of the parameters are given in the text.

In fig. 4 we show a comparison of a numerically measured amplitude growth with a rate calculated from eq. (19). The comparison is at $q = 0.02$, which is small enough for the linear stability analysis to apply, and for $P^{\mathrm{s}} = P^{\mathrm{f}} = 0.03$. One sees in the figure the linear growth (the higher order term is negligible) with a coefficient compatible with the numbers that appear in (19). Though simulations show that there is a critical wavelength, we were unable to measure it from the growth rate of a perturbation. For $q = 0.05$, we could not make the perturbative amplitude small enough for linear stability analysis to apply. The reason is that noise generates additional modes at many wavelengths which overwhelm any small perturbation. Higher q means increased sensitivity to momentum density fluctuations of the gas which act over small time scales compared to the times one would need follow the growth of a perturbation.

We start our simulations with a perturbation of the interface in the shape of a cosine curve, at the center of the channel, of width about half the channel width. The instability persists and grows as a finger (cf. fig. 5). A finger tip instability starts developing after about a 1000 time steps. Such tip splitting has been observed in experiment and in numerical simulations. In fig. 6 we show a case where after 1500 time steps a three-lobe instability develops at the tip. Similar results have been obtained by Meiburg and Homsy [7], using a Lagrangian vortex method, and adding noise to their algorithm. We also note that the shape of the finger depends sensitively on the amplitude of the cosine perturbation. For smaller amplitudes, there is a narrowing of the stem and appearance of side fingers which influence the behavior at the tip.

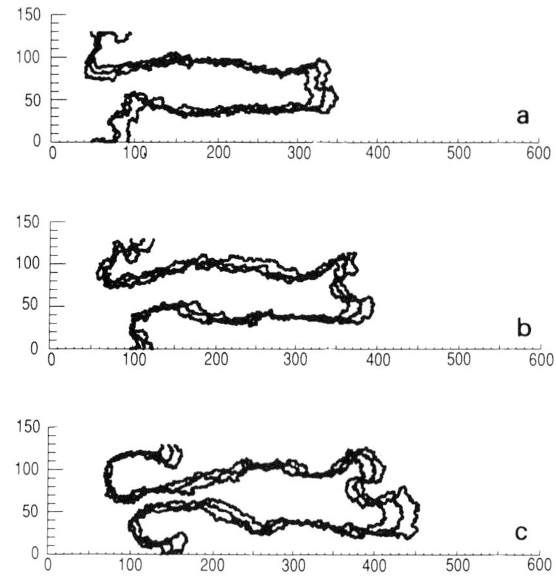

Fig. 5. Growth of a finger instability as a function of time in a channel of width 150 and length 600 (in lattice units). The direction along the length corresponds to the x axis. The initial perturbation is a cosine curve of width 75 and amplitude equal to 200. Other values are $P^s = P = 0.03$. The difference between curves represents 200 time steps. The rightmost portion and shape of the finger are obtained after (a) 1200, (b) 1800, and (c) 2400 time steps.

7. Discussion and conclusions

We clearly observe a fingering instability of our interface. Because of noise, intrinsic to the algorithm, the finger itself becomes unstable to tip splitting. As already said, the physics depends on the value of the capillary number, defined above. Instead, let us use the parameter d_0 defined by

$$d_0 = \left(\frac{\lambda_c}{W}\right)^2 = 4\pi^2 C_a^{-1}, \qquad (21)$$

where λ_c is the critical wavelength above which the system is linearly unstable. For small values of d_0 tip splitting occurs. Small values are ob-

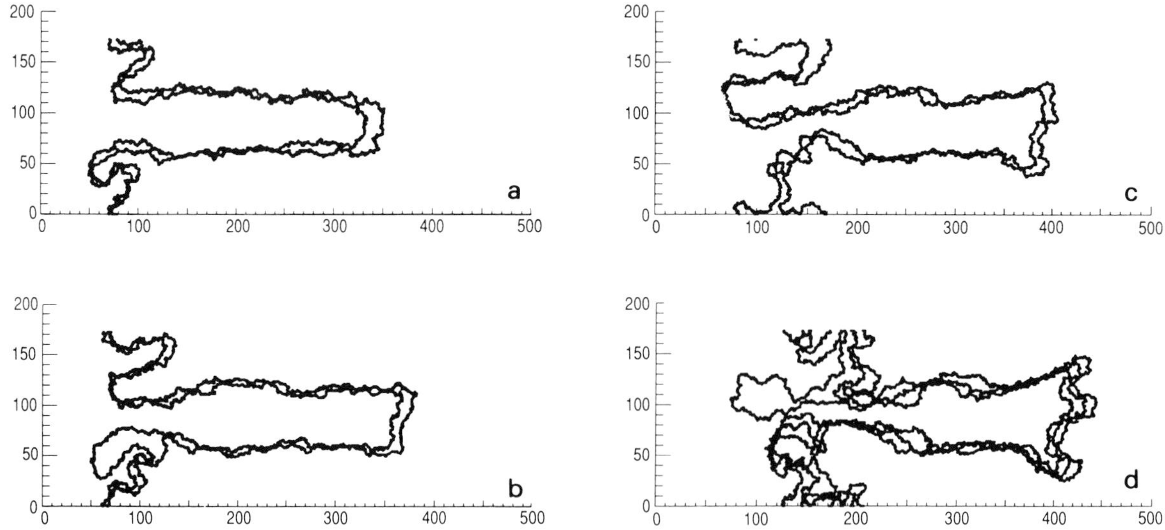

Fig. 6. Growth of a finger instability as a function of time in a channel of width 200 and length 800 (in lattice units). The initial perturbation is a cosine curve of width 85 and amplitude 200 in terms of lattice units. The rightmost position and shape of the finger are obtained after (a) 800, (b) 1200, (c) 1600 and (d) 2200 time steps. The time difference between successive finger shapes is 200 time steps.

tained in the limit of small surface tension, when the capillary number becomes large. The instability, which is nonlinear, can be prevented from taking place either experimentally or numerically by stabilizing the tip curvature.

In ref. [7] tip splitting happens for d_0 around 0.01, and the shape of the finger as it develops depends strongly on noise. Tip splitting also occurs earlier in time, the smaller the value of d_0. Our findings are in agreement with these results. The same phenomena are observed in an approach based on the implementation of Laplace's equation through random walkers [8]. Here again, when d_0 falls below 0.05, the finger instability develops. It turns out that for the cases we have considered d_0 is of the order of several hundredths. The reason d_0 is small is that the critical wavelength itself is small. The latter is because our measured surface tension is, and because the body force cannot be diminished at will because it helps to control density gradients. It is no surprise therefore that we witness tip splitting during simulation runs. Given the physical character of the interface for which we developed our algorithm, we are unable to reach high enough values of d_0 for the finger itself to remain stable.

Let me summarize. We have been successful in developing within the framework of lattice gas hydrodynamics an interface algorithm, which leads to Saffman–Taylor-type instabilities between two immiscible fluids. This is interesting insofar as the lattice is a microscopic model built upon binary variables, and the phenomenon itself is a typical, complex macroscopic phenomenon based on Darcy's law. The lattice gas approach is however afflicted with disadvantages that – although they occur in experiment – prevent a quantitative study. The main one is noise, which is intrinsic to the algorithm and which leads to tip splitting. Another one is the gas compressibility, which we believe tends to enlarge the finger. Thus, though we have met the challenge we set for ourselves, we cannot claim we have resolved the issue of interface instability in lattice gas hydrodynamics. Better algorithms must be possible.

Acknowledgements

I wish to acknowledge the contributions of my co-authors to the work described here, K. Balasubramanian, W.F. Saam and in particular D. Burgess, who worked successfully on the interface algorithm.

This work was supported by Department of Energy grant number DE-F-G02-88ER13916A000.

References

[1] K. Balasubramanian, F. Hayot and W.F. Saam, Phys. Rev. A 36 (1987) 2248.

[2] D. Burgess, F. Hayot and W.F. Saam, Phys. Rev. A 38 (1988) 3589.

[3] D. Burgess, F. Hayot and W.F. Saam, Phys. Rev. A 39 (1989) 4695.

[4] D. Burgess and F. Hayot, Phys. Rev. A 40 (1989) 5187.

[5] P.G. Saffman and G.I. Taylor, Proc. R. Soc. London Ser. A 245 (1958) 312.

[6] D. Rothman, Geophysics 53 (1988) 509;
D. Rothman and J. Keller, J. Stat. Phys. 52 (1988) 1119.

[7] E. Meiburg and G.M. Homsy, Phys. Fluids 31 (1988) 429.

[8] D. Bensimon, L.P. Kadanoff, S. Liang, B.I. Shraiman and Ch. Tang, Rev. Mod. Phys. 58 (1986) 977;
S. Liang, Phys. Rev. A 33 (1986) 2663.

Physica D 47 (1991) 72–84
North-Holland

LATTICE GAS AUTOMATA FOR FLOW THROUGH POROUS MEDIA

Shiyi CHEN, Karen DIEMER, Gary D. DOOLEN, Kenneth EGGERT,
Castor FU, Semion GUTMAN and Bryan J. TRAVIS
Center for Nonlinear Studies, Los Alamos National Laboratory, Los Alamos, NM 87545, USA

Received 17 November 1989

Lattice gas hydrodynamic models for flows through porous media in two and three dimensions are described. The computational method easily handles arbitrary boundaries and a large range of Reynolds numbers. Darcy's law is confirmed for Poiseuille flow and for complicated boundary flows. Multiply connected pore structures similar to actual sandstone with fixed fractal dimension and porosity are generated. Permeability as a function of fractal dimension and porosity is calculated and compared with results of other methods and experiments.

1. Introduction

Lattice gas hydrodynamics [1] has recently received considerable attention because of its speed and simplicity. This method not only describes macroscopic phenomena using ensemble averages, but it also provides microscopic detail which is important for improving our understanding of volume-averaged parameters used in large-scale simulations of flow through porous media.

Porous flow problems are usually solved by volume-averaged approaches. For example, flow characteristics are often described by Darcy's law, which prescribes a linear relation between the pressure gradient and the flux. With these approaches, one can study macroscopic properties of flows. However, applications such as enhanced oil recovery and the analysis of contaminant transport rely on understanding the functional relationship between the volume-averaged parameters used in field scale problems and the microscopic processes occurring at the pore scale. Because of the very complicated boundaries in pore structures, a macroscopic framework cannot provide exact solutions of flow through porous media and diffusion in pores. Lattice gas methods have been shown to have the capability of solving these problems.

The objective of this paper is to test lattice gas flow past complicated boundaries and to verify that this method reproduces Darcy's law for low Reynolds number flows through porous media. We present some quantitative results and estimate the utility of the lattice gas model. In section 2, we will briefly review lattice gas models and discuss the mean free path and viscosity of the models. In section 3, we describe a procedure for generating pore structures with prescribed fractal dimension and porosity for a finite system size. We also give a definition of permeability and verify Darcy's law for three-dimensional channel flows. Typical velocity and pressure distributions at the pore scale for flow through porous media are presented in section 4. Permeability as a function of fractal dimension and porosity are also presented here for two-dimensional flows. Three-dimensional results will be given in section 5. In section 6, a short summary is provided.

2. Lattice gas models for hydrodynamics in two and three dimensions

2.1. Lattice gas automata for two dimensions

We will use seven-bit models for two-dimensional problems. For simplicity, we now describe the FHP-II model [2], but most simulations in this

0167-2789/91/$03.50 © 1991 – Elsevier Science Publishers B.V. (North-Holland)

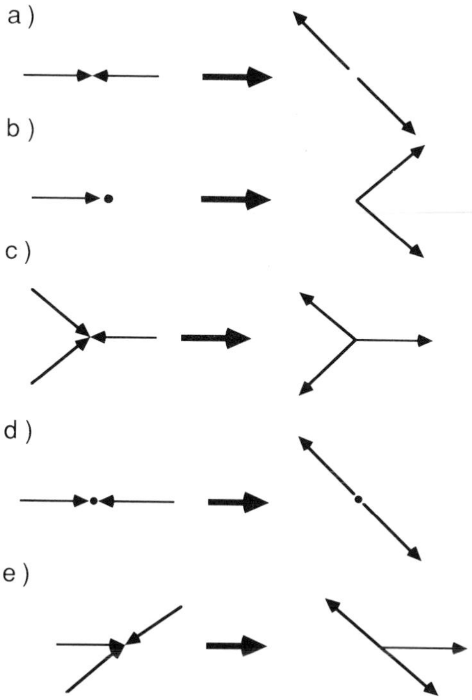

Fig. 1. The collision rules for the 7-bit lattice gas model. The particles have unit mass and unit speed. The left side refers to the state before the collision. The right side refers to the state after the collision.

paper use a seven-bit nondeterministic model B [3]. A triangular lattice is used with six unit velocity particles and a rest particle allowed at each site. Particles have unit mass and their updating is described by two operations, collision and streaming. See fig. 1 for the collision rules which conserve mass and momentum. The microscopic equation for the above two operations is

$$N_a(\boldsymbol{x} + \boldsymbol{e}_a, t+1) = N_a(\boldsymbol{x}, t) + \Omega_a, \qquad (1)$$

where N_a is the particle number at site $\boldsymbol{x}$ at time t with momentum $\boldsymbol{e}_a$ ($a = 1, \ldots, 6$), and Ω_a denotes the time rate-of-change of particle number due to collisions.

We define macroscopic parameters, local mean density, n, and mean momentum, $n\boldsymbol{u}$, as follows:

$$n = \sum_a f_a, \qquad (2)$$

$$n\boldsymbol{u} = \sum_a f_a \boldsymbol{e}_a, \qquad (3)$$

where f_a is a neighborhood average of $N_a(\boldsymbol{x}, t)$.

After an expansion in powers of velocity u and expansion about equilibrium, Boltzmann theory gives the following Navier–Stokes equations [1]:

$$\frac{\partial n}{\partial t} + \nabla \cdot n\boldsymbol{u} = 0, \qquad (4)$$

$$\frac{\partial (n u_\alpha)}{\partial t} + \frac{\partial n g(n)\, u_\alpha u_\beta}{\partial x_\beta} = -\frac{\partial p^*}{\partial x_\beta} + \frac{\partial (\mu\, \partial u_\alpha / \partial x_\beta)}{\partial x_\beta}, \qquad (5)$$

where n is the fluid density, u_α is the velocity and μ is the shear viscosity. The coefficient of the convective term, $g(n)$, should be unity to simulate a physical continuum system. We note that $g(n)$ only depends on density and the geometry of a lattice gas model when the collision transition probabilities satisfy semi-detailed balance. For the 7-bit lattice gas model, we have $g(n) = 7(7-2n)/12(7-n)$. The pressure is given by $p^* = \frac{3}{7}n(1 - \frac{5}{6}g(n)u^2)$. The dependence of the pressure on the macroscopic velocity causes unphysical oscillations, which can be overcome by controlling the initial density [4] or by using multispeed lattice gas models [5,6].

2.2. Lattice gas automata for three dimensions

The lattice gas model used in this paper for three-dimensional calculations was originally proposed by d'Humieres, Lallemand and Frisch [7]. It is a three-dimensional projection of a four-dimensional single speed model. A four-dimensional single-speed model is required because no three-dimensional single-speed lattice model yields a stress tensor which is isotropic to fourth order in the velocity.

The face-centered hypercubic lattice (FCHC) used in our simulations is the simplest lattice to meet the required (icosahedral) symmetry conditions. The FCHC lattice is the set of all points in the four-dimensional integer lattice for which the sum of the coordinates is even. Each lattice site has 12 nearest neighbors that are a distance $\sqrt{2}$ away. Particle collisions at each site can involve up to 24 particles and still conserve mass and momentum. The velocity of a particle at site

$\boldsymbol{x}$ is defined as $\boldsymbol{e}_a$ ($a = 1, 2, ..., 24$). The streaming operation consists of moving each particle to its neighbor at $\boldsymbol{x} + \boldsymbol{e}_a$. In order to simplify the computation, we impose a periodic condition in the fourth dimension, leading to a modified model, called the pseudo-FCHC model. After projecting the physical quantities to three-dimensional space, we obtain eqs. (4) and (5). In this model, $g(n) = 2(24 - 2n)/3(24 - n)$ and the pressure is given by $p^* = \frac{1}{2}n(1 - g(n)u^2)$. This three-dimensional projection contains particles with speed 1 and $\sqrt{2}$.

2.3. Viscosity and mean free path

The viscosity depends on the collision rules. In order to increase the Reynolds number and decrease the mean free path, Hénon [8] introduced optimized collision rules which maximize R_*, where

$$R_* = \frac{g(n)}{\nu} c_s = \frac{Re}{ML}.$$

ν is kinematic viscosity, μ/n, and c_s is the sound speed. Because computation time is proportional to R_*^4, it is important that R_* be as large as possible. In fig. 2, we present R_* versus density per direction (reduced density), d, for the two-dimensional seven-bit nondeterministic model B. The solid line indicates the analytic results and diamonds indicate the computer results for Couette flow.

The collision table we use for the three-dimensional lattice gas model has $R_*^{\max} = 6.21$ at density $n = 7.64$. In fig. 3, we give the theoretical results of R_* versus reduced density.

The mean free path is an important parameter for calculations of flow through porous media because the equations of lattice gas hydrodynamics are only valid for space scales much larger than the mean free path. Since we are interested in the fine scale flow properties in porous media, we must require that the pore space scale is large enough to describe the hydrodynamics.

We use the following definition for the mean free path, λ, for a particle in the lattice gas model,

$$\lambda = \sum_k \left(\frac{\langle |\boldsymbol{e}_k| \rangle}{\sum_i j_i^{(k)} f_i^k} \right) n, \tag{6}$$

where $\langle |\boldsymbol{e}_k| \rangle$ is the macroscopic ensemble-averaged speed for type-k particles. f_i^k is the type-i collision frequency per cell at each time step for type-k particles. $j_i^{(k)}$ is the number of type-k particles involved in collisions of type i. The collision frequency for each particle of type-k particles is $n^{-1} \sum_i j_i^{(k)} f_i^k$. In fig. 4, we give the mean free path

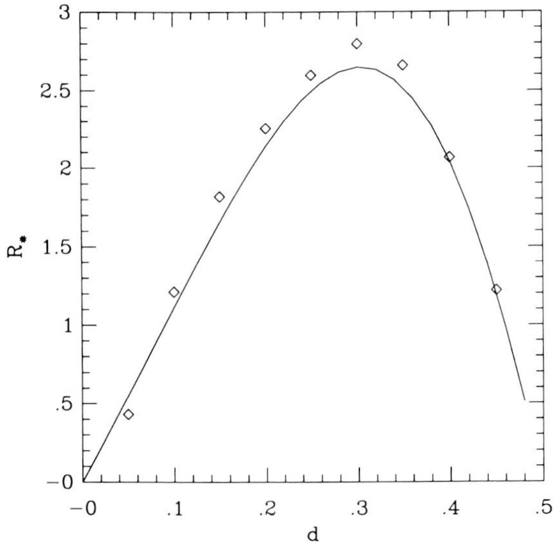

Fig. 2. R_* versus reduced density for the two-dimensional seven-bit nondeterministic model B.

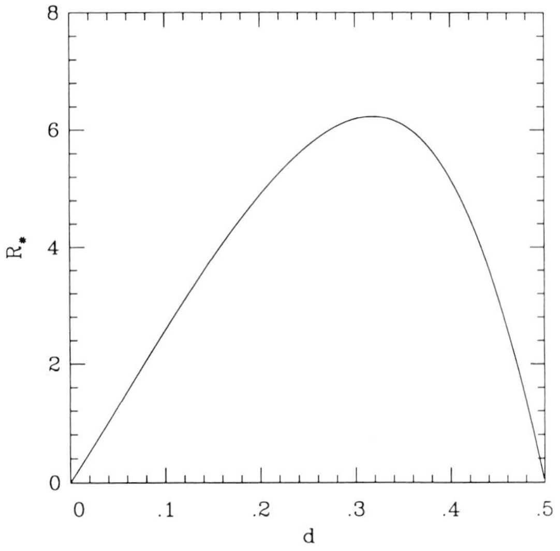

Fig. 3. R_* versus reduced density for the three-dimensional FCHC model.

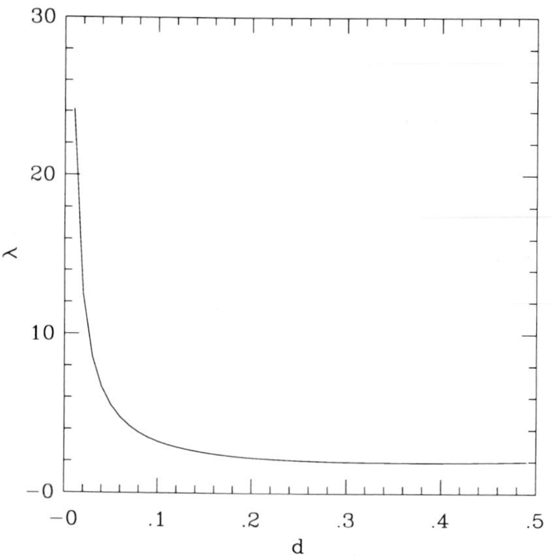

Fig. 4. Mean free path per particle versus reduced density for the two-dimensional model.

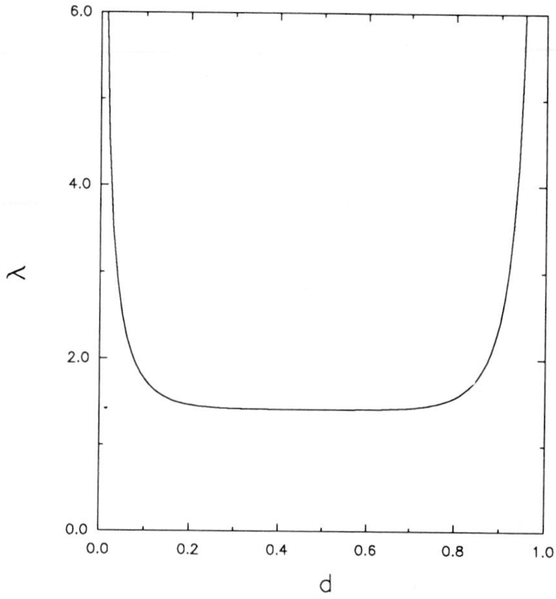

Fig. 5. Mean free path versus reduced density for the three-dimensional lattice gas model.

versus density for the seven-bit nondeterministic model B [3] in two dimensions.

Speed in the four-dimensional model has a single magnitude; however, the speed in three dimensions has two magnitudes, 1 for particles with nonzero speeds in the fourth dimension, such as (1, 0, 0, 1), and $\sqrt{2}$ for particles with zero speed in the fourth dimension, such as (1, 0, 1, 0). The mean free path versus density is given in fig. 5.

2.4. *Implementation of the non-slip boundary and pressure-gradient condition*

The non-slip condition at boundaries A–B and A′–B′ in fig. 6 is easily implemented by reversing the velocity of the particles which hit the boundaries.

There are two ways to create a pressure gradient between $x = 0$ and $x = L_x$. The first one is to use a uniform forcing condition at the inlet. There, we can use a prescribed uniform flipping rate, which produces a net momentum increment in the flow direction. For example, we can change a rest particle to a particle with a momentum in the flow direction. Because the pressure change is linearly proportional to the forcing rate, the pressure change can be prescribed. Another method we use in this paper is to maintain a high pressure at the

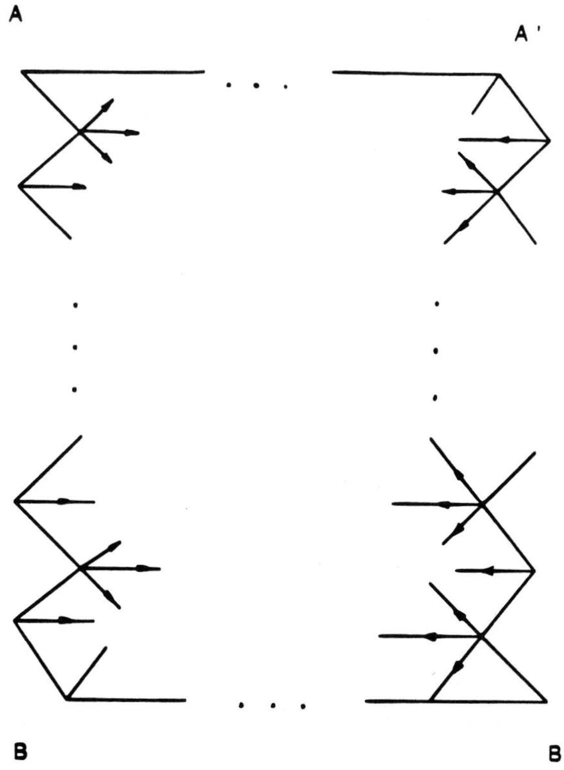

Fig. 6. The bounce-back boundary collision rules used in the wall cells. This condition guarantees zero velocity at the wall.

inlet and a low pressure at the outlet by using different densities, say $n = n_1$ for $x = 0$ and $n = n_2$ for $x = L_x$ where $n_1 > n_2$. The flow is in the x-direction. Because the pressure is linearly proportional to density for small velocities, we have a pressure drop between the $x = 0$ plane and the $x = L_x$ plane. Simulations show an equivalence between these two methods for channel flow.

Statistical methods are used to obtain macroscopic quantities. A time average or spatial average is required. We are interested in the detailed flow structure for arbitrarily complicated boundaries and the calculation of permeability using steady state measurements, so we use a time average for steady flows to reduce the microscopic noise.

3. Fractal dimension, porosity and permeability

3.1. The generation of multiply connected pore structures

We now describe a procedure to create 2D and 3D structures having complex internal connectivity. The basic concept is to distribute features (either grains or open spaces) randomly in a region, using an appropriate number density distribution function. Grain and pore sizes for rocks and solids frequently follow a log-normal distribution. However, it has been observed [9] that many rocks have a fractal distribution of features at small scales (less than approximately one hundred microns). In a fractal medium, the number density m of features of size L is given by

$$m(L) = AL^{2-D}, \tag{7}$$

where D is the fractal dimension and A is a coefficient which depends on the material. In this paper, we restrict ourselves to media with fractal characteristics. Our model generator can, however, describe a log-normal distribution for large features and a fractal law distribution for small-to-intermediate scales. The following procedure is used:

Step 1. Pick a set of feature sizes. The sizes must lie in the range of validity of (7). Typically, each successive size is picked by dividing the previous size by a factor of two.

Step 2. Determine the number density for each feature size using eq. (7). Calculate $\Delta = \triangle x = \triangle y = \triangle z = 1/m(L)$. Divide the region of interest into subdomains of size $N\triangle x$, $N\triangle y$ and $N\triangle z$, where the value of N is supplied by the user. The overall dimensions of the region must be whole multiples of $N\Delta$. The expected number, $E(L)$, of feature size L in each subdomain is N^3.

Step 3. Within the specified subdomains, the expected number of features of size L are distributed randomly. Feature shapes can be chosen as cubes or spheres. The largest feature is distributed first. Smaller features can overlap the larger ones previously distributed. We restrict the locations of features to the subdomains. This restricts the possible realizations, mainly by eliminating those realizations with extreme variations due to large-scale clumping of features. The structure obtained should be characteristic of the material to be described. The values of $\triangle x$, $\triangle y$ and $\triangle z$ are usually considerably larger than the corresponding L; so the restriction to placement within a subdomain is not a strong one.

Step 4. The nature of any feature, i.e., whether it represents a solid grain or an open region, is determined probabilistically. Initially, all lattice sites are set to grain. An input quantity b, ranging between 0 and 1, is used to control the final porosity. A feature is declared a grain if

$$\mathrm{Min}(1, \mathrm{int}(\mathrm{rand}/b)) = 1, \tag{8}$$

otherwise, it represents open space. In eq. (8), 'Min' is the minimum operation, 'rand' is a random number between 0 and 1, and 'int' is the greatest integer function. This procedure produces a structure with porosity close to b. In three dimensions, structures with at least one path through the system are obtained for porosities as low as 10%. In two dimensions, connected systems are produced for porosities above 50–60%.

We generated the porous structure shown in plate I using a feature size spectrum of (32, 17, 10, 5,..., given in terms of number of lattice sites), a fractal dimension of 2.75, and the factor $A = 0.425$. Total porosity of the sample was calculated to be 0.404. The yellow color represents the rock and the dark blue represents the fluid field. We plot only half of the lattice points used for computations.

3.2. Darcy's law and permeability in channel flow

Darcy's law plays a central role in flow through porous media, requiring the following linear relation between the pressure gradient and the flux:

$$q = -\frac{K}{\mu}\frac{\mathrm{d}p}{\mathrm{d}x}. \tag{9}$$

Here, q is the volume flow rate per unit area, K is the permeability, μ is the dynamic viscosity of the fluid and $\mathrm{d}p/\mathrm{d}x$ is the pressure gradient. Permeability defined by Darcy's law is similar to conductivity defined in an electrical conductor. Permeability is an inverse measure of friction generated by the pore structure.

Darcy's law is valid for flows with Reynolds numbers less than one, in which the nonlinear convective term in the Navier–Stokes equation is not important. For some simple cases, permeability can be obtained analytically; for example flow between parallel planes and flow in a circular pipe. The flow between parallel plates will give a permeability dependent only on the channel width and independent of the pressure gradient and kinematic viscosity [10],

$$K = \tfrac{1}{12}h^2, \tag{10}$$

where h is the channel width.

We measure permeability in this paper, as have others [13–16], using a steady-state lattice gas method, i.e., for prescribed pressure gradient and pore structure, we measure the steady-state flux to obtain the permeability using eq. (9). For our lattice gas simulations, we produce a pressure gradient by holding a fixed density at the inlet and a lower fixed density at the outlet. If the density change is sufficiently small, compressibility effects are negligible.

Three-dimensional channel flows are calculated for a given pressure drop between the inlet and outlet boundaries. A typical channel velocity distribution is given in fig. 7. The average density for this calculation is $d = 0.3$ with a shear viscosity of

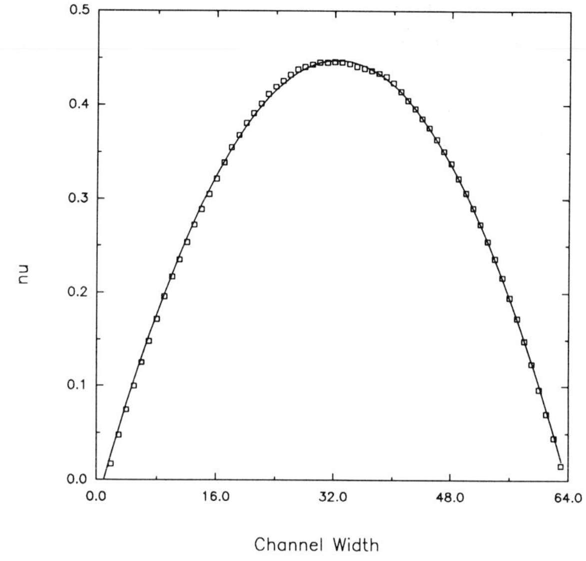

Fig. 7. Averaged velocity distribution for three-dimensional channel flow. The solid line is the parabolic curve from analytical solution and the squares indicate simulation results.

0.35. The mean free path is about 1.4 lattice units. Hence, we can use the channel widths of 10, 20, 30, 40, 50 to study the wall effect. A nonslip condition is applied on the walls ($z = 0$ and $z = h$). For all calculations, we initialize the macroscopic velocity to be zero. After 10000 time steps to relax to a

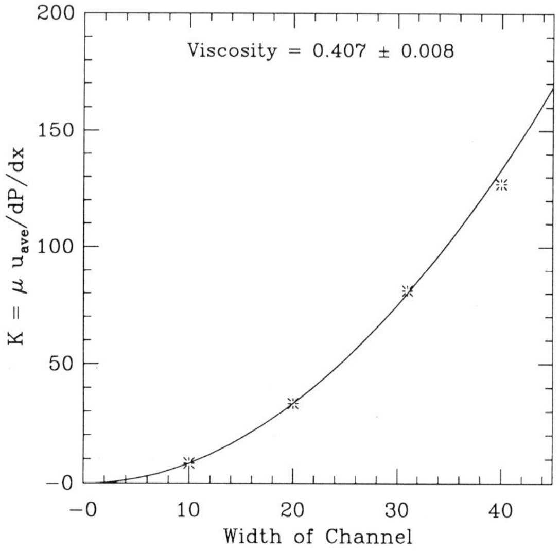

Fig. 8. Permeability versus channel width for three-dimensional Poiseuille flow. The solid line indicates theoretical parabolic results; stars indicate simulation results.

steady state, we average an 10000 additional time steps to calculate the net particle flux in each y–z plane for the given x value. From the flux, we calculate the permeability as shown in fig. 8. We see that the simulations agree well with the expected parabolic analytical results.

3.3. Modified Darcy's law for finite Reynolds numbers

In flow through porous media, the convective term becomes increasingly important as Reynolds numbers increase above unity; and, consequently, Darcy's law becomes less accurate. Eq. (9) is then replaced by the Forchheimer equation:

$$-\frac{\mathrm{d}p}{\mathrm{d}x} = \frac{\nu}{K} q + \rho \beta q^2, \tag{11}$$

where β is a flow coefficient. It was recently observed [11] that the permeability K and coefficient β are, in fact, functions of several variables characterizing the fluid and the rock matrix.

A model geometry used for examining variability of K and β in the Reynolds number range from 1 to 10 is presented in plate II. The porous medium is a network of straight channels intersecting at 45°. The ratio of the channel width to the pore radius is 1:3. The whole structure is oriented in such a way that the incoming flow makes an angle of 22.5° with respect to the channels from the left-hand side. The seven-bit (collision saturated nondeterministic model A) [3] is used. A lattice of 288 × 254 sites was used. We use 16 × 254 sites for the constant pressure inlet and outlet. We use 24000 time steps to obtain a statistical equilibrium, and then average the pressure distribution for 6000 time steps. The pressure distribution for one case is shown in plate II.

To obtain K and β, pressure and flow velocity u were computed as averaged values across the void space in vertical cross sections. The coefficient K was chosen to minimize the integral

$$\int_0^{L_x} \left(\frac{\mathrm{d}p}{\mathrm{d}x} + \frac{\nu}{K} u \right)^2 \mathrm{d}x,$$

and coefficient β was computed to minimize

$$\int_0^{L_x} \left(\frac{\mathrm{d}p}{\mathrm{d}x} + \frac{\nu}{K} u + \rho \beta u^2 \right)^2 \mathrm{d}x.$$

In figs. 9a and 9b, we show the simulation results of K and β as a function of Reynolds number. We see that permeability K changes slightly with the Reynolds number, while β is quite sensitive to the Reynolds number. Permeability depends only on geometry, while β directly depends on flow velocity.

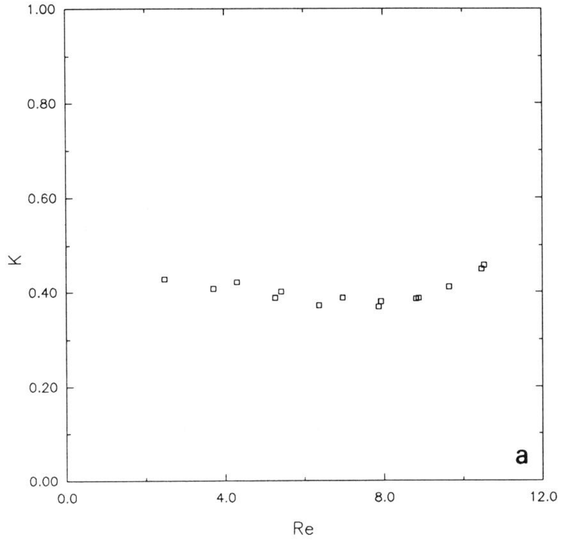

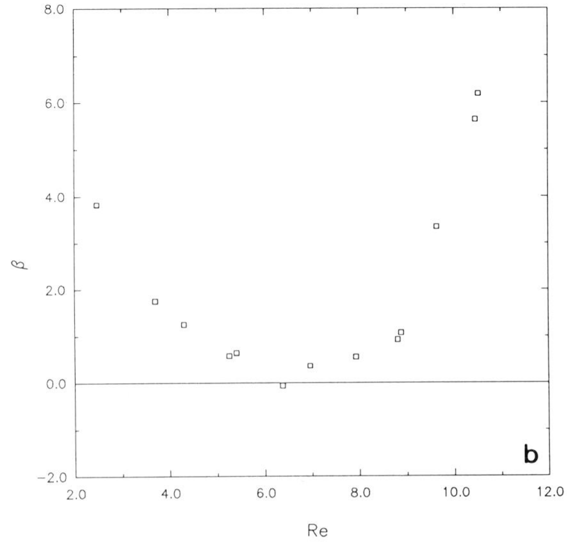

Fig. 9. (a) Permeability versus Reynolds number for the geometry in plate II. (b) Coefficient β in eq. (11) versus Reynolds number for the geometry in plate II.

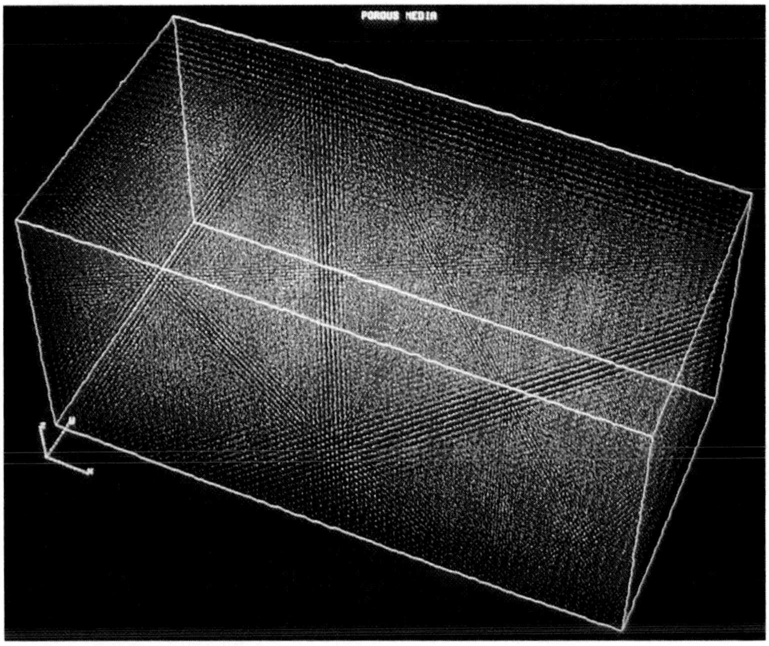

Plate I. A typical rock structure generated by the fractal generator with prescribed dimension and porosity.

Plate II. Network geometry used to study the modified Darcy's law. Red represents high pressure, yellow represents medium pressure, and light blue represents low pressure.

4. Lattice gas simulations for 2D flows

In this section, we study the relationship between permeability, porosity, and fractal dimension. Several simulations have been performed. The medium is generated with given porosity and fractal dimension in a square two-dimensional region. For all calculations, we use a nonslip condition at the boundaries and the net flow is in the x direction.

In fig. 10, we present a typical velocity vector field. To establish steady state, we first run for 50000 time steps and then average over the following 20000 times steps. We also have run another 20000 time steps to verify that the system has achieved a steady state. We found that the fluid particles prefer to go through a few primary channels, while most flow regions are blocked by the rock or flow slowly because the channel width is small. One advantage of the lattice gas method is that velocity fields are readily obtained.

We present the relation between the permeability and porosity in fig. 11. We find that the trend agrees with results in fig. 8. As expected, an increase of porosity provides a monotonic increase in permeability. Fig. 12 presents the relation between permeability and fractal dimension holding porosity constant.

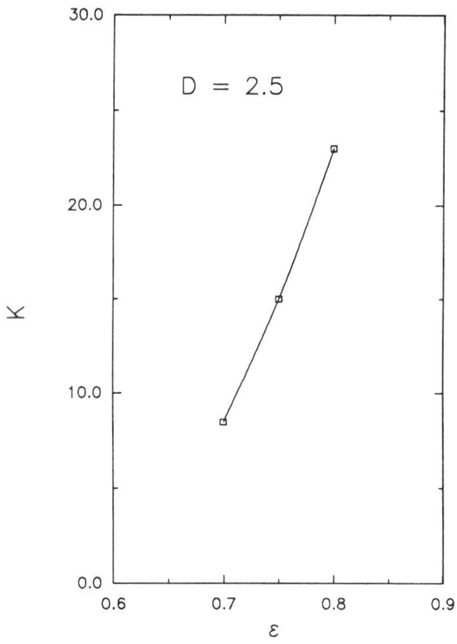

Fig. 11. Permeability versus porosity for fixed fractal dimension, $D = 2.5$.

Scientists at the Petroleum Recovery Research Institute in Socorro, New Mexico, are using a technology for etching into glass the actual pore structures taken from photographs of thin sections of

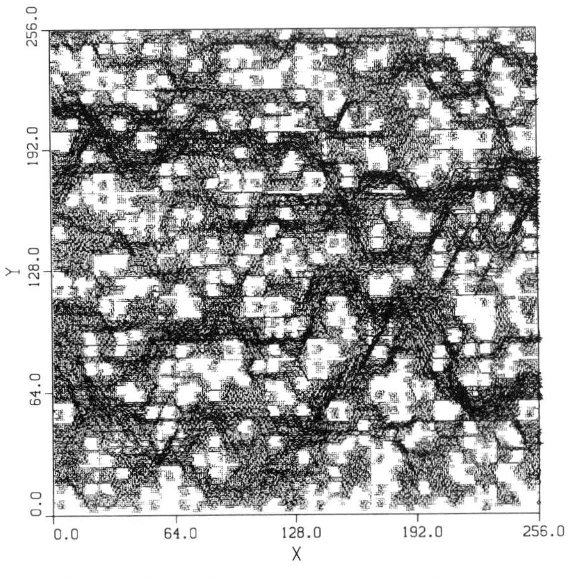

Fig. 10. A typical lattice gas velocity vector field for a two-dimensional flow.

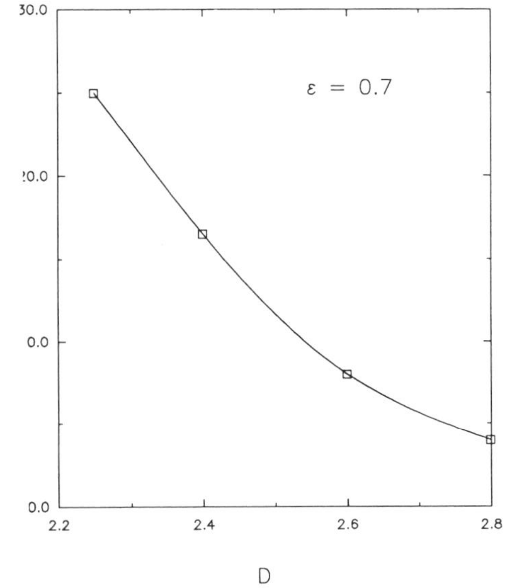

Fig. 12. Permeability versus fractal dimension for fixed porosity, $\epsilon = 0.7$.

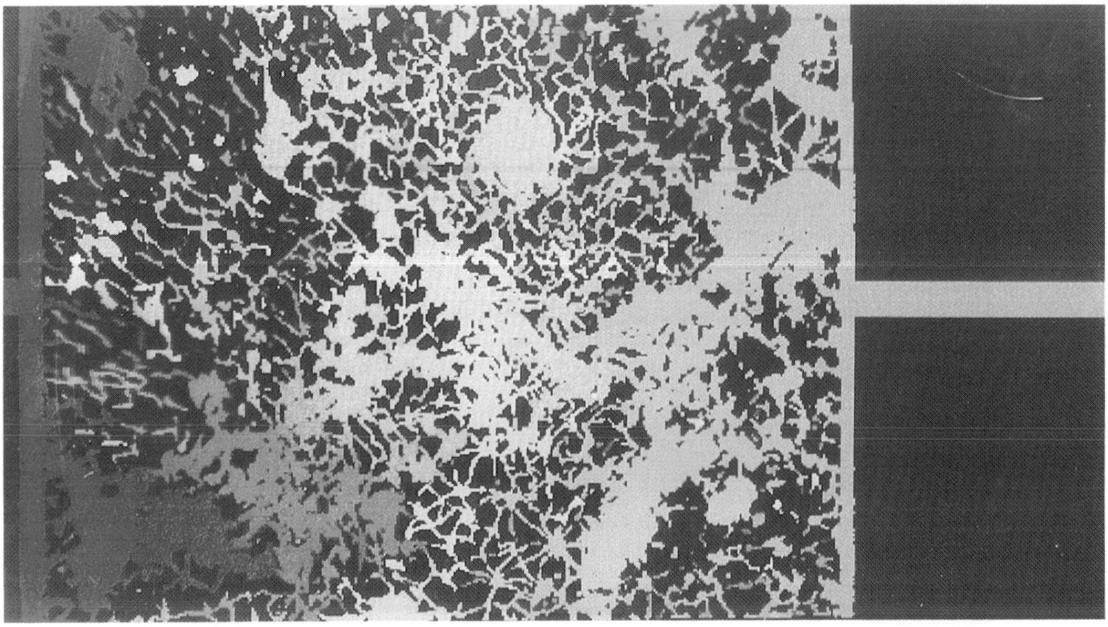

Plate III. Pressure distribution for an etched glass micro-model. Red represents high pressure, yellow represents low pressure and dark blue represents the experimental porous medium.

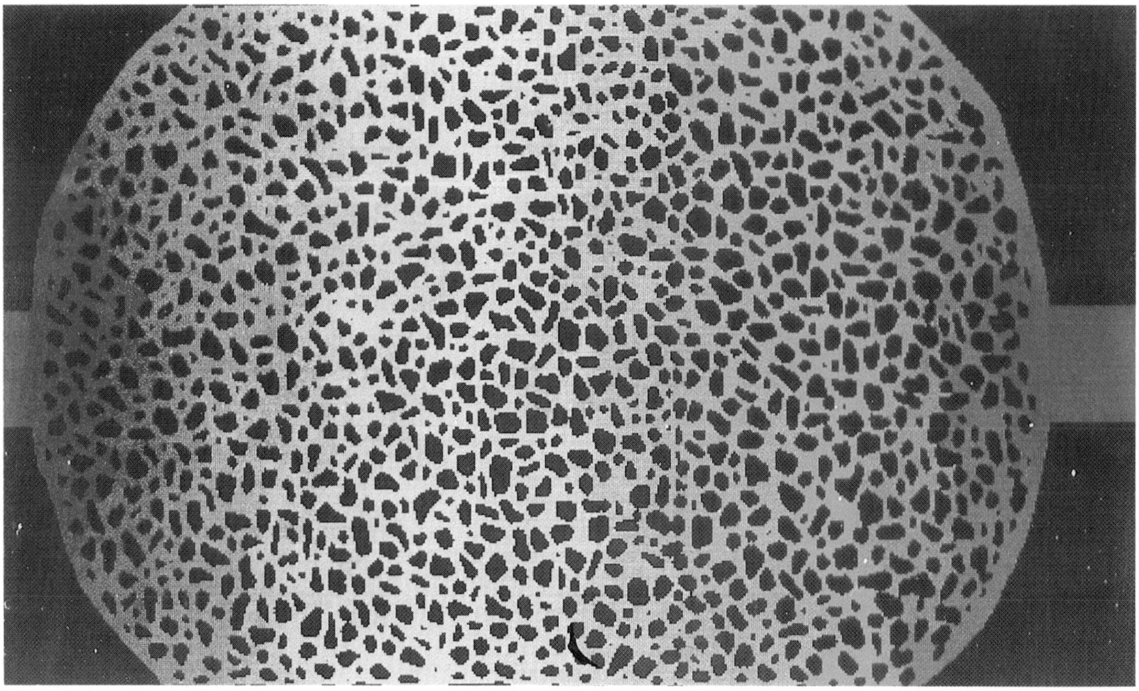

Plate IV. Pressure distribution for an etched glass micro-model. Red represents high pressure, yellow represents medium pressure, and blue represents low pressure.

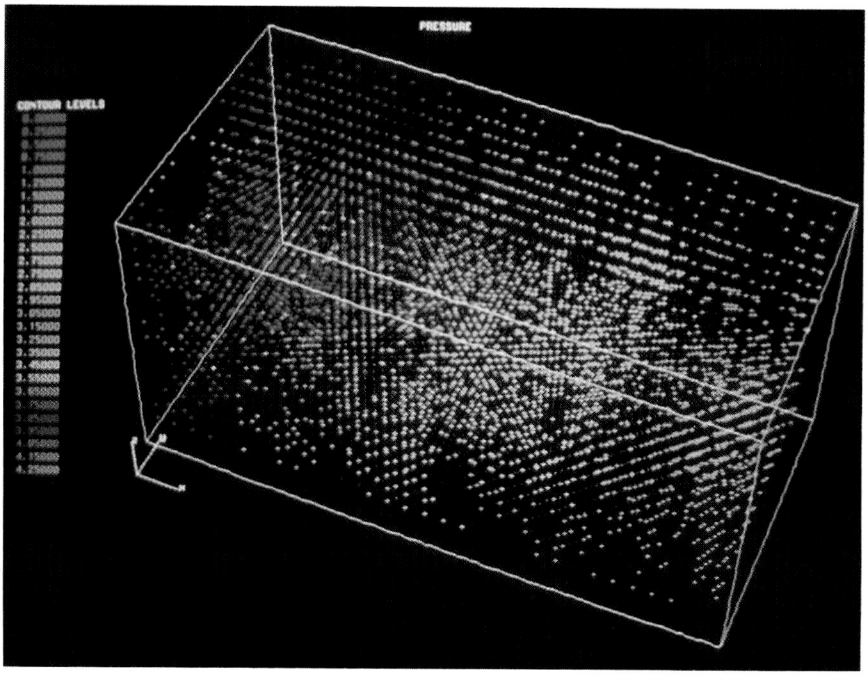

Plate V. Pressure distribution in three dimensions for the medium in plate I.

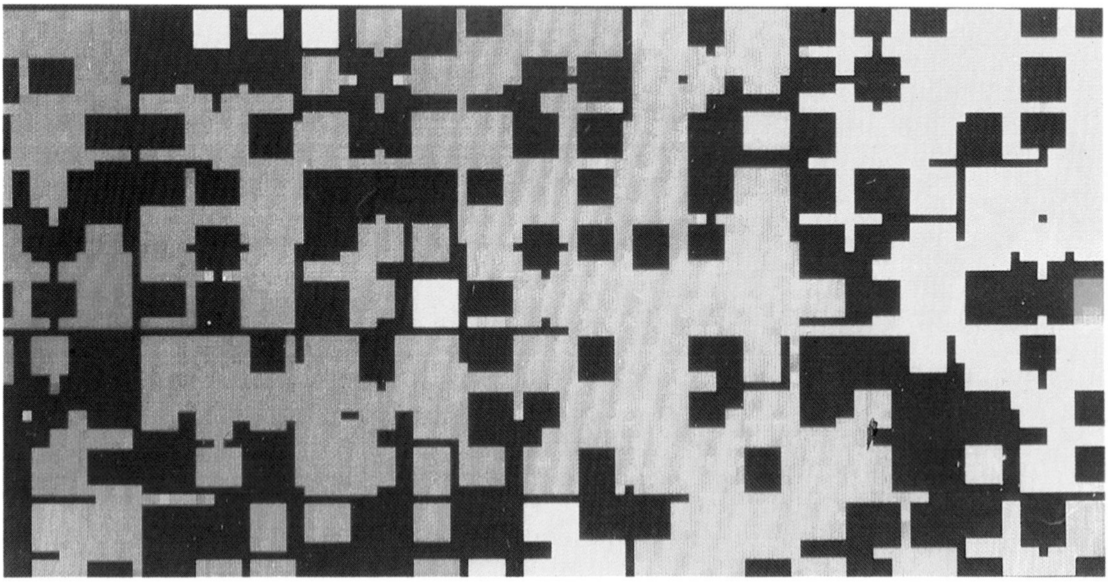

Plate VI. Pressure distribution for x–y cross section at $z = 32$.

sandstone and rock types. Images of these etched glass micro-models are digitized and used directly in our calculations. Plates III and IV show two examples of the etched glass experiments. The calculated pressure distributions are presented in plates III and IV. Dark blue represents the medium and the other colors represent pressures. In plots not shown here, we observe a linear pressure drop across the sample in plate III. Experimental measurements of the micro-model permeability for the image in plate IV is approximately 23 darcy. Our lattice gas simulation produced a permeability of 21.9 darcy. Agreement between simulation and experiment is within the 10% uncertainty of the experiment.

5. Lattice gas results for 3D flows

The programs used in simulations in this paper are generalizations of those developed for the initial three-dimensional hydrodynamic simulations [12]. Details of the code were strongly influenced by the Cray-XMP architecture, and the intrinsic simplicity of the algorithm allowed a comparatively large amount of programming time to be invested in careful optimization.

Each lattice site is represented by a 24-bit number. The updating algorithm involves two basic steps: first, a routine which allows particles at the same site to interact with each other, and second, a routine to move particles from one site to the next.

The first routine is implemented using table look-up. With 24 bits per site, this would normally require a table with 16 million entries. We utilized hole–particle duality (the complement of a 24-bit number, has the complement of the tabular value for its tabular value) and we also packed two entries in each Cray word. This table look-up procedure also incorporates the effects of the media itself: a bounce-back operation is implemented if the site is at the boundary.

The particle-moving routine was written in Cray assembly language for efficiency. It is called three times per time step, once for each dimension. The routine was designed to move particles in both the forward and backward directions simultaneously to reduce memory traffic, allowing these routines to operate at a rate of 40 million sites per second on a single head of a Cray-XMP. The codes are easily adaptable to multiple processor operation.

The system used in the following calculations has 126 lattice cells in the x direction, 63 lattice cells in the y direction and 63 lattice cells in the z direction. We apply the pressure drop in the x direction. The porosity used in this paper is representative of values found in natural materials.

In plate IV, we show the three-dimensional pressure distribution in the pore structure. The color represents the value of pressure. The left-hand side of the plot has red indicating high pressure and the right-hand side has blue representing low pressure. The left-side pressure is 3.6, and the right-side pressure is 3.4. The Reynolds number for this run is near 1.0. After averaging the pressure over the y–z cross section for fixed x, we find that the pressure drop along x is almost linear. A linear pressure drop for homogeneous porous media is expected at small flow velocity.

To see the more detailed structure of the pressure distribution, we plot a typical cross section of plate V in plate VI. Plate VI shows the x–y cross-section–pressure distribution for $z = 32$ (middle plane). Because the structure of the rock is three-dimensional, the pressure distribution for this cross section is quite different from the two-

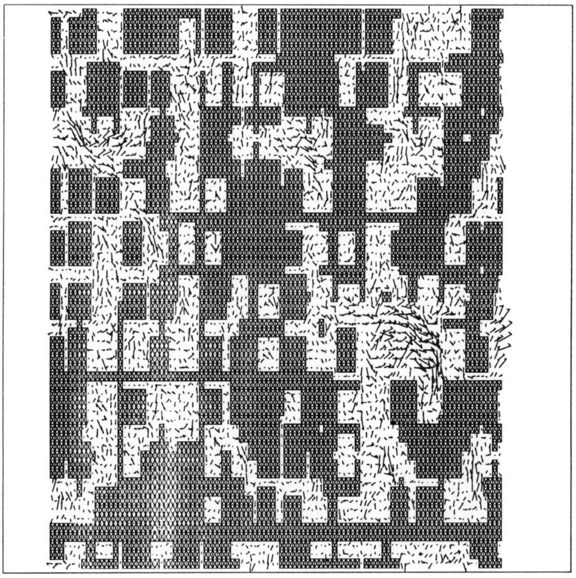

Fig. 13. x–y cross-section–velocity distribution at $z = 32$.

dimensional results shown before. In three dimensions, we often see a discontinuous pressure gradient. A continuous change of pressure actually occurs in three-dimensional simulations. However, visualization along planar sections gives the appearance of pressure discontinuities.

In fig. 13, we present the velocity distribution for the x–y cross section at $z = 32$. This velocity distribution is what is normally expected.

6. Summary

In this paper, we have successfully studied flow through porous media using lattice gas methods. We have generated media with fixed fractal dimension and porosity allowing us to simulate physical pore structures. Simulations for complicated boundaries show reasonable velocity and pressure distributions. The dependence of permeability upon porosity and fractal dimension in two-dimensional flows agree qualitatively with observations. Excellent agreement with experiment is obtained for one flow geometry.

We use time averaging to reduce noise. The lattice gas method is faster than traditional numerical schemes because of the efficiency with which boundary conditions can be implemented. Moreover, lattice gas automata can simulate a wide range of Reynolds numbers, while methods, such as the Laplace equation for flow through porous media, is restricted to very small Reynolds number. The lattice gas method appears most promising for simulating flow through porous media.

Acknowledgements

We thank Peter Ford, Brosl Hasslacher and Tsutomu Shimomura for essential discussions. NASA AMES provided the three-dimensional plotting software. We are also grateful to the Petroleum Recovery Research Institute at Socorro, New Mexico and Dr. F. Kovarick for supplying us with their micro-model experimental results. This work was supported by the US Department of Energy at Los Alamos National Laboratory, by DARPA under grant DPP88-50, and by NASA Innovative Research Program under grant NAGW 1648.

References

[1] U. Frisch, B. Hasslacher and Y. Pomeau, Phys. Rev. Lett. 56 (1986) 1505.

[2] D. d'Humieres and P. Lallemand, Complex Systems **1** (1987) 599.

[3] K. Diemer, K. Condie, S. Chen, T. Shimomura and G. Doolen, Density and velocity dependence of Reynolds numbers for several lattice gas models, in: Lattice Gas Methods for Partial Differential Equations, ed. G.D. Doolen (Addison-Wesley, Reading, MA, 1989) pp. 137–178.

[4] S. Chen, Z. She, L.C. Harrison and G.D. Doolen, Optimal initial condition for lattice gas hydrodynamics, Phys. Rev. A 39 (1989) 2725.

[5] S. Chen, H. Chen and G.D. Doolen, How the lattice gas model for the Navier–Stokes equation improves when a speed is added, Complex Systems 3 (1989) 243.

[6] K. Molvig, P. Donis, J. Myczkowski and G. Vichniac, Removing the discrete artifacts in 3D lattice gas fluids, in: Discrete Kinetic Theory, Lattice Gas Dynamics and Foundations of Hydrodynamics, ed. R. Monaco (World Scientific, Singapore, 1989) p. 409.

[7] D. d'Humieres, P, Lallemand and U. Frisch, Europhys. Lett. 2 (1986) 291.

[8] M. Henon, Complex Systems 1 (1987) 762.

[9] C.E. Krohn, J. Geophys. Res. 93 (1988) 3297.

[10] F.A.L. Dullien, Porous Media, Fluid Transport and Pore Structure (Academic Press, New York, 1979).

[11] M. Tiss, R.D. Evans, The measurement and correlation of the non-Darcy flow coefficients in consolidated porous media, preprint.

[12] T. Shimomura, G.D. Doolen, B. Hasslacher and C. Fu, Los Alamos Science Special Issue 15 (1987) 201–210.

[13] K. Balasubramanian, F. Hayot and W. F. Saam. Darcy's law from lattice-gas hydrodynamics, Phys. Rev. A. 36 (1987) 2248–2253.

[14] F. Hayot, Fingering instability in a lattice gas, Physica D 47 (1991) 64–71; these Proceedings.

[15] D.H. Rothman, Macroscopic laws for immiscible two-phase flow in porous media: results from numerical experiments, J. Geophys. Res. (1990), in press.

[16] S. Succi, E. Foti and M. Gramignani, Flow through geometrically irregular media with the lattice gas automata, MECCANICA, to be published (1990).

Physica D 47 (1991) 85–96
North-Holland

A LIQUID–GAS MODEL ON A LATTICE

Cécile APPERT[a], Daniel H. ROTHMAN[b] and Stéphane ZALESKI[a]
[a]*Laboratoire de Physique Statistique*[1], *Ecole Normale Supérieure, 24 Rue Lhomond, 75231 Paris Cedex 05, France*
[b]*Department of Earth, Atmospheric, and Planetary Sciences, Massachusetts Institute of Technology, Cambridge, MA 02139, USA*

Received 20 February 1990

We describe a triangular lattice model able to undergo a liquid–gas transition. The model is obtained by adding an attractive force to the Frisch–Hasslacher–Pomeau gas in the form of non-local interaction. Several types of interactions are suggested and their properties are discussed. When the attractive forces are strong enough the model decomposes into a dense and a light phase. The equation of state of the model is analogous to a van der Waals equation. The theoretical prediction of the equation of state, obtained using a Boltzmann or factorization assumption, agrees well with numerical observations. The isotropy of the model is tested by a numerical computation of the two-dimensional power spectrum.

1. Introduction

Lattice gas models [1, 2] provide an interesting method for the numerical solution of the Navier–Stokes equations. As this Conference shows, the advent of lattice gases has led to many applications, not only in hydrodynamics, but also in statistical mechanics. Among the many possible variants of lattice gas rules are those which allow the simulation of flows containing interfaces. A great advantage of the lattice gas method for interface flow is that interfaces can be created spontaneously in the model and need not be followed explicitly. In contrast, traditional methods of solution can be relatively complicated; these methods must use markers to keep track of the position of the interfaces or use variable grids [3].

Several lattice gases with interfaces have been suggested. An early model incorporated reactive collisions between species of particles [4, 5]. Another model obtained interfaces between miscible fluids by choosing minimally diffusive collision rules [6]. Yet another approach [7], perhaps closer the traditional spirit of ref. [3], explicitly follows the motion of an interface.

In refs. [8–10] a model with two species of particles and two thermodynamic phases was presented. This model was the first example of a phase transition in a lattice gas. Other models of phase transitions have since been introduced, including one involving a transition in the viscosity [11] and another yielding a phase transition in a ternary fluid [12]. In general, we use the name "immiscible lattice gases" (ILG) for the multiple-species models that yield phase-separation transitions.

In this paper we discuss a new model, called a "liquid–gas" (LG) model, that can undergo a phase-separation transition with only a single species of particles [13]. The transition produces a separation between a dense phase and a light phase. The dense phase is made of frequently interacting particles, while the light phase is made of non-interacting particles. This phase transition is similar to the commonly observed liquid–gas transition.

[1]Associated with CNRS and Université Paris VI and Paris VII.

0167-2789/91/$03.50 © 1991 – Elsevier Science Publishers B.V. (North-Holland)

Simple models for the liquid–gas transition have been suggested for a long time, and our model is far from being new in this respect. The Ising model with a conserved number of particles may be viewed as such a model, but it has a clear disadvantage because it does not conserve momentum. Thus hydrodynamic behavior may not be observed. The Kadanoff–Swift model [14], which conserves momentum, is also based on Ising interactions. However, it has never led to simulations. Recently a lattice model which conserves momentum and separates into a dense phase and a light phase was introduced by Chen et al. [15]. It differs from the model presented here by the nature of the dense phase. In our model most particles in the dense phase are moving particles. In the Chen et al. model the dense phase is mainly made of bound pairs at rest. Thus the hydrodynamic properties of this model should be very different from ours.

There are at least two distinct motivations for the study of LG models. First, it is important to have a simple phase-transition model where momentum is conserved. In particular the conservation laws which a model obeys determine its universality class for dynamical properties [16]. The fluid mechanics of mixtures of widely different densities provides the other motivation. The creation of water waves by the wind is only one such problem. In the ILG models the densities of all phases are equal. Thus the LG models are a first step toward more complicated multiphase models.

In this article, we discuss only the questions concerning the equation of state and the isotropy of LG models. The various possible applications of LG models as well as their other properties, such as viscosity, non-linear hydrodynamics and surface tension, remain to be investigated.

2. Definitions

We recall the definition of the model and its variants, already partially given in ref. [13]. The following notation for the points on a triangular lattice will be useful:

$$\boldsymbol{x}_{ij} = \left(i + \tfrac{1}{4}\left[(-1)^j - 1\right], \tfrac{1}{2}\sqrt{3}\, j\right).$$

There may be up to seven particles on each site of the lattice, one pointing in each of the six directions and one at rest. The particles move with unit velocities $\boldsymbol{c}_j$ defined by

$$\boldsymbol{c}_j = (\cos[2\pi(j-1)/6], \sin[2\pi(j-1)/6])$$

or stay at rest with velocity $\boldsymbol{c}_0 = 0$. The configuration at a given site $\boldsymbol{x}_{ij}$ is given by a Boolean vector $s = (s_0, s_1, \ldots, s_6)$. The particles evolve in three steps: they propagate from one site to the next, collide locally, and interact with distant particles. The long-range interactions are the novel feature of the model. They are designed to simulate a central attractive force.

To be more precise, define a streaming operator S as in [2]. It propagates particles according to their velocity. Define also a collision operator C. It takes particles on a site and redistributes momentum among them. We used the “FHP-III” collision operator, which achieves minimal viscosity. Finally we define non-local interaction operators I_n and I_n'. The evolution operator from time n to $n+1$ is

$$E_n = C \circ I_n \circ I_n' \circ S. \tag{1}$$

The indices n are meant to be mod 3. The three operators I_n are deduced from each other by $2\pi/3$ rotations. These operators act on pairs of sites. The way in which pairs are chosen will be described below. The operators exchange momentum between the two ends of the pair, but leave the total momentum invariant. This is done by rearranging the directions of the particles on each site. We shall describe two variants of this exchange, the simplified model and the maximal interaction model.

The simplified model is identical to the one described in ref. [13]. The operator I_n is obtained by the composition of several elementary operators:

$$I_1 = \mathbf{I}_1^{(a)} \circ \mathbf{I}_1^{(b)} \circ \ldots \circ \mathbf{I}_1^{(e)}.$$

The operators $\mathbf{I}_1^{(a)}$, etc. are shown in fig. 1. They attempt to take a particle in one of the directions in full lines and to put it in one of the directions in dashed lines. This is not always possible: a direction in full lines may be empty or a direction in dashed lines may be occupied. The operator then leaves the configuration unchanged. In other words, it reduces to identity. Finally the operators I'_n are obtained from I_n by a π rotation around $\boldsymbol{x}_{00}$.

To describe the pairs it is convenient to work with the special case $n = 1$. The pairs of sites are chosen so that all sites are visited in one application of the operator I_1. Thus the set of all pairs covers the lattice. In mathematical terms, the pairs are a partition which we shall call P_1. The pairs in P_1 are all horizontal: $p_{kl} = (\boldsymbol{x}_a, \boldsymbol{x}_b) = (\boldsymbol{x}_{2k,l}, \boldsymbol{x}_{2k+r,l})$ and r is an odd number called the range of the interaction. P_n is deduced from P_1 by a $(n-1)\pi/3$ rotation.

We now turn to the maximal interaction model. Let $\boldsymbol{g} = \Sigma_i s_i \boldsymbol{c}_i$ be the momentum on site s. Let the two end sites of a pair be s_a and s_b, and let $\boldsymbol{g}_a$ and $\boldsymbol{g}_b$ be the corresponding momenta. The number of particles will be $m_a = \Sigma_i s_{ai}$, $m_b =$

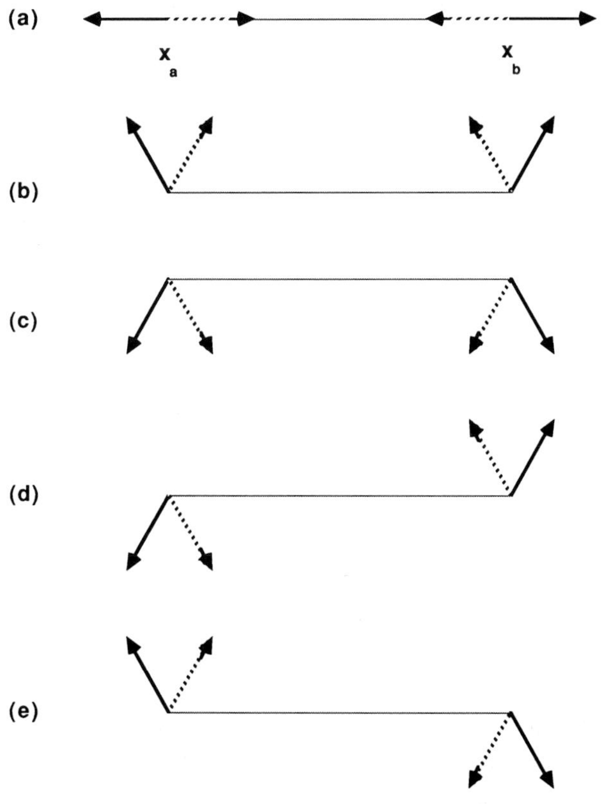

Fig. 1. A pictorial description of the interaction operators $\mathbf{I}_{1p}^{(m)}$. Each operator takes particles in the directions indicated by the full lines and puts them in the directions indicated by the dashed lines. The particles in the other directions are untouched.

$\Sigma_i s_{bi}$. We still keep the above partition in pairs. The momenta after the interaction are $\boldsymbol{g}'_a, \boldsymbol{g}'_b$. The momentum exchanged in the interaction is

$$\boldsymbol{t} = \boldsymbol{g}'_b - \boldsymbol{g}_b = \boldsymbol{g}_a - \boldsymbol{g}'_a .$$

$|t|$ is the quantity we attempt to maximize, with the constraint that the total momentum, particle number on each site, and transverse momentum on each site should be conserved. Let the two sites of the pair be $\boldsymbol{x}_a, \boldsymbol{x}_b$ with $\boldsymbol{x}_b - \boldsymbol{x}_a = r\boldsymbol{c}_n$ at time step $n \pmod 3$. Let the radial momentum components be $\xi_a = \boldsymbol{g}_a \cdot \boldsymbol{c}_n$, $\xi_b = \boldsymbol{g}_b \cdot \boldsymbol{c}_n$ and the transverse components be η_a, η_b and $\xi = \xi_a + \xi_b$. We must leave ξ, η_a, η_b, m_a, and m_b invariant. Consider the class $\mathscr{G}(\xi, \eta_a, \eta_b, m_a, m_b)$ containing all configurations $\sigma' = (s'_a, s'_b)$ having the same invariants. In this class there is at least one configuration which maximizes $-\boldsymbol{t} \cdot \boldsymbol{c}_n$. When several possible outputs exist, a choice must be made. A possible rule is to effect that choice randomly. Another is to choose arbitrarily the output configuration. We took that second option for reasons of simplicity. Yet another possibility would be to choose among the tied outcomes to optimize some other feature of the model, such as viscosity. We have not considered the hydrodynamics of the model in this work so we have not used the latter possibility.

Other variants of the model may be obtained by a change in the order in which operators are composed. For instance a possible rule is

$$E_n = C \circ I_n \circ S \tag{2}$$

for $0 \le n < 3 \pmod 6$ and

$$E_n = C \circ I'_n \circ S \tag{3}$$

for $3 \le n \bmod 6$. In this rule the operators cover all pairs of the lattice in 6 time steps instead of 3, and the strength of the interaction is reduced. Care must be taken to preserve isotropy when the order of operators is chosen. For instance the ordering:

$$E = C \circ I_1 \circ I_2 \circ I_3 \circ I'_1 \circ I'_2 \circ I'_3 \circ S$$

yields catastrophic results, because interactions in direction $\boldsymbol{c}_1$ play a special role.

3. Numerical simulations

Numerical simulations of these models are made using table lookup algorithms for the operators. The operator $C \circ I_n$ may be represented by a single table. Each site is coded in a single machine word. This makes the streaming step S of the algorithm more expensive than the collision and propagation steps. Thus applying several interaction operators in a single time step as in (1) is inexpensive. Packing two interaction operators in a single time step halves the period of the evolution operator from 6 time steps to 3; shorter periods are less likely to influence the hydrodynamic time scale of the model.

Typically a simulation would start with a homogeneously filled lattice. For certain values of the initial density, the lattice separates in two phases, which may be distinguished by their density. The light phase has reduced density $d = \langle m \rangle / 7$ smaller than 0.1 while the dense phase has reduced density around 0.5 in typical cases. In the light phase the interactions seldom result in momentum transfer, while in the dense phase much of the momentum transfer comes from interactions. Thus the two phases may be seen as a liquid and a gas. The transition to a liquid–gas mixture occurs for ranges $r \ge 3$ for the maximal interaction model and for $r \ge 5$ in the simplified model. Thus the maximal model is more efficient in bringing about the transition.

An interesting feature of the observed patterns is the asymmetry between the two phases. Fig. 2 shows the patterns observed for $r = 3$ at intermediate times and for two values of the initial density. One may see that the patterns are dominated by bubbles of gas in a connected liquid for

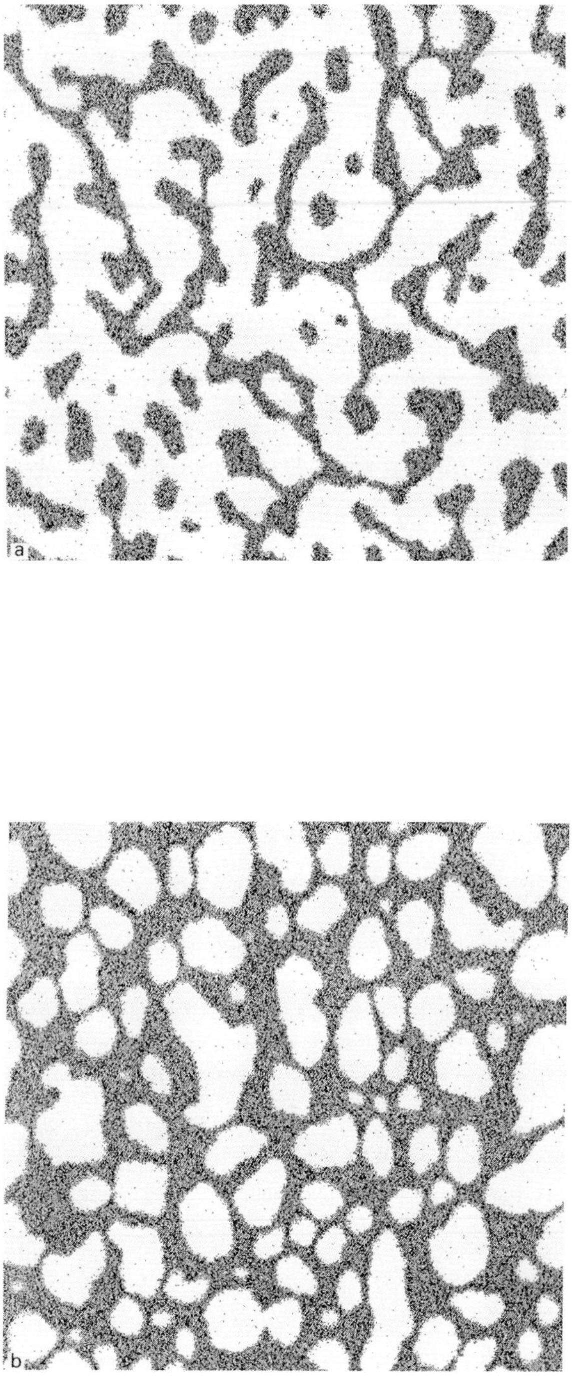

Fig. 2. The patterns obtained for range $r = 3$ at $t = 270$. The maximal interaction model is used. The lattice size is 480×480. (a) Initial density $d = 0.2$, (b) initial density $d = 0.3$.

initial density $d = 0.3$ while they show elongated droplets for $d = 0.2$. When the interaction is stronger, the droplets of liquid are connected by thin filaments. Fig. 3 shows the time evolution for $r = 5$ and $d = 0.2$ and fig. 4 is the same for $d = 0.3$. Filaments appear in both cases. They probably persist a long time because of the viscosity of the liquid phase. The absence of filaments at $r = 3$ may be explained as follows. The lattice is then closer to a critical state. Thus stronger fluctuations and a longer correlation length may blur and break filaments.

4. Equation of state

An equation of state may be predicted by using a factorization assumption as in the Boltzmann equation. Define the momentum transfer tensor $\boldsymbol{P}_{\alpha\beta}$ as the amount of momentum transferred both by interactions and particle propagations. Let $\pi_j = \langle |\boldsymbol{t}_j| \rangle$ be the average momentum transferred in interaction I_j. π_j' is defined as the average momentum transferred in interaction I_j'. The momentum transferred during the first evolution E_1 is

$$\boldsymbol{P}^1_{\alpha\beta} = \sum_i N_i^1 c_{i\alpha} c_{i\beta} - \tfrac{1}{2} r \pi_1 c_{1\alpha} c_{1\beta} - \tfrac{1}{2} r \pi_1' c_{4\alpha} c_{4\beta} .$$

α and β are indices for the two space directions, and $\boldsymbol{c}_i = (c_{i1}, c_{i2})$. N_i^t is the average number of particles at time t: $N_i^t = \langle s_i \rangle_t$. An isotropic configuration may be obtained when the average momentum $\langle \boldsymbol{g} \rangle$ vanishes uniformly in the fluid. Then

$$\boldsymbol{P}_{\alpha\beta} = p \delta_{\alpha\beta} .$$

If we assume that the π_i are all equal, we obtain after time averaging the equation of state

$$p = 3\rho/7 - r\pi(\rho), \tag{4}$$

where $\rho = 7d$ is the density, $\pi(\rho) = \tfrac{1}{2}(\pi_1 + \pi_1')$. The function $\pi(\rho)$ may be computed easily once

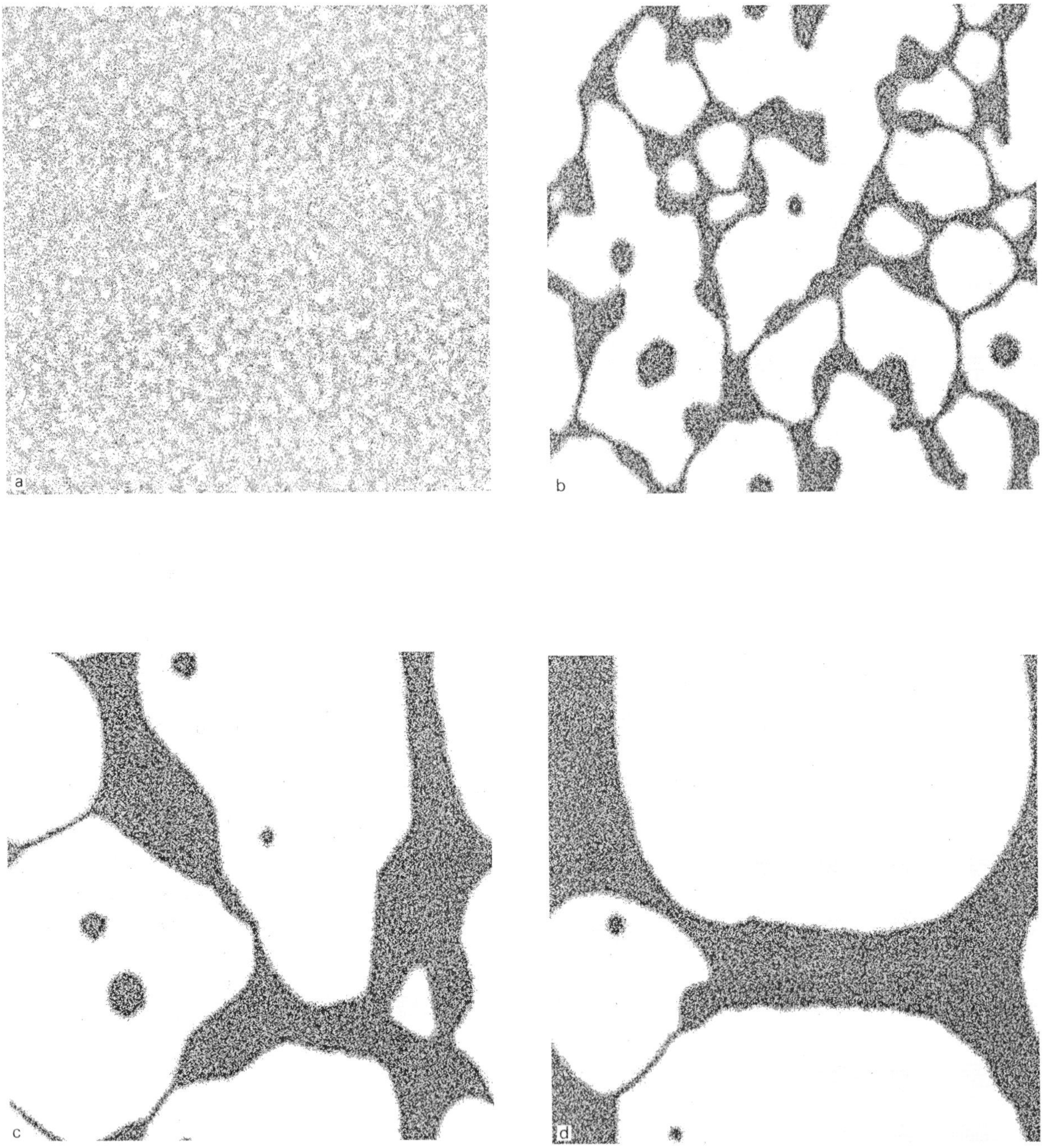

Fig. 3. The patterns obtained for range $r = 5$. Initial density is $d = 0.2$. The maximal interaction model is used. (a) $t = 10$, (b) $t = 270$, (c) $t = 1250$, (d) $t = 5120$.

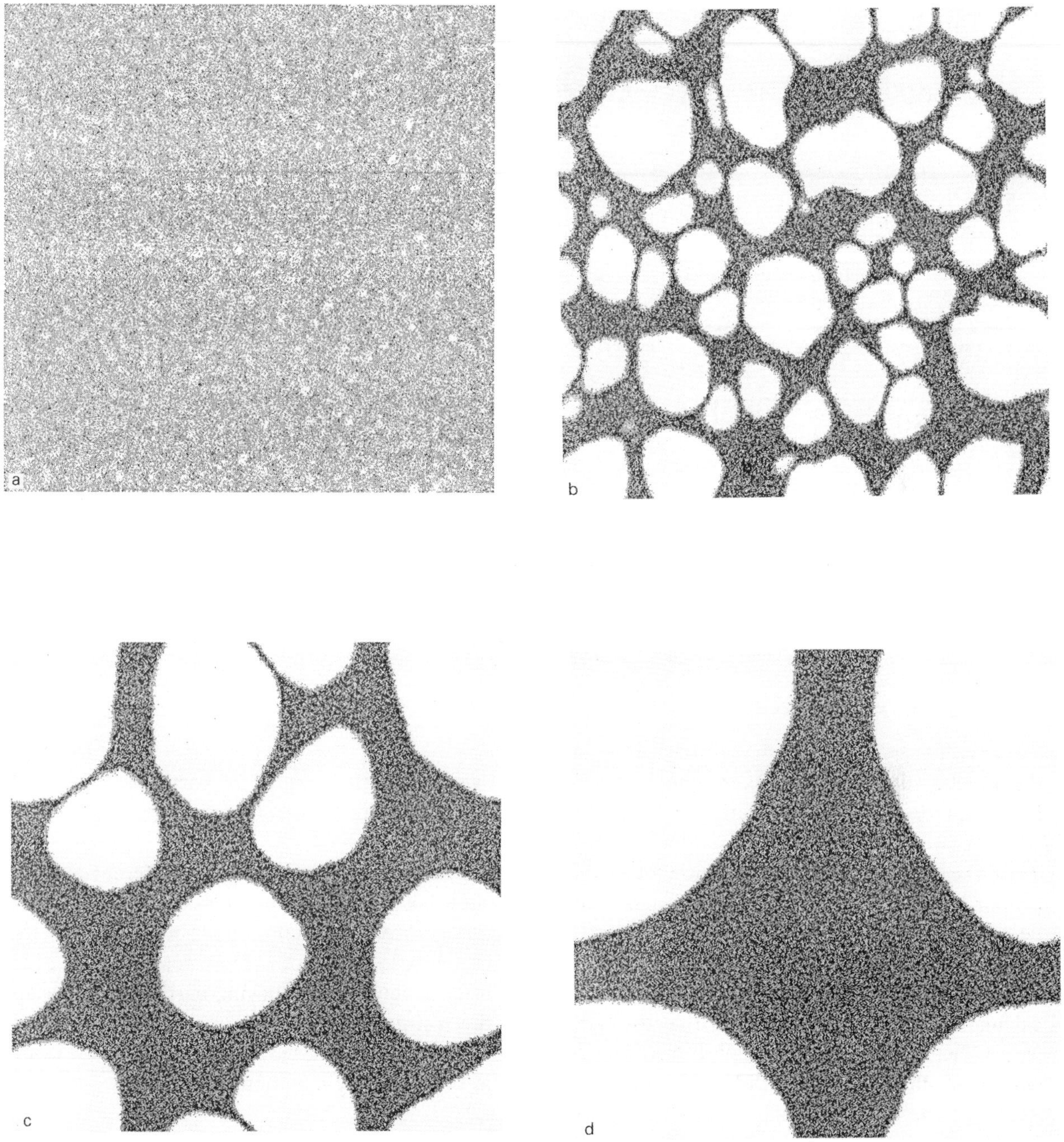

Fig. 4. The patterns obtained for range $r = 5$ with initial density $d = 0.3$ and maximal interaction model. (a) $t = 10$, (b) $t = 270$, (c) $t = 1250$, d) $t = 5120$.

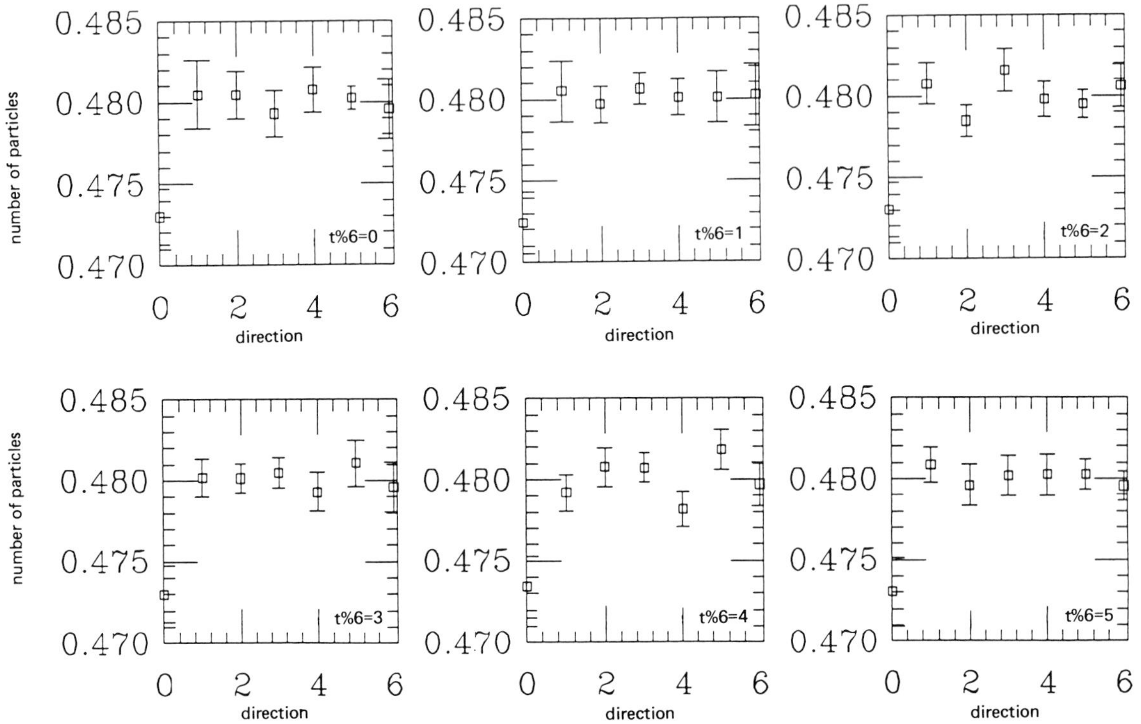

Fig. 5. The measured densities N_i. Averages are taken for different values of time n mod 6.

the precise interaction rules are specified. One important result of the theoretical calculation is that the function $\pi(\rho)$ does not depend on r. Thus the correction to the equation of state is proportional to the range r. In particular it may be of arbitrary magnitude.

The theory used to predict $\pi(\rho)$ was described in ref. [13]. In what follows we just outline the calculation. We start by assuming that the lattice is in a factorized state. We shall discuss the case where the fluid is at rest with $\langle \boldsymbol{g} \rangle = 0$. We let $A(\sigma, \sigma'; j)$ be the transition rate from the pair σ to σ' from interaction I_j. For the models described above, A is 1, 0 or $\frac{1}{2}$. Before the interaction I_1 the probability of seeing a state s is

$$W_0(s) = d^m (1-d)^{(7-m)}. \tag{5}$$

This assumption has been verified numerically. In fig. 5 we show probability densities which are averages of the number of particles $N_i = \langle s_i \rangle$ in each direction i for the simplified model. To investigate the effect of the time dependence of the evolution operator E_n, six different averages were taken at time steps $n = k$ (mod 6). There is no significant difference either between time steps or between directions. However, there is a difference between the moving-particle densities and the rest-particle density.

As shown in ref. [13] the assumption (5) results in the following equation of state:

$$\pi(\rho) = \frac{1}{2}\Big(\sum_{\sigma} A(\sigma, \sigma'; 1)\, (\xi_b - \xi_b')\, d^m (1-d)^{(14-m)} + \sum_{\sigma'} A(\sigma', \sigma''; 4)\, (\xi_b' - \xi_b'')\, W_{1b}(s_a') W_{1a}(s_b') \Big), \tag{6}$$

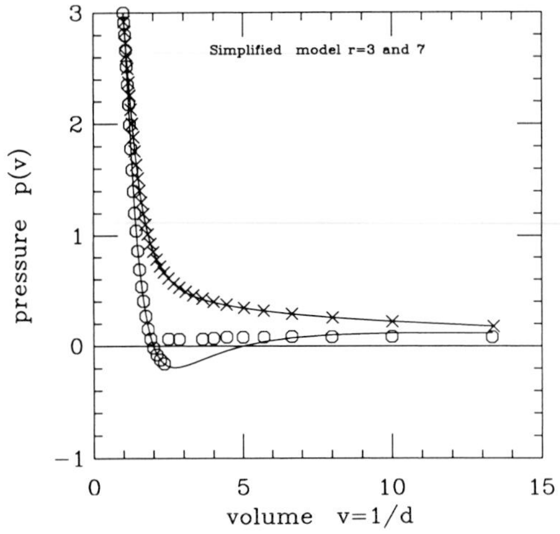

Fig. 6. The equation of state of the simplified model. Full lines are theoretical results. Squares are numerical results for $r=3$ and crosses are for $r=7$. A transition is seen for $r=7$ but not for $r=3$. When the two phases coexist, the pressure is constant but for finite size effects.

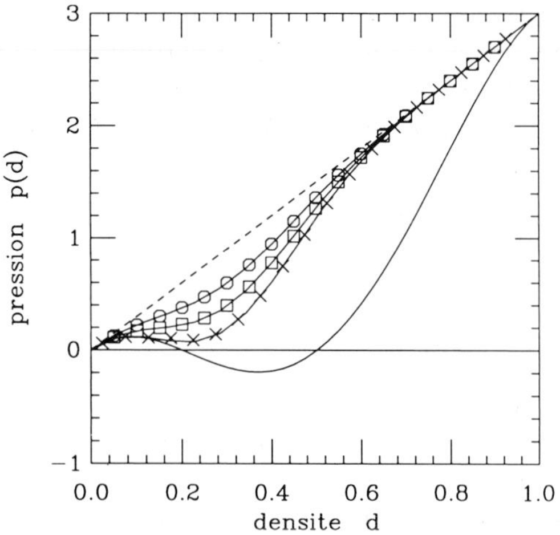

Fig. 8. Equation of state for the variant of the model defined in the test. (○) $r=3$, (□) $r=5$, (×) $r=7$. The lowest full line is the simplified model for $r=7$, and is shown for comparison.

where

$$W_{1a}(s'_a) = \sum_{\sigma, s'_b} A(\sigma, \sigma'; 1)\, W_0(s_a)\, W_0(s_b) \qquad (7)$$

and $\sigma = (s_a, s_b)$, $\sigma' = (s'_a, s'_b)$, $m = \sum_{i=0}^{6} s_{ai} + \sum_{i=0}^{6} s_{bi}$. As was already shown in ref. [13], the expression above allows the prediction of the equation of state of the lattice gas with very good accuracy in the case of the simplified model (fig. 6 shows the pressure as a function of volume per particle, $1/d$). The optimal model yields worse results (figs. 7 and 8). One possible reason is that

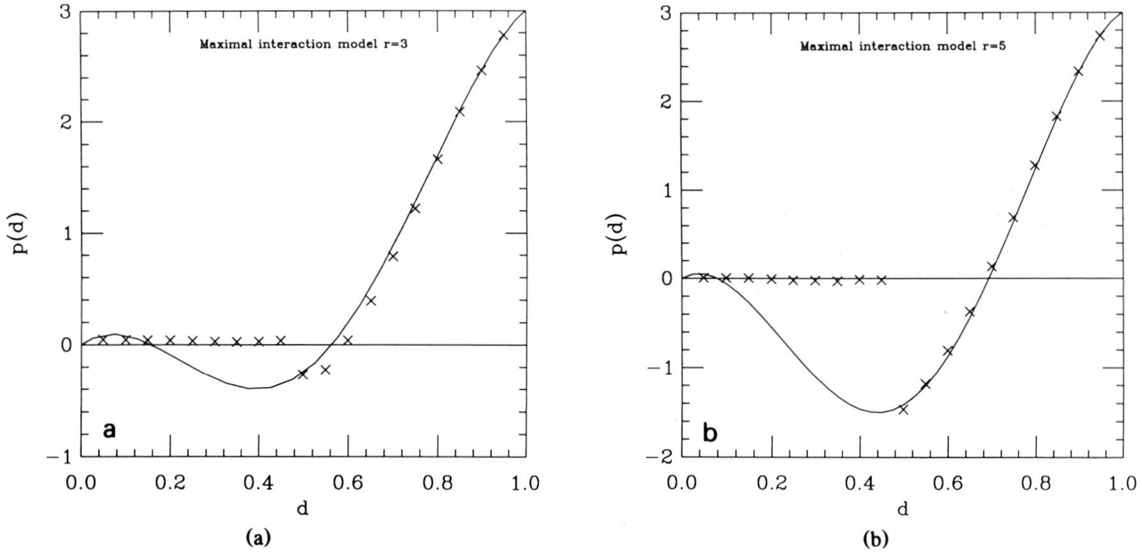

Fig. 7. Equation of state of the maximal interaction model with (a) $r=3$, (b) $r=5$.

expression (5) does not incorporate the difference between the moving- and rest-particle densities.

To illustrate how the equation of state might be adjusted, we also investigated a variant which we shall call the "low-density" model. In the low-density model the interactions are tuned so that they are effective only at low density. To achieve this, the interactions are performed just as in the simplified model. However, on each site the interaction is performed only if $\max(m_a, m_b) < 4$ where m_a, m_b are defined as above. If the latter condition is not satisfied no interactions are performed. This rule reduces the strength of the attractive force when the density goes above $\rho = 3$. In fig. 8 we show the resulting equation of state, and the theoretical prediction.

5. Isotropy and structure functions

A remarkable feature of the patterns of figs. 3 and 4 is the appearance of filaments. A first impression is that these patterns may not be isotropic. Anisotropy would indeed be an undesirable feature of the model. Thus we have performed a systematic study of the model's isotropy.

Isotropy may be quantitatively studied by Fourier transformation of the density field. Specifically, we have computed the discrete two-dimensional power spectrum

$$S(\boldsymbol{k}, t) = \frac{1}{N} \left| \sum_{\boldsymbol{x}} \mathrm{e}^{-\mathrm{i}\boldsymbol{k}\cdot\boldsymbol{x}} \left[\rho(\boldsymbol{x}, t) - \bar{\rho} \right] \right|^2, \qquad (8)$$

where $\rho(\boldsymbol{x}, t)$ is the number density at time t at a site with coordinates given by $\boldsymbol{x}$, $\bar{\rho} = 7d$ is the density averaged over the entire lattice, and $\boldsymbol{k}$ is the discrete wavenumber. Computations were performed with the maximal interaction model on a lattice of size $N \times N$, with $N = 400$, $\bar{\rho} = 2.1$, and $r = 5$.

If the model were isotropic, one would obtain a circularly symmetric $S(\boldsymbol{k}, t) = S(k, t)$ with $k = |\boldsymbol{k}|$. Figs. 9a, 9b, and 9c show contour plots of spectra computed at time steps 31, 62, and 93, each averaged over an ensemble of 50 independent numerical experiments. The level curves for the largest value of $S(\boldsymbol{k}, t)$ are noisy and therefore

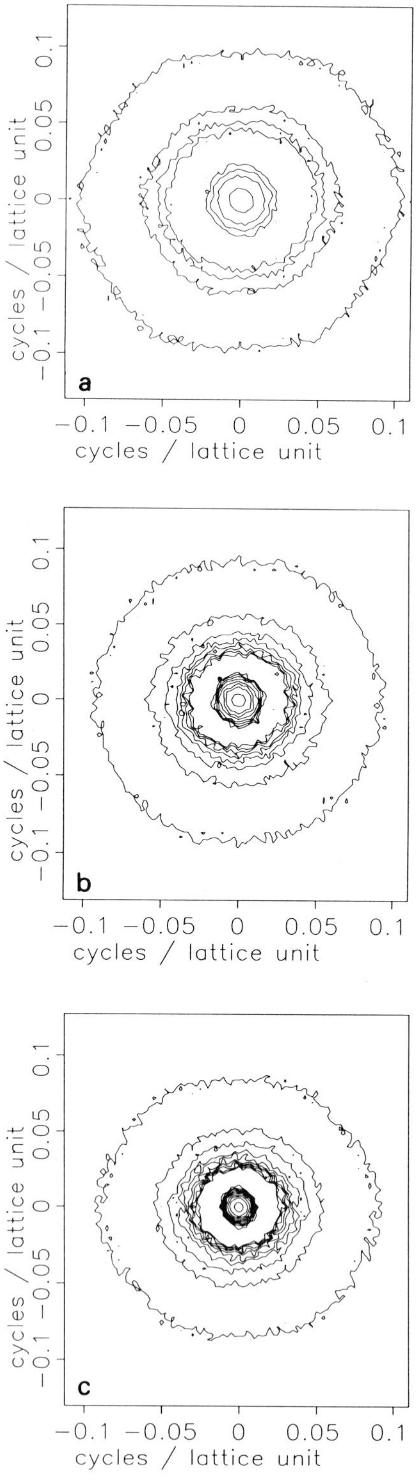

Fig. 9. 2D power spectra of the model. The scaling of the wavenumber axes has been corrected for the $2/\sqrt{3}$ factor between them. (a) Time step 31, (b) time step 62, (c) time step 93.

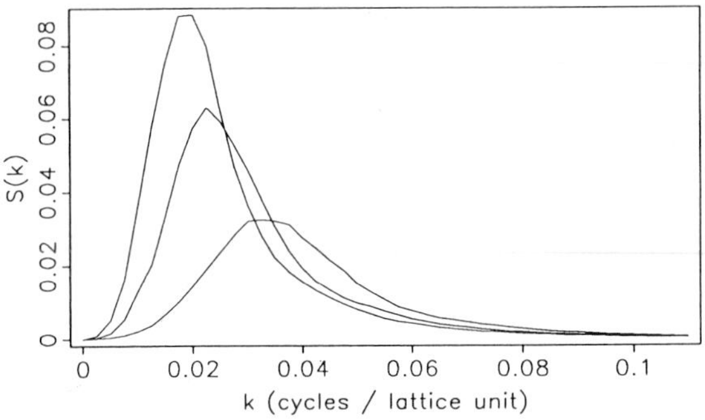

Fig. 10. The circularly averaged structure function $S'(k)$. The maximum of each spectrum occurs at a positive value of k which decreases with time.

clipped to enhançe the clarity of the figure. In each figure, but especially in fig. 9a, one can detect the hexagonal symmetry at large $\boldsymbol{k}$, corresponding to small-scale anisotropy. The spectra at small $\boldsymbol{k}$ appear more closely isotropic. We thus conclude, to within the accuracy of our measurements, that the long-wavelength behavior of the model is isotropic. A study of the isotropy of the surface tension would be necessary to make more definitive conclusions.

Assuming that the spectra are indeed isotropic, we have computed the circularly averaged spectra $S'(k,t)$ corresponding to the three spectra in fig. 9. These "structure functions" are shown in fig. 10. One interesting feature of the structure functions is their self-similarity. Define $k_{\mathrm{m}}(t)$ as the wavenumber at which $S'(k,t)$ is maximum. Fig. 11 shows a graph of $S'(k,t)/S'(k_{\mathrm{m}}(t))$ as a function of $k/k_{\mathrm{m}}(t)$ for the three unscaled structure functions of fig. 10. The three scaled functions in fig. 11 are close to being on top of each other, especially the spectra computed at time steps 62 and 93. In fig. 11 we have also compared the results of our computations to a theoretical form suggested in ref. [17].

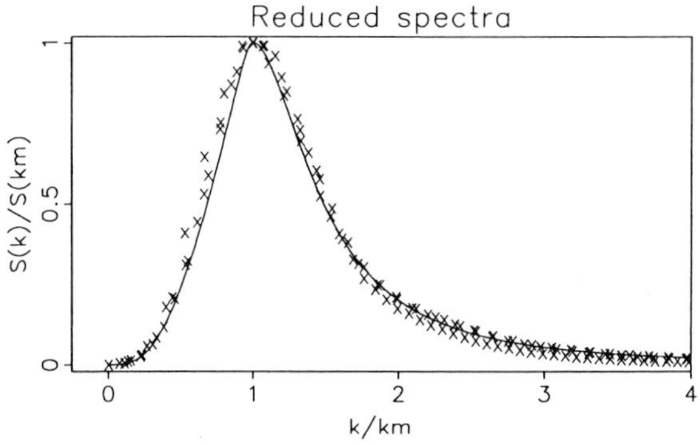

Fig. 11. The scaled structure function of the model. The full line is a fit to ref. [17].

6. Conclusions

We have described a lattice-gas model that simulates a liquid–gas transition. Theoretical predictions of the model's equation of state, obtained via a Boltzmann approximation, agree well with numerical experiments. These predictions are successful in a variety of models. The observed patterns are clearly different from the spinodal decomposition patterns observed in simpler models without momentum conservation.

Numerical computations of power spectra indicate that the hydrodynamic behavior of the model is isotropic. Future studies will further quantify the hydrodynamic properties of the model in addition to illustrating applications to the study of multiphase flows with greatly contrasting densities.

Acknowledgements

S.Z. thanks the Mathematical Disciplines Institute of The University of Chicago, where part of this work was performed. This work was supported in part by NSF Grant EAR-8817027, NATO Travel Grant 891061, and by the sponsors of the MIT Porous Flow Project.

References

[1] U. Frisch, B. Hasslacher and Y. Pomeau, Phys. Rev. Lett. 56 (1986) 1505.

[2] U. Frisch, D. d'Humières, B. Hasslacher, P. Lallemand, Y. Pomeau and J-P. Rivet, Complex Systems 1 (1987) 648.
[3] J.M. Hyman, Physica D12 (1984) 396.
[4] D. d'Humières, P. Lallemand and G. Searby, Complex Systems 1 (1987) 633.
[5] P. Clavin, P. Lallemand, Y. Pomeau and J. Searby, J. Fluid Mech. (1988) 437.
[6] M. Bonetti, A. Noullez and J.-P. Boon, in: Cellular Automata and the Modeling of Complex Physical Systems, eds. P. Manneville, N. Boccara, G. Vichniac and R. Bidaux (Springer, Berlin, 1989) p. 239.
[7] D. Burgess, F. Hayot and W.F. Saam, Phys. Rev. A 38 (1988) 3589; A 39 (1989) 4695.
[8] D.H. Rothman and J.M. Keller, J. Stat. Phys. 52 (1988) 1119.
[9] D.H. Rothman, in: Discrete Kinetic Theory, Lattice-Gas Dynamics, and Foundations of Hydrodynamics, ed. R. Monaco (World Scientific, Singapore, 1989) p. 286.
[10] D.H. Rothman and S. Zaleski, J. Phys. (Paris) 50 (1989) 2161.
[11] D.H. Rothman, J. Stat. Phys. 56 (1989) 517.
[12] A. Gunstensen and D. Rothman, A lattice-gas model for three immiscible fluids, Physica D 47 (1991) 47–52, these Proceedings.
[13] C. Appert and S. Zaleski, Phys. Rev. Lett. 64 (1990) 1.
[14] L. Kadanoff and J. Swift, Phys. Rev. A 165 (1968) 310.
[15] H. Chen, S. Chen, G.D. Doolen, Y.C. Lee and H.A. Rose, Multi-thermodynamic phase lattice gas automata incorporating interparticle potentials, Los Alamos preprint (1988).
[16] P.C. Hohenberg and B.I. Halperin, Rev. Mod. Phys. 49 (1977) 435.
[17] P. Fratzl and J.L. Lebowitz, Universality of scaled structure function in quenched systems undergoing phase separation, Acta Metall., in press.

Physica D 47 (1991) 97–111
North-Holland

LATTICE GAS MODELS FOR NONIDEAL GAS FLUIDS

Shiyi CHEN, Hudong CHEN, Gary D. DOOLEN, Y. C. LEE, H. ROSE
Center for Nonlinear Studies and T-Division, Los Alamos National Laboratory, Los Alamos, NM 87545, USA

and

Helmut BRAND
Center for Nonlinear Studies, Los Alamos National Laboratory,
and Department of Physics, University of Essen, W-43 Essen 1, Germany

Received 20 March 1990

A lattice gas model with a nonideal gas equation of state is presented. Transitions between the solid and gas phase are described. Computer simulations of applications of this model to shock waves are discussed. Generalization of this model to liquid crystal flow is also outlined.

1. Introduction

Since Frisch, Hasslacher and Pomeau (FHP) [1] proposed the lattice gas automata for solving Navier–Stokes equations, much theoretical and computational progress has been made [2–5]. Lattice gas automata offer several advantages over most numerical schemes: first, lattice gas methods approximate physical systems using parallel algorithms, making the design of a parallel dedicated lattice gas machine possible; second, a lattice gas computer program is much simpler than traditional numerical schemes; third, memory is used with maximum efficiency; fourth, because particle number, momentum and energy are conserved exactly during each scattering process, there is no roundoff error in the model and the scheme is unconditionally stable.

The collision processes for most lattice gas models only happen for some particular configurations in a same site, such as two- or three-body collisions. During the advection process, particles move along the lattice directions to their nearest neighbors. Thus, particles in a lattice gas behave like classical particles constrained to move on a lattice with an interaction potential identical to hard spheres with zero radii. The equation of state of these lattice gas models approximates an ideal gas law.

Many interesting physical phenomena occur for systems which have the properties of a nonideal gas. The simplest model in classical thermodynamics is the Ising model, in which only nearest-neighbor interactions are used. Recently, we introduced nearest-neighbor interaction potentials for particles in an FHP-like model [3]. It is interesting that this lattice gas model approximates the Navier–Stokes equations and its equation of state differs considerably from that of an ideal gas. Computer simulations of this model show a first-order phase transition below the critical point. There is a coexistence region for temperatures below this critical point.

In this paper, we present a detailed investigation of this model and discuss applications, including the shock wave simulations, and an extension of this model to two-dimensional liquid-crystal systems. In section 2, we describe the model. In section 3, we discuss the properties of the model. Section 4 presents some detailed applications. In section 5, we present the model for a nematic liquid crystal and the results of some of our simulations. A short summary appears in section 6.

2. Lattice gas model for nonideal gas fluids

To model a system with a nonideal gas equation of state and with a realistic energy equation [6,7], a lattice gas model must have more than one speed. In addition, particle interactions must include a potential energy, which can depend on density and temperature.

The original 2D FHP lattice gas model [1] consists of identical unit-mass particles on a hexagonal lattice and allows six possible particle states along the six different linear momentum directions. An exclusion rule is imposed so that no more than one particle at a given site can have the same momentum state. If we use $N_a(\boldsymbol{x})$ $(a = 1, \ldots, 6)$ to denote the particle occupation in state a at site $\boldsymbol{x}$, then $N_a = 0$ or 1. For simplicity, the collision rules between moving particles in this paper only include 2R, 2L, 3S and 4S in the notation of Wolfram [8]. In addition to the moving particles, the new lattice gas model allows at each site $\boldsymbol{x}$ another kind of particle (a "bound pair" with zero momentum), with occupation $N_0(\boldsymbol{x})$ $(= 1, 0)$. The mass of a bound pair is twice as that of a free particle. Note that this zero speed particle is not the same as a rest particle in the traditional seven-bit lattice gas model, where an additional rest particle is allowed with the same mass as a moving particle, but with no interactions between rest particles at different sites [9]. Here we have introduced a square-well potential energy between nearest-neighbor bound pairs:

$$\epsilon(r) = -\epsilon_0 \qquad r = c$$
$$\quad = 0 \qquad \text{otherwise},$$

where $\epsilon_0 = \text{const.} \geq 0$ and c is the distance between lattice sites. The constant can be set equal to unity without loss of generality, if one rescales the temperature. A bound pair only has nonzero potential energy with those bound pairs at its six nearest-neighbor sites. Free-particle interactions remain unchanged from the original FHP model. Bound pairs possess a total potential energy, $E = -\frac{1}{2}\epsilon_0 \sum_{\boldsymbol{x},a} N_0(\boldsymbol{x}) N_0(\boldsymbol{x} + \hat{\boldsymbol{e}}_a)$, which varies according to the distribution of the bound pairs. A transition between bound pairs and free pairs (with opposite momentum) is included. The ratio of the probabilities for the system to change from one state to another is proportional to $\exp(-\beta\Delta E)$, where ΔE is the potential energy difference between the two states and β is the reciprocal temperature defined for the canonical ensemble. The transition process between free particles and bound pairs is a Markov process. It can be shown that the canonical ensemble is the equilibrium invariant measure for this lattice gas system with temperature $1/\beta$ [10].

In this simplest model, the following transitions are allowed: (i) a pair of oppositely directed free particles may form a bound pair with zero net momentum, and (ii) a bound pair may become two oppositely directed free particles. The potential energy change associated with a binding transition at a site is $\Delta E(\boldsymbol{x}) = -\epsilon_0 \sum_{i=1}^{6} N_0(\boldsymbol{x} + \hat{\boldsymbol{e}}_i)$. With constant temperature everywhere, the updating rules are specified as follows: to avoid multiple transitions, the lattice is divided into three independent sublattices, each of them a hexagonal lattice with a lattice constant, $\sqrt{3}\,c$; particles on the same sublattice are separated by more than one lattice unit and hence do not mutually interact. At each time step, the updating of the system associated with the transition process is done in parallel in three steps, each step involving only one sublattice. A binding probability $\phi = \lambda \exp(-\beta\Delta E)/[1 + \exp(-\beta\Delta E)]$ $(\lambda \leq 1)$ is assigned at each site of a sublattice. The unbinding probability, $\bar{\phi}$, for bound pairs is $1-\phi$. A transition is not allowed if it leads to a state which has more than one particle per microstate. For example, for $N_0 = 1$, $\bar{\phi}$ is sampled and, if successful, one of the three paired momentum directions is chosen with equal probability. An unbinding is allowed only if there are no free particles occupying the chosen pair of directions. For $N_0 = 0$, one of the three paired momentum directions is chosen with equal probability. If the chosen pair of the free particle states is occupied, ϕ is sampled and, if successful, a binding occurs such that the pair of free particles form a bound pair and N_0 becomes 1. For fixed β, $\lambda = 1$ leads to the shortest time for the system to reach equilibrium. Streaming and elastic collision processes also occur at each time step. The FHP model is the special case with $\beta = -\infty$.

The microdynamical evolution of this simple system is described by the following set of local kinetic equations:

$$N_a(\boldsymbol{x} + \hat{\boldsymbol{e}}_a, t+1) = N_a(\boldsymbol{x}, t) + \Lambda_a + \Pi_a,$$
$$a = 1, \ldots, 6,$$
$$N_0(\boldsymbol{x}, t+1) = N_0(\boldsymbol{x}, t) + \Pi_0, \quad (1)$$

where Λ_a represents the usual FHP contribution from pure elastic collisions for the free particles [2,8]. Π_a $(a = 0, \ldots, 6)$ is the additional contribution from the transition processes:

$$\Pi_a = \mathcal{B}_a^+[1 - N_a(\boldsymbol{x}, t)] - \mathcal{B}_a N_a(\boldsymbol{x}, t),$$
$$a = 0, \ldots, 6, \quad (2)$$

where $\mathcal{B}_a^+$ and $\mathcal{B}_a$ (= 0, or 1) are the creation and annihilation operators for N_a due to the transition processes, which depend on the particle occupations at site $\boldsymbol{x}$ and the configuration of the bound pairs at the six nearest-neighbor sites. The detailed form of Π_a guarantees that the particle occupation for each state is either 0 or 1. From the explicit expressions for $\mathcal{B}_a$ and $\mathcal{B}_a^+$ [3], one can show that the conservation of mass and momentum satisfies: $\sum_{a=1}^{6} \Pi_a + 2\Pi_0 = 0$ and $\sum_{a=1}^{6} \hat{\boldsymbol{e}}_a \Pi_a = 0$.

To derive an approximate equation of state, we use the mean field approximation which assumes: (i) there are no correlations between particles at the same site and same time, $\langle N_a(\boldsymbol{x}, t)\, N_b(\boldsymbol{x}, t)\rangle = \langle N_a(\boldsymbol{x}, t)\rangle \langle N_b(\boldsymbol{x}, t)\rangle$, $a \neq b$, where $\langle\ \rangle$ represents the ensemble average; and (ii) homogeneous particle distributions $\langle N_a(\hat{\boldsymbol{x}})\rangle = \langle N_a(\boldsymbol{x}')\rangle$. It can be shown using an H theorem [2] that the equilibrium distribution will have the form

$$f_a = \frac{1}{1 + \exp(\alpha + \gamma \hat{\boldsymbol{e}}_a \cdot \boldsymbol{u})}, \quad a = 1, \ldots, 6, \quad (3)$$

and

$$f_0 = \frac{1}{1 + \exp(2\alpha - \beta\varepsilon)}, \quad (4)$$

where the $f_a = \langle N_a \rangle$ $(a = 1, \ldots, 6)$ represent the free-particle equilibrium distributions and $f_0 = \langle N_0 \rangle$ represents the bound-pair distribution. $\boldsymbol{u}$ is the velocity and ε is the average potential energy per bound pair. α and β are Lagrange multipliers, which are functions of $\boldsymbol{u}$, ρ and ε, and can be determined from the conservation of mass and momentum. Density and velocity are defined by

$$\rho = \sum_a f_a + 2f_0,$$
$$\rho \boldsymbol{u} = \sum_a \hat{\boldsymbol{e}}_a f_a. \quad (5)$$

Using the Chapman–Enskog expansion, the hydrodynamic equations for this model are

$$\partial\rho/\partial t + \nabla \cdot \rho\boldsymbol{u} = 0,$$
$$\partial_t(\rho\boldsymbol{u}) + \nabla \cdot (\rho g(\rho)\boldsymbol{u}\boldsymbol{u})$$
$$= -\nabla(p + \rho h(\rho)\boldsymbol{u}^2) + \nu\nabla^2(\rho\boldsymbol{u}), \quad (6)$$

where $g(\rho) = \rho(\rho_{\mathrm{f}} - 3)/\rho_{\mathrm{f}}(\rho_{\mathrm{f}} - 6)$, ρ_{f} is the free particle density and p is the kinetic pressure determined from the equation of state, which is equal to one-half of the free-particle density, $\rho_{\mathrm{f}} = \sum_{a=1}^{6} f_a$. The velocity-dependent term vanishes as the macroscopic velocity approaches zero. The explicit form of $h(\rho)$ can be derived by substituting the first-order velocity expansion of f_a and f_0 into the stress tensor $\Pi_{ij} = \sum f_a(\hat{\boldsymbol{e}}_a)_i(\hat{\boldsymbol{e}}_a)_j$.

In fig. 1, we show the mean field approximation pressure indicated by the solid lines for four reciprocal temperatures (0.2, 0.6, 1.0 and 1.2). V is

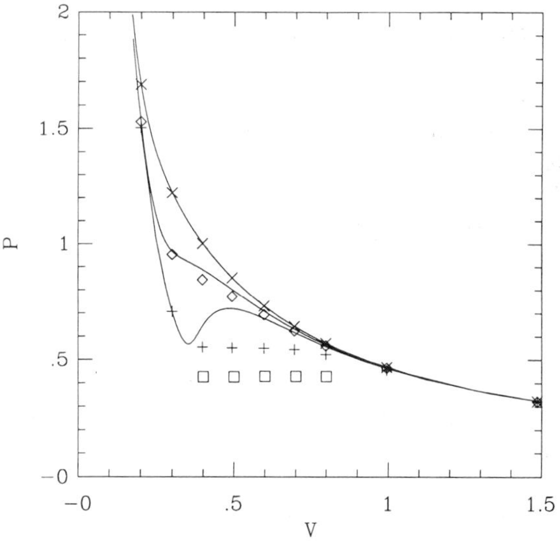

Fig. 1. Pressure versus specific volume for four different reciprocal temperatures ($\beta = 0.2, 0.6, 1.0$ and 1.2). The solid line represents the mean field approximation. The computer results are shown by $\times$, $\diamond$, $+$ and $\square$ respectively.

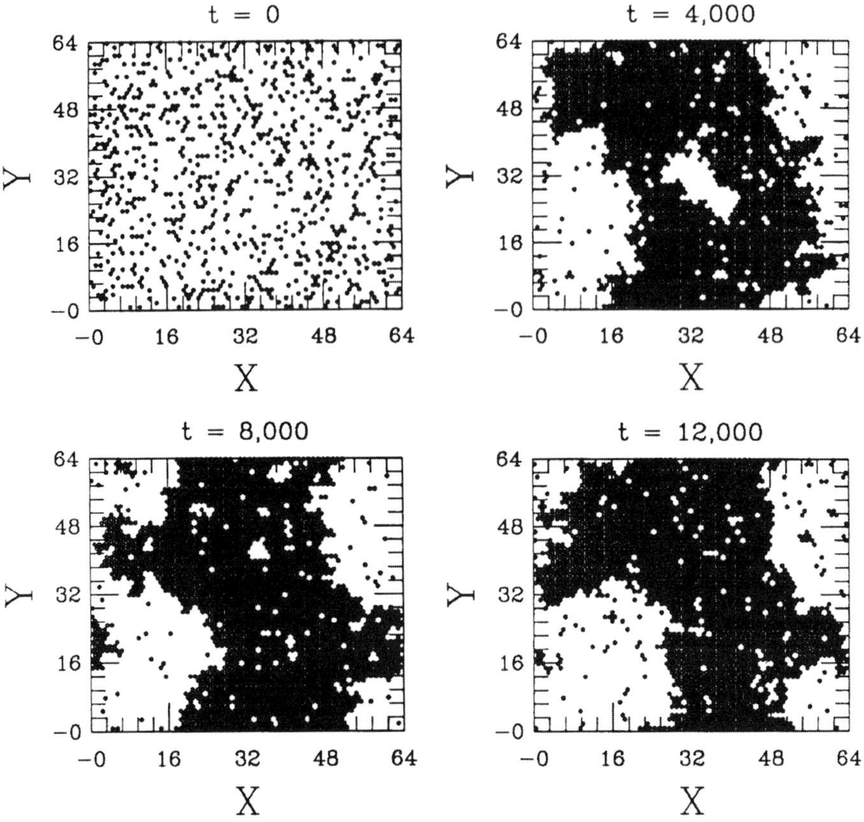

Fig. 2. Instantaneous distribution of bound pairs at time $t = 0$, 4000, 8000 and 12000 in the two-phase coexistence region with $V = 0.5$ and $\beta = 1.2$.

the specific volume, $1/\rho$. The simulation was done for zero velocity in a system with 64×64 lattice sites. We also used system sizes of 128×128 sites and 256×256 sites to test the system-size dependence. The simulations indicated that there is less than a 1% change for a system size bigger than 64×64 sites. The $\times$, $\diamond$, $+$ and $\square$ in fig. 1 represent simulation results for different reciprocal temperatures, $\beta = 0.2, 0.6, 1.0$ and 1.2. These simulations clearly show a critical point for $1 < \beta < 1.2$ where $\partial p/\partial V = 0$. In fig. 2, we present the bound-pair distributions for $\beta = 1.2$ and density $\rho = 2$ at times $t = 0$, 4000, 8000, and 12000. The initial condition for the bound-pair density is 0.2 with a random spatial distribution. The same bound-pair density and pressure exists between $t = 8000$ and $t = 12000$, but there is a difference in their spatial distributions. The spatially homogeneous pressure and spatially inhomogeneous bound pair distribution strongly suggests the coexistence of two phases. Also we noted that the bound pairs tend to coalesce at low temperature.

In fig. 3, we present $g(\rho)$ versus ρ at equilibrium for four values of the inverse temperature: $-\infty$ (FHP model), 0.2, 0.6 and 1.0. It is interesting that for reciprocal temperature near 0.6, there is a large range of densities, all with approximately the same value for $g(\rho)$. This is an important property, which can be used for simulating compressible flows with large density variations.

3. Properties of the model

3.1. *A mapping between our model and the Ising model for static properties*

Using a simple mapping, it can be shown that the invariant measure of the bound pairs is equiv-

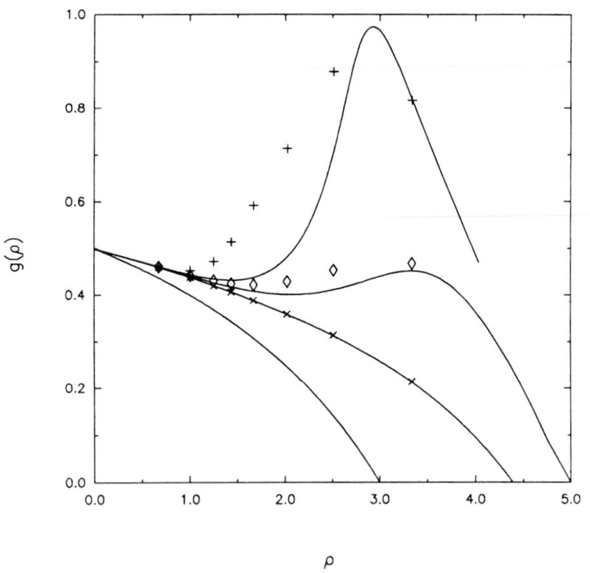

Fig. 3. The convection coefficient $g(\rho)$ versus average density per cell. The solid lines represent the mean field results and $\times$, $\diamond$, and $+$ represent the simulation results for $\beta = 0.24, 0.6$ and 1.0. The lowest solid line gives the analytical results for $\beta = -\infty$ (FHP lattice gas model).

alent to that of the Ising model on a triangular lattice with a prescribed external magnetic field, which depends on density and temperature.

In order to simplify the problem, let us define the phase space, Γ, as the set of all possible assignments of the configuration of the lattice L: $S(\cdot) = (N_i(\boldsymbol{x}) = 0,1;\ i = 0,1,\ldots,b;\ \boldsymbol{x} \in L)$, where $i = 0$ indicates a bound pair and b is the number of distinguishable moving particle states per site. The probability at time step t for a given configuration $P(t, S(\cdot))$ satisfies

$$P(t, S(\cdot)) \geq 0, \qquad \sum_{S(\cdot)\in\Gamma} P(t, S(\cdot)) = 1. \tag{7}$$

Since our model is a Markov process, the Liouville equation for $P(t, S(\cdot))$ can be written as

$$P(t+1, TS'(\cdot)) = \sum_{S(\cdot)\in\Gamma} A(S(\cdot) \to S'(\cdot))\, P(t, S(\cdot)), \tag{8}$$

where T is the advection operator,

$$TN_i(\boldsymbol{x}) \to N_i(\boldsymbol{x} + \hat{\boldsymbol{e}}_i), \quad i = 1,\ldots,b, \tag{9}$$

and $A(S(\cdot) \to S'(\cdot))$ is the transition probability from configuration $S(\cdot)$ to $S'(\cdot)$, which satisfies the the total mass and momentum conservation:

$$\sum_i (N_i' - N_i)\theta_i A(N \to N'|\boldsymbol{x}) = 0,$$

$$\sum_i (N_i' - N_i)\hat{\boldsymbol{e}}_i A(N \to N'|\boldsymbol{x}) = 0, \quad \forall N, N', \boldsymbol{x},$$

$$\theta_i = 0,1, \quad i = 1,\ldots,b. \tag{10}$$

At equilibrium, the system is translationally and rotationally invariant, so we can write $A(S(\cdot) \to S'(\cdot))$ in a formal factorized form: $A(S(\cdot) \to S'(\cdot)) = \prod_{\boldsymbol{x}\in L} A(N \to N'|\boldsymbol{x})$ and from the transition rules discussed in section 2, we can write the following generalized detailed-balance relation:

$$\frac{A(S(\cdot) \to S'(\cdot))}{A(S'(\cdot) \to S(\cdot))} = \exp\{[E(S(\cdot)) - E(S'(\cdot))]\beta\}, \tag{11}$$

where $E(S(\cdot))$ is the total potential energy for a given configuration $S(\cdot)$:

$$E = -\tfrac{1}{2}\epsilon_0 \sum_{\boldsymbol{x},i} N_0(\boldsymbol{x})\, N_0(\boldsymbol{x} + \hat{\boldsymbol{e}}_i). \tag{12}$$

The following form is obtained for the time-independent solution of the Liouville equation (8) using eqs. (9), (10) and (11) for systems with total mass and momentum conservation:

$$P(S(\cdot)) = -\frac{\exp[E(S(\cdot))\beta]}{\Omega}\, \delta\Big(N_T - \sum_{\boldsymbol{x}\in L}\sum_{i=0}^{b} \theta_i s_i(\boldsymbol{x})\Big) \times \delta\Big(\sum_{\boldsymbol{x}\in L}\sum_{i=0}^{b} \hat{\boldsymbol{e}}_i s_i(\boldsymbol{x})\Big), \tag{13}$$

where $s_i(\boldsymbol{x})$ is the occupation number for a momentum state at $\boldsymbol{x}$ and where Ω is the partition function:

$$\Omega = \sum_{S(\cdot)\in\Gamma} \exp[E(S(\cdot))\beta]\, \delta\Big(N_T - \sum_{\boldsymbol{x}\in L}\sum_{i=0}^{b} \theta_i s_i(\boldsymbol{x})\Big) \times \delta\Big(\sum_{\boldsymbol{x}\in L}\sum_{i=0}^{b} \hat{\boldsymbol{e}}_i s_i(\boldsymbol{x})\Big). \tag{14}$$

In the thermodynamic limit, we replace the δ-functions by $\exp(-\mu N - \boldsymbol{\gamma}\cdot\boldsymbol{\pi})$, where μ and $\boldsymbol{\gamma}$ are chosen to satisfy mass and momentum constraints: $\langle N\rangle = N_{\rm T}$ and $\langle \pi\rangle = 0$, where $N_{\rm T}$ is the total particle number in the system and N and $\boldsymbol{\pi}$ are the total mass and total momentum for a given configuration. We obtain a probability distribution which factorizes the bound-pair and free-particle dependence:

$$P(S(\cdot)) = \frac{\exp[-E(S(\cdot))\beta - N(s(\cdot))\mu - \boldsymbol{\pi}(s(\cdot))\cdot\boldsymbol{\gamma}]}{\Omega}. \tag{15}$$

The reduced bound-pair probability is

$$P_0(S(\cdot)) = \frac{\exp[-E_0(S(\cdot))\beta - N_T(S_0(\cdot))\mu]}{\Omega_0} = P_0(S_0(\cdot)), \tag{16}$$

since $E_0(S(\cdot)) = E_0(S_0(\cdot))$, where

$$N_T(S_0(\cdot)) = 2\sum_{\boldsymbol{x}\in L} N_0(\boldsymbol{x}),$$

$$\Omega_0 = \sum_{S(\cdot)\in\Gamma} \exp[-E_0(S_0(\cdot))\beta - N_0(S_0(\cdot))\mu]$$

and $\mu = \mu(\beta,\rho)$.

In order to compare with the Ising model, we introduce the following transformation,

$$N_0(\boldsymbol{x}) = \tfrac{1}{2}[s(\boldsymbol{x}) + 1], \tag{17}$$

where $s(\boldsymbol{x}) = 1, -1$ for $N_0 = 1, 0$, respectively. The potential energy for the total system can be written as

$$E = -\tfrac{1}{8}\epsilon_0 \sum_{\boldsymbol{x},i} s(\boldsymbol{x})s(\boldsymbol{x}+\hat{\boldsymbol{e}}_i) - \tfrac{1}{4}\epsilon_0 b \sum_{\boldsymbol{x}} s(\boldsymbol{x}), \tag{18}$$

and the probability distribution has the form

$$P_0(S_0(\cdot)) = \frac{1}{\Omega_0}\exp\Big(-\tfrac{1}{8}\beta\epsilon_0 \sum_{\boldsymbol{x},i} s(\boldsymbol{x})\, s(\boldsymbol{x}+\hat{\boldsymbol{e}}_i)\Big) \times \exp\Big(-\beta B_0(\beta,\rho)\sum_{\boldsymbol{x}} s(\boldsymbol{x})\Big), \tag{19}$$

which is the exact Ising probability distribution

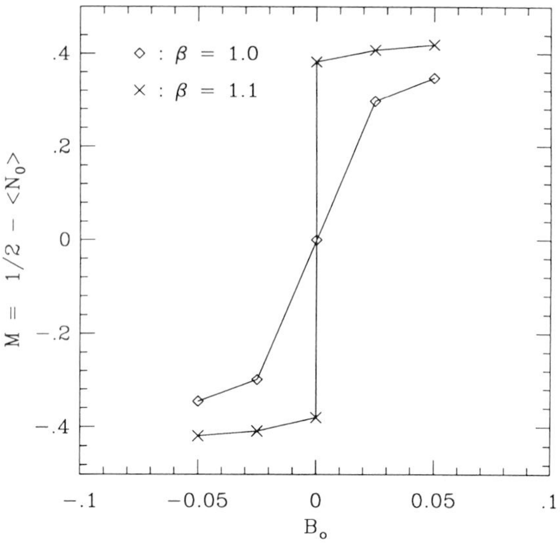

Fig. 4. Magnetization versus magnetic field for the Ising model (from lattice gas simulation without FHP-type collision) for the hexagonal lattice. The simulation confirms that the critical reciprocal temperature is near 1.09.

in an external magnetic field with $J = \frac{1}{2}\epsilon_0$ and $B_0 = \frac{1}{4}\epsilon_0 b + B_0'$. At the critical point, $B_0' = -\frac{1}{4}\epsilon_0 b$.

For the two-dimensional triangular Ising model, the critical reciprocal temperature, $\beta_{\rm c}$, is

$$\beta_{\rm c} = \frac{\log 3}{2J}. \tag{20}$$

Setting $J = \frac{1}{2}$ for our model, we obtain a critical reciprocal temperature $\beta_{\rm c} = \log 3 \approx 1.0986$. In fig. 1, we see that the critical reciprocal temperature is near 1.1. In fig. 4, we present the Ising model results for magnetization $M = \frac{1}{2} - \langle N_0\rangle$ for a lattice gas system with an external magnetic field B_0', simply by removing the collision process from the lattice gas code. A sharp jump in magnetization is observed at the critical temperature.

3.2. Sound speed and measurements

The sound speed is defined by $\partial p/\partial\rho$ for lattice gas systems. Because we place the lattice gas system in contact with a uniform-temperature heat bath, the sound speed is the isothermal sound speed. It is possible to measure this sound speed directly as the propagation speed of a density fluctuation in the long-wavelength and low-frequency limit. In order to do that, we define the following Laplace–Fourier transformation:

$$\tilde{\rho}(\boldsymbol{k}, s) = \int_0^\infty \mathrm{d}t \, \exp(-st) \times \int_{-\infty}^{\infty} \exp(\mathrm{i}\boldsymbol{k}\cdot\boldsymbol{r}) \, \rho(\boldsymbol{r}, t) \, \mathrm{d}\boldsymbol{r}, \tag{21}$$

where $s = \epsilon + \mathrm{i}\omega$.

We assume that the density and velocity in eq. (6) can be separated into a constant part and fluctuating part, $\rho = \rho_0 + \delta\rho$ and $\boldsymbol{u} = \delta\boldsymbol{u}$, where we have also assumed that the constant part of the velocity is zero. We can obtain a dispersion relation using (21) and eliminating higher-order fluctuations in (6),

$$s\,\delta\tilde{\rho}(\boldsymbol{k}, s) + \mathrm{i}\boldsymbol{k}\cdot\tilde{\boldsymbol{j}}(\boldsymbol{k}, s) = \delta\tilde{\rho}(\boldsymbol{k}, 0),$$
$$(s + \nu k^2)\,\tilde{\boldsymbol{j}}(\boldsymbol{k}, s) + c_s^2\,\mathrm{i}\boldsymbol{k}\delta\tilde{\rho} = \tilde{\boldsymbol{j}}(\boldsymbol{k}, 0). \tag{22}$$

Here, $\boldsymbol{j} = \rho_0\delta\boldsymbol{u}$ and $c_s^2 = \partial p/\partial\rho$. After simple manipulation, one obtains the following relation:

$$\frac{\langle\delta\rho(\boldsymbol{k}, s)\,\delta\rho^*(\boldsymbol{k}, 0)\rangle}{\langle\delta\rho(\boldsymbol{k}, 0)\,\delta\rho^*(\boldsymbol{k}, 0)\rangle} = \frac{s + \nu k^2}{s(s + \nu k^2) + c_s^2 k^2}. \tag{23}$$

Defining the density correlation function as

$$S(\boldsymbol{k}, \omega) = \int_{-\infty}^\infty \mathrm{d}t \, \exp(-\omega t) \times \int_{-\infty}^{\infty} \exp(\mathrm{i}\boldsymbol{k}\cdot\boldsymbol{r}) \, F(r, t) \, \mathrm{d}\boldsymbol{r}, \tag{24}$$

where $F(|\boldsymbol{r} - \boldsymbol{r}'|, t - t') = \langle\rho(\boldsymbol{r}, t)\,\rho(\boldsymbol{r}', t')\rangle$. We have

$$\frac{S(\boldsymbol{k}, \omega)}{S(\boldsymbol{k}, 0)} = 2\mathrm{Re}\lim_{\epsilon\to 0}\frac{\langle\delta\rho(\boldsymbol{k}, s)\,\delta\rho^*(\boldsymbol{k}, 0)\rangle}{\langle\delta\rho(\boldsymbol{k}, 0)\,\delta\rho^*(\boldsymbol{k}, 0)\rangle},$$

which has peaks at $\omega = c_s k, -c_s k$, for small wavenumbers.

The sound speed of the lattice gas systems has been determined by measuring the density correlation function for $k = 2\pi/L_x$ ($k = 0$ represents the constant density mode) in a system with 128×128 lattice sites. The y-direction space average has been used to replace the ensemble average. Initially zero velocity and constant density are imposed on the system. After the system approaches equilibrium, we run an additional $T = 16384$ time steps to obtain a time series for the Fourier transformation and use this to obtain the density correlation power spectrum. In fig. 5, we show the unnormalized power density spectrum, $\langle S(\omega)\,S^*(\omega)\rangle$, for the FHP-I model (dash–dotted line), which has a peak at $\omega = 91.8 \times 2\pi/T$, which corresponds to a sound speed of 0.707, accurate within 2%. The same method has been used for the new model (solid line in fig. 1). Except near the critical point, the density spectrum always has a peak at $c_s k$, where c_s is the sound speed. Because of the long-range correlation, the measurement of sound speed is not accurate near the critical point. We present typical sound speed measurements for the density $\rho = 2.42$. In fig. 6, we give the sound speed as a function of the reciprocal temperature. We see that the sound speed decreases toward zero as β approaches the critical reciprocal temperature of 1.1. The existence of small sound speeds is important for supersonic flows, as we will discuss in section 4.

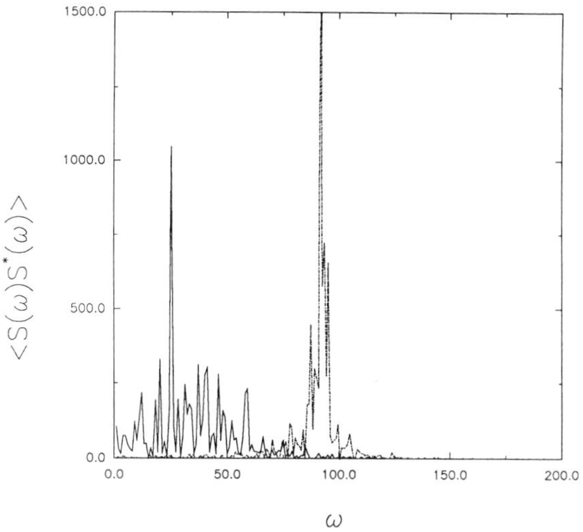

Fig. 5. The power spectrum for wavenumber $k = 2\pi/L$: the dash–dotted line represents the FHP result and the solid line represents the present model, both at density $\rho = 2.42$. The frequency of the peak is related to the sound speed of the system by $\omega = c_s k$.

3.3. Viscosity measurements

The measurement of kinematic viscosity has been implemented using forced channel flow be-

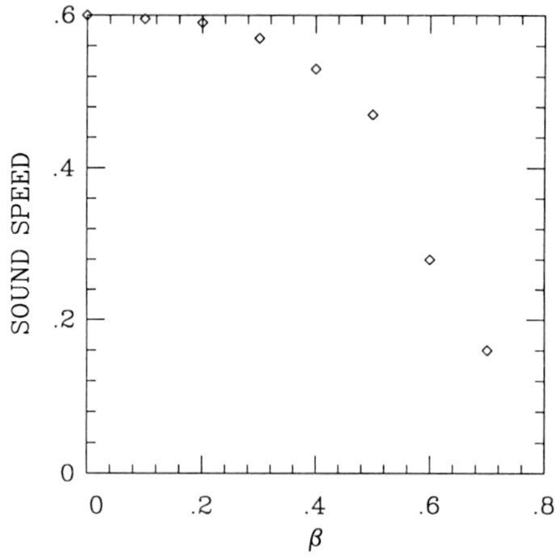

Fig. 6. The sound speed versus the reciprocal temperature β for $\rho = 2.42$. The sound speed decreases with decreasing temperature.

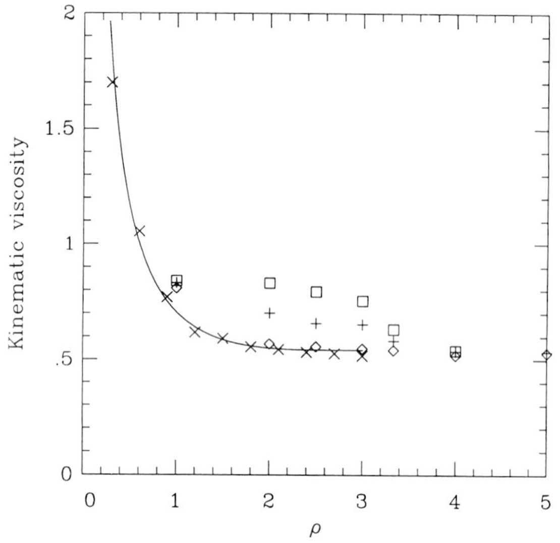

Fig. 7. Simulation results for kinematic viscosity versus density for $\beta = -\infty, 0.6, 1.0$, and 1.2 ($\times, \diamond, +$ and $\square$). The solid line represents the analytical result for the FHP-II lattice gas.

tween two parallel plates (Poisseuille flow). The forcing scheme is similar to that used by Kadanoff et al. [11]. To maintain energy conservation, no transitions are allowed between moving particles and bound pairs during the forcing process. A system size of 256×64 sites was used. The boundary condition in the x-direction is periodic and the y-direction is non-slip (bounce back). The viscosity of the system is given by,

$$\nu = \frac{3}{32 j_{\max}} f L_y^2, \tag{25}$$

where $j_{\max}$ is the averaged x-direction momentum along the centerline of the channel; f is the forcing for each time step per unit area and L_y is the lattice size in the y-direction. After equilibration, a time average is used to obtain a parabolic x-direction momentum distribution. In fig. 7, measurements of viscosity versus density are given for several reciprocal temperatures, β. The solid line describes $\beta = -\infty$ (FHP). The $\times, \diamond, +$ and $\square$ represent $\beta = -\infty, 0.6, 1.0$ and 1.2, respectively. The viscosity increases with increasing β (i.e. with a decrease of temperature) for a given density in the coexistence region ($1 < \rho < 3$). Moreover, a higher density is associated with a lower viscosity for a fixed temperature. This occurs because the mean free path decreases for the system with a reduced density in the regime between zero and one-half. In the very low density region, two moving particles having a head-on configuration will have a small probability. This causes the moving particles and bound pairs to have almost the same temperature distribution, as can be seen in the P–V diagram in fig. 1.

The total density as a function of distance from one wall for different β are presented in fig. 8 for the case $\rho = 2$. This density is in the domain of coexistence (fig. 1). We find that the density distribution across the channel is neither constant nor symmetric for low temperature. For β higher than the critical value, the system still behaves like the FHP lattice gas model. If the velocity is always less than 0.1, then the nonphysical velocity-dependent term in the equation of state should not seriously affect the distribution of particles in space. Hence, the nonuniform density distribution across the channel comes from a variation of temperature. When we reduce the temperature, the bound pairs concentrate at the center of the channel because of the increased chance of a head-on collision. This phenomenon is observed in experimental solid–gas or solid–liquid flows [12]. The density variation across the channel does not affect the momentum profile in the simulation. The normalized momentum profiles for different β (even for coexistence region) are still parabolic. But the

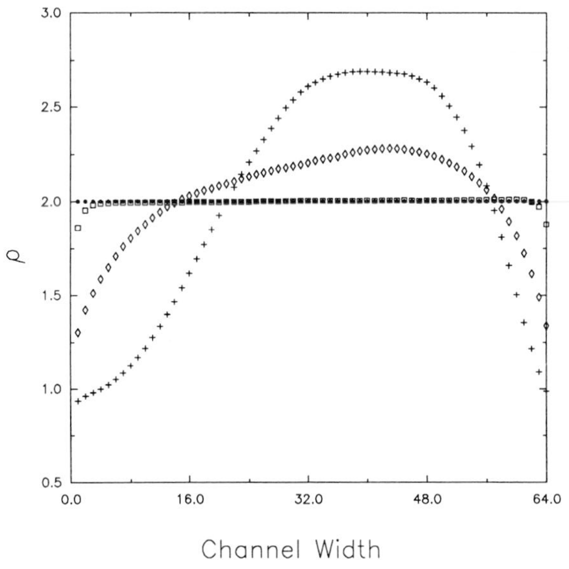

Fig. 8. The total density distribution as a function of channel width for forced channel flow for the case $\rho = 2$. The solid circle, square, diamond and plus signs represent $\beta = -\infty, 0.6, 1.0$ and 1.2, respectively.

velocity has changed due to the definition in eq. (5). In fig. 9, we show a typical velocity distribution from our simulation (□) along channel direction y for a density $\rho = 2$, and $\beta = 1.0$ compared with a parabolic profile (solid line). Nonsymmetric velocity is observed. The phenomenon of the change of velocity distribution with a change of void ratio has also been observed in other two-phase flow experiments [13].

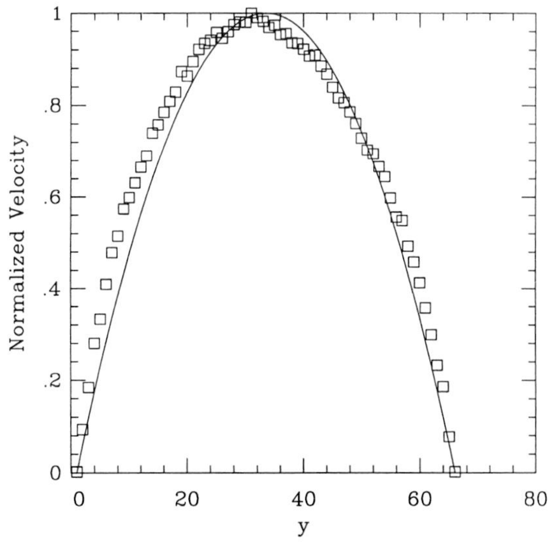

Fig. 9. The normalized channel flow velocity distribution along the y-direction for $\beta = 1.0$. Squares represent the present lattice gas simulation results and the solid line is a parabolic profile.

4. One-dimensional shock waves

The model discussed in section 3 can be consistently applied to supersonic flows because the sound speed is very small when the system is near the critical point, as shown in subsection 3.2. The hydrodynamic equation (6) is obtained assuming a small macroscopic velocity expansion. The sound speed in the FHP-I lattice gas is $1/\sqrt{2}$. Hence, it is not consistent to use the original lattice gas model to simulate a flow with Mach number bigger than 1. Also, because the convection coefficient, $g(\rho)$, in Navier–Stokes equation is density dependent for previous models, it will not consistent to rescale $g(\rho)$ at different density. The present model provides the possibility of keeping $g(\rho)$ almost constant over a large density region as shown in section 2.

An important phenomenon in supersonic flows is shock wave formation: the discontinuous macroscopic solution for the Navier–Stokes equation and the microscopic transition from a uniform upstream flow to a uniform downstream flow. The main difference between our model and the usual hydrodynamic discontinuous solution is that the shock wave phenomena usually depend on temperature or entropy. Our model has a constant temperature. Thus only pressure, density and velocity fields can be measured. In this paper, we are interested in one-dimensional shock waves, particularly the piston problem [14]. We initialize the flow with uniform velocity along the x-direction and place a fixed wall on the right. The fluid is compressed when it strikes the wall and forms a wave moving upstream (in the negative x-direction). To obtain one-dimensional results, we use periodic boundary conditions in the y-direction and average over the y-direction. The fluid velocity in the region between the wave interface and the wall is zero. We define the Mach number of the system to be the ratio of the upstream (incoming) velocity to the sound speed of the fluid. The Mach number can be varied by changing the incoming velocity. Usually shocks are studied in the reference frame of the shock. Then all quantities are time-independent. Because we are interested in the shock wave structure and the speed of the shock, we use a laboratory frame of reference.

In the FHP lattice gas, the sound speed does not depend on density and temperature. The propagation speed of a density discontinuity will move at the speed of sound. The numerical simulations for a six-bit lattice gas model have demonstrated the existence of a macroscopic discontinuous solution for the FHP lattice gas and confirm that the interface propagation speed is the sound speed.

Measurements of the shock speed have been carried out for our phase transition lattice gas model with a density of 2.42 and $\beta = 0.7$. The measured sound speed for this material is 0.16. This simulation was done in a system with 1024×1024 lattice sites. A constant velocity boundary at $x = 0$ was sampled each time step from the equilibrium distributions (3) and (4). The initial condition is the uniform velocity field, sampled again from eqs. (3) and (4). The propagation speed of the interface in the present model is lower than the FHP model. The $g(\rho)$ variation is large for a system with high Mach number. Hence, the present model has much better Galilean invariance for weak shock flows. For example, for the system with Mach number 1.2, we have $g_2(\rho = 2.49) = 0.473$; the downstream density is $\rho = 3.8$, which gives $g_1 = 0.475$, where subscript 2 refers to the incoming density and subscript 1 refers to the high density region. If the Mach number is 1.875, $g_2(\rho = 2.42) = 0.442$ and $g_1(\rho = 4.25) = 0.313$. We see that the first case has approximately the same $g(\rho)$ before and after the shock wave, while there is a big change in $g(\rho)$ for the second case. The propagation speed of the shock interface also depends strongly on Mach number. The first example has a propagation speed of 0.375 and the second example has a value of 0.401. Using the conservation of mass and the measured density, we have a propagation speed of 0.38 in the first case. In fig. 10, we present the shock interface for time steps from $t = 0$ to $t = 1000$ for every 100 time steps for a system with $\rho_2 = 2.42$ and $p_2 = 1.62$. Even though there is some noise because of the finite space average in the y-direction, we see that the propagation speed is still a constant.

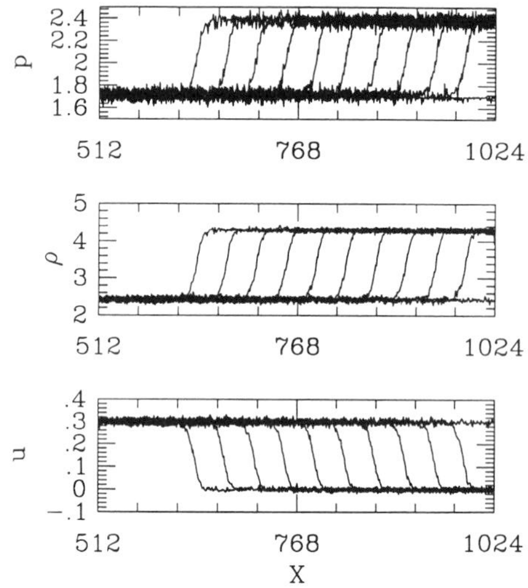

Fig. 10. Evolution of the discontinuous interfaces for (a) pressure; (b) density and (c) velocity from $t = 0$ to $t = 1000$. Time interval is 100.

The structure of the shock wave is an interesting problem for statistical mechanics reasons. A continuum description, such as the Navier–Stokes equation, is limited to the macroscopic information and cannot describe the interface structure in detail. In fig. 11, we present typical pressure and density profiles for a normal shock wave at

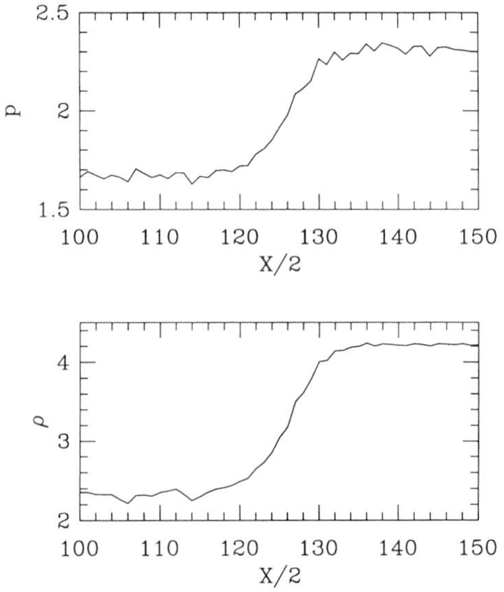

Fig. 11. Instantaneous simulation pressure (a) and density (b) profiles for a normal shock wave with Mach number 1.875.

time $t = 2000$ for a Mach number of 1.875. The detailed structure is very similar to other simulations [15]. A two-cell x-direction average has been used to smooth the noise in the simulation. The mean free path is defined by the ratio of the mean microscopic velocity to the collision frequency. For our model and the density in this calculation, the mean free path is about 3 lattice lengths [5]. The transition thickness is about 10 mean free paths in our simulation. Also from our simulations, we find that the shock thickness depends on Mach number. A strong shock produces a short shock thickness.

The relation between pressure, density and velocity before and after a shock interface is given by the Rankine–Hugoniot relation [14], which is derived assuming conservation of mass, momentum and energy. Energy conservation for our case is trivial because our system has a constant temperature. The viscosity term in the momentum equation is nonzero only in the transition region. Including the dependence on $g(\rho)$ and $h(\rho)$ in (6), we have the following relations:

$$\rho u = \text{const.}$$

$$p + \rho(g(\rho) + h(\rho))u^2 = \text{const.} \tag{26}$$

valid throughout the shock in the frame of reference of the shock.

Because of the complicated form of $g(\rho)$ and $h(\rho)$, it is difficult to get a simple expression for the Rankine–Hugoniot relation. In fig. 12, we show the simulation results for the relation between p_2/p_1 and ρ_2/ρ_1 for the incoming velocities of $0.2, 0.25, 0.3, 0.35$ and 0.37. There is a 3% error in the measured data due to noise. The crosses in the plot represent for the analytical solution of a perfect gas from the formula

$$\frac{p_2}{p_1} = \left(\frac{\gamma+1}{\gamma-1}\frac{\rho_2}{\rho_1} - 1\right) \Big/ \left(\frac{\gamma+1}{\gamma-1} - \frac{\rho_2}{\rho_1}\right).$$

The diamonds represent the lattice gas simulation. Because these two systems are different, one should not expect exact agreement, but there is qualitative agreement. It will be interesting to work out a scaling relation between different materials for the Rankine–Hugoniot relation. If one can find it, then all simulations for realistic systems can be carried out using lattice gas simulations.

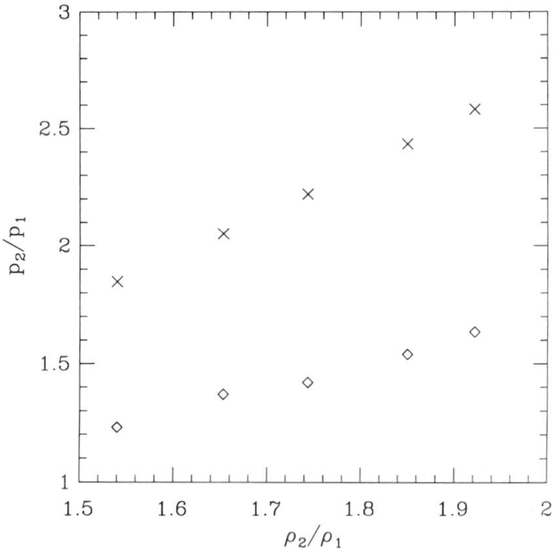

Fig. 12. The Rankine–Hugoniot relation for perfect gases (×) and a similar relation for the present model from computer simulation (◇).

5. An extension of the model: order phase transition

The difference between the present lattice gas model and previous FHP-type lattice gas models is, so far, only the existence of the interaction potentials between bound-pair particles at nearest-neighbor sites. Another interaction can be added which depends on the orientation of particles. For this extension, a directional degree of freedom must be introduced. Suppose that, in addition to having a spatial coordinate, $\boldsymbol{x}$, and linear momentum, $\hat{\boldsymbol{e}}_a$, we let all particles have an orientational degree of freedom. For simplicity, we limit ourselves to the simplest model which allows bound-pair particles to have an orientation which is independent of the linear momentum and interaction potentials. Hence, the bound-pair particles have anisotropic properties. This requires an additional degree of freedom, the molecular orientation θ, which can vary continually from $0°$ to $360°$ in two-dimensional space. We restrict θ to be a multiple of $60°$ because of the hexagonal lattice. The general interaction Hamiltonian between two particles at different sites and orientations can be represented by $H = H(\theta_{ij}, r)$. Here θ_{ij} is the relative angle between particles i and j and r is the distance between them. In order to extend the model to liquid crystals, we restrict the Hamiltonian to be an even function of θ. Physically this restriction is related to the indistinguishability of head and tail of the macroscopic properties of liquid crystals. We have designed lattice rules which allow a transition between an orientationally ordered state and a disordered state. This corresponds to the nematic–isotropic transition. There are two types of potentials for liquid crystals [16]: separable and nonseparable potentials. The separable potential requires that the Hamiltonian with a dependence on θ and r can be written as a summation of products of a function of angle times a function of the distance: $H(\theta, r) = \sum_n H_n(\theta) H_n'(r)$. It has been proved [17,18] rigorously that no long-range order can exist in systems with separable potentials. In contrast, for nonseparable interactions, the existence of the true long-range order is not excluded. In this paper, we only study the separable case and take $H'(r)$ to be the same as in the model discussed earlier. There are many possibilities for $H(\theta)$ [16,19]. For D-dimensional problems, it is interesting to study the following interaction potential:

$$H(\theta) = \frac{D\cos^2(\theta) - 1}{D - 1}. \tag{27}$$

For this Hamiltonian, a randomly distributed orientation will give zero average for $\langle H(\theta) \rangle$. A completely parallel orientation will have $\langle H(\theta) \rangle = 1$. Here $\langle \; \rangle$ is the ensemble average. When the interactions between orientations are isotropic, it is possible to see two-phase transitions of the Kosterlitz–Thouless type [20]. The low-temperature ordered phase is characterized by quasi-long-range order, with a power-law decay for the correlation function, $g_2(r)$, defined as follows:

$$g_2(r) = \langle \cos\{2[\theta(r) - \theta(0)]\} \rangle, \tag{28}$$

where $\theta(r) - \theta(0)$ is the angular difference between an orientation at a given position and the orientation at distance r. The $\langle \; \rangle$ can be replaced by a summation over all space. In order to test the effect of the position interaction potential $H(r)$ on $H(\theta)$, we introduce the parameter χ, the ratio of $H(r)$ to $H(\theta)$. Then $H(r, \theta)$ in this paper ($D = 2$) is

$$H(r, \theta) = \chi H(r) + 2\cos^2(\theta) - 1, \tag{29}$$

where $H(r) = \epsilon_0 N_0(\boldsymbol{x}) \sum_{i=1}^{6} N_0(\boldsymbol{x} + \hat{\boldsymbol{e}}_i)$ and θ is the relative orientation angle between the nearest-neighbor bound pairs.

We define the order parameter for the system to be the ensemble average of $H(\theta)$. All the operations in section 2 (collision, streaming and transition) will apply. The transition will have to account for binding energy. Because of the indistinguishability of head and tail, we only need to calculate ΔE_i ($i = 1, 2, 3$) for three possible orientations. Mean field theory results can be obtained, but in this paper we only present simulation results. The lattice gas simulation has been done for a system with a 64×64 lattice sites. A nine-bit lattice gas code was developed containing six moving states and three bound orientation states. The equilibration process from an arbitrary initial condition depends strongly on density and temperature. Usually, when the system is not highly oriented, the time to approach equilibrium is relatively short. If the system is near $H(\theta) = 1$, then the equilibrium time is very long.

In fig. 13, we present a plot of $\langle H(\theta) \rangle$ versus β for density, $\rho = 2.4$ (fig. 13a) and $\rho = 4.8$ (fig. 13b) for $\chi = 0, 0.5$ and 1.0 denoted by $\square$, $\times$ and $\diamond$, respectively. We see that the transition is not the first-order where there would be a jump in the order parameter. Rather, a continuous behavior is found, where one has a big change of the value of the order parameter over a small internal of β values. The continuum transition is more likely for a higher density system for the same χ and more likely for a high temperature (small β) for the same density system. This is the same as the first-order phase transition as we have seen for bound pairs in sections 2 and 3.

In plate I, we show snapshots of the orientational equilibrium distribution for a system at different temperatures, but with the same density, $\rho = 4.8$, at time $t = 10000$ after a random initial orientation. The color coding represents the orientation. Yellow represents the x-direction orientation, red represents the orientation having an angle 60° with x-direction; and black represents the orientation having an angle 120° with x-direction. The pattern formation depends on the initial condition. As is clear from the plots, there is a increase

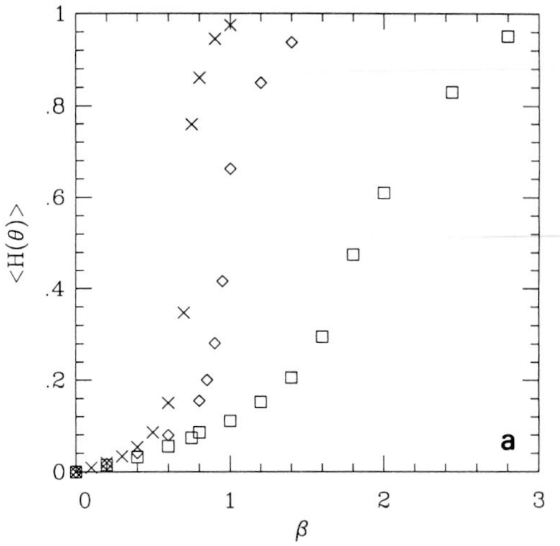

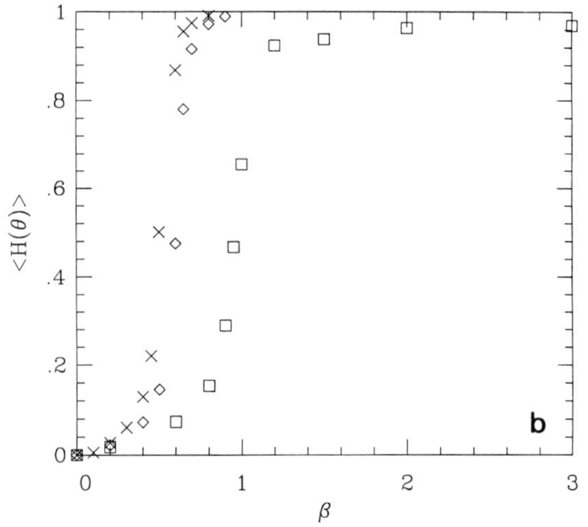

Fig. 13. The nematic order parameter $\langle H(\theta)\rangle$ as a function of β for $\chi = 0$ (□), 0.5 (◇) and 1.0 (×): for density $\rho = 2.4$ (a) and $\rho = 4.8$ (b). Note that orientational order increases as χ is increased for fixed density, and as the density is increased for fixed χ.

of the alignment of the orientation with increasing β.

A simulation for determining the spatial dependence of the angular correlation was run which used $\rho = 4.8$, $\chi = 0.5$ and $\beta = 1.2$. A typical correlation function $g_2(r)$ versus spatial separation r is presented in fig. 14 in log–log form at $t = 20000$ after a random initial condition. This time is long enough to allow the system to relax to equilibrium. We found that the spatial decay rate is not exponential. Also we observe that for r smaller than 7,

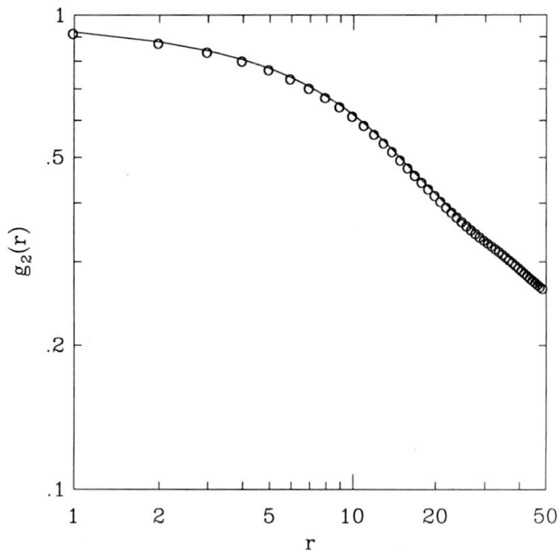

Fig. 14. Log–log plot of the correlation function $\langle\cos\{2[\theta(r) - \theta(0)]\}\rangle$ (circles) versus spatial separation. We can see that there is an algebraic decay over the range from $r = 20$ to 50. Here $\rho = 4.8$, $\chi = 0.5$ and $t = 20000$.

the correlation decays faster than algebraic, while $20 \leq r \leq 50$, we observe a linear decay (an algebraic decay in the linear coordinate), this indicates a quasi-long-range order. A transition occurs between these two regions. The result has also be observed by others [18]. The dependence of $g_2(r)$ on density ($\rho = 1.8, 2.4$ and 4.8) is shown in fig. 15 (with $\chi = 0.5$). At lower density, only short-range correlations appear. The dependence of $g_2(r)$ on temperature ($\beta = 0.4, 0.6$ and 1.0) is shown in fig. 16 for a system with $\chi = 0.5$ and density $\rho = 4.8$. The small β (high temperature) behavior is related to short-range correlations which exponentially decay. This agrees with Frenkel's results using Monte Carlo simulations [18]. The traditional correlations include $g_{2l}(r)$ which is defined as $g_{2l}(r) = \langle\cos\{2l[\theta(r) - \theta(0)]\}\rangle$. Because of our definition of orientation for a hexagonal lattice, these high-order correlations are equivalent to $g_2(r)$.

6. Concluding remarks

In this paper, we have studied a lattice gas model for fluids with an equation of state corresponding to a nonideal gas. Mean field theory

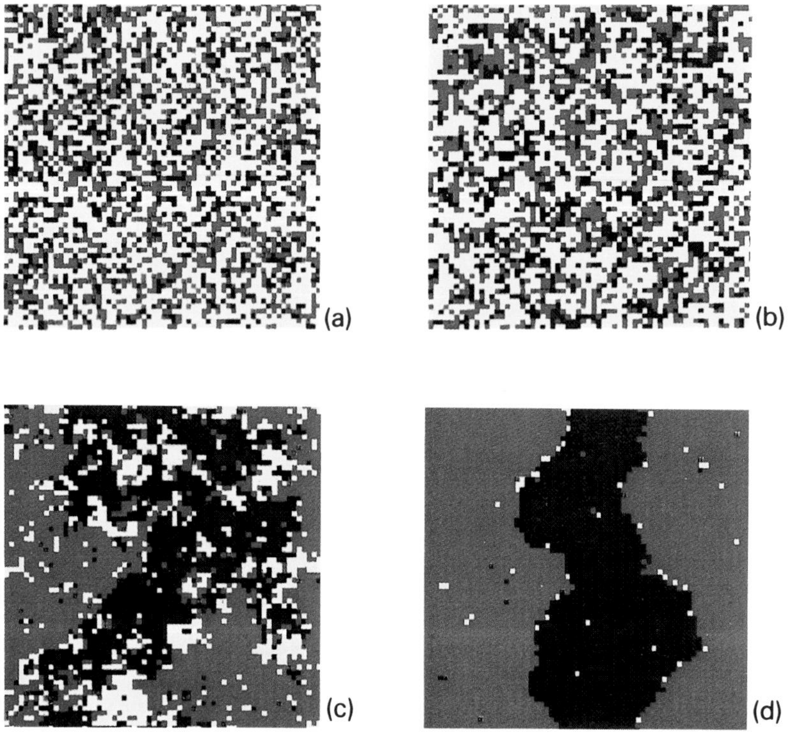

Plate I. Snapshots of configurations for a system of 64×64 lattice sites at density $\rho = 4.8$ at reciprocal temperatures $\beta = 0$ (a), $\beta = 0.4$ (b), $\beta = 0.6$ (c), and $\beta = 0.8$ (d) for $t = 10000$ after a random initial distribution. Note the increase in local orientational order as β increases. The yellow color represents the particle with orientation pointing to the x-direction, red represents the particle with an orientation having a $60°$ angle with the x-axis and black represents the particle with an orientation having a $120°$ angle with the x-axis.

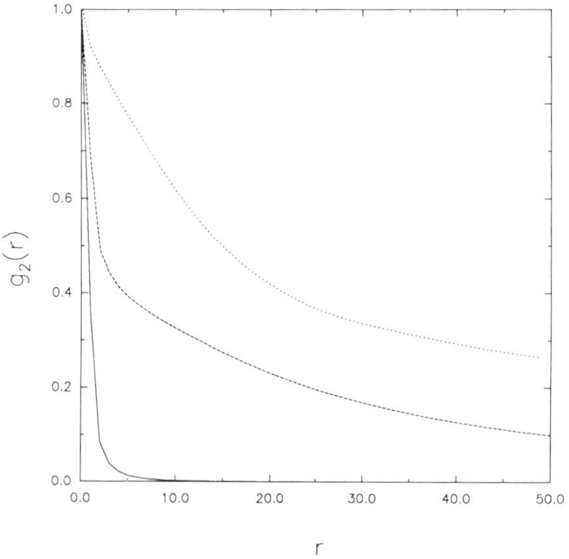

Fig. 15. The correlation function versus spatial separation for the system with $\beta = 1.0$: $\rho = 1.8$ (solid line), $\rho = 2.4$ (dashed line) and $\rho = 4.8$ (dotted line). Note that a system with a small density has only a short-range correlation.

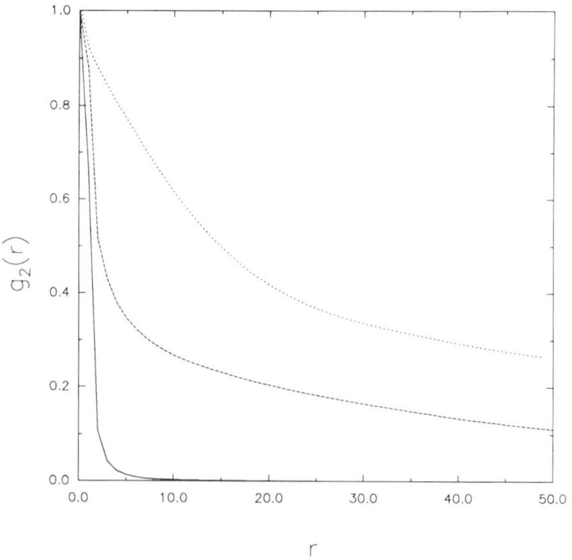

Fig. 16. The correlation function versus spatial separation for a system with density $\rho = 4.8$ and reciprocal temperatures of $\beta = 0.4$ (solid line), $\beta = 0.6$ (dashed line) and $\beta = 1.0$ (dotted line).

and simulations of this model strongly suggest the existence of a phase transition. Application of this model to two-phase flow in channels, to one-dimensional shock waves and to oriental phase transitions have shown promising results. There are several interesting points worth mentioning. First, this model allows one to specify an arbitrary, spatially dependent temperature. This feature can be used for research on the pattern formation, such as the formation of snowflakes. Second, the convection coefficient, $g(n)$, in this model depends on temperature. For a large range of density, this coefficient can be a constant. This is a desirable property to be used to maintain the Galilean invariance of the lattice gas. Third, the nonideal gas equation of state and phase transition properties have been produced by nearest-neighbor interactions between bound pairs. This long-range interaction naturally introduces correlations between bound pairs at low temperature. For simulating realistic materials, a long-range interaction may be necessary.

It is possible to extend the present studies to model liquid-crystal hydrodynamics and other complex fluids. Results in this paper have demonstrated the capability of simulating a large number of rod-like particles. In order to obtain a constitutive equation which couples orientation and velocity, a variable associated with angular momentum must be introduced and collision rules that conserve angular momentum are required.

Acknowledgements

We thank B. Hasslacher, L. Lam and W.H. Matthaeus for helpful discussions. This work was supported by the US Department of Energy at Los Alamos National Laboratory, by DARPA grant DPP88-50, and by the NASA Innovative Research Program under grant NAGW 1648. H.R.B. thanks the Deutsche Forschungsgemeinschaft for support of his work.

References

[1] U. Frisch, B. Hasslacher and Y. Pomeau, Phys. Rev. Lett. 56 (1986) 1505.

[2] U. Frisch et al., Complex Systems 1 (1987) 649.

[3] H. Chen, S. Chen, G.D. Doolen, Y.C. Lee and H.A. Rose, Phys. Rev. A 40 (1989) 2850.

[4] D.H. Rothman and J.M. Keller, J. Stat. Phys. 52 (1989) 1119.

[5] S. Chen, K. Diemer, G.D. Doolen, K. Eggert, C. Fu and B. Travis, Lattice gas automata for flow through porous media, Physica D 47 (1991) 72–84.

[6] S. Chen, H. Chen and G.D. Doolen, Complex Systems 3 (1989) 243.

[7] K. Molvig, P. Donis, J. Myczkowski and G. Vichniac, Removing the discrete artifacts in 3D lattice gas fluids, in: Discrete Kinetic Theory, Lattice Gas Dynamics and Foundation of Hydrodynamics, ed. R. Monaco (World Scientific, Singapore, 1989) p. 409.

[8] S. Wolfram, J. Stat. Phys. 45 (1986) 471.

[9] D. d'Humieres and P. Lallemand and G. Searby, Complex Systems 1 (1986) 598.

[10] N. Metropolis et al., J. Chem. Phys. 21 (1953) 1087.

[11] L.P. Kadanoff, G. McNamara and G. Zanetti, Complex Systems 1 (1986) 790.

[12] Y.Y. Hsu and R.W. Graham, Transport Processes in Boiling and Two-Phase Systems (Hemisphere Publishing Corporation, New York, 1976).

[13] G.F. Hewitt and N.S. Hall-Taylor, Annular Two-Phase Flow (Pergamon Press, New York, 1970).

[14] H.W. Liepmann and A. Roshoko, Elements of Gasdynamics (Wiley, New York, 1958).

[15] G.A. Bird, Molecular Gas Dynamics (Oxford Univ. Press, Oxford, 1976).

[16] M.J.P. Gingras, P.C.W. Holdsworth and B. Bergersen, Europhys. Lett. 9 (1989) 539.

[17] J.P. Straley, Phys. Rev. A 4 (1971) 675.

[18] D. Frenkel and R. Eppenga, Phys. Rev. A 31 (1985) 1776.

[19] E. Vives and A. Planes, Phys. Rev. A 38 (1988) 5391.

[20] M. Kosterlitz and D.J. Thouless, J. Phys. C 6 (1973) 1181.

CHAPTER 4

REACTIONS AND DIFFUSION

Physica D 47 (1991) 115–123
North-Holland

A LATTICE GAS AUTOMATA MODEL FOR HETEROGENEOUS CHEMICAL REACTIONS AT MINERAL SURFACES AND IN PORE NETWORKS

J.T. WELLS[a], D.R. JANECKY[b] and B.J. TRAVIS[c]
[a]*Department of Geological Science AJ-20, University of Washington, Seattle, WA 98195, USA*
[b]*Isotope Geochemistry Group, Los Alamos National Laboratory, Los Alamos, NM 87545, USA*
[c]*Geoanalysis Group, Los Alamos National Laboratory, Los Alamos, NM 87545, USA*

Received 15 January 1990

A lattice gas automata (LGA) model is described which couples solute transport with chemical reactions at mineral surfaces and in pore networks. Chemical reactions and transport are integrated into a FHP-I LGA code as a module so that the approach is readily transportable to other codes. Diffusion in box calculations are compared to finite element Fickian diffusion results and provide an approach to quantifying space–time ratios of the models. Chemical reactions at solid surfaces, including precipitation/dissolution, sorption, and catalytic reaction, can be examined with the model because solute diffusion and mineral surface processes are all treated explicitly. The simplicity and flexibility of the LGA approach provides the ability to study the interrelationship between fluid flow and chemical reactions in porous materials, at a level of complexity that has not previously been computationally possible.

1. Introduction

Chemical reaction processes are often spatially distributed in both natural and engineered systems. Many of these systems also involve intrinsically coupled hydrologic and chemical processes. In natural systems, a detailed description of mineral–fluid interaction is critical to our understanding of a wide variety of geochemical processes, including weathering, diagenesis, and hydrothermal alteration. There are a variety of practical geochemical problems which demand sophisticated modeling capabilities. In petroleum reservoirs, issues of interest include oil migration, secondary recovery processes, and the evolution and/or manipulation of porosity. Similarly, reactions during ground water flow and contaminant migration are important in environmental systems.

As chemical reaction models have become increasingly comprehensive, it has been recognized that chemical reactions between rocks and aqueous solutions are heterogeneous on both spatial and temporal scales, and that new modeling approaches are required to better understand many pressing issues. Interrelated factors such as rock texture, mineral distribution, and pore or fracture network geometry all affect processes of mineral dissolution, deposition, and mass transport. For bulk reactions on large scales of tens to thousands of meters, models of geochemical processes have provided significant insights using averaged or phenomenological descriptions of the permeability, fluid flow fields, mineral distributions, and fluid composition [1]. There are also models for reactions on the molecular scale [2]. Between these two scales, however, there is a paucity of general and flexible models of fluid–rock interaction. This intermediate scale is particularly important to a quantitative understanding of both geochemical and flow processes in natural porous media because it is precisely the scale at which most detailed analytical, experimental, and descriptive methods provide information. In addition, describing coupled flow fluid and chemical reactions has proven to be especially difficult using conventional modeling techniques [3–6]. This is particularly true for transient problems in

0167-2789/91/$03.50 © 1991 – Elsevier Science Publishers B.V. (North-Holland)

which fluid flow and/or mineral reaction rates vary as a function of time.

Rothman [7] and Travis et al. [8] examined potential applications of lattice gas automata (LGA) methods to geological problems by formulating models to simulate flow through complex porous media. We are investigating LGA models for simulating coupled solute transport and chemical reactions at mineral surfaces. Primary advantages of this approach are that coupled fluid flow–chemical reaction processes are implicitly included, and that processes can be simulated which are heterogeneous in both space and time. Thus, LGA calculations appear to be ideally suited to the analysis of systems in the critical intermediate spatial scale range. The addition of chemical reactions and solute transport to LGA models of porous flow, as described here, constitutes a powerful new tool for the computational analysis of complex geochemical processes. Of particular interest are the pore-scale description of porous media and information about spatial and temporal heterogeneities in flow and reaction.

2. Lattice gas automata hydrodynamic model

The core of our approach is a LGA hydrodynamic model. Such models are being developed and applied to the investigation of 2- and 3-D hydrodynamic processes [9–11]. The chemical reaction algorithm is designed to be integrated into the hydrodynamic codes as a relatively self-contained module. All calculations presented in this paper use the minimal collision 2D FHP-I model of Frisch, Hasslacher, and Pomeau [12]. In the FHP-I model, fluid particles travel on a hexagonal symmetric lattice. Up to six particles can occupy any node, but exclusion rules allow only one particle travelling in any one direction to occupy a node. This model has been shown to provide a numerical solution to the two-dimensional Navier–Stokes equations, and thus is capable of modeling real systems [12, 13].

LGA collision rules satisfy conservation of momentum and particle number at every node on a lattice. In an FHP-I model, two types of collisions are recognized, head-on two-particle collisións and symmetric three-particle collisions. Particles involved in a collision are simply rotated 60°. For two-particle collisions, particles rotate clockwise half of the time and counterclockwise the other half of the time. Particle–wall collisions result in the 180° rotation of the particle, maintaining a no-slip boundary condition.

3. Chemical reaction sub-models

The simplest models for chemical processes in porous media involve diffusion in the solvent occupying pores, and reactions at solid surfaces. Our predominant interest is in systems involving solutes in an aqueous solvent; however, chemical processes in oil or gas saturated systems or even two-phase systems [14] can also be examined using the LGA method. Surface reactions may involve significant mass transfer via dissolution and/or precipitation, which modifies the pore network structure and thus the hydrodynamic environment. When the components of interest are at minor or trace concentrations, sorption and desorption reactions become important and the pore network may be either fixed or variable. Catalytic reactions of solutes at solid surfaces could also be modeled using this approach. To provide the flexibility to model such processes, the chemical state of the system is described in parallel to the hydrodynamic state. Thus, the concentration of each chemical species or component is stored for every fluid and solid node. Calculation of chemical processes is integrated with the flow calculations at each time step after movement of the particles on the lattice and before the analysis of collision results. Such an approach is not limited to one solute species or immiscible species. In addition, as noted above, the approach is directly portable between LGA models.

3.1. Diffusion

Solutes are transported down a concentration gradient by diffusion. To simulate diffusion in an LGA model, a concentration of a hypothetical solute is assigned to each open space node. If the mass of solute or changes in the mass of solute are small relative to the total mass of solution, only minor perturbations of unit momentum for the lattice particles are involved, which can be ignored. This approach to solution composition tracking is similar to the energy tracking approach of Sero-Guillaume and Bernardin [15]. Each particle is assigned a concentration equal to that of the node at which it is residing. During each time step, particles travel to new nodes and the concentration of those nodes is assigned the concentration of the arriving particle. If more than one particle arrives at a node during a time step, the resulting concentration at that node becomes the mean concentration of all particles (fig. 1). During the next time step, all particles carry the mean concentration to their next destination. This method of simulating diffusion contrasts with models for inter-diffusion of different types or species of particles [16, 17], in several ways. Primarily, extension of the approach to multicomponent diffusion is fairly straightforward. In addition, the rate of diffusion can potentially be varied for any solute by merely transferring increments of concentration between nodes by the flow of LGA particles.

This approach has been examined by performing a set of calculations of diffusion in a closed box (fig. 2). The initial concentration field consists of a step function in which the concentration equals 32 arbitrary units in the left half of the box, and 8 units in the right half. Mean concentration profiles at different time steps for the LGA calculation are in good agreement with profiles calculated using a finite different technique (fig. 3). Each point on the observed profiles constitutes the average value of each column in the box. This calculation was performed with an average density of 1.2 particles per node, or a reduced density (particles per direction per node) of 0.2.

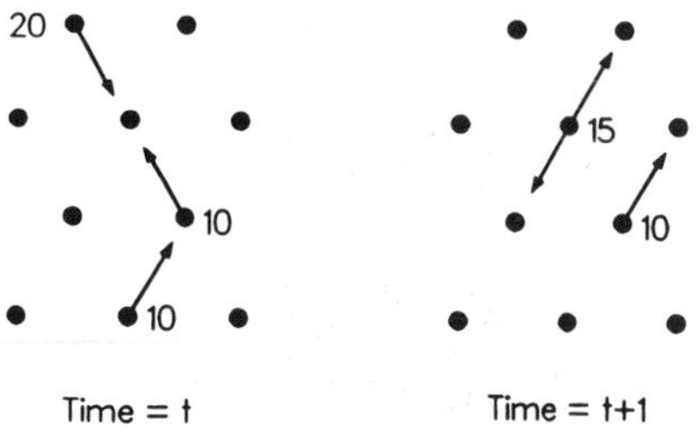

Fig. 1. Diffusion process for a hypothetical solute in the LGA solvent. Concentrations are associated with nodes, and carried by particles which mix when multiple particles reside on a node.

The overall rate of diffusion is a function of the particle density and frequency of interactions (fig. 4). Diffusion coefficients are strongly dependent on component mass on the lattice, which increases linearly with particle density. However, when observed diffusion coefficients per node are normalized for particle density, the strong influence of mean free path [18] as a function of reduced density is evident. In addition, diffusion profiles tend to be smoother and less noisy as particle densities and the number of nodes in the box increase.

The LGA model has no intrinsic temporal or spatial scale; however, the rate of diffusion may be used to delineate the space–time relationships of models applied to real physical problems. For instance, if we are interested in diffusion of ions in an aqueous solution, we can choose a diffusion coefficient of 10^{-9} m^2 s^{-1}, which is an appropriate value for several ions of geological importance such as Ca^{2+}, Cl^-, and Fe^{3+} [19]. By choosing either the nodal dimension or the length of the time step, the other quantity can be determined by comparing LGA results to Fickian diffusion calculations. For example, at a reduced density of 0.2, if we assume the LGA model time step is 1 s and the diffusion coefficient above, the nodal spacing is 0.002 m.

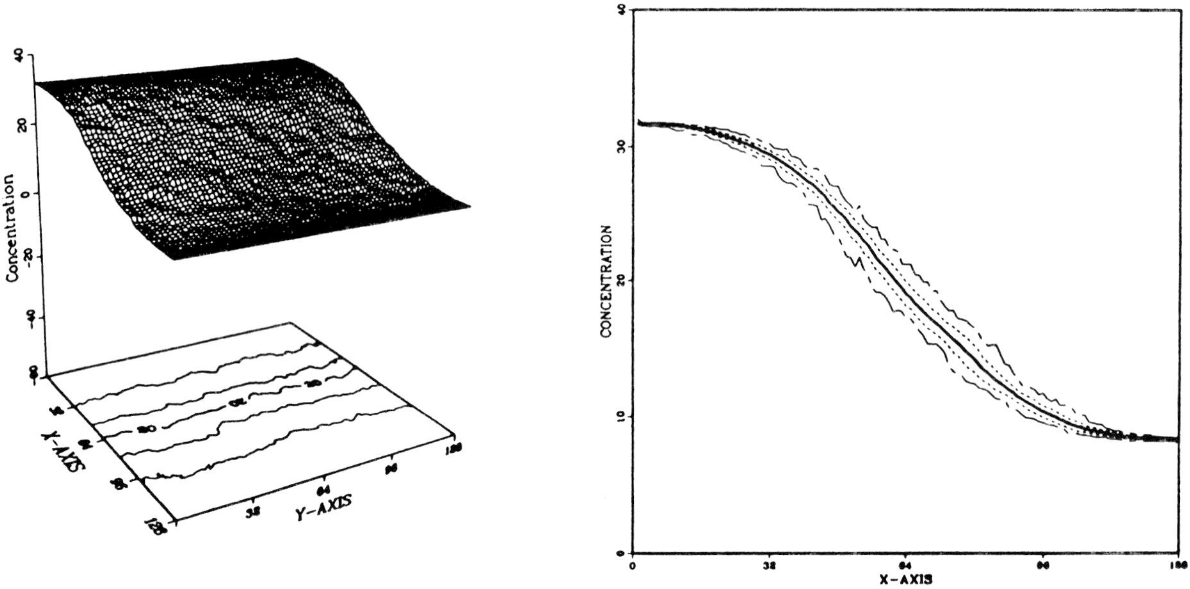

Fig. 2. Diffusion profiles at 1600 steps for a 128 × 128 box and 0.2 reduced density with initial concentration step of 32 to 8 units. A concentration surface is illustrated in (a), while (b) shows the average profile (solid line), one standard deviation from average for each row (dotted lines), and extrema for each row (dashed lines).

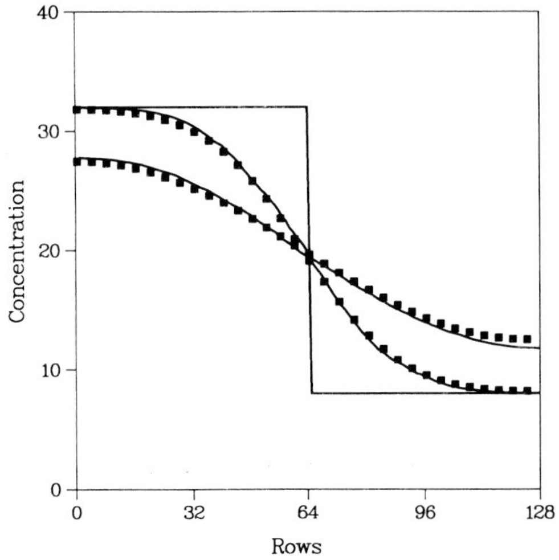

Fig. 3. Comparison of diffusion profiles for 0, 1000 and 5000 steps with a reduced density of 0.2 (solid lines) with results from finite element calculation of Fickian diffusion (solid squares).

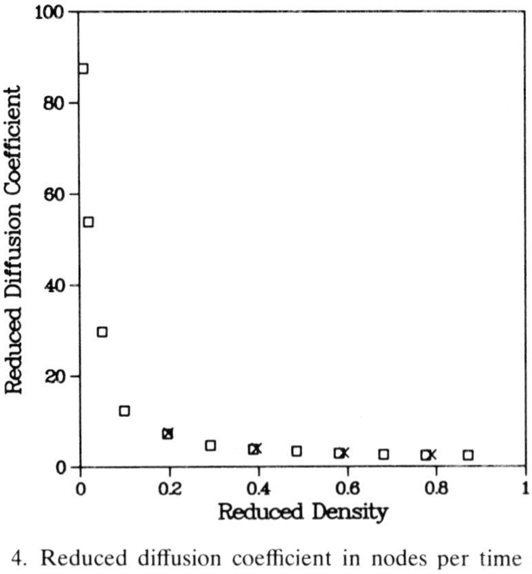

Fig. 4. Reduced diffusion coefficient in nodes per time step per unit reduced density relative to average density of lattice particles per direction per node for 128 × 128 (□) and 256 × 256 (×) boxes.

3.2. Surface reactions

Surface reactions involving dissolution and precipitation are simulated by allowing wall nodes to serve as sources or sinks for mass of a dissolved component. Whenever a particle collides with a wall, a unit of mass may be exchanged, thus increasing or decreasing the local concentration in solution depending upon the saturation state of the fluid (fig. 5). A probability of mass transfer

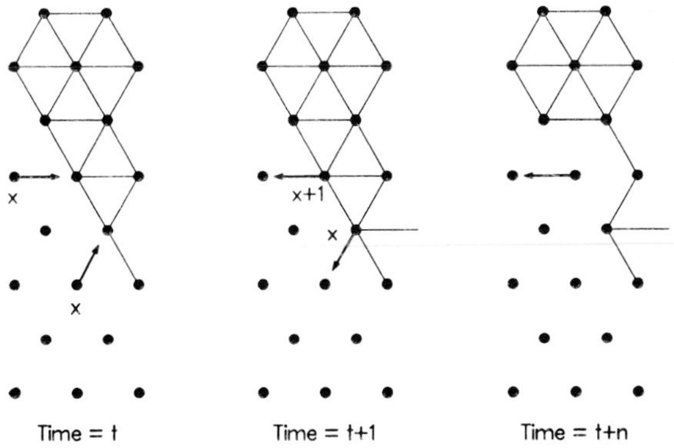

Fig. 5. Surface reaction processes for undersaturated lattice particles colliding with solid surface (a), rebounding in the next time step (b) with one particle having incremented its concentration, and after many concentration transfers to solution, the transformation of a solid node into a solution node (c).

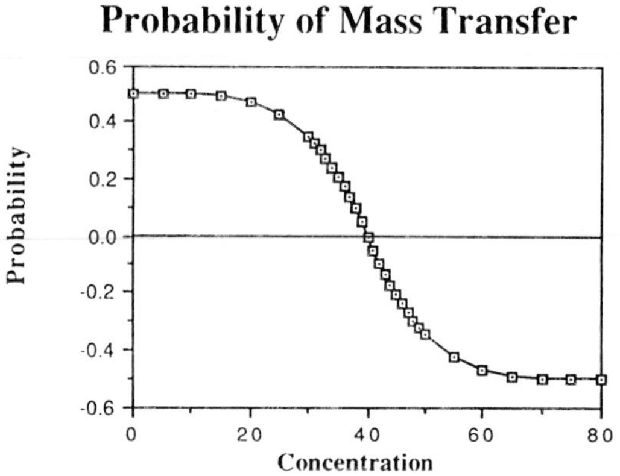

Fig. 6. Representative probability function for dissolution equilibrium, and precipitation as a function of concentration. The curve corresponds to a transition state theory rate law.

is defined for each collision, which is a function of the saturation state of the mineral (fig. 6). The maximum probability can be adjusted to match the reaction rate of the mineral of interest. The shape of the probability function in fig. 6 was designed to be consistent with transition state theory, in which the rate of reaction is a function of the degree of disequilibrium between fluid and mineral [20]. The probability function is also consistent with thermodynamic theory, in that the probability of mass transfer is zero at equilibrium.

After some large number of mass transfer events in an undersaturated fluid, a wall node becomes an open space node, simulating dissolution (fig. 5). Similarly, after some number of supersaturated mass transfer events, an open space node becomes a wall node, simulating mineral precipitation. The number of events required to create or destroy a wall node depends on the relative concentrations per unit volume in the solvent and solid phases.

Many mineral dissolution reactions are spatially heterogeneous. For example, crystallographic defects are sites of excess strain energy which constitute regions especially favorable to dissolution [21]. Such heterogeneous surface reactions can be simulated with the LGA model by increasing the probability of mass transfer in linear zones of a mineral. Fig. 7a shows the early stages of a dissolution calculation with a central zone of enhanced probability of mass transfer. Over the course of this simulation the entire wall recedes slightly, but the central region dissolves more readily, resulting in an etch pit (figs. 7b and 7c). The local solution concentration within the etch pit varies as a function of the rates of both dissolution and diffusion to the bulk solution.

Implicit integration of both dissolution and solute transport processes by the LGA model is a major advantage in simulating natural processes. Thus, variations in etch pit development and morphology can be examined for transport-controlled dissolution, in which the rate-limiting step is the diffusion of the dissolved component away from a mineral surface. Other modeling approaches have been limited to treating only the mineral surface [21], which precludes examination of diffusion control on etch pit formation and surface dissolution. Figs. 7b and 7c illustrate the development of etch pits in simulations that are diffusion-controlled to different extents. In fig. 7c, the local solution concentration within the etch pit is elevated over the concentration in the bulk fluid. With increased dissolution rate of the defect material (fig. 7c), the local concentra-

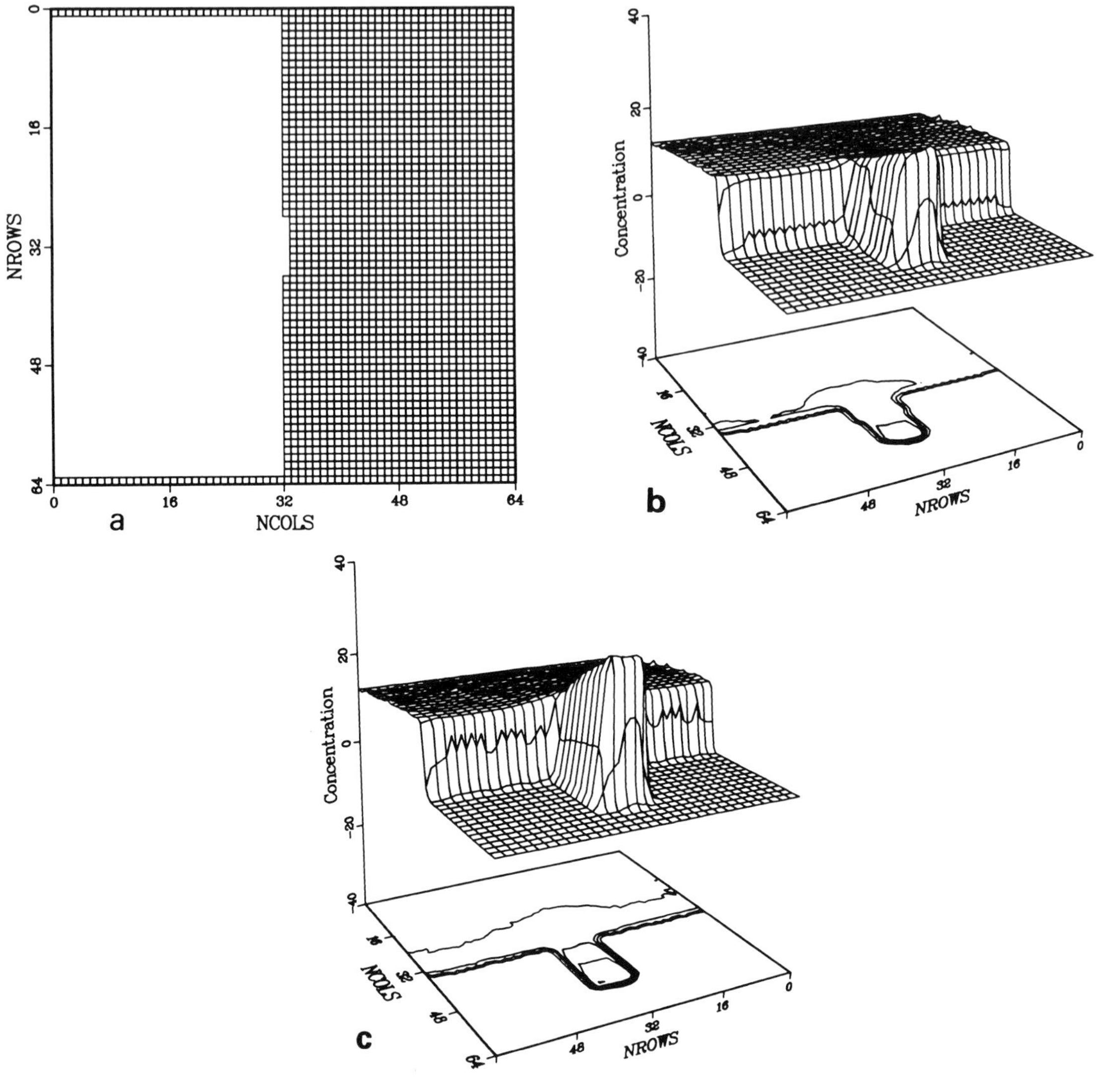

Fig. 7. Static dissolution experiment for a hypothetical single-component solid into a solution with an initial concentration of 12 and a reduced density of 0.3. An 8-node wise line defect in the center of the solid phase dissolves more rapidly to create an etch pit. The system configuration is shown at 100 time steps in (a), along with concentration profiles for surface-controlled dissolution (b) and transport-controlled dissolution (c). Enhanced probabilities of mass transfer for the defect zone are 0.4 (b) and 0.85 (c).

tion in the etch pit increases. As equilibrium is approached, the local probability of mass transfer decreases and continued dissolution becomes influenced by the ability of the solute to diffuse out of the etch pit. In addition, morphology of the etch pit is influenced by transport-controlled processes, with restriction of dissolution to the defect zone as diffusion becomes more important.

Extension of these models to multiple solid phases for a single component or sorption of a trace component on surfaces is straightforward. Reactions among multiple species in solution and reactions involving multi-component solid phases will involve expansion to multiple-component tracking arrays and formulation of an efficient approach to reactions between components.

4. Coupled fluid flow and chemical reactions

Numerical models that describe coupled fluid flow and multi-component chemical reactions in geological environments have proven very difficult to implement. The transient form of the flow–reaction equations are particularly computationally unwieldy. The LGA approach is a viable alternative to the more conventional finite difference and finite element methods because it can be applied to systems, such as porous solids, with complex boundary conditions between solid and fluid phases. It does not rely on averaged values for porosity and permeability, and it is well

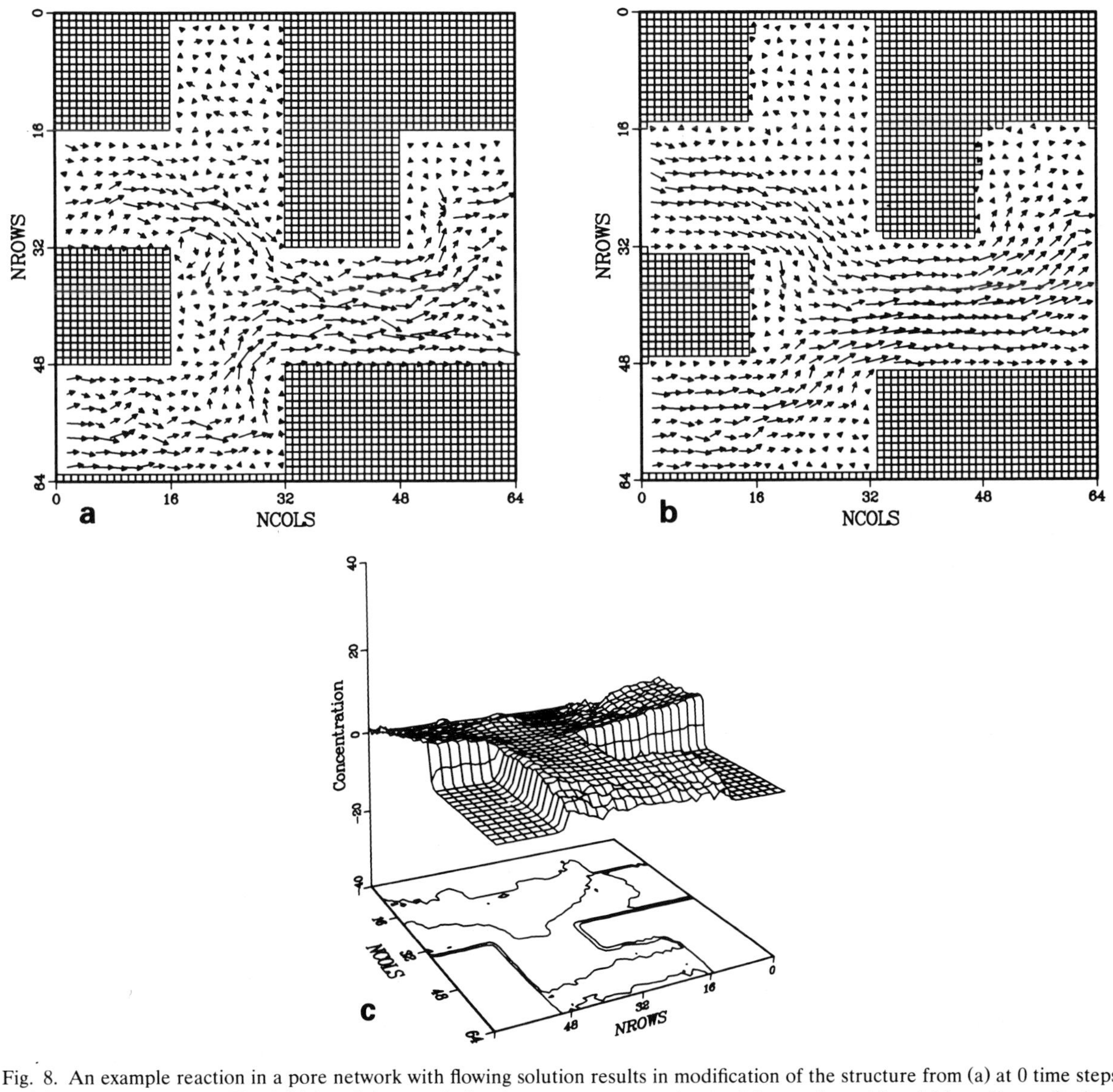

Fig. 8. An example reaction in a pore network with flowing solution results in modification of the structure from (a) at 0 time steps to (b) at 400 time steps. The initial and input solution has zero concentration and interaction with soluble walls results in a heterogeneous concentration field and dissolution pattern (c).

suited for the analysis of spatial and temporal heterogeneities in solute transport and reactions, as discussed above.

Rothman [7] demonstrated the potential of LGA methods for modeling flow through porous geologic media. Our representations of chemical reactions and diffusion can be added to the LGA model of flow through porous media to give a fully coupled model of water–rock reactions and fluid flow. An example of a simple porous flow model is shown in fig. 8a. In this figure, the vectors represent the local fluid velocity, averaged over seven adjacent nodes and over 100 time steps. The concentration field that results from the reaction of a wall mineral with an initially undersaturated fluid is clearly inhomogeneous (fig. 8c). The concentration within the principle flow channel is relatively low, whereas the concentration in dead-end pores and around obstructions is relatively high. Dissolution within the pore network is illustrated in fig. 8b. This type of simulation has important implications for increasing our understanding of the relationship between porosity and permeability in dynamic systems in which the porous structure of a rock is altered by mineral dissolution and growth.

The effect of reaction and transport heterogeneities on the overall evolution of fluid composition and rock porosity and permeability is not well understood at this time, but is certainly a function of fluid velocity, rock composition, and the geometry of the porous network. The LGA approach has the great potential to aid in investigations of these phenomena.

5. Summary

We have presented a lattice gas automata model which couples solute transport with chemical reactions at mineral surfaces and in pore networks. The model can treat both surface-controlled and transport-controlled mineral reactions because solute diffusion, fluid flow, and mineral surface detachment processes are all treated explicitly. It is particularly useful for the analysis of porous media problems in which fluid flow and mineral reactions are heterogeneous in space and time. The simplicity and flexibility of this lattice gas automata approach provides the ability to study the interrelationship between fluid flow and chemical reactions in porous materials at a level of complexity that has not been computationally possible in the past.

Acknowledgements

This work is supported by funding from the Institute of Geophysics and Planetary Physics (Los Alamos National Laboratory, University of California) and advanced concepts funding from the Geosciences Program, US DOE Office of Basic Energy Science.

References

[1] R.L. Bassett and D.C. Melchior, in: Chemical Modeling of Aqueous Systems II, eds. D.C. Melchior and R.L. Bassett (Am. Chem. Soc., Washington, DC, 1990) p. 1.
[2] A.C. Lasaga and G.V. Gibbs, Phys. Chem. Minerals 14 (1987) 107.
[3] P.C. Lichtner, Geochim. Cosmochim Acta 52 (1988) 143.
[4] J.J. Ague and G.H. Brimhall, Econ. Geol. 83 (1989) 506
[5] C.I. Steefal and A.C. Lasaga, in: Chemical Modeling of Aqueous Systems II, eds. D.C. Melchior and R.L. Bassett (Am. Chem. Soc., Washington, DC, 1990) p. 212.
[6] G.L. Carnahan, in: Chemical Modeling of Aqueous Systems II, eds. D.C. Melchior and R.L. Bassett (Am. Chem. Soc., Washington, DC, 1990) p. 234.
[7] D.H. Rothman, Geophys. 53 (1988) 509.
[8] B.J. Travis, K.G. Eggert, S.Y. Chen and G. Doolen, EOS 69-44 (1988) 1193.
[9] D. d'Humieres, P. Lallemand and U. Frisch, Europhys. Lett. 2 (1986) 291.
[10] D. d'Humieres and P. Lallemand, Complex Systems 1 (1987) 297.
[11] U. Frisch, D. D'Humieres, B. Hasslacher, P. Lallemand, Y. Pomeau and J.-P. Rivet, Complex Systems 1 (1987) 649.
[12] U. Frisch, B. Hasslacher and Y. Pomeau, Phys. Rev. Lett. 56 (1986) 1505.

[13] S. Wolfram, J. Stat. Phys. 45 (1986) 471.
[14] D.H. Rothman and J.M. Keller, J. Stat. Phys. 52 (1988) 1119.
[15] O.E. Sero-Guillaume and D. Bernardin, Eur. J. Mech. B/Fluids 9 (1989) 1.
[16] J.P. Boon and A. Noullez, in: Discrete Kinematic Theory, Lattice Gas Dynamics, and Foundations of Hydrodynamics, ed. R. Monaco (World Scientific, Singapore, 1989) p. 309.
[17] B. Boghosian and C.D. Levermore, in: Discrete Kinematic Theory, Lattice Gas Dynamics, and Foundations of Hydrodynamics, ed. R. Monaco (World Scientific, Singapore, 1989) p. 44.
[18] K. Diemer, K. Hunt, S. Chen, T. Shimomura and G. Doolen, in: Lattice Gas Methods for Partial Differential Equations, ed. G.D. Doolen (Addison-Wesley, Reading, MA, 1990) p. 137.
[19] E.H. Oelkers and H.C. Helgeson, Geochim. Cosmochim. Acta 52 (1988) 63.
[20] A.C. Lasaga, J. Geophys. Res. 89 (1984) 4009.
[21] A.C. Lasaga and A.E. Blum, Geochim. Cosmochim. Acta 50 (1986) 2363.

Physica D 47 (1991) 124–131
North-Holland

DETERMINISTIC LATTICE MODEL FOR DIFFUSION-CONTROLLED CRYSTAL GROWTH

Fong LIU and Nigel GOLDENFELD
Department of Physics and Materials Research Laboratory, University of Illinois at Urbana-Champaign, Urbana, IL 61801, USA

Received 15 January 1990

An efficient lattice model is developed to study the late stages of diffusion-controlled crystal growth. We establish the existence of a dense branching morphology and its relation to diffusion-limited aggregation. We find a clear morphological transition from kinetic-effect-dominated growth to surface-tension-dominated growth, marked by a difference in the way growth velocity scales with undercooling. We also study the evolution of interfacial instability and find a scaling behaviour for the interface power spectra, indicating the non-linear selection of a unique length scale.

1. Introduction

Dendritic crystal growth [1] is a well-known example of pattern formation in non-equilibrium systems where simple mechanisms may generate rather complex structures due to intrinsic dynamic instabilities. It is related to other pattern formation phenomena like viscous fingering in Hele-Shaw cells [2], electrodeposition [3], and diffusion-limited aggregation (DLA) [4].

The simplest case is diffusion-controlled crystal growth from a supercooled melt, where the primary physical process is diffusion of the latent heat released at the liquid–solid interface as the crystal grows. This process is described by the model [5]

$$\partial_t u = D \nabla^2 u, \tag{1.1}$$

$$v_n = -D \nabla u \cdot \boldsymbol{n} |_{\mathrm{sol}}^{\mathrm{liq}}, \tag{1.2}$$

$$u_s = \Delta - d_0(\theta)\, \kappa - \beta(\theta)\, v_n. \tag{1.3}$$

Here $u(\boldsymbol{x}, t)$ is the dimensionless temperature field, D is the diffusion constant, v_n is the velocity of the interface along normal $\boldsymbol{n}$, u_s is the temperature at the interface, Δ is the dimensionless undercooling, which is the driving force for crystal growth, $d_0(\theta)$ is the anisotropic capillary length, κ is the curvature of the interface, and $\beta(\theta)$ is an anisotropic kinetic coefficient. The angle θ refers to the angle between $\boldsymbol{n}$ and a crystal axis. The second term on the r.h.s. of boundary condition (1.3) accounts for equilibrium surface tension effects, while the last term, $-\beta v_n$, reflects the fact that the interface is away from local thermodynamic equilibrium.

This moving boundary problem can, in principle, be attacked naively by directly discretizing the partial differential equations (PDE). A front tracking method is usually required to locate the interface at all times, a task demanding increasing number of grid points due to interfacial instabilities. This poses serious difficulty for programming and computational efficiency, a difficulty shared by many hydrodynamic simulations. Partly for this reason, the interesting regime of time-dependent non-linear growth has never been explored.

In this paper, we develop an efficient lattice model for diffusion-controlled crystal growth. The discrete spatial patterns at different times are related by deterministic injective maps. Here rather than using an algorithm derived by discretizing the PDE, we model the natural growth process directly in terms of discrete dynamical systems. This method of modelling, referred to as the cell-dynamical scheme (CDS), has been successfully used in studying phase separation and recently applied to hydrodynamics systems [6]. Our lattice model is independent of the PDE model eqs. (1.1)–(1.3). In fact both are phenomenological descriptions of the natural phenomena. We believe that they are in the same universality class, in the sense that both will yield the same macroscopic (large scale) properties of crystal growth. In the following sections, the lattice model is introduced, and used to study the morphological aspects of the late stages of crystal growth.

2. Lattice model of crystal growth

Consider a crystal growing into a supercooled liquid on a two-dimensional square lattice. Lattice sites labeled with index i are assigned two time-dependent variables: a continuous temperature field u_i^n and a binary phase field $\phi_i^n \in \{0, 1\}$ with superscript n being the discrete time-step label. $\phi_i = 1$ (0) if site i is in the solid (liquid) state. The crystal grows by continuous updating u_i, ϕ_i according to the following rules[#1]:

(a) At time step n, all perimeter sites (liquid sites with at least one *nearest* solid neighbor) are identified. Solidification occurs only on perimeter sites. The remelting from solid back to liquid is forbidden.

(b) A perimeter site will crystallize ($\phi_i = 0 \to 1$) if it satisfies the condition

$$u_i \leq \Delta + \lambda\left(\sum_{j \in \mathrm{Nbrs}} \omega_j \phi_j - 6 \right), \tag{2.1}$$

where j is summed over both nearest and next-nearest neighbors of i with different weight factor ω_j. We choose $\omega_j = 2$ for nearest neighbors and $\omega_j = 1$ for next-nearest neighbors.

(c) Let all perimeter sites that satisfy condition (2.1) change to solid. Then add latent heat to those sites by increasing the temperature by an amount l_a:

$$u_i \to u_i + l_a. \tag{2.2}$$

(d) In this step, heat diffusion is performed before going to time step $n + 1$. The temperature field u_i is updated using the explicit finite difference scheme

$$u_i \to u_i + \frac{D}{m}(\langle u \rangle - u_i), \tag{2.3}$$

where $m > 1$ is an integer and $\langle \ \rangle$ denotes average over nearest and next-nearest neighbors: $\langle u \rangle = \frac{1}{6}\Sigma_{j \in \mathrm{nn}} u_j + \frac{1}{12}\Sigma_{j \in \mathrm{nnn}} u_j$. Here the diffusion constant D is taken to be identical in both liquid and solid. Physically, the temperature field relaxes on a longer time scale than the phase field does, so (2.3) is repeated m times. The continuum limit (1.1) is recovered as $m \to \infty$.

In this model, the physical processes of latent heat generation and diffusion are modelled by step (c) and (d) respectively. Note that the temperature and phase fields are coupled through eq. (2.1), which corresponds to boundary condition (1.3) with λ playing the role of capillary length. This can be seen by noting that the term $-(\Sigma \omega_j \phi_j - 6)$ approximates the local interface curvature, similar to the algorithms in DLA simulations [8] where interface curvature is calculated by counting the number of solid particles in a box centered at the interface. In addition, the anisotropy and non-equilibrium kinetic effects are also implicitly present in our model, a consequence of the underlying square lattice and spatio-temporal discreteness.

Our model is deterministic by construction, in contrast to DLA-type simulations where stochastic random walkers are used. However, the effect of noise can be incorporated rather conveniently; either by choosing random initial conditions or by

[#1] Our model is based upon that developed by Packard in ref. [7]; note that in the earlier model of Packard, the rules do not correspond to the correct boundary condition for diffusion-controlled crystal growth.

adding a stochastic term to eq. (2.1). The latter mimics persistent noise in real crystal growth experiments. Condition (2.1) is replaced by

$$u_i \leq \Delta[1 + \delta\eta(i,n)] + \lambda\left(\sum_{j \in \mathrm{Nbrs}} \omega_j \phi_j - 6\right), \quad (2.4)$$

where $\eta(i,n)$ is a random number uniformly distributed in $[-1, 1]$ and $\delta < 1$ is the noise amplitude.

The parameters D, m, λ and l_a merely determine the length, time and temperature scales of the simulation and may be set to convenient values and kept fixed. The undercooling Δ is the control parameter that we may adjust. A weakness of the model is that the strength of anisotropy and kinetic effects are not quantitatively controllable. We will assume that the dependences are included implicitly in Δ.

For the sake of computational efficiency, all arithmetic operations are performed in integers rather than floating-point numbers. Further, the most time-consuming part of the simulation can be parallelized. With this model, we are able to monitor simultaneously both the evolution of interfaces and the temperature fields to very late stages of growth, something which has never been achieved using traditional methods.

3. Numerical results

Our simulations are carried out in both circular and planar geometries. By varying the control parameters, the model exhibits a wide variety of patterns that have been observed in experiments. Fig. 1 is an example of a dendritic crystal. Gray levels in the picture reflect the temperature scale. Some of our numerical results are summarized below.

3.1. Morphological transitions

We first simulate crystal growth from the center of a square lattice with a random initial solid seed. We show in fig. 2 structures grown under

Fig. 1. Dendritic crystal pattern grown with the lattice model. Grey levels represent the temperature value. Parameters are $D = 4.0$, $m = 5$, $\lambda = 0.015$, $l_a = 1$, $\Delta = 0.195$ and $\delta = 0.1$.

fixed capillary length and diffusion constant, but different undercooling Δ. We observe a clear morphological transition from a dense dendritic structure along symmetry axes (fig. 2a) at relatively high undercooling $\Delta > 0.125$, to a less branching diagonal structure (figs. 2c, 2d) at smaller undercooling $\Delta < 0.125$.

In between the two morphologies, there is a regime of almost isotropic growth (fig. 2b) where the branching or tip-splitting events occur in a seemingly random direction. This morphology is identified as the dense branching morphology (DBM) [9], and discussed later in the text.

This transition from diagonal to axial growth results from the competition between surface tension and kinetic anisotropies. At low undercooling when the interface is close to local thermodynamic equilibrium, interface relaxation is determined by energetics considerations. Hence, the surface tension anisotropy causes the crystal to grow along the diagonal directions. At higher undercooling when growth is fast, local equilibrium is violated. The kinetic effect wins over

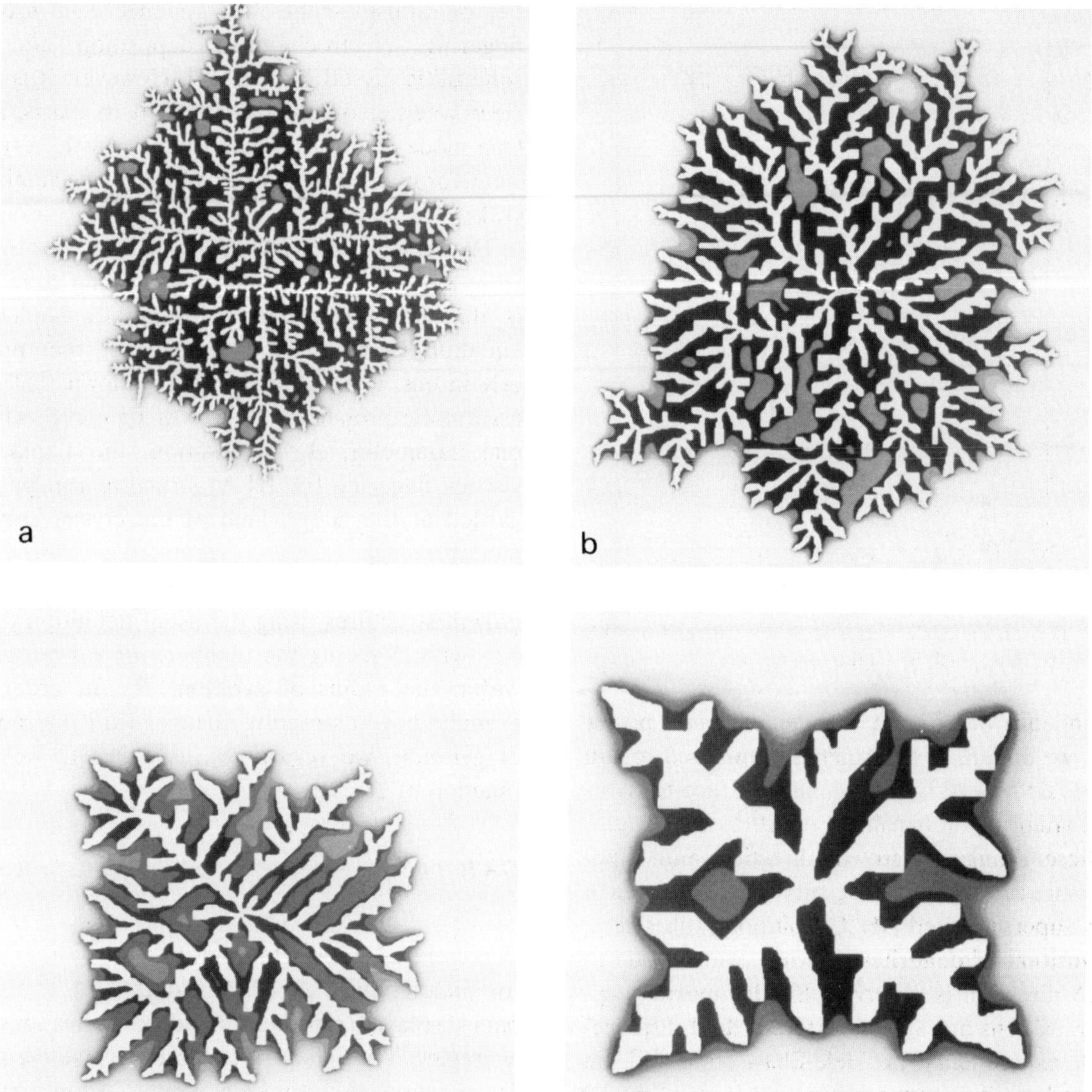

Fig. 2. The morphological transition from axial to diagonal growth as undercooling Δ is decreased. All crystals are grown at $\lambda = 0.01$, $\delta = 0.2$ but with different Δ. (a) $\Delta = 0.15$, the cluster has 85 171 solid particles; (b) $\Delta = 0.125$, with 144 522 particles; (c) $\delta = 0.1$, 94 072 particles; and (d) $\Delta = 0.05$, 47 305 particles. The scale of picture (d) has been doubled.

surface tension and the growth is parallel to the axes.

Beside the explicit visual demonstration of fig. 2, the presence of a morphology transition also manifests itself in the way that the growth velocity scales with the undercooling. In fig. 3 the growth velocity v is plotted as a function of undercooling Δ. The dependence of v on Δ changes abruptly at $\Delta \sim 0.125$, the same value at which the morphological transition is found by direct examination of fig. 2. The inset of fig. 3 is the same plot on logarithmic axes. Fitting the

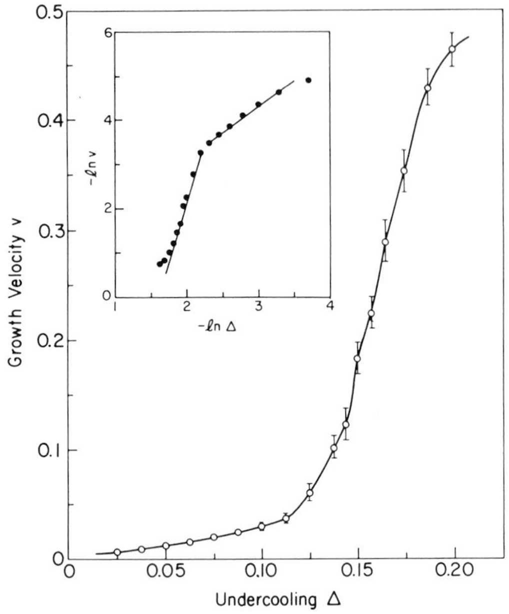

Fig. 3. Growth velocity v versus Δ at $\lambda = 0.01$ and $\delta = 0.2$. Inset is the same graph in natural log–log plot.

dependence of v on Δ with *approximate* power laws, we obtain, in the kinetics-dominated region (large Δ) $v \sim \Delta^{5.5}$ and in the surface-tension-dominated region (small Δ) $v \sim \Delta^{1.3}$.

These change of growth direction and exponents are reminiscent of results on crystal growth from supersaturated NH_4Cl solution [10]. There, a transition of dendrites growing along $\langle 110 \rangle$ to $\langle 111 \rangle$ direction is observed. Similar morphological transitions are also reported in fluid displacement experiments in Hele-Shaw cells [9], in electrodeposition experiments and in a model calculation [11].

3.2. *Asymptotic growth morphology*

In this section, we investigate the asymptotic ($t \to \infty$) properties of growth and address the question: what is the asymptotic growth structure in the absence of crystalline anisotropy? Is it a ramified DLA-like structure or the so-called dense branching morphology (DBM) – a branching structure with stable circular envelope and radial symmetry? DBMs are observed in a number of diffusion-controlled systems, e.g., viscous fingering, electrochemical deposition and in spherulitic crystal growth [12]. However, it is not clear whether or not the DBM, if its exists, is a true mode of diffusion-controlled growth, a transient, or a result of other physical mechanisms [13].

The crystal growth system differs from the Hele-Shaw system in that the governing equation is the diffusion equation instead of Laplace's equation. By rescaling the length and time properly in eqs. (1.1)–(1.3), it can be shown that the asymptotic properties of growth only depend on one parameter, Δ. In addition, the Laplacian viscous fingering (or DLA) structure can be regarded as the $\Delta \to 0$ limit of the crystal growth model.

We measure the fractal dimension D_f of radial growth structures using the usual definition $N \sim R_g^{D_f}$ with N being the number of solid particles within the radius of gyration, R_g. In order to examine how a structure changes with the radius of gyration, we regard D_f as a slowly varying function of R_g, i.e.,

$$D_f(R_g) = \frac{d(\log N)}{d(\log R_g)}. \tag{3.1}$$

Fig. 4 shows the fractal dimension D_f versus R_g for undercooling $\Delta = 0.15$ and $\Delta = 0.175$. Error bars displayed represent standard derivations of averaging. Within statistical accuracy, we see that D_f increases monotonically with R_g from a more or less DLA-like value 1.7 to the asymptotic value of the DBM $D_f = 2$.

Our simulations suggest that the asymptotic morphology for crystal growth is a DBM ($D_f = 2$). On the other hand, the asymptotic structure of Laplacian viscous fingering (as the $\Delta \to 0$ limit of crystal growth) is that of DLA with $D_{DLA} \approx 1.7$. This statement regarding the connection between DBM and DLA can be summarized in terms of a scaling hypothesis for cluster size

$$N(R, \Delta) = R^{D_{DLA}} f(R/\xi(\Delta)), \tag{3.2}$$

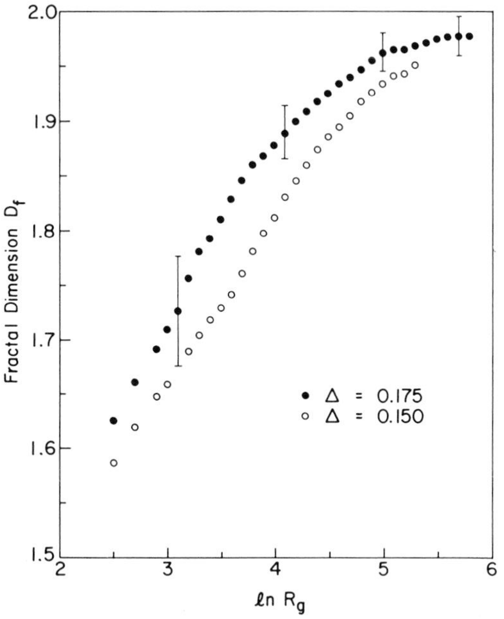

Fig. 4. Fractal dimension D_f of the pattern varies with radius of gyration R_g under different undercoolings. Typical error bars are shown on a few data points only.

where the function f satisfies $f(x) \sim x^{2-D_{DLA}}$ as $x \to +\infty$ and $f(0) = \text{const.}$; ξ is a characteristic length at which the structure crosses over from DLA to DBM.

We must point out that, strictly speaking, a DBM with an ideal circular envelope only exists at zero anisotropy. This is satisfied at about $\Delta \sim 0.125$ (see fig. 2b), where the competition between kinetic and surface tension anisotropies balances and gives approximate isotropy. The reason that DBMs are not observed as ubiquitously as are anisotropic dendrites is the inevitable presence of crystalline anisotropy.

3.3. *Interfacial structure*

In this section we analyze the dynamical time evolution of the interfacial instabilities. Simulations are performed in rectangular cells starting from a horizontal substrate of linear dimension $L = 600$. The initial interface profile is a random linear superposition of different Fourier modes. Fig. 5 shows a typical growth pattern at $\Delta = 0.1625$.

Inspecting fig. 5, we observe that the development of instability passes through a short transient and then quickly settles down to a seemingly steady growth with a characteristic length scale:

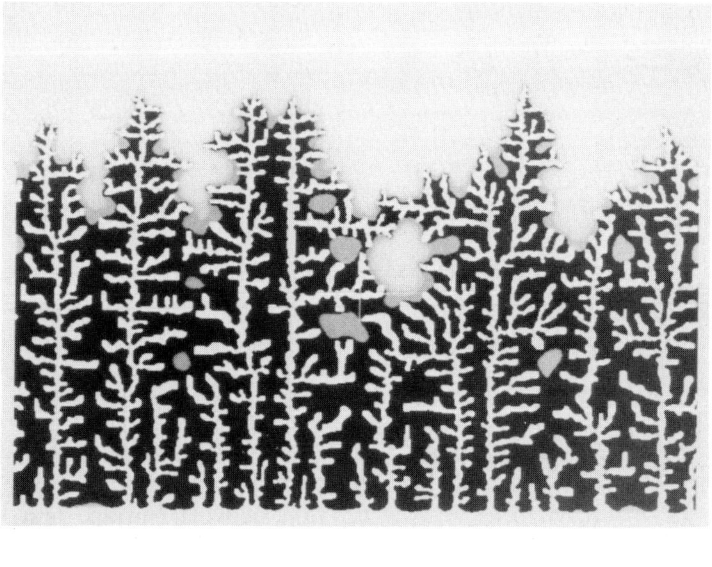

Fig. 5. Growth pattern in planar geometry. $\Delta = 0.1625$, $\lambda = 0.01$ and $\delta = 0.2$. Periodic boundary condition is used.

spacings between sidebranches and individual dendrites are constant in time. Interfacial structures can be well probed quantitatively by their power spectrum. In our case, we first eliminate overhangs of the interface by projecting it onto the substrate, resulting in a projected interface $h(x,t)$. The power spectrum can then be calculated

$$P(q,t) = \frac{1}{L^2}\left|\sum_{j=1}^{L} h(x_j,t)\exp(iqx_j)\right|^2, \qquad (3.3)$$

where q is the wavenumber.

Power spectra for $\Delta = 0.1625$ from $t = 500$ to $t = 1700$ are presented in the inset of fig. 6. To check that growth has reached the steady state, we further scale the power spectra by the interface width $w^2(t) = (1/L)\Sigma_{j=1}^{L}[h(x_j,t) - \bar{h}]^2$. After scaling, the power spectra for different times collapse nicely onto a single curve, as shown in fig. 6.

Two features can be identified from the power spectrum $P(q,t)$. The first is the selection of a unique length scale from the continuum of perturbations with all possible wavelengths. This non-linear selection mechanism, insensitive to short-wavelength fluctuations or noise, is totally different from the dynamical selection principle found in steady state problems [5].

Another feature evident from fig. 6 is that $P(q) \sim q^{-2}$ for large q: the short-wavelength fluctuations have the well-known spectrum of a random surface under thermal capillary roughening. This result demonstrates that, in contrast to the case of steady state problems, where the short-length-scale surface tension strongly affects the global shape of the needle crystal by behaving as a singular perturbation, here the short wavelength fluctuations are irrelevant to the selection of a large length scale during growth. Similar scaling behaviour of the interface power spectra has been observed in a spin Monte Carlo study [14] as well as in boundary integral type simulations of dendritic growth [15].

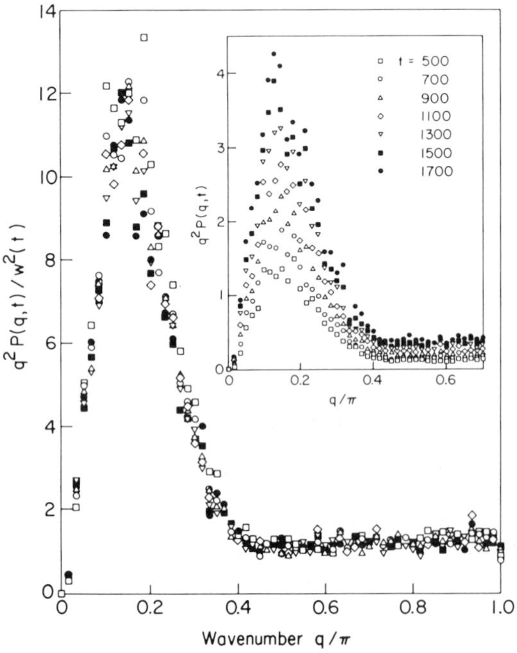

Fig. 6. Scaled power spectra $P(q,t)$ of the growing interface. Different symbols represent times from $t = 500$ to $t = 1700$ in 200 intervals. Unscaled power spectra are plotted in the inset.

3.4. Summary

We have demonstrated the usefulness of discrete cell dynamical system modelling through an efficient lattice model of diffusion-controlled crystal growth. This type of approach provides an alternative to the traditional continuum PDE approach. The key idea is the renormalization group philosophy: different microscopic models may lead to the same macroscopic phenomenology. This is in the same spirit as the modelling of macroscopic phenomena by cellular automata [16] and coupled map lattices [17].

To be specific, our deterministic lattice model provides an efficient way of studying the crystal growth process, especially the complex structures which emerge at late stages of growth. We find a morphological transition induced by the competition between surface and kinetics anisotropies. Under weak anisotropy, we recognize a DBM as the asymptotic structure and make the connection with the structure resulting from the DLA. Our study also reveals a non-linear wavelength selection distinct from that in steady state problems.

In the future, we hope to develop systematic methods for generating non-traditional numerical schemes, using the renormalization group [18]. In this way, the goal of optimizing these numerical methods can be achieved. Furthermore, we would like to be able to determine which features of a dynamical process are universal, and thus can be captured by coarse-grained simulations along the lines presented here, and which features cannot be successfully modelled using the CDS approach.

Acknowledgements

We are grateful to N. Packard for helpful discussion and for convincing us of the utility of good graphics. We thank the MRL Center of Computation in the University of Illinois for the use of their facilities and Jennifer Shannon for help with the graphics. This work is supported by grant NSF-DMR-86-12860 and NSF-DMR-87-01393. One of us (N.D.G.) gratefully acknowledges the support of the Alfred P. Sloan Foundation.

References

[1] J.S. Langer, Rev. Mod. Phys. 52 (1980) 1;
D.A. Kessler, J. Koplik and H. Levine, Adv. Phys. 37 (1988) 255.

[2] D. Bensimon, L.P. Kadanoff, S. Liang, B.I. Shraiman and C. Tang, Rev. Mod. Phys. 58 (1986) 977.

[3] Y. Sawada, A. Dougherty and J.P. Gollub, Phys. Rev. Lett. 56 (1986) 1260;
D. Grier, E. Ben-Jacob, R. Clarke and L.M. Sander, Phys. Rev. Lett. 56 (1986) 1264.

[4] See F. Family and D.P. Landau, eds., Kinetics of Aggregation and Gelation (North-Holland, Amsterdam, 1984).

[5] N.D. Goldenfeld, in: Physicochemical Hydrodynamics, ed. M. Velarde (Plenum Press, New York, 1988) p. 547.

[6] Y. Oono and S. Puri, Phys. Rev. Lett. 58 (1987) 836; Phys. Rev. A 38 (1988) 434;
S. Puri and Y. Oono, Phys. Rev. A 38 (1988) 1542;
Y. Oono and A. Shinozaki, Sci. Form, in press.

[7] N. Packard, in: Proceedings of the First International Symposium for the Science on Form, Tsukuba University (1985).

[8] T. Vicsek, Phys. Rev. Lett. 53 (1984) 2281; Phys. Rev. A 32 (1985) 3084.

[9] E. Ben-Jacob, G. Deutscher, P. Garik, N.D. Goldenfeld and Y. Lereah, Phys. Rev. Lett. 57 (1986) 1903.

[10] S-K. Chan, H.H. Reimer and M. Kahlweit, J. Cryst. Growth 32 (1976) 303.

[11] E. Ben-Jacob, P. Garik, D. Grier and T. Muller, Superlattices Microstructures 3 (1987) 599.

[12] N.D. Goldenfeld, J. Cryst. Growth. 84 (1987) 601, and references therein.

[13] S.E. May and J.V. Maher, Phys. Rev. A 40 (1989) 1723.

[14] R. Harris and M. Grant, J. Phys. A 23 (1990) 567.

[15] D. Jasnow and J. Viñals, Phys. Rev. A 40 (1989) 3864.

[16] S. Wolfram, Theory and Applications of Cellular Automata (World Scientific, Singapore, 1986).

[17] K. Kaneko, Prog. Theor. Phys. 72 (1984) 480.

[18] N.D. Goldenfeld, O. Martin, Y. Oono and F. Liu, Anomalous dimension and the renormalization group in a non-linear diffusion process, Phys. Rev. Lett. 64 (1990) 136;
N.D. Goldenfeld, O. Martin and Y. Oono, in: Proceedings of the Third Nobeyama Workshop on Supercomputing and Experiments in Fluid Dynamics, ed. S. Orszag, J. Sci. Comp. 4 (1989) 355.

Physica D 47 (1991) 132–158
North-Holland

REACTIVE LATTICE GAS AUTOMATA

Anna LAWNICZAK
Department of Mathematics and Statistics, University of Guelph, Guelph, Ontario, Canada N1G 2W1

David DAB
Faculté des Sciences, C.P. 231, Université Libre de Bruxelles, 1050 Bruxelles, Belgium

Raymond KAPRAL
Chemical Physics Theory Group, Department of Chemistry, University of Toronto, Toronto, Ontario, Canada, M5S 1A1

and

Jean-Pierre BOON
Faculté des Sciences, C.P. 231, Université Libre de Bruxelles, 1050 Bruxelles, Belgium

Received 15 Februry 1990

A probabilistic lattice gas cellular automaton model of a chemically reacting system is constructed. Microdynamical equations for the evolution of the system are given; the continuous and discrete Boltzmann equations are developed and their reduction to a generalized reaction–diffusion equation is discussed. The microscopic reactive dynamics is consistent with any polynomial rate law up to the fourth order in the average particle density. It is shown how several microscopic CA rules are consistent with a given rate law. As most CA systems, the present one has spurious properties whose effects are shown to be unimportant under appropriate conditions. As an explicit example of the general formalism a CA dynamics is constructed for an autocatalytic reactive scheme known as the Schlögl model. Simulations show that in spite of the simplicity of the underlying discrete dynamics the model exhibits the phase separation and wave propagation phenomena expected for this system. Because of the microscopic nature of the dynamics the role of internal fluctuations on the evolution process can be investigated.

1. Introduction

Nonlinear chemically reacting systems exhibit many different types of spatial and temporal behavior, for example, chemical oscillations and waves [1,2]. Typically the analysis of such phenomena is based on a reaction–diffusion equation. This macroscopic description will be adequate if the phenomena of interest occur on sufficiently long distance and time scales and fluctuations do not play an important role. The macroscopic reaction–diffusion equation has its basis in an underlying molecular dynamics which is necessarily quite complex for a chemically reacting system. In this article we explore a class of microscopic models for nonlinear reaction–diffusion systems. We adopt the probabilistic lattice gas automaton approach [3,4] where space, time and particle velocities are discrete. In spite of this simple microscopic description the macroscopic behavior of the automaton is consistent with that obtained from the reaction–diffusion equation and, furthermore, the lattice gas analysis allows one to explore the role of internal fluctuations on the dynamics of this nonlinear system.

A number of simplifications of the microscopic dynamics is made in the construction of the cellular automaton model. Normally we shall be interested in reactions taking place in a solvent, so interactions of the chemical species with the solvent must also be taken into account. A complete description of the reactive and nonreactive dynamics would then entail a consideration of several species. However, it is possible to construct a much simpler model that still preserves the features of the full system dynamics. The only aspect of the solvent dynamics that is of interest is the fact that collisions with the reacting molecules can change the velocities of these species. If the solvent is in excess then collisions between solvent molecules will occur frequently and will serve to maintain the velocity distributions of these species close to equilibrium. On the other hand, the system can be constrained to lie far from equilibrium by fixing the concentrations of some species; without loss of generality we consider this to be the case for all species except one, say X. In view of these considerations we may ignore the details of the dynamics of all species but X, as well as the solvent and focus solely on the dynamics of X.

In qualitative terms the cellular automaton model is constructed in the following way. Space is made discrete by restricting the dynamics of the X species to take place on a square lattice. In addition to the X species we assume there exist "ghost" particles which serve to account for the presence of other species and solvent molecules. The X molecules can undergo two types of collisions: elastic and reactive. Elastic collisions are modeled by assuming that the particles undergo a random walk generated by local random rotations at a node of the lattice. Reactive collisions consist of random species changes in accord with the kinetics of the system and random rotations to simulate the effect of velocity changes that occur as a result of the reaction. Particles move from node to node with unit velocity. This basic model consists of the single X species with four discrete unit velocities. We note that in this ghost particle description (i) the microscopic reactive dynamics does not satisfy mass conservation, and (ii) the fact that the concentration of the ghost species is constant may influence the observed fluctuations in the X species.

We give a precise mathematical formulation of the lattice gas cellular automaton rules and the microdynamical equations for the Boolean fields describing the system state. Space and time scaling transformations are considered in order to investigate the kinetic regime where the automaton can be reduced to a continuous space and time Boltzmann equation. We also present a reduction of the linearized Boltzmann equation to a generalized reaction–diffusion equation. A discrete lattice Boltzmann equation is derived and the spurious invariants inherent in most lattice gas models [5] are discussed. Automaton simulations are presented for the application of this class of automata to a specific reaction–diffusion scheme: the Schlögl model [6].

2. Cellular automaton model

The cellular automaton model for a chemically reacting system [7,8] can be described in formal terms as follows: Particles of species X move on a square lattice $\mathcal{L}_\epsilon \subset \mathrm{Z}^2$, which is a square centered on the origin with sides equal to the integer part of ϵ^{-1} and periodic boundary conditions. At each node, labeled by the discrete vector $\boldsymbol{r}$, there are four cells labeled by an index i defined modulo four. The cells are associated with the unit velocity vectors $\boldsymbol{c}_i$ connecting the node to its four nearest neighbors. (We assume i increases counterclockwise and 1 corresponds to the positive direction of the x-axis.)

An exclusion principle forbids two particles to be at the same node with the same velocity; therefore, each cell $(\boldsymbol{r}, i)$ has only two states coded with a Boolean variable $s_i(\boldsymbol{r})$:

$$s_i(\boldsymbol{r}) = 1 \quad \text{occupied,} \qquad = 0 \quad \text{unoccupied.} \tag{2.1}$$

A configuration of particles at a node $\boldsymbol{r}$ at time k is described by a random vector $\boldsymbol{\eta}(k, \boldsymbol{r}) = \langle \eta_i(k, \boldsymbol{r}) \rangle_{i=1}^4$ with values in a state space S of all ($2^4 = 16$) four bit words; i.e.,

$$S = \{ s = \langle s_i \rangle_{i=1}^4 \colon s_i = 0 \text{ or } 1 \text{ for } i = 1, \ldots, 4 \}. \tag{2.2}$$

A configuration of the lattice $\mathcal{L}_\epsilon$ at time k is described by a Boolean field

$$\boldsymbol{\eta}(k) = \{\boldsymbol{\eta}(k, \boldsymbol{r}) \colon \boldsymbol{r} \in \mathcal{L}_\epsilon\}, \tag{2.3}$$

with values in a phase space $\Gamma = S^{\mathcal{L}_\epsilon}$ of all possible assignments $s(\cdot)$:

$$s(\cdot) = \left\{ s(\boldsymbol{r}) = \langle s_i(\boldsymbol{r})\rangle_{i=1}^4 \colon \boldsymbol{r} \in \mathcal{L}_\epsilon \right\}. \tag{2.4}$$

The evolution of the system occurs at discrete time steps and the cellular automaton updating rule can be defined on the Boolean field (2.3). At each node at each time step the updating rule consists of propagation followed by collisions, which may be either elastic or reactive. For generality we assume that collisions occur with probability p (thus, no collision with probability $(1-p)$).

The lattice gas automaton rule consists of a product $R \circ C \circ P$ of three basic operators: rotation R, chemical transformation C and propagation P. We first describe these building blocks of the lattice gas dynamics and then express the updating rules in terms of the microdynamical equations for the Boolean fields.

2.1. Propagation

During the propagation step each particle moves in the direction of its velocity from its cell to a nearest-neighbor node. The velocity of each particle is conserved during this operation. We denote the propagation operator by P and the configuration in the ith direction after the propagation step by η_i^P. Thus,

$$P\colon \eta_i^P(k, \boldsymbol{r}) = \eta_i(k-1, \boldsymbol{r} - \boldsymbol{c}_i). \tag{2.5}$$

2.2. Rotation

Particles change their velocities as a result of random rotations R. The rotation operations are local and thus only involve particles at a given node; clearly rotations do not change the number of particles at a node. At each node, independently of the others, the configuration of the particles is rotated by $\pi/2$ or $-\pi/2$ with probability $1/2$. In more formal terms we let

$$\{\xi_{\pi/2}(k, \boldsymbol{r}) \colon \boldsymbol{r} \in \mathcal{L}_\epsilon,\ k = 1, 2, \ldots\} \tag{2.6}$$

be a sequence of independent copies of random variables $\bar{\xi}_{\pi/2}$ such that the probabilities Pr satisfy

$$\Pr(\bar{\xi}_{\pi/2} = 1) = \Pr(\bar{\xi}_{\pi/2} = 0) = 1/2. \tag{2.7}$$

If η_i^R denotes a configuration in the ith direction after a rotation at a node we have

$$R\colon \eta_i^R = \xi_{\pi/2}\eta_{i+3} + (1 - \xi_{\pi/2})\eta_{i+1}, \quad i = 1, \ldots, 4, \tag{2.8}$$

where the subscript addition is modulo four. We may also write this equation in the form

$$\eta_i^R = \eta_i + \xi_{\pi/2}(-\eta_i + \eta_{i+3}) + (1 - \xi_{\pi/2})(-\eta_i + \eta_{i+1}) \equiv \eta_i + \Delta_i^R(\eta), \tag{2.9}$$

which defines the collision term $\Delta_i^R(\eta)$ (indeed the rotation operator redistributes the velocities as collisions with solvent particles would) that takes on the values $\{-1, 0, 1\}$. In (2.8) and (2.9) we have dropped the dependence of η_i on k and $\boldsymbol{r}$. We adopt this convention in the sequel when confusion is unlikely to occur.

Rotations can also be described in terms of a probability matrix. A particular rotation at a node can be defined by an input state $s = \langle s_1, s_2, s_3, s_4\rangle$ and an output state $s' = \langle s_1', s_2', s_3', s_4'\rangle$ with an associated probability $R(s, s')$. For example, for the rotation $s = \langle 1, 0, 1, 1\rangle \to s' = \langle 1, 1, 0, 1\rangle$ the associated probability is

$$R(\langle 1,0,1,1\rangle;\langle 1,1,0,1\rangle) = R(s;s') = 1/2. \tag{2.10}$$

It is convenient to define a rotation matrix for all s and s', even if the rotation rules do not provide for an actual transition between s and s'; for such transitions $R(s;s')$ can simply be set to zero. Note also that for an input state that does not change, we have $R(s;s') = 1$ for $s = s'$. In this way the entire set of collision rules can be neatly encoded into a single 16×16 rotation matrix $\mathbf{R}$ with elements $R(s;s')$.

2.3. Chemical transformation

In the chemical transformation step at each node, independently of the others, particles are randomly created or annihilated in reactions of the type $\alpha X \to \beta X$ with the net reaction probabilities $P_{\alpha\beta} = P_{\alpha\beta}(\epsilon)$ $(\alpha, \beta = 0, \ldots, 4)$ regardless of their velocity state. The diagonal elements $P_{\alpha\alpha}$ of the transition probability matrix $\mathbf{P} = [P_{\alpha\beta}]$ correspond to nonreactive events and we may write

$$P_{\alpha\alpha} = 1 - \sum_{\beta \neq \alpha} P_{\alpha\beta}. \tag{2.11}$$

The off-diagonal elements of $\mathbf{P}$ for which $\alpha > \beta$ correspond to reactive transitions where particles are destroyed while if $\alpha < \beta$ particles are created.

Next we describe the chemical transformations in terms of the microdynamical variables η. Let

Table 1
Rotation probability matrix $\mathbf{R}$.

	0000	1000	0100	0010	0001	1100	1010	1001	0101	0011	0110	1110	1101	1011	0111	1111
0000	1															
1000			1/2		1/2											
0100		1/2		1/2												
0010			1/2		1/2											
0001		1/2		1/2												
1100								1/2			1/2					
1010									1							
1001						1/2				1/2						
0101							1									
0011								1/2			1/2					
0110						1/2				1/2						
1110													1/2		1/2	
1101												1/2		1/2		
1011													1/2		1/2	
0111												1/2		1/2		
1111																1

$$\{\xi_i^{\alpha\beta}(k,\boldsymbol{r})\colon i=1,\dots,4;\ \boldsymbol{r}\in\mathcal{L}_\epsilon,\ k=1,2,\dots\},\quad \alpha\neq\beta,\quad 0\le\alpha,\beta\le 4 \tag{2.12}$$

be independent random sequences, independent of the sequence (2.6), of identically distributed independent Bernoulli-type random variables with distributions determined by the following conditions. Let $n_{\alpha\beta}$ denote the number of all possible distinct outcomes of a reaction $\alpha X\to\beta X$. For each $\alpha=0,\dots,4$ we introduce a set $R(\alpha)$ of all allowed reactions different from the identity; i.e., $R(\alpha)=\{\tau\colon \tau\neq\alpha,\ 0\le\tau\le 4\}$, and for $\beta\in R(\alpha)$ let $R(\alpha\beta)=R(\alpha)-\{\beta\}$. We now introduce random variables which govern chemical transformations and define distributions of the random variables $\xi_i^{\alpha\beta}$. We introduce indices $i_k\in\{1,\dots,4\}$ and let $q=|\alpha-\beta|$; the following products of random variables may be defined for a distinct set $\{i_1,\dots,i_q\}$ of velocity indices:

$$\gamma^{\alpha\beta}(i_1\cdots i_q)=\prod_{l=1}^{q}\xi_{i_l}^{\alpha\beta}\ \sideset{}{'}\prod_{k\neq 1,\dots,q}(1-\xi_{i_k}^{\alpha\beta})\prod_{\tau\in R(\alpha\beta)}\prod_{m=1}^{4}(1-\xi_m^{\alpha\tau}), \tag{2.13}$$

where the prime on the product indicates that only distinct (distinct from each other and the $i_1,\dots,i_q$) i_k values are considered. The probability distributions associated with these random variables are

$$\Pr(\gamma^{\alpha\beta}(i_1\cdots i_q)=1)=P_{\alpha\beta}/n_{\alpha\beta},\quad \Pr(\gamma^{\alpha\beta}(i_1\cdots i_q)=0)=1-P_{\alpha\beta}/n_{\alpha\beta}. \tag{2.14}$$

Hence the product of random variables $\gamma^{\alpha\beta}(i_1\cdots i_q)$ in (2.13) ensures that a reaction $\alpha X\to\beta X$ will occur with the desired probability $P_{\alpha\beta}$. We also introduce the following products of Boolean fields labeled by the number of particles α in the initial state of the reaction:

$$Q^\alpha(\eta;i_1\cdots i_\alpha)=\prod_{l=1}^{\alpha}\eta_{i_l}\ \sideset{}{'}\prod_{k\neq 1,\dots,\alpha}(1-\eta_{i_k}), \tag{2.15}$$

where again the prime refers to distinct values of i_k. This product of Boolean fields ensures that there are α particles at a node with velocities $i_1,\dots,i_\alpha$.

The local chemical transformation operator can now be easily written in terms of these functions since the configuration after chemical transformation will depend on the initial configuration determined by Q^α and on the probability of chemical transformation given that configuration which is governed by $\gamma^{\alpha\beta}$. If η_i^C denotes a configuration in the ith direction after chemical transformation, the chemical transformation operator takes the form

$$\begin{aligned}
C\colon \eta_i^C &= \sum_{\alpha=1}^{4}\sideset{}{'}\sum_{i_2,\dots,i_\alpha}Q^\alpha(\eta;ii_2\cdots i_\alpha)\Bigg\{(1-\eta_i)\sum_{\beta=0}^{\alpha-1}\sideset{}{'}\sum_{j_2,\dots,j_q=i_2}^{i_\alpha}\gamma^{\alpha\beta}(ij_2\cdots j_q)\\
&\quad+\eta_i\Bigg[1-\sum_{\beta=0}^{\alpha-1}\sideset{}{'}\sum_{j_2,\dots,j_q=i_2}^{i_\alpha}\gamma^{\alpha\beta}(ij_2\cdots j_q)\Bigg]\Bigg\}\\
&\quad+\sum_{\alpha=0}^{3}\sideset{}{'}\sum_{j_1,\dots,j_\alpha}Q^\alpha(\eta;j_1\cdots j_\alpha)\Bigg\{(1-\eta_i)\sum_{\beta=\alpha+1}^{4}\sideset{}{'}\sum_{i_2,\dots,i_q}\gamma^{\alpha\beta}(ii_2\cdots i_q)\\
&\quad+\eta_i\Bigg[1-\sum_{\beta=\alpha+1}^{4}\sideset{}{'}\sum_{i_2,\dots,i_q}\gamma^{\alpha\beta}(ii_2\cdots i_q)\Bigg]\Bigg\}.
\end{aligned} \tag{2.16}$$

This lengthy form exposes the structure of η_i^C. The two factors involving sums on α correspond to processes for which the number of particles at a node decreases ($\alpha=1,\dots,4$) or increases ($\alpha=0,\dots,3$).

The first term in the curly brackets of each factor arises from reactive events while the second factor corresponds to non-reactive events. The Q^α factors ensure that the configuration at a node corresponds to one with α particles. In the sums on the j_k one should recall that the primes also imply that the j_k's are distinct from i.

This form may be simplified if we use the properties:

$$Q^\alpha(\eta; i_1 \cdots i_\alpha)(1-\eta_i) = (1-\delta_{i,i_1})Q^\alpha(\eta; i_1 \cdots i_\alpha), \tag{2.17}$$

$$Q^\alpha(\eta; i_1 \cdots i_\alpha)\eta_i = \delta_{i,i_1} Q^\alpha(\eta; i_1 \cdots i_\alpha), \tag{2.18}$$

and

$$\eta_i = \sum_{\alpha=1}^{4} \sum_{i_2,\ldots,i_\alpha}{}' Q^\alpha(\eta; i i_2 \cdots i_\alpha). \tag{2.19}$$

We then obtain

$$\begin{aligned} \eta_i^C = \eta_i &+ \sum_{\alpha,\beta(\alpha<\beta)} \sum_{i_1,\ldots,i_\alpha}{}' Q^\alpha(\eta; i_1 \cdots i_\alpha) \sum_{j_2,\ldots,j_q}{}' \gamma^{\alpha\beta}(i j_2 \cdots j_q) \\ &- \sum_{\alpha,\beta(\alpha>\beta)} \sum_{i_2,\ldots,i_\alpha}{}' Q^\alpha(\eta; i i_2 \cdots i_\alpha) \sum_{j_2,\ldots,j_q=i_2}^{i_\alpha}{}' \gamma^{\alpha\beta}(i j_2 \cdots j_q)\,, \end{aligned} \tag{2.20}$$

where the last two terms correspond to processes where the particle number increases or decreases at a node, respectively. Again $i_1 \cdots i_\alpha$ are distinct from i as well as each other.

Finally, we define a chemical transformation term $\Delta_i^C(\eta; \alpha, \beta)$ by comparison of

$$\eta_i^C = \eta_i + \sum_{\alpha,\beta(\alpha\neq\beta)} \Delta_i^C(\eta; \alpha, \beta), \tag{2.21}$$

with (2.20). The C operator yields the property

$$\begin{aligned} \Delta_i^C(\eta; \alpha, \beta) &= -1 \quad \text{or 0 if } \alpha > \beta, \\ &= 1 \quad \text{or 0 if } \alpha < \beta. \end{aligned} \tag{2.22}$$

As in the case of rotations, chemical transformations can be described by a chemical transformation probability matrix. For given input and output states each reaction has an associated probability which we write as $C(s; s')$. For example, for the reaction $s = \langle 1,1,0,0\rangle \to s' = \langle 1,1,1,0\rangle$, the associated probability is

$$C(\langle 1,1,0,0\rangle; \langle 1,1,1,0\rangle) = C(s; s') = P_{23}/2. \tag{2.23}$$

The chemical transformation matrix $\mathbf{C}$ with elements $C(s; s')$ is defined as follows:

$$\begin{aligned} C(s; s') &= P_{|s|,|s'|}/n_{|s|,|s'|} \qquad \text{for } s \neq s', \\ &= P_{|s|,|s'|} \qquad \text{for } s = s', \end{aligned} \tag{2.24}$$

where $|s|$ denotes the number of particles in a state s.

2.4. The transformation RC

The automaton model involves the product $D = R \circ C$ of the local chemical transformation C and rotation R operators, and it is useful to consider some properties of this transformation before describing the full automaton dynamical equations. It follows directly from (2.8) and (2.20) that the result of this sequence of transformations is

$$\eta_i^D = \xi_{\pi/2}\eta_{i+3}^C + (1-\xi_{\pi/2})\eta_{i+1}^C, \tag{2.25}$$

or using (2.21)

$$\eta_i^D = \xi_{\pi/2}\eta_{i+3} + (1-\xi_{\pi/2})\eta_{i+1} + \sum_{\alpha,\beta(\alpha\neq\beta)} \left[\xi_{\pi/2}\Delta_{i+3}^C(\eta;\alpha,\beta) + (1-\xi_{\pi/2})\Delta_{i+1}^C(\eta;\alpha,\beta)\right]. \tag{2.26}$$

Using (2.9) and defining

$$\Delta_i^C = \sum_{\alpha,\beta(\alpha\neq\beta)} \left[\xi_{\pi/2}\Delta_{i+3}^C(\eta;\alpha,\beta) + (1-\xi_{\pi/2})\Delta_{i+1}^C(\eta;\alpha,\beta)\right], \tag{2.27}$$

we have

$$\eta_i^D = \eta_i + \Delta_i^R(\eta) + \Delta_i^C(\eta) \equiv \eta_i + \Delta_i^D(\eta). \tag{2.28}$$

An alternative expression can be derived starting with (2.25) and using (2.16) in place of (2.20).

The dynamical description given above for the RC transformation process can be recast into a probability matrix form. Each reaction has an associated probability $D(s;s')$ that depends on the input and output states. Since a reactive collision consists of a chemical transformation followed by a rotation, and these operations are independent, we have

$$D(s;s') = \sum_{s''\in S} R(s;s'')\,C(s'';s'), \tag{2.29}$$

which defines the matrix $\mathbf{D}$. We note that $\mathbf{R}$ and $\mathbf{C}$ commute; indeed the order of the rotation and chemical transformation steps is irrelevent because, apart from particle creation or annihilation, the transformation $\mathbf{C}$ does not affect the velocity distribution.

It is convenient to rewrite $\mathbf{D}$ as the sum of the rotation probability matrix and another matrix that is proportional to the chemical transformation probabilities $P_{\alpha\beta}(\epsilon)$. To this end we may write $\mathbf{C}$ as the sum of a unit matrix $\mathbf{1}$ and a matrix $\mathbf{C}^{\mathrm{c}}$:

$$\mathbf{C} = \mathbf{1} + \mathbf{C}^{\mathrm{c}}. \tag{2.30}$$

Application of $\mathbf{R}$ yields

$$\mathbf{D} = \mathbf{R} + \mathbf{D}^{\mathrm{c}}. \tag{2.31}$$

This decomposition of $\mathbf{D}$ corresponds to the breakup of Δ_i^D into Δ_i^R and Δ_i^C contributions in (2.28) and is especially useful in carrying out scaling transformations since all dependence on the reactive probabilities resides in one term.

We may give a clear physical interpretation of the automaton collision dynamics by considering a slightly different decomposition of the $\mathbf{C}$ matrix. Recalling that the diagonal elements of $\mathbf{C}$ correspond to the case where no chemical transformation occurs, we may split $\mathbf{C}$ into diagonal and off-diagonal matrices,

$$\mathbf{C} = \mathbf{C}^{\mathrm{d}} + \mathbf{C}^{\mathrm{o}}. \tag{2.32}$$

Then $\mathbf{D}$ may be written as

$$\mathbf{D} = \mathbf{R}\mathbf{C}^{\mathrm{d}} + \mathbf{R}\mathbf{C}^{\mathrm{o}} \equiv \mathbf{D}^{\mathrm{el}} + \mathbf{D}^{\mathrm{ch}}. \tag{2.33}$$

These manipulations show that the probability matrix $\mathbf{D}$ can be split into elastic 'el' and reactive 'ch' parts. The elastic collision process is simply a rotation of the local particle configuration; the reactive collision process is the result of a chemical transformation process, where the number of particles at the node changes in accord with the collision rules, followed by a rotation of the resulting configuration in order to mimic the change in velocity state that accompanies the chemical transformation process.

Finally, utilizing the chemical transformation probability matrix the collision operator Δ_i^D can be written in a form involving a sum over all possible collisions. Let

$$\{\xi_{s;s'}(k,\boldsymbol{r}): (s;s') \in S \times S\}, \quad \boldsymbol{r} \in \mathcal{L}_\epsilon, \quad k = 1,2,\ldots \tag{2.34}$$

be independent sequences of independent copies of Boolean random variables $\bar{\xi}_{s;s'}$ such that for each $(s;s') \in S \times S$

$$\begin{aligned}\bar{\xi}_{s;s'} &= 1 \text{ with probability } D(s;s') && \text{for allowed collisions,}\\ &= 0 \text{ with probability } 1 - D(s;s') && \text{otherwise.}\end{aligned} \tag{2.35}$$

Thus for each input $s \in S$

$$\sum_{s'} \xi_{s;s'} = 1 \tag{2.36}$$

with probability one. It is easy to see that after a collision a configuration in the ith direction, η_i^D, can be expressed as follows:

$$\eta_i^D = \sum_{s,s'} s_i' \xi_{s;s'} \prod_{j=1}^{4} \eta_j^{s_j} (1-\eta_j)^{1-s_j}. \tag{2.37}$$

The factor s_i' ensures the presence of a particle in cell i after collision; the various factors in the product over the index j ensure that before the collision the pattern of the η_j's matches that of the s_j's. Using (2.36) and (2.37) and the identity

$$\sum_{s} s_i \prod_{j=1}^{4} \eta_j^{s_j} (1-\eta_j)^{1-s_j} = \eta_i, \tag{2.38}$$

we can rewrite Δ_i^D in the form:

$$\Delta_i^D(\eta) = \sum_{s,s'} (s_i' - s_i) \xi_{s;s'} \prod_{j=1}^{4} \eta_j^{s_j} (1-\eta_j)^{1-s_j}. \tag{2.39}$$

2.5. Microdynamical equations

The cellular automaton rule consists of successive applications of the transformations described above: propagation followed by collision, $R \circ C \circ P$. The microdynamical equations for the automaton follow directly from the microdynamical equations for these transformations. Let

$$\{\xi_p(k,\boldsymbol{r}): r \in \mathcal{L}_\epsilon,\ k = 1,2,\ldots\} \tag{2.40}$$

be a sequence, independent of the sequences (2.6) and (2.12), of independent copies of the random variable $\bar{\xi}_p$ such that

$$\Pr(\bar{\xi}_p = 1) = p, \quad \Pr(\bar{\xi}_p = 0) = 1 - p, \tag{2.41}$$

where $p = p(\epsilon)$. If we let

$$\tilde{\eta}_i = \eta_i^P(k, \boldsymbol{r}) = \eta_i(k-1, \boldsymbol{r} - \boldsymbol{c}_i), \tag{2.42}$$

then from the definition of the cellular automaton updating scheme we have

$$\eta_i(k, \boldsymbol{r}) = (1 - \xi_p)\tilde{\eta}_i + \xi_p \tilde{\eta}_i^D. \tag{2.43}$$

for each $\boldsymbol{r} \in \mathcal{L}_\epsilon$, $k = 1, 2, \ldots$. Then, using (2.28) we can write the microdynamical equations for the cellular automaton as

$$\eta_i(k, \boldsymbol{r}) = \tilde{\eta}_i + \xi_p \left\{ \Delta_i^R(\tilde{\eta}) + \Delta_i^C(\tilde{\eta}) \right\}. \tag{2.44}$$

2.6. Liouville equation and mean values

The evolution of the lattice gas can also be described in terms of a probability distribution $P_\mu(k; s(\cdot))$ [4], which gives the probability of occurrence of a configuration $s(\cdot)$ at time k with the initial distribution (i.e., at $k = 0$) $\mu = \mu_\epsilon$. Since the random variables $\xi_{s;s'}(k, \boldsymbol{r})$ defined in (2.34) are independent, the entire Boolean field $\{\eta(k)\colon k = 0, 1, \ldots\}$ defined in (2.3) is a Markov process with transition probabilities

$$P_\mu(k-1, s(\cdot); k, s'(\cdot)) = \prod_{\boldsymbol{r} \in \mathcal{L}_\epsilon} D(s^P(\cdot); s'(\cdot)), \tag{2.45}$$

where for each $\boldsymbol{r} \in \mathcal{L}_\epsilon$, $s^P(\boldsymbol{r}) = \langle s_i(\boldsymbol{r} - \boldsymbol{c}_i)\rangle_{i=1}^4$, and the Liouville equation, which is actually the Chapmann–Kolmogorov equation for this Markov process becomes

$$P_\mu(k, s'(\cdot)) = \sum_{s(\cdot)\in\Gamma} \prod_{\boldsymbol{r} \in \mathcal{L}_\epsilon} D(s^P(\cdot); s'(\cdot)) P_\mu(k-1, s(\cdot)), \tag{2.46}$$

Eq. (2.46) expresses the fact that the probability at time k of a given configuration $s'(\cdot)$ is the sum of the probabilities at $k - 1$ of all possible configurations $s(\cdot)$ times the transition probability.

The introduction of the probability distribution P_μ and its corresponding expectation E_μ enables one to define physically interesting average quantities. The mean population in the ith direction is

$$N_i^\epsilon(k, \boldsymbol{r}) = E_\mu(\eta_i(k, \boldsymbol{r})), \tag{2.47}$$

while the local density is given by

$$\rho^\epsilon(k, \boldsymbol{r}) = \sum_{i=1}^4 N_i^\epsilon(k, \boldsymbol{r}), \tag{2.48}$$

and the mean velocity by

$$\begin{aligned} \boldsymbol{u}^\epsilon(k, \boldsymbol{r}) &= (\rho^\epsilon(k, \boldsymbol{r}))^{-1} \sum_{i=1}^4 \boldsymbol{c}_i N_i^\epsilon(k, \boldsymbol{r}) \qquad &&\text{if } \rho^\epsilon(k, \boldsymbol{r}) \neq 0, \\ &= 0 &&\text{otherwise.} \end{aligned} \tag{2.49}$$

Of course N_i^ϵ, ρ^ϵ and $\boldsymbol{u}^\epsilon$ depend on μ, which we drop to simplify the notation.

3. Boltzmann equation and macroscopic law

It is interesting to consider the conditions under which the microdynamical cellular automaton equations can be reduced to a continuous space and time kinetic (Boltzmann) equation for the local average particle number. The macroscopic chemical rate law and reaction–diffusion equation are also derived from the lattice Boltzmann equation.

3.1. Scaling and kinetic regime

To investigate this regime we must consider the limiting behavior of the mean quantities N_i^ϵ and ρ^ϵ when $\epsilon \to 0$. Consider the expected value of (2.44). Since the random variables $\{\xi_p(k,\boldsymbol{r}),\ \xi_{\pi/2}(k,\boldsymbol{r}) \colon \boldsymbol{r} \in \mathcal{L}_\epsilon\}$ are independent of the evolution of the cellular automaton up to time $k-1$, this implies that

$$N_i^\epsilon(k,\boldsymbol{r}) = \tilde{N}_i^\epsilon + p(\epsilon)\Big(\mathcal{C}_i^R(\tilde{N}^\epsilon) + \tfrac{1}{2} \sum_{\alpha,\beta(\alpha\neq\beta)} E_\mu(\Delta_{i+1}^C(\tilde{\eta};\alpha,\beta) + \Delta_{i+3}^C(\tilde{\eta};\alpha,\beta))\Big), \tag{3.1}$$

where $\tilde{N}_i^\epsilon = N_i^\epsilon(k-1, \boldsymbol{r}-\boldsymbol{c}_i)$ and for any vector $w = \langle w_i \rangle_{i=1}^4$

$$\mathcal{C}_i^R(w) = \tfrac{1}{2}(w_{i+1} + w_{i+3} - 2w_i). \tag{3.2}$$

If we factorize $E_\mu(\Delta_i^C(\tilde{\eta};\alpha,\beta))$ for every α and β at all times, then the expected occupancy N_i^ϵ at time k would satisfy the lattice Boltzmann equation

$$N_i^\epsilon(k,\boldsymbol{r}) = \tilde{N}_i^\epsilon + p(\epsilon)[\mathcal{C}_i^R(\tilde{N}^\epsilon) + \mathcal{C}_i^C(\tilde{N}^\epsilon)], \tag{3.3}$$

where

$$\mathcal{C}_i^C(\tilde{N}^\epsilon) = \sum_{\alpha,\beta(\alpha\neq\beta)} [\mathcal{C}_{i+1}^C(\tilde{N}^\epsilon;\alpha,\beta) + \mathcal{C}_{i+3}^C(\tilde{N}^\epsilon;\alpha,\beta)]. \tag{3.4}$$

We may write (3.4) more explicitly as

$$\sum_{\alpha,\beta(\alpha\neq\beta)} \mathcal{C}_i^C(N;\alpha,\beta) = \frac{1}{2}\Big(\sum_{\alpha=0}^{3} \sum_{i_1,\dots,i_\alpha}{}' Q^\alpha(N; i_1 \cdots i_\alpha) r_\alpha - \sum_{\alpha=1}^{4} \sum_{i_2,\dots,i_\alpha}{}' Q^\alpha(N; i i_2 \cdots i_\alpha) f_\alpha \Big), \tag{3.5}$$

where the coefficients r_α and f_α are expected values of the random variables:

$$E_\mu\Big(\sum_\beta \sum_{j_2,\dots,j_q}{}' \gamma^{\alpha\beta}(i j_2 \cdots j_q)\Big) = \sum_{\beta=0}^{\alpha-1} \binom{\alpha-1}{\beta} \frac{P_{\alpha\beta}}{n_{\alpha\beta}} \equiv f_\alpha \qquad (1 \le \alpha \le 4),$$

$$E_\mu\Big(\sum_\beta \sum_{j_2,\dots,j_q}{}' \gamma^{\alpha\beta}(i j_2 \cdots j_q)\Big) = \sum_{\beta=\alpha+1}^{4} \binom{4-\alpha-1}{\beta-\alpha-1} \frac{P_{\alpha\beta}}{n_{\alpha\beta}} \equiv r_\alpha \qquad (0 \le \alpha \le 3). \tag{3.6}$$

Since factorization in (3.3) does not hold exactly unless the system is in equilibrium and in a state that strictly satisfies the Boltzmann hypothesis, N_i^ϵ as computed from this equation will not give the true expected occupancies of the process. Nevertheless, we suppose that under some conditions on the initial state which are made precise below, the automaton inherits the limiting behavior of (3.3) in some space–time scaling regimes corresponding to small values of ϵ. Next we study the kinetic regime where (3.3) converges to the continuous space–time Boltzmann equation [9].

The kinetic regime is characterized by a Knudsen number (the ratio of the mean free path and the typical distance on which densities vary) of order one. The mean free path is of order ϵ. We choose the initial distribution μ_ϵ as a product distribution such that

$$E_\mu(\eta_i(0;\boldsymbol{r})) = N_i(\epsilon\boldsymbol{r}), \tag{3.7}$$

where $N_i(\boldsymbol{x})$, $\boldsymbol{x} \in [-1,1]^2$ is a non-negative smooth function independent of ϵ bounded by one with periodic boundary conditions. For $\boldsymbol{r} \in \mathcal{L}_\epsilon$ a microscopic point, $\boldsymbol{x} = \epsilon \boldsymbol{r}$ is the corresponding macroscopic point. In this way we impose a density profile which varies on the average on a distance of order ϵ^{-1} on the lattice. Since the space scale varies as ϵ^{-1} and there are finite velocities in the system we must also rescale time. In particular the microscopic time k corresponds to the macroscopic time $t = \epsilon k$. In this space–time regime, if we suppose that the reaction probabilities $P_{\alpha\beta}(\epsilon)$ are constant and that $p(\epsilon) = \epsilon p$ for some constant p, then with some conditions on the expected values the limit $\epsilon \to 0$ leads to the continuous space–time Boltzmann equation:

$$\frac{\partial N_i(t,\boldsymbol{x})}{\partial t} + \boldsymbol{c}_i \cdot \nabla N_i(t,\boldsymbol{x}) = p\left[\mathcal{C}_i^R(N) + \mathcal{C}_i^C(N)\right] \equiv p\mathcal{C}_i(N). \tag{3.8}$$

Details concerning the derivation of this equation are given in the appendix.

3.2. Rate law

The results of section 3.1 also provide a route to the continuity equation and chemical rate law for the total density. Summation of (3.8) over the velocity index i gives the continuity equation for the local particle density $\rho = \sum_{i=1}^4 N_i$:

$$\frac{\partial \rho(t,\boldsymbol{x})}{\partial t} + \nabla \cdot \boldsymbol{u}(t,\boldsymbol{x})\rho(t,\boldsymbol{x}) = 2p \sum_{i=1}^{4} \sum_{\alpha,\beta(\alpha\neq\beta)} \mathcal{C}_i^C(N;\alpha,\beta). \tag{3.9}$$

After simplification (3.9) can be written in the form

$$\frac{\partial \rho(t,\boldsymbol{x})}{\partial t} + \nabla \cdot \boldsymbol{u}(t,\boldsymbol{x})\rho(t,\boldsymbol{x}) = p\Bigg(\kappa_0 - \kappa_1\rho(t,\boldsymbol{x}) + \kappa_2 \sum_{i<j} N_i(t,\boldsymbol{x})\, N_j(t,\boldsymbol{x})$$

$$- \kappa_3 \sum_{l<j<i} N_i(t,\boldsymbol{x}) N_j(t,\boldsymbol{x}) N_l(t,\boldsymbol{x}) + \kappa_4 \prod_{m=1}^{4} N_m(t,\boldsymbol{x})\Bigg), \tag{3.10}$$

where

$$\kappa_n = \sum_{\alpha=0}^{n} \binom{n}{\alpha} (-1)^\alpha \sum_{\beta\neq\alpha} (\beta-\alpha) P_{\alpha\beta}. \tag{3.11}$$

Finally, if the system is spatially homogeneous and in equilibrium in velocity space so that $N_i(t) = \rho(t)/4$ we obtain the quartic chemical rate law:

$$\frac{\mathrm{d}\rho(t)}{\mathrm{d}t} = p\left(\kappa_0 - \kappa_1\rho(t) + \frac{3}{8}\kappa_2\rho^2(t) - \frac{1}{16}\kappa_3\rho^3(t) + \frac{1}{256}\kappa_4\rho^4(t)\right). \tag{3.12}$$

With $\rho = \rho_\mathrm{s} + \delta\rho$ and $d = \rho_\mathrm{s}/4$, the linearized version of this rate law is

$$\frac{\mathrm{d}\delta\rho(t)}{\mathrm{d}t} = -p\left(\sum_{m=0}^{3} \binom{3}{m} (-1)^m \kappa_{m+1} d^m\right) \delta\rho(t), \tag{3.13}$$

which allows one to identify the chemical relaxation time as

$$\tau_\mathrm{ch}^{-1} = p \sum_{m=0}^{3} \binom{3}{m} (-1)^m \kappa_{m+1} d^m. \tag{3.14}$$

3.3. Linearized Boltzmann equation and the macroscopic law

In this subsection we consider the linearized version of the Boltzmann equation (3.8) and derive a generalized reaction–diffusion equation for the automaton along with autocorrelation function expressions for the transport coefficients. If we suppose that the system is close to a homogeneous steady state we can linearize the Boltzmann equation about this state. Letting $N_i(t,\boldsymbol{x}) = N_i^{\mathrm{s}} + \phi_i(t,\boldsymbol{x})$ with $N_i^{\mathrm{s}} = \rho_s/4 = d$ we have

$$\partial_t \phi_i(t,\boldsymbol{x}) + \boldsymbol{c}_i \cdot \nabla \phi_i(t,\boldsymbol{x}) = p \sum_j A_{ij} \phi_j(t,\boldsymbol{x}), \tag{3.15}$$

where

$$A_{ij} = \left(\frac{\partial C_i(N)}{\partial N_j} \right)_{\mathrm{s}}, \tag{3.16}$$

where the subscript 's' refers to $N_k = N_k^{\mathrm{s}}$. The matrix $\mathbf{A}$ may be written as a sum of elastic (rotation) and reactive collision contributions in view of the decomposition of the Boltzmann collision operator in (3.8):

$$A_{ij} = \left(\frac{\partial C_i^R(N)}{\partial N_j} \right)_{\mathrm{s}} + \left(\frac{\partial C_i^C(N)}{\partial N_j} \right)_{\mathrm{s}} = A_{ij}^R + A_{ij}^C. \tag{3.17}$$

The elements of these cyclic 4×4 matrices follow directly from the definitions in (3.2), (3.4), (3.5) and (3.17). The elements of the first row of the elastic collision (rotation) matrix are

$$A_{11}^R = -1, \quad A_{12}^R = A_{14}^R = \tfrac{1}{2}, \quad A_{13}^R = 0. \tag{3.18}$$

Similarly the elements of the first row of the linearized reactive collision matrix are

$$\begin{aligned} A_{11}^C = A_{13}^C &= (-r_0 + r_1) + (3r_0 - 5r_1 + 2r_2 + f_1 - f_2)d \\ &+ (-3r_0 + 7r_1 - 5r_2 + r_3 - 2f_1 + 4f_2 - 2f_3)d^2 \\ &+ (r_0 - 3r_1 + 3r_2 - r_3 + f_1 - 3f_2 + 3f_3 - f_4)d^3, \end{aligned} \tag{3.19}$$

$$\begin{aligned} A_{12}^C = A_{14}^C &= (-r_0 + \tfrac{1}{2}r_1 - \tfrac{1}{2}f_1) + (3r_0 - 4r_1 + r_2 + 2f_1 - 2f_2)\, d \\ &+ \left(-3r_0 + \tfrac{13}{2}r_1 - 4r_2 + \tfrac{1}{2}r_3 - \tfrac{5}{2}f_1 + 5f_2 - \tfrac{5}{2}f_3\right) d^2 \\ &+ (r_0 - 3r_1 + 3r_2 - r_3 + f_1 - 3f_2 + 3f_3 - f_4)\, d^3. \end{aligned} \tag{3.20}$$

The remaining rows of these matrices can be obtained by cyclic permutations of the first row, which follows from the lattice symetries.

The linearized Boltzmann equation may be written compactly in matrix form as

$$\partial_t \boldsymbol{\phi}(t,\boldsymbol{x}) = \mathbf{L}\boldsymbol{\phi}(t,\boldsymbol{x}), \tag{3.21}$$

where

$$L_{ij} = -\delta_{ij}\boldsymbol{c}_i \cdot \nabla + pA_{ij}. \tag{3.22}$$

Given this Boltzmann-level description of the automaton we next derive an equation for the density of particles at position $\boldsymbol{x}$ at time t,

$$\rho(t, \boldsymbol{x}) = \sum_i N_i(t, \boldsymbol{x}). \tag{3.23}$$

It is convenient to consider the deviation of the density from its steady-state value since this is related to the perturbed distribution function by

$$\delta\rho(t, \boldsymbol{x}) = \rho(t, \boldsymbol{x}) - \rho_s = \sum_i \phi_i(t, \boldsymbol{x}). \tag{3.24}$$

In order to carry out this reduction of the Boltzmann equation we introduce a matrix operator $\mathbf{P}$ that projects an arbitrary vector-valued function onto configuration space:

$$\mathbf{P}\boldsymbol{f} = \tfrac{1}{4}\mathbf{U}\boldsymbol{f} = \tfrac{1}{4}\boldsymbol{u}\sum_i f_i, \tag{3.25}$$

where $\mathbf{U}$ is a matrix whose elements are unity and $\boldsymbol{u}$ is a column vector whose elements are unity. Note that $\mathbf{P} = \mathbf{P}^2$ and that

$$\mathbf{P}\boldsymbol{\phi} = \tfrac{1}{4}\delta\rho(t, \boldsymbol{x})\,\boldsymbol{u}. \tag{3.26}$$

Defining the complementary projector as $\mathbf{Q} = 1 - \mathbf{P}$ and using standard projection operator methods [10] we can obtain an equation for the time development of $\delta\rho(t, \boldsymbol{x})$:

$$\partial_t \delta\rho(t, \boldsymbol{x}) = \tfrac{1}{4}\sum_i [\mathbf{PL}\boldsymbol{u}]_i\,\delta\rho(t, \boldsymbol{x}) + \int_0^t \mathrm{d}t'\,\tfrac{1}{4}\sum_i [\mathbf{PL}\exp\{\mathbf{QL}(t-t')\}\,\mathbf{QL}\boldsymbol{u}]_i\,\delta\rho(t', \boldsymbol{x}) + \sum_i [\mathbf{PL}\exp(\mathbf{QL}t)\,\mathbf{Q}\boldsymbol{\phi}(0, \boldsymbol{x})]_i. \tag{3.27}$$

The coefficient of $\delta\rho$ in the first term on the right-hand side may be directly evaluated to give the chemical relaxation time defined in (3.14):

$$\tfrac{1}{4}\sum_i (\mathbf{PL}\boldsymbol{u})_i = \tau_{\mathrm{ch}}^{-1}. \tag{3.28}$$

The memory kernel is related to the velocity autocorrelation function and hence the diffusive propagation of the local density. Evaluation of the matrix products yields for this term the expression

$$\tfrac{1}{4}\sum_i [\mathbf{PL}\exp(\mathbf{QL}t)\,\mathbf{QL}\boldsymbol{u}]_i = \partial_\alpha \tfrac{1}{4}\sum_j\sum_l c_{j\alpha}[\exp(\mathbf{QL}t)]_{jl} c_{l\beta}\partial_\beta \equiv \partial_\alpha \mathcal{K}_{\alpha\beta}(t)\partial_\beta. \tag{3.29}$$

The last term on the right-hand side of (3.27) is the initial condition term; it will vanish if the initial value of $\boldsymbol{\phi}$ is such that there is equilibrium in velocity space. In this circumstance the generalized macroscopic reaction–diffusion equation for the cellular automaton takes the form

$$\partial_t \delta\rho(t, \boldsymbol{x}) = -\tau_{ch}^{-1}\delta\rho(t, \boldsymbol{x}) + \int_0^t \mathrm{d}t'\,\partial_\alpha \mathcal{K}_{\alpha\beta}(t-t')\partial_\beta \delta\rho(t', \boldsymbol{x}). \tag{3.30}$$

In the limit of small spatial gradients we can neglect the spatial dependence in the $\mathbf{L}$ matrix in the exponent in $\mathcal{K}$ and the velocity autocorrelation function takes the form

$$\mathcal{K}_{\alpha\beta}(t) = \tfrac{1}{8}\sum_j\sum_l \boldsymbol{c}_j \cdot [\exp(p\mathbf{QA}t)]_{jl}\boldsymbol{c}_l \delta_{\alpha\beta} \equiv K(t)\delta_{\alpha\beta}. \tag{3.31}$$

This expression for the velocity autocorrelation function may be easily evaluated by considering the eigenvalues and eigenvectors of the projected matrix $\mathbf{QA}$. These are

$$v_1 = \begin{pmatrix} 1 \\ 1 \\ 1 \\ 1 \end{pmatrix}, \quad v_2 = \begin{pmatrix} 0 \\ 1 \\ 0 \\ -1 \end{pmatrix}, \quad v_3 = \begin{pmatrix} 1 \\ 0 \\ -1 \\ 0 \end{pmatrix}, \quad v_4 = \begin{pmatrix} 1 \\ -1 \\ 1 \\ -1 \end{pmatrix}, \tag{3.32}$$

corresponding to the eigenvalues $\lambda_1 = 0, \lambda_2 = \lambda_3 = -1$ and $\lambda_4 = -2 + \gamma$, respectively, where

$$\gamma = (r_1 + f_1) - 2(r_1 - r_2 + f_1 - f_2)d + (r_1 - 2r_2 + r_3 + f_1 - 2f_2 + f_3)d^2. \tag{3.33}$$

Using these results and the fact that $K(t)$ can be written compactly in terms of $\boldsymbol{v}_2$ we find

$$K(t) = \tfrac{1}{4}\boldsymbol{v}_2^{\mathrm{T}} \exp(p\mathbf{QA}t)\, \boldsymbol{v}_2 = \tfrac{1}{2}\exp(-pt), \tag{3.34}$$

where the superscript T on the eigenvectors refers to the transpose. So the generalized reaction–diffusion equation takes the form

$$\partial_t \delta\rho(t, \boldsymbol{x}) = -\tau_{\mathrm{ch}}^{-1}\delta\rho(t, \boldsymbol{x}) + \tfrac{1}{2}\nabla^2 \int_0^t \mathrm{d}t' \exp[-p(t - t')]\, \delta\rho(t', \boldsymbol{x}). \tag{3.35}$$

The diffusion coefficient is of course the time integral of the velocity autocorrelation function, and we have

$$D = \int_0^\infty \mathrm{d}t\, K(t) = \frac{1}{2p}. \tag{3.36}$$

A number of features of these results merit discussion. The singular reactive contribution τ_{ch}^{-1} arises from the instantaneous nature of the chemical interconversion process [11]. Note also that there is no reactive part in the memory kernel in (3.34) due to the fact that the reaction does not perturb the velocity distribution: reaction is equally likely from any velocity state.

The evolution operator appearing in the velocity autocorrelation function does contain elastic and reactive collision parts; however, the reactive collisions do not affect the time development of this function since only the $\boldsymbol{v}_2$ or $\boldsymbol{v}_3$ eigenvectors of $\mathbf{QA}$ contribute to $K(t)$. It is easy to see that $K(t)$ is independent of the reactive part of $\mathbf{QA}$ in view of the fact that

$$\mathbf{QA}\boldsymbol{v}_2 = \mathbf{QA}\boldsymbol{v}_3 = \mathbf{QA}^{\mathrm{el}}\boldsymbol{v}_2 = \mathbf{QA}^{\mathrm{el}}\boldsymbol{v}_3. \tag{3.37}$$

This feature arises from the fact that, as noted above, the reaction does not perturb the velocity distribution in our simple model reaction. If reaction occurred by an activated process this would give rise to a perturbation of the velocity distribution and the reestablishement of the equilibrium distribution via elastic collisions would lead to memory effects due to reactive events. From this it follows that the rate law (3.12) is exact for this Boltzmann model since there are no memory effects due to the coupling of the reaction to the velocity distribution. The result $D = 1/2p$ is expected in view of the nature of the random walk that the particles undergo.

4. Discrete Boltzmann approximation

In section 3, the limiting procedure needed to reduce the microdynamical cellular automaton equations to a continuous space and time Boltzmann equation was described; an appropriate scaling of space, time and collision rate $p(\epsilon)$ was required. In CA simulations, the system is treated at a microscopic level and some of the hypotheses on which the reduction to the continuous Boltzmann equation (3.8) rests must be partly relaxed. In this section we consider the discrete kinetic Boltzmann equation which is a better approximation than (3.8) for finite collision rate models ($\lim_{\epsilon\to 0} p(\epsilon) \neq 0$). Exact solutions and macroscopic phenomenological constants are computed and the connection with the continuous Boltzmann equation is discussed.

4.1. Discrete Boltzmann equation

For the sake of simplicity, we restrict ourselves to a subset of the class of cellular automata defined in section 2 and consider cellular automata with a collision probability $p(\epsilon)$ equal to one. In this case, at each time step, each particle is rotated by $\pi/2$ or $-\pi/2$, and each node can be the locus of a chemical reaction whose outcome depends on the configuration and the transition probability matrix $P_{\alpha\beta}$. In order to discuss the validity of the discrete Boltzmann equation and to compare the results obtained in the preceding sections to those given below we introduce the following scaling of the transition probability matrix

$$P_{\alpha\beta} = p^{c} P^{0}_{\alpha\beta} \quad \text{for } \alpha \neq \beta, \tag{4.1}$$

where the diagonal elements $P_{\alpha\alpha}$ are again given by (2.11).

In the non-reactive limit $p^{c} = 0$ and each particle performs a random walk on the lattice. These random walks are not independent since when two or more particles are simultaneously present on a node, they are rotated by the same angle in order to preserve the exclusion principle. Except for this weak dependence between trajectories due to the exclusion principle all elastic collisions are independent and a kinetic Boltzmann equation for the average local number of particles can be derived.

In the presence of reactive transformations collisions can no longer be assumed to be independent. For example, particles which have reacted to produce a new particle can collide reactively again with this new particle on nearby nodes. However, if p^{c} is sufficiently small, reactive transitions are rare events and we can again assume independence between collisions. In this approximation, we can factorize the expected value of the microdynamical equation (2.44) (with the scaled reactive transition probabilities) so that the expected occupancy N_i at time k satisfies a lattice Boltzmann equation similar to (3.3):

$$N_i(k, \boldsymbol{r}) = \tilde{N}_i + [\mathcal{C}_i^R(\tilde{N}) + p^c \mathcal{C}_i^C(\tilde{N})] = \mathcal{C}_i^{p^c}(N). \tag{4.2}$$

This discrete Boltzmann equation can also be conveniently rewriten as

$$N_i(k+1, \boldsymbol{r} + \boldsymbol{c}_i) = N_i(k, \boldsymbol{r}) + \{\mathcal{C}_i^R(N(k, \boldsymbol{r})) + p^c \mathcal{C}_i^C(N(k, \boldsymbol{r}))\}. \tag{4.3}$$

4.2. Linearized Boltzmann equation

In this subsection, we consider the linearized version of the lattice Boltzmann equation (4.3). As in section 3.3, if we suppose that the system is close to an homogenous steady state we can linearize the collision term $[\mathcal{C}_i^R(N_i) + p^c \mathcal{C}_i^C(N_i)]$ about this state. Setting $N_i(k, \boldsymbol{r}) = N_i^s + \phi_i(k, \boldsymbol{r})$ as was done in section 3.3, we have the linearized discrete Boltzmann equation

$$\phi_i(k+1, \boldsymbol{r} + \boldsymbol{c}_i) = \phi_i(k, \boldsymbol{r}) + \sum_j A_{ij} \phi_j(k, \boldsymbol{r}), \tag{4.4}$$

where the matrix $\mathbf{A}$ is defined as in (3.16) with $\mathcal{C}_i$ replaced by $\mathcal{C}_i^{p^c}$. This equation can be written in a more compact form:

$$\phi_i(k+1, \boldsymbol{r} + \boldsymbol{c}_i) = \sum_j B_{ij} \phi_j(k, \boldsymbol{r}), \tag{4.5}$$

with

$$\mathbf{B} = \mathbf{I} + \mathbf{A}^R + p^c \mathbf{A}^C. \tag{4.6}$$

To solve the set of linear finite difference equations (4.5), it is convenient to work in Fourier space. We define

$$\phi_i^{(q,\omega)}(k, \boldsymbol{r}) = \phi_i^{(0)} \exp(\mathrm{i}\boldsymbol{q} \cdot \boldsymbol{r} + \omega k), \quad i = 1, \ldots, 4. \tag{4.7}$$

Introducing this form into (4.5) we obtain a linear algebraic system for the $\phi_i^{(0)}$'s:

$$\sum_{j=1}^{4} M_{ij} \phi_j^{(0)} = 0, \quad \forall i = 1, \ldots, 4, \tag{4.8}$$

with

$$M_{ij} = \begin{pmatrix} B_{11} - \mathrm{e}^{\omega}\mathrm{e}^{q_x} & B_{12} & B_{11} & B_{12} \\ B_{12} & B_{11} - \mathrm{e}^{\omega}\mathrm{e}^{q_y} & B_{12} & B_{11} \\ B_{11} & B_{12} & B_{11} - \mathrm{e}^{\omega}\mathrm{e}^{-q_x} & B_{12} \\ B_{12} & B_{11} & B_{12} & B_{11} - \mathrm{e}^{\omega}\mathrm{e}^{-q_y} \end{pmatrix}. \tag{4.9}$$

Non-trivial solutions follow from the condition det $\mathbf{M} = 0$. For the restricted set of cellular automata with $p(\epsilon) = 1$ considered here, the solution of this equation yields

$$\mathrm{e}^{\omega_1} = 0, \tag{4.10}$$

$$\mathrm{e}^{\omega_2} = 0, \tag{4.11}$$

$$\mathrm{e}^{\omega_3} = B_{11}[\cos(q_x) + \cos(q_y)] + \sqrt{B_{11}^2[\cos(q_x) - \cos(q_y)]^2 + 4B_{12}^2 \cos(q_x)\cos(q_y)}. \tag{4.12}$$

$$\mathrm{e}^{\omega_4} = B_{11}[\cos(q_x) + \cos(q_y)] - \sqrt{B_{11}^2[\cos(q_x) - \cos(q_y)]^2 + 4B_{12}^2 \cos(q_x)\cos(q_y)}, \tag{4.13}$$

These solutions are interpreted in the next section.

4.3. Linearized discrete Boltzmann equation and macroscopic law

In order to make the connection between the discrete Boltzmann equation (4.4) and the expected macroscopic reaction–diffusion behavior we now consider the set of solutions (4.10)–(4.13) in the macroscopic limit.

First consider the two solutions (4.10)–(4.11). Their degeneracy is a manifestation of the singularity of the collision matrix $\mathbf{B}$,

$$|\mathbf{B}| = 0, \tag{4.14}$$

when $p(\epsilon) = 1$. Actually, this singularity is independent of the chemical transition probabilities and follows from a singularity of the elastic collision matrix (here we include the general $p(\epsilon)$ dependence for clarity),

$$|\mathbf{I} + p(\epsilon)\mathbf{A}^R| = 0, \tag{4.15}$$

when $p(\epsilon) = 1$.

From a physical point of view, this singularity reflects the fact that two nodes with symmetrical configurations obtained by reflection in space produce the same distribution of post-collisional configurations. In other words, when $p(\epsilon) = 1$, all the information about the sign of the velocity of a particle is destroyed in one collision. In accordance with this one-step total decay the matrix $\mathbf{B}$ has an non-trivial kernel. A basis of this kernel is given by $\boldsymbol{v}_2$ and $\boldsymbol{v}_3$ defined in (3.32), which are the pure antisymmetric states in the directions (2,4) and (1,3), respectively. In the macroscopic limit the decay time is of order ϵ and these solutions are unobservable.

We next consider the expression (4.12), and its expansion in the limit of small gradients ($\lim_{\boldsymbol{q}\to 0}$). To first significant order in $\boldsymbol{q}$, we have

$$e^{\omega} = 2(B_{11} + B_{12}) - \tfrac{1}{2}(B_{11} + B_{12})\boldsymbol{q}^2. \tag{4.16}$$

The relation between the relaxation time τ_{ch}^{-1} (3.14), and the matrix elements B_{ij} is

$$2(B_{11} + B_{12}) = 1 - \tau_{\text{ch}}^{-1}, \tag{4.17}$$

and the expansion of the logarithm of (4.16) to lowest order in $\boldsymbol{q}^2$ yields

$$\omega = \ln(1 - \tau_{\text{ch}}^{-1}) - \tfrac{1}{4}\boldsymbol{q}^2, \tag{4.18}$$

which is the dispersion relation for the linearized reaction–diffusion equation

$$\partial_t \delta\rho(t, \boldsymbol{x}) = -\tilde{\tau}_{\text{ch}}^{-1}\delta\rho(t, \boldsymbol{x}) + \tilde{D}\nabla^2\delta\rho(t, \boldsymbol{x}), \tag{4.19}$$

where

$$\tilde{D} = \tfrac{1}{4}, \tag{4.20}$$

and

$$\tilde{\tau}_{\text{ch}}^{-1} = \ln(1 - \tau_{\text{ch}}^{-1}). \tag{4.21}$$

As expected we obtain a reaction–diffusion mode. However, the diffusion constant $\tilde{D}$ as well as the chemical relaxation time $\tilde{\tau}_{\text{ch}}$ differ from their predicted values obtained in section 3 by terms arising from the discrete nature of the model. If chemical reactions become rare events (i.e., $p^{\text{c}} \to 0$), then, $\tau_{\text{ch}}^{-1} \to 0$ and $\tilde{\tau}_{\text{ch}}^{-1}$ converges to the continuous limit value (3.14). Similarly, if elastic collisions become infrequent (i.e., $p \to 0$), it can be shown that $\tilde{D}$ converges to the diffusion coefficient (3.36) obtained from the continuous Boltzmann equation.

Expansion of (4.13) with respect to $\boldsymbol{q}$ yields

$$e^{\omega} = 2(B_{11} - B_{12}) - \tfrac{1}{2}(B_{11} - B_{12})\boldsymbol{q}^2. \tag{4.22}$$

The corresponding mode will not be analyzed in detail here. We note that as a result of a spurious diffusive invariant when $p(\epsilon) = 1$ the time decay of this mode and that of the reaction–diffusion mode may be of the same order; however, it will be show in section 6 how this mode may be filtered out of the CA simulations so that the reaction–diffusion behavior alone survives.

5. Spurious properties

The CA considered here possess spurious invariants [5], i.e., quantities which are conserved under the CA rules and which may interfer with the reaction–diffusion mode. In this section, we describe two spurious properties and consider their effects on the reaction–diffusion mode.

5.1. Checkerboard parity

The square lattice CA exhibits a "checkerboard parity" [12]: two particles separated by an odd distance in the Manhattan metric (that is two particles on differently colored nodes when the lattice nodes are painted as a checkerboard) will never interact. This property does not exist for other lattice geometries, for example the triangular lattice used in 2D lattice gas hydrodynamics. It will also vanish if the particles have a certain probability to remain on a node and not propagate, or if they can propagate over more than one node in one time step (in this case the exclusion principle must be relaxed).

The "checkerboard parity" property splits the cellular automaton universe into two totally independent subsystems corresponding to alternate colors on the checkerboard at alternate time steps. Therefore, when ensemble averaging should be avoided one must select one of the subsystems; for instance, such a selection must be made when spontaneous symmetry breaking occurs, as will be illustrated in section 7. It should also be noted that the boundary conditions may affect the "checkerboard parity" property since, for example, periodic boundary conditions may connect the two subsystems.

5.2. Diffusion spurious invariant

Consider the model in the absence of reactive transformations. In this case the total number of particles is conserved, and the model exhibits a pure diffusive mode corresponding to this conservation law. More precisely, the diffuson mode reflects the existence of an eigenvalue equal to one in the rotation matrix (cf. (4.15)). If chemical transformations are incorporated the total number of particles is no longer invariant, and the diffusion mode becomes the reaction–diffusion mode characterized by the dispersion relation (4.18).

Next consider now the model when $p(\epsilon) = 1$. In the absence of reactive transitions, as already mentioned, each particle is rotated by $\pm\pi/2$ at each time step. As a result, a particle with velocity in the directions $(1,3)$ at time k will have velocity in the directions $(2,4)$ at time $k+1$, and as a consequence the total number of particles in the directions $(1,3)$ or $(2,4)$ are conserved quantities at even times, and interchange with each other at odd times. It should be noticed that the spurious conservation law disappears when $p(\epsilon) \neq 1$. It will also vanish if the rotation transformation described in section 2.2 is modified so that node configurations are rotated by $\pi/2$, $-\pi/2$, or π with probability $1/3$.

According to the new conservation law that appears when $p(\epsilon) = 1$, the rotation matrix has an eigenvalue -1 and the model exhibits a new mode that survives on large time scales. When reactions are taken into account this spurious mode is characterized by the dispersion relation (4.22). In the homogeneous limit the decay of this mode is determined only by the chemical transition probabilities and may be of the same order as the decay of the reaction–diffusion mode, hence it can interfere with the latter mode.

In CA simulations a model with $p(\epsilon) = 1$ can be useful in order to save CPU time and to reduce the amount of memory needed. In such simulations, the spurious mode (4.22) can be eliminated by choosing an initial lattice configuration which does not excite this mode. Even if excited the spurious mode is generally filtered out in measurements of the total number of particles since the mode corresponding to particle number is approximately orthogonal to the spurious mode.

6. Chemical rate laws and microscopic dynamics

In the preceding sections we have seen that the CA model as formulated for the square lattice with the exclusion principle gives rise to a macroscopic chemical rate law with the general form of a quartic polynomial (3.12). Underlying this macroscopic law are microscopic reactions of the type $\alpha X \to \beta X$ occurring with probability $P_{\alpha\beta}$. There are several microscopic CA dynamics that can lead to a given macroscopic rate law; we examine this feature below.

The relationship between the coefficients of the powers of the density in the rate law (3.12) and the microscopic reaction probabilities was given in (3.11). From this it follows that

$$\sum_{\beta\neq\alpha}(\beta-\alpha)P_{\alpha\beta}=\sum_{n=0}^{\alpha}\binom{\alpha}{n}(-1)^n\kappa_n=\nu_\alpha,\quad \alpha=0,\ldots,4. \tag{6.1}$$

The relations (6.1) and (2.11) and the assumption that

$$0\leq P_{\alpha\beta}\leq 1,\quad \alpha,\beta=0,\ldots,4, \tag{6.2}$$

provide the conditions that the chemical transformation transition probabilities $P_{\alpha\beta}$ must satisfy in order give the rate law. From (2.11) and (6.1) it follows that for $\alpha=0,\ldots,3$

$$P_{\alpha,\alpha+1}=-\sum_{\beta\neq\alpha,\alpha+1}(\beta-\alpha)P_{\alpha\beta}+\nu_\alpha, \tag{6.3}$$

$$P_{\alpha\alpha}=1+\sum_{\beta\neq\alpha,\alpha+1}(\beta-\alpha-1)P_{\alpha\beta}-\nu_\alpha, \tag{6.4}$$

and for $\alpha=4$

$$P_{43}=\sum_{\beta\neq 3,4}(\beta-4)P_{4\beta}-\nu_4, \tag{6.5}$$

$$P_{44}=1-\sum_{\beta\neq 3,4}(\beta-3)P_{4\beta}+\nu_4. \tag{6.6}$$

Eqs. (6.3)–(6.6) constitute a set of constraints on the $P_{\alpha\beta}$ for a given set of macroscopic rate coefficients. It is of course obvious that the five ν_α along with the conditions on the diagonal elements (2.11) are not sufficient to determine all the elements of $\mathbf{P}$. One may choose, for instance, for each $\alpha=0,\ldots,3$, $P_{\alpha\beta}$ for $\beta\neq\alpha,\alpha+1$ and for $\alpha=4$, $P_{4\beta}$ for $\beta\neq 3,4$ as independent parameters along with the ν_α for $\alpha=0,\ldots,4$ to determine the complete structure of $\mathbf{P}$. Clearly a number of microscopic models characterized by $\mathbf{P}$ can be constructed that are consistent with a given macroscopic chemical rate law. Below we illustrate this flexibility by considering a specific example.

6.1. Schlögl model

The Schlögl model [6] is a simple example of an autocatalytic chemical reaction composed of the following elementary steps:

$$A\overset{k_0}{\rightarrow}X,\quad X\overset{k_1}{\rightarrow}A,\quad 2X+B\overset{k_2}{\rightarrow}3X,\quad 3X\overset{k_3}{\rightarrow}2X+B. \tag{6.7}$$

This reaction scheme corresponds to a cubic rate law for species X:

$$\mathrm{d}\rho/\mathrm{d}t=-k_3\rho^3+k_2b\rho^2-k_1\rho+k_0a, \tag{6.8}$$

and yields to the following reaction–diffusion equation

$$\partial\rho/\partial t=-k_3\rho^3+k_2b\rho^2-k_1\rho+k_0a+\nabla^2\rho. \tag{6.9}$$

The system can be constrained to lie far from equilibrium by fixing the concentrations a and b of the A and B species respectively and these concentrations can be incorporated in the definitions of the rate constants. Thus if we treat A, B and the solvent molecules as "ghost" species as described in section 1 we need only focus on the dynamics of the X species. The rate law (6.8) can be written in the form of (3.12) if we make the identifications: $0=\kappa_4$, $k_3=p\kappa_3/16$, $k_2b=3p\kappa_2/8$, $k_1=p\kappa_1$ and $k_0=p\kappa_0$.

The Schlögl model clearly involves elementary reactions that either increase or decrease the number of particles by 1. Hence in the construction of the automaton dynamics we restrict attention to reactions of the type $\alpha X \to (\alpha \pm 1)X$. Not all of these elementary reactions appear in the Schlögl mechanism. However, since up to four particles can reside at a node the manner in which the state of a node changes can be due to several possible elementary reactions. For instance, if the number of particles at a node changes from 4 to 3 it cannot be due to an independent reaction like $4X \to 3X$ since this is not one of the elementary Schlögl reactions (the rate law is a cubic polynomial), but on the lattice processes like $3X \to 2X$, etc. can lead to an effective transformation equivalent to $4X \to 3X$, etc. if one particle is considered as reactively passive. So chemical transformations on the lattice cannot be identical to the elementary steps of the phenomenological model, and conversely, there is a number of sets of lattice elementary transformations which are compatible with the phenomenological rate law. Hence we have a manifestation of the abovementioned flexibility in microscopic description. Next we work out the details of the construction of microscopic models consistent with the cubic Schlögl rate law.

From (3.11) and the restriction to reactions of the form $\alpha X \to (\alpha \pm 1)X$ it follows that

$$P_{01} = \kappa_0, \quad (P_{10} - P_{12}) = -\kappa_0 + \kappa_1, \quad (P_{21} - P_{23}) = -\kappa_0 + 2\kappa_1 - \kappa_2,$$

$$(P_{32} - P_{34}) = -\kappa_0 + 3\kappa_1 - 3\kappa_2 + \kappa_3, \quad P_{43} = -\kappa_0 + 4\kappa_1 - 6\kappa_2 + 4\kappa_3. \tag{6.10}$$

The five relations in (6.10) do not fix the eight transition probabilities corresponding to reactions that increase or decrease the particle number by one at a node. If a reaction does not appear in the mechanism (6.7) it either need not be included or must have a probability that is related to that of one of the reactions in the mechanism.

Given these general principles we may for example take $P_{12} = P_{34} = 0$ and $P_{23} = \kappa_2$. We term this choice model (a) and find for the reaction probabilities:

$$P_{01} = \kappa_0, \quad P_{10} = \kappa_1 - \kappa_0, \quad P_{21} = 2\kappa_1 - \kappa_0, \quad P_{23} = \kappa_2,$$

$$P_{32} = \kappa_3 - 3\kappa_2 + 3\kappa_1 - \kappa_0, \quad P_{43} = 4\kappa_3 - 6\kappa_2 + 4\kappa_1 - \kappa_0. \tag{6.11}$$

Other choices are equally acceptable. As another example we may set $P_{12} = P_{21} = 0$ and let $P_{32} = \kappa_3$ and obtain model (b):

$$P_{01} = \kappa_0, \quad P_{10} = \kappa_1 - \kappa_2, \quad P_{23} = \kappa_2 - 2\kappa_1 + \kappa_0, \quad P_{32} = \kappa_3,$$

$$P_{34} = 3\kappa_2 - 3\kappa_1 + \kappa_0, \quad P_{43} = 4\kappa_3 - 6\kappa_2 + 4\kappa_1 - \kappa_0. \tag{6.12}$$

Additional models may be constructed in this manner. In fact, if one is simply interested in macroscopic cubic kinetics then an even larger number of microscopic models is possible since there is no need to single out the reactions in (6.7). In the next section we discuss the results of simulations of the CA Schlögl model (a).

7. Cellular automaton simulations

Simulations of the cellular automaton evolution have been performed for a microdynamics corresponding to the Schlögl model described in section 6.1. This is one of the simplest autocatalytic reaction schemes giving rise to bistable steady states and it exhibits a wide variety of interesting spatio-temporal behavior.

7.1. Schlögl model phenomenology

The cubic Schlögl rate law (6.8) yields three real homogeneous steady states ($x_1 \leq x_3 \leq x_2$) for certain values of the parameters ($k_0, \ldots, k_3, a, b$); two of these (x_1 and x_2) are temporally stable while x_3 is unstable. The existence of this bistability gives rise to interesting phenomena when the spatially distributed system is considered.

The Schlögl reaction–diffusion equation (6.9) has the form of the time-dependent Ginzburg–Landau equation for a non-conserved order parameter; hence, a variety of phase separation processes can be investigated [13,14]. For example we can imagine fixing the parameters on a surface in parameter space that corresponds to equally stable deterministic bistable steady states. An initial random configuration with mean concentration corresponding to the unstable steady state x_3 will evolve into macroscopic domains of the two stable phases with mean concentrations x_1 and x_2. Long-time domain growth is governed by domain wall curvature effects [15]. A finite system will evolve to the homogeneous x_1 or x_2 phase with equal probability. This long-time domain growth regime is endowed with simple scaling behavior.

If the parameters are such that the system is in the bistable region but the deterministic states are no longer equivalent, the growth occurs by a nucleation process. For instance, nuclei of the more stable phase in a sea of the less stable phase will grow provided their radii are greater than the critical radius for the selected parameters.

The effects of fluctuations on such processes can be explored in CA simulations of the type reported below.

7.2. Simulations

The cellular automaton numerical simulations presented here have been performed with $p(\epsilon) = 1$ (rotation occurs at each time step) for the Schlögl model with the particular choice (6.10) for the chemical transition matrix (Schlögl model (a)). Like in standard lattice gas simulations, the Boolean field $\eta(k)$ describing the state of the system at a time k is represented by a matrix of four-bit words, each of which describes the state of a node. In this representation, the propagation step of the dynamics consists in moving bits from each matrix element to adjacent ones. Rotations and chemical transformations were carried out by means of a look-up table giving the different output configurations along with their respective probabilities for each possible input configuration at a node. The choice of the actual output configuration, among all the possible post-collisional configurations was determined for each node at each time step using the microdynamical description of section 2. Note that the use of a look-up table, which is particularly efficient from a computational point of view, is only possible if the number of node configurations is not too large. This is one reason for the introduction of the exclusion principle (2.1) in the model.

In order to simulate a reaction–diffusion equation, the initial lattice configuration must be chosen to represent a given initial concentration field. Since the concentration does not uniquely determine the velocity distribution, there are several velocity distributions that are compatible with a given initial concentration field. We have considered only initial states where the velocity is selected from a uniform distribution since for this particular choice the spurious mode (4.22) is not strongly excited.

7.3. Experiment 1

A first set of simulations was done in order to check if the CA is capable of reproducing the qualitative features of the phase separation behavior of the Schlögl reaction–diffusion equation (6.9). For various parameter values in the bistable region the system was prepared with a mean density corresponding to that of the deterministic homogeneous unstable state x_3 (as predicted by (6.8)) by randomly occupying each node, independent of its neighbors, with a probability selected to yield this mean density.

A typical evolution of one of the checkerboard subsystems starting from this initial condition is shown in fig. 1. Due to inhomogeneities in the initial configuration as well as to internal fluctuations as a result of the microscopic nature of the system there is a complex initial evolution where domains of the two stable states form (see panels for $t = 2$, 40 and 80 in fig. 1). After this initial stage the system is roughly separated into domains of different phases whose evolution is determined by domain wall curvature as noted above (see panels for $t = 100$, 300 and 1000 in fig. 1). Eventually one phase will completely invade the other and the system will become homogeneous. The figure shows that the cellular automaton exhibits qualitative behavior in agreement with the above phenomenology.

Note that the two checkerboard subsystems constitute different realizations of the evolution and therefore must be considered separately in order to display the symmetry breaking and domain formation.

7.4. *Experiment 2*

A second set of simulations was carried out in order to compare the homogeneous stable steady-state concentration obtained from the cellular automaton simulation to that predicted from the rate law (6.8). For various parameter values of the Schlögl model the system was prepared in an homogeneous state with average concentration x_2. Starting from this initial configuration, the system was allowed to relax for a sufficiently long time (i.e., long compared to the characteristic decay time of the autocorrelation of the total number of particles in the system). The concentration of species X was then computed by spatio-temporal averaging over the entire system and time.

Fig. 2 shows the ratio of the observed concentration to the stable steady-state concentration predicted from the rate equation (6.8) for various parameter values. Significant discrepancies between the rate law predictions and the experimental values are observed which reflect the fact that the cellular automaton does not behave as a Boltzmann model. The relation between the non-Boltzmann effects and the chemical reaction rate is studied in the following experiment.

7.5. *Experiment 3*

The simulations of experiment 2 were repeated with rescaled chemical transition probabilities as given by the transformation (4.1) The effect of this scaling is to decrease the probability of reactions by a factor p^c. This scaling modifies the chemical rate law (6.8) but does not change its steady states. From a physical point of view, the transformation (4.1) controls the time scale separation between the elastic collisions that govern the diffusive properties of the model and the reactive collisions that are responsible for changes in the number of particles. Fig. 3 shows the ratio of the observed concentration to the stable steady-state concentration predicted by the rate law for various values of the scaling parameter p^c. As the chemical reactions become infrequent compared to elastic collisions the discrepancies decrease.

Similar behavior has been obtained using enhanced diffusion instead of a decreased chemical rate [7] (the diffusion was increased by modifying the time evolution operators as $(R \circ P)^n \circ C$ where n is an integer). An infinite diffusion coefficient has also been modeled by considering a cellular automaton where the rotation step was replaced by a stirring step which completely mixes the system at each time step. For this model, under the same experimental conditions, measured and predicted values of the steady state concentration agree to 5 digits [7].

In the light of this experiment, we propose an interpretation of the discrepancies in terms of reactive recorrelations: particles which have reacted can collide reactively again on nearby nodes before local diffusive relaxation can take place. It is likely that this effect is enhanced by the autocatalytic nature of the reactions that occur on the lattice.

This interpretation appears more clearly in a model where only one reaction is considered, the reaction $2X \to 3X$, and where rotation occurs at each time step ($p(\epsilon) = 1$). Assume a large finite system of volume

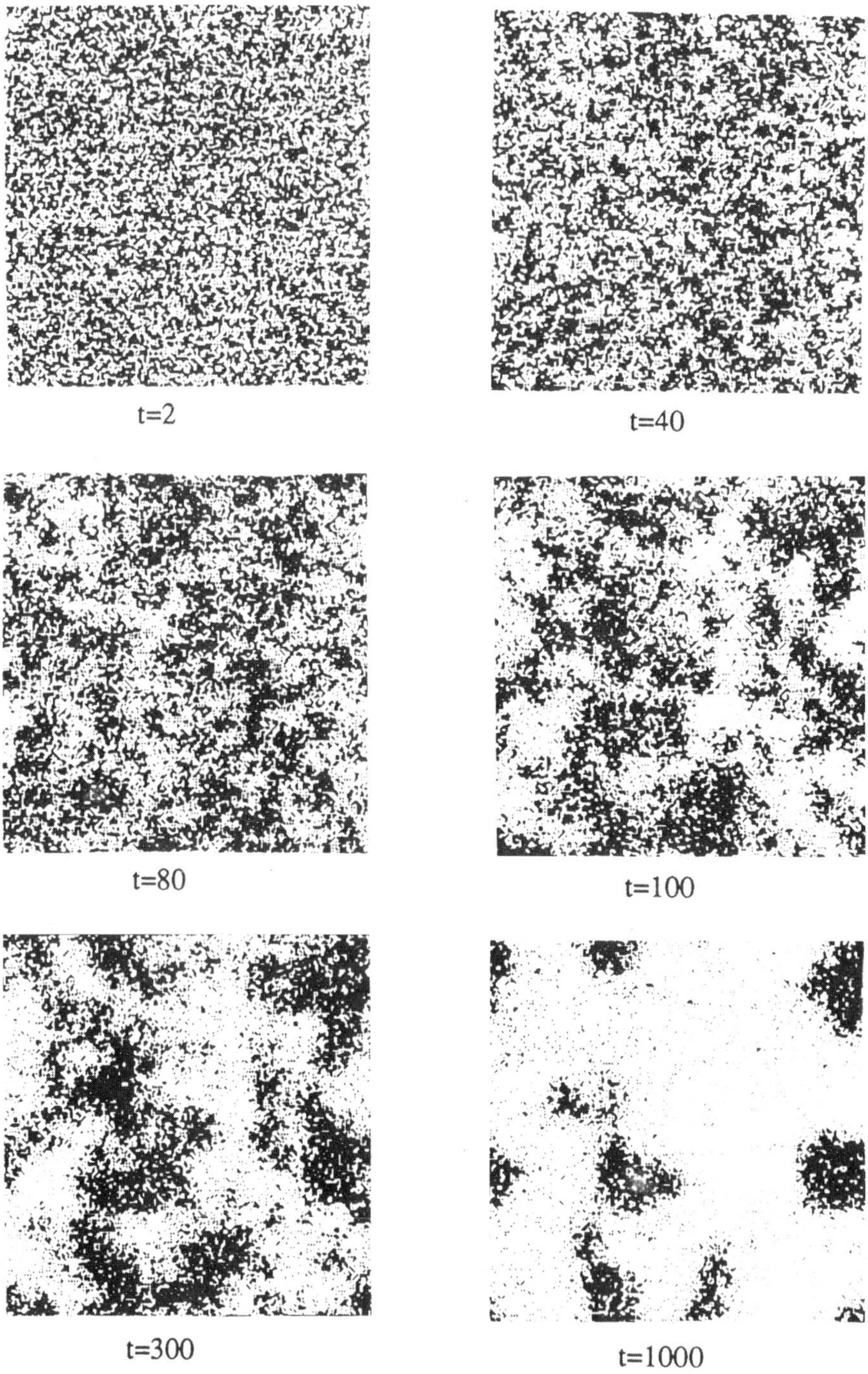

Fig. 1. Simulation of domain formation for the Schlögl model (a) with $\kappa_0 = 0.002$, $\kappa_1 = 0.039$, $\kappa_2 = 0.19$, and $\kappa_3 = 0.49$. Universe size: 256×256 nodes. Black dots represent nodes occupied by more than one particle.

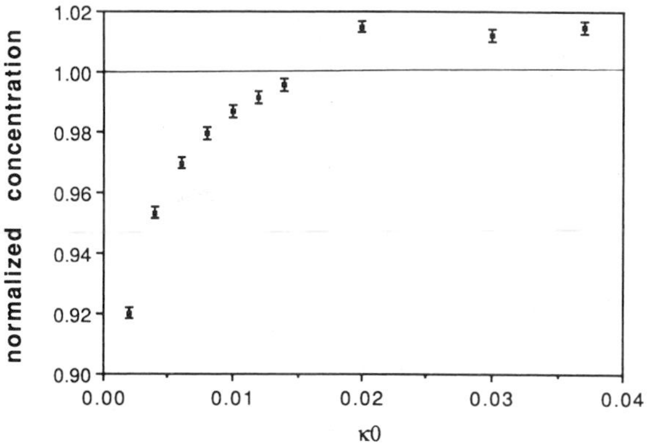

Fig. 2. Ratio of the observed steady-state concentration to the prediction of the phenomenological equation (6.8) as a function of κ_0. (64×64 nodes, periodic boundary conditions, $\kappa_1 = 0.039$, $\kappa_2 = 0.019$, $\kappa_3 = 0.49$, relaxation time=5000, averaging over 100000 time steps).

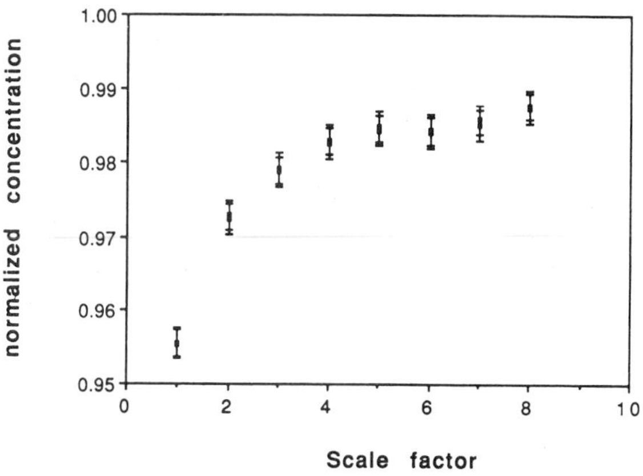

Fig. 3. Ratio of the observed steady-state concentration to the prediction of the phenomenological equation (6.8) as a function of the scale factor $(p^c)^{-1}$ (64×64 nodes, periodic boundary conditions, $\kappa_0 = 0.002$, $\kappa_1 = 0.039$, $\kappa_2 = 0.019$, $\kappa_3 = 0.49$, relaxation time=5000, averaging over 100000 time steps).

(area) V with periodic boundary conditions, and an initial state with only two particles in the system (on the same checkerboard subsystem). If one waits a sufficiently long time the two particles will collide and will be able to react. Assuming a reaction probability equal to one, the two particles will produce a third particle. Now the lattice configuration is very special: all the particles of the system (three) are on the same node. Starting from this configuration there is a probability close to one that two particles will again collide on nearby nodes in a few time steps producing a new particle and giving rise to another special configuration: three particles on the same node and a fourth on a nearby node. As shown by computer simulations, after the first collision an avalanche process generaly starts and the system is no longer governed by the reaction–diffusion equation. We observe that such cascade processes depend on the constant supply of ghost particles. In real systems the densities of other species that participate in the autocatalytic reactions may themselves be locally depleted or enhanced thereby influencing the growth of the number of X particles.

Assume now that the reaction probability is of the order of V^{-1}. In this case, when the two initial particles collide reactively and produce a third particle the configuration will again be very special. However the probability that the system will take advantage of this particular configuration to produce a fourth particle is close to zero as the three particles have time to diffuse in the whole system before a new reactive collision occurs. As a result, the reaction rates must be small enough if the occurrence of a chemical reaction is to be governed only by the local concentration, and not by special configurations that follow reactive events [16–18].

8. Concluding remarks

We have shown that a minimal CA scheme can be constructed to model a general class of reaction–diffusion systems for which we have developed a statistical mechanical lattice theory and performed CA simulations which support the validity of this new approach to the study of reactive systems. We note that some specific reaction models have been studied earlier using discrete models [19,20].

One of the main results of the present work is the observation that the validity of the phenomenological description of reaction–diffusion phenomena rests on a Boltzmann hypothesis: we find that when reactive collisions occur frequently deviations from the phenomenological prediction become important. These non-Boltzmann effects are interpreted in terms of reactive recorrelations. Obviously such "ring correlations" should be size dependent as indicated by the limiting cases considered in section 7.5; they must also depend on the dimension of the system: the larger the dimension (e.g. passing from 2D to 3D) the lower the recorrelation probability. Size and dimension effects (including reaction–diffusion on a fractal lattice) are now being further explored and correlation function measurements are being performed. Along the same lines another interesting problem is the existence of reaction–diffusion long-time correlations (conceptually similar to the persistence of hydrodynamically induced correlations [21]).

An important question has been the subject of an unsettled debate: what is the real origin of pattern formation in actual reaction–diffusion systems? It is obvious that fluctuations play a crucial role; however it not clear whether they are the only responsible factor. Indeed experimental "artefacts" (such as defects and boundaries) also come into play for triggering space- and time-dependent structures. Since the latter parameters can be controlled at will in model systems, the lattice gas automaton approach should prove very valuable to establish the relative importance of intrinsic fluctuations and extrinsic factors in reaction–diffusion systems. In order to pursue studies related to the above applications one must first examine the nature of the microscopic fluctuations in these reacting lattice gas models and determine to what extent they mimic those in real reacting systems.

On the basis of this approach, the basic scheme developed here can be extended and further generalized. The extension to three-dimensional reaction–diffusion systems is straightforward by passing from the square lattice to the cubic lattice. The inclusion of additional reactive species X, Y, Z, ... could be done without major difficulty, since the specification of the particular properties of the components can be implemented by exploiting the techniques used for "colored" lattice gas automata [3]. For instance with two species, one can treat the Brusselator model [22] and therefrom extend the analysis to a set of coupled Brusselators, whose coupling is the source of a rich variety of behaviors [23]. This program is presently in progress. The application of the basic reactive CA model to the widely used triangular lattice (FHP model [4]) allows extension to a class of reaction–diffusion equations with a source term of higher polynomial order (up to order 6). However the actual virtue of the lattice generalization is rather in that the symmetry of the triangular lattice conforms to the isotropy required for hydrodynamics [4], thereby allowing for the investigation of the coupling between reactive processes and hydrodynamic phenomena [24]. Further generalization along these lines would involve the three-dimensional versions of the model via the FCHC lattice [25] in order to explore the fascinating aspects of laminar, chaotic and turbulent reactive flows.

Acknowledgements

J.P.B. and D.D. acknowledge support by the Fonds National de la Recherche Scientifique (FNRS, Belgium) and the work of R.K. and A.L. was supported in part by grants from the Natural Sciences and Engineering Research Council of Canada. Part of the work reported here was supported by European Community Grant SC1-0212.

Appendix A

Scaling limit and the Boltzmann equation

In this appendix we provide additional details concerning the reduction of the automaton dynamics to a continuous space–time Boltzmann equation in the limit $\epsilon \to 0$. Two assumptions on the nature of the expected values are required for this reduction.

(i) With the initial distribution as given in (3.6) we expect that local equilibrium will hold; i.e., there exist functions $N_i(t, \boldsymbol{x})$ such that

$$E_\mu(\eta_i(k, \boldsymbol{r})) \mapsto N_i(t, \boldsymbol{x}), \tag{A.1}$$

where $\epsilon \to 0$, $\epsilon k \to t$ and $\epsilon \boldsymbol{r} \to \boldsymbol{x}$.

(ii) We also expect that for all $\boldsymbol{x} \in [-1, 1]^2$ and $t \geq 0$

$$\lim_{\epsilon \to 0} \sup_{F \in \mathcal{F}} \left| E_\mu \left(\prod_{i \in F} \eta_i(k, \boldsymbol{r} - \boldsymbol{c}_i) \right) - \prod_{i \in F} N_i(\epsilon k, \epsilon \boldsymbol{r}) \right| = 0, \tag{A.2}$$

where $\mathcal{F}$ denotes the set of all subsets of $\{1, \ldots, 4\}$ and $\lim_{\epsilon \to 0}$ denotes the limit as $\epsilon k \to t$ and $\epsilon \boldsymbol{r} \to \boldsymbol{x}$.

The derivation of the Boltzmann equation (3.7) can be carried out as follows. Let $g \in C_0^1([0, \infty))$ (without loss of generality $g(0) = 0$), and $f \in C_0^1(T^2)$. Consider

$$\int_0^\infty \iint \mathrm{d}t \, \mathrm{d}\boldsymbol{x} \, f(\boldsymbol{x}) g'(t) N_i(t, \boldsymbol{x}) = \lim_{\epsilon \to 0} \epsilon^2 \sum_{k=1}^{\infty} \sum_{\boldsymbol{r} \in \mathcal{L}_\epsilon} f(\epsilon \boldsymbol{r}) g(\epsilon k) \left[N_i(\epsilon(k-1), \epsilon \boldsymbol{r}) - N_i(\epsilon k, \epsilon \boldsymbol{r}) \right]. \tag{A.3}$$

From (3.1) and making use of the two assumptions (A.1) and (A.2) we have

$$\lim_{\epsilon \to 0} \epsilon^2 \sum_{k=1}^{\infty} \sum_{\boldsymbol{r} \in \mathcal{L}_\epsilon} f(\epsilon \boldsymbol{r}) g(\epsilon k) \{ [N_i(\epsilon(k-1), \epsilon \boldsymbol{r}) - N_i(\epsilon(k-1), \epsilon(\boldsymbol{r} - \boldsymbol{c}_i))] - \epsilon p [C_i^R(\tilde{N}) + C_i^C(\tilde{N})] \}$$

$$= -\lim_{\epsilon \to 0} \epsilon^2 \sum_{k=1}^{\infty} \sum_{\boldsymbol{r} \in \mathcal{L}_\epsilon} [f(\epsilon \boldsymbol{r} + \epsilon \boldsymbol{c}_i) - f(\epsilon \boldsymbol{r})] \, g(\epsilon k) N_i(\epsilon k, \epsilon \boldsymbol{r}) - \int_0^\infty \iint \mathrm{d}t \, \mathrm{d}\boldsymbol{x} \, f(\boldsymbol{x}) g(t) p [C_i^R(N) + C_i^C(N)]$$

$$= \int_0^\infty \iint \mathrm{d}t \, \mathrm{d}\boldsymbol{x} \, f(\boldsymbol{x}) g(t) \left\{ \boldsymbol{c}_i \cdot \nabla N_i(t, \boldsymbol{x}) - p [C_i^R(N) + C_i^C(N)] \right\}, \tag{A.4}$$

which completes the proof. In (A.4) $\tilde{N}_i = N_i(\epsilon(k-1), \epsilon(\boldsymbol{r} - \boldsymbol{c}_i))$.

References

[1] G. Nicolis and F. Baras, eds., Chemical Instabilities (Reidel, Dordrecht, 1984);
C. Vidal and A. Pacault, eds., Nonequilibrium Dynamics in Chemical Systems (Springer, Berlin, 1984).

[2] Oscillations and Traveling Waves in Chemical Systems, eds. R.J. Field and M. Burger (Wiley, New York, 1985).

[3] G. Doolen, ed., Lattice Gas Methods for Partial Differential Equations, (Addison–Wesley, Reading, MA, 1990).

[4] U. Frisch, D. d'Humières, B. Hasslacher, P. Lallemand, Y. Pomeau and J.P. Rivet, Complex Systems 1 (1987) 648.

[5] D. d'Humières, Y.H. Qian and P. Lallemand, in: Discrete Kinetic Theory, Lattice Gas Dynamics and Foundations of Hydrodynamics, eds. I.S.I. Monaco and R. Monaco (World Scientific, Singapore, 1989) pp. 102–113.

[6] F. Schlögl, Z. Phys. 253 (1972) 147;
F. Schlögl, C. Escher and R.S. Berry, Phys. Rev. A 27 (1983) 2698.

[7] D. Dab and J.-P. Boon, in: Cellular Automata and Modeling of Complex Physical Systems, ed. P. Manneville (Springer, Berlin, 1989) pp. 257–273.

[8] D. Dab, A. Lawniczak, J.-P. Boon and R. Kapral, Phys. Rev. Lett. 64 (1990) 2462.

[9] A. De Masi, R. Esposito, J.L. Lebowitz and E. Presutti, Commun. Math. Phys. 125 (1989) 127;
A. De Masi, R. Esposito, J.L. Lebowitz and E. Presutti, in: Discrete Kinetic Theory, Lattice Gas Dynamics and Foundations of Hydrodynamics, eds. I.S.I. Monaco and R. Monaco (World Scientific, Singapore, 1989);
A. De Masi, N. Ianiro, A. Pellegrinotti and E. Presutti, in: Studies in Statistical Mechanics, Vol. 11, eds. J.L. Lebowitz and E. Montroll (North-Holland, Amsterdam, 1984) p. 123.

[10] R. Zwanzig, in: Lectures in Theoretical Physics, Vol. 3 (Wiley, New York, 1961) p. 135;
H. Mori, Prog. Theor. Phys. 33 (1965) 423.

[11] R. Kapral, Adv. Chem. Phys. 48 (1981) 71 .
[12] C.D. Levermore and B.M. Boghosian, in: Cellular Automata and Modeling of Complex Physical Systems, ed. P. Manneville (Springer, Berlin, 1989) pp. 118–129.
[13] J.D. Gunton, M. San Miguel and P.S. Sahni, in: Phase Transitions and Critical Phenomena, Vol. 8, eds. C. Domb and J.L. Lebowitz (Academic, New York, 1983) p. 26.
[14] G.L. Oppo and R. Kapral, Phys. Rev. A. 36 (1987) 5810.
[15] S.M. Allen and J.W. Cahn, Acta Metall. 27 (1979) 1085.
[16] D. Ben Avraham, J. Stat. Phys. 48 (1987) 315; Science 241 (1988) 1620.
[17] M.A. Burschka, C.A. Doering and D. Ben Avraham, Phys. Rev. Lett. 63 (1989) 700.
[18] B. J. West, R. Kopelman and K. Lindenberg, J. Stat. Phys. 54 (1989) 1429.
[19] P. Grassberger, Z. Phys. B 47 (1982) 365.
[20] H.S. Berryman and D.R. Franceschetti, Phys. Lett. A 136 (1989) 348.
[21] A. Noullez and J.-P. Boon, Physica D 47 (1991) 212–215, these Proceedings;
M.A. van der Hoef and D. Frenkel, Physica D 47 (1991) 191–197, these Proceedings.
[22] I. Prigogine and R. Lefever, J. Chem. Phys. 48 (1968) 1695.
[23] I. Schreiber and M. Marek, Phys. Lett. A 91 (1982) 263;
I. Stuchl and M. Marek, J. Chem. Phys. 77 (1982) 2956.
[24] P. Clavin, P. Lallemand, Y. Pomeau and G. Searby, J. Fluid Mech. 188 (1989) 437;
V. Zehnle and G. Searby, J. Phys. (Paris) 50 (1989) 1083.
[25] D. d'Humières, P. Lallemand and U. Frisch, Europhys. Lett. 2 (1986) 291–297.

Physica D 47 (1991) 159–168
North-Holland

A STOCHASTIC CELLULAR AUTOMATON SIMULATION OF THE NON-LINEAR DIFFUSION EQUATION

Leesa BRIEGER[a] and Ernesto BONOMI[b]
[a]*Department of Mathematics, Ecole Polytechnique Fédérale de Lausanne, CH-1015 Lausanne, Switzerland*
[b]*GASOV Group, Ecole Polytechnique Fédérale de Lausanne, CH-1015 Lausanne, Switzerland*

Received 15 January 1990

In this article we investigate a cellular automaton simulation of the non-linear diffusion equation. The diffusion coefficient characterizing the equation is used to locally bias the random walks of particles on a square lattice. Emphasis is placed on respecting the massively parallel nature of the automaton model, while also correctly simulating the macroscopic behavior described by the equation. The result, a highly parallel algorithm, presents an interesting possibility for parallel and dedicated machines.

1. Introduction

The model of Brownian motion furnished by the random walk of a particle on a square lattice is governed by a simple local rule: displace the particle with equal probability in any one of the four directions on the lattice. The same rule, applied individually to each member in a population of particles and with an exclusion principle permitting at most one particle per site, simulates Brownian motion in the population. Such a model, discrete in space and in time, in which local or microscopic rules imposed on the individuals drive the macroscopic behavior of the system, is a cellular automaton. This is the basis for our stochastic automaton model of diffusion.

We follow the macroscopic evolution of the model by means of the balance equations (master equations), which derive from the description of the probability that a site is occupied by a particle. This probability is governed by the microscopic rules of evolution of the automaton. Depending on the rules of the model, the balance equations take the form of finite difference approximations of well-known macroscopic equations of mathematical physics, in which case the average behavior of the cellular automaton model respects these equations. The savoir faire consists in defining the right microscopic rules to correctly approximate the partial differential equations of interest (in this presentation, the diffusion equation), thereby defining a "non-numerical" tool for approximately solving them. The automaton then simulates the discretized differential equations, reproducing in space and time their solutions affected by the statistical noise which follows from the fluctuations of the microscopic dynamics (the configuration of the automaton at a given instant is a typical realization of the probabilistic behavior described by the balance equations). This noise can be attenuated in accordance with the law of large numbers.

The advent of parallel and dedicated machines renders cellular automata potentially powerful tools of simulation. To optimize the benefit of a parallel machine for a given problem, it is in general necessary to design an algorithm organized into concurrent tasks in such a way as to optimize the mapping between the logical structure of the problem and the architecture of the machine. In cellular automaton models, the evolution of the system takes place by discrete time steps on the basis of a local function, identical at all sites, which can be evaluated at all the sites simultaneously. Thus the model itself is finely grained, and the correspondence with massively parallel machines is immediate.

In section 2, we present the details of the model and demonstrate how the master equations are derived from a given set of rules that govern the evolution of the system. In section 3, we examine three alternative implementations and the corresponding master equations. In the first implementation, the asynchronous automaton, a single particle at a time is randomly chosen from the population and allowed to move. In the second implementation, a synchronous model, the entire population moves simultaneously, defining a parallel algorithm. Unless care is taken in the definition of the transition rules, however, this synchronicity of the model can induce an interaction between particles due to the competition for free sites. Then the resulting master equations have the form of the discretized diffusion equation perturbed by an error term. We examine this more closely in an investigation of the law of Fourier. Finally, in the third implementation, we arrive at an algorithm which is synchronous and parallel and simulates correctly the diffusion equation.

In section 4, we present the automaton model for simulating the non-linear diffusion equation. We remark that one application of this equation is to describe fluid flow in a porous medium, which takes place via various transport mechanisms, including capillary flow and molecular diffusion [1]. These mechanisms, combined and modelled as effective diffusion, give to the description of fluid flow the non-linear form of Fick's equation of diffusion,

$$\frac{\partial u}{\partial t} = \operatorname{div}(\mathscr{D} \operatorname{grad} u), \tag{1}$$

in which the effective diffusion coefficient $\mathscr{D}$ is a function of u, the solution. Assuming that such a diffusion coefficient, characterizing the fluid–matrix system, can be theoretically or experimentally determined, we implement it in the automaton model by using it to alter the particles' displacement probabilities. The result is a model which simulates inhomogeneous, non-linear diffusion. We compare the results of this model to a finite element solution of the equation for a given diffusion coefficient $\mathscr{D}(u)$.

2. The one-particle model

The random walk of a particle on a square lattice defines a probability distribution, $P(r_{ij}, t_n)$, discrete in time and space, which is just the probability that position r_{ij} is occupied by the Brownian particle at time t_n (time step n). The process is characterized by the mesh size Δx and the time interval between subsequent events, Δt. $P(r_{ij}, t_n)$ evolves diffusively on the lattice, and as Δx and Δt approach zero, keeping the ratio $D = \Delta x^2/4\,\Delta t$ constant, $P(r_{ij}, t_n)$ approaches $P(r, t)$, the solution of the continuous diffusion equation (Fick's law of diffusion) [2]:

$$\frac{\partial P(r,t)}{\partial t} = D\nabla^2 P(r,t), \tag{2}$$

in which D is the diffusion coefficient and ∇^2 indicates the Laplacian operator. The diffusive character of the displacement rule is seen clearly in the master equations for the model. We let P_r^n denote the probability of occupation by the particle at site $r = r_{ij}$, at time step n. The process is Markovian and the probability that the particle occupies site r at time t_{n+1} is given as a function of the neighboring probabilities at time t_n by the following expression:

$$P_r^{n+1} = \sum_{q \in \mathscr{N}(r)} p_q^n W_{qr}^n. \tag{3}$$

$\mathscr{N}(r)$ is the five-site neighborhood of r consisting of r and its nearest neighbors on the lattice, and W_{qr}^n denotes the conditional probability of moving from site q to site r at time t_n. $W_{qr}^n = 0$ whenever

q and r are not in the same neighborhood, and the following conservation condition (conservation of mass) is respected:

$$\sum_{r \in \mathcal{N}(s)} W_{sr}^n = 1. \tag{4}$$

Solving for W_{rr}^n in (4), we use this to rewrite (3) and obtain the following description for the evolution of P:

$$P_r^{n+1} - P_r^n = \sum_{\substack{q \in \mathcal{N}(r) \\ q \neq r}} P_q^n W_{qr}^n - P_r^n \sum_{\substack{q \in \mathcal{N}(r) \\ q \neq r}} W_{rq}^n. \tag{5}$$

In the one-particle random walk model, the conditional probabilities for $q \in \mathcal{N}(r)$ are the following:

$$W_{qr}^n = \tfrac{1}{4}(1 - \delta_{qr}), \tag{6}$$

Consequently, expression (5) takes the form

$$P_r^{n+1} - P_r^n = \frac{1}{4}\left(\sum_{\substack{q \in \mathcal{N}(r) \\ q \neq r}} P_q^n - 4P_r^n \right), \tag{7}$$

which is a stable explicit finite difference representation of the diffusion equation (2) [3], using the central difference approximation to the second derivative and in which

$$D = \Delta x^2 / 4\,\Delta t.$$

Thus the random walk model reproduces the finite difference approximation of the diffusion equation (2) on the grid. As the discretization goes to zero, with D constant, the finite difference solution converges to the solution of the continuous equation.

3. Alternative implementations and parallelism

In our model we adapt the stochastic rule to a population of particles. Each particle chooses at random one of the directions on the square lattice, targeting one of its four nearest neighbors. Respecting an exclusion principle which permits at most one particle per site, a particle moves to its targeted site if this site is free and simply does not move if the site it has targeted is already occupied. This is the strategy adopted in the following models.

In the asynchronous model, particles are treated one at a time, chosen at random from the population on the N sites of the configuration. The evolution of probability P is given by (5), due to the Markovian nature of the process, where the time interval from t_n to t_{n+1} marks the occurrence of a single event, i.e. the stochastic displacement rule applied at a single site. Here the probability of transition of a particle from q to r at time t_n is as follows:

$$W_{qr}^n = \frac{1}{4N}(1 - P_r^n), \quad q \in \mathcal{N}(r), q \neq r. \tag{8}$$

Substituting these probabilities into (5), we derive the following balance equation:

$$P_r^{n+1} - P_r^n = \frac{1}{4N}\left(\sum_{\substack{q \in \mathcal{N}(r) \\ q \neq r}} P_q^n - 4P_r^n \right). \tag{9}$$

This is just the finite difference form of the diffusion equation (2) with

$$D = \Delta x^2 / 4N\Delta t.$$

We conclude that in the asynchronous automaton, the particle population diffuses on the lattice, and the model reproduces the solution of the discrete approximation of diffusion equation (2), Fick's equation. This implementation does not have the advantage of being parallel, and we consider a synchronous alternative.

In the synchronous model, all particles move simultaneously and a time step Δt represents one configuration update. If each particle's random walk is governed independently, then conflicts between particles competing for a single free site must be resolved. In our implementation, when more than one particle attempt to move to a single open site, only one of the competing particles, chosen randomly from among them, is allowed to move to the unoccupied site, and the others do not move. This induces interactions between particles beyond just the immediate neighborhood. To characterize the resulting correlations, we use the transition probabilities which can be written as follows:

$$W_{qr}^n = \tfrac{1}{4}(1 - P_r^n) A_{qr}^n, \quad q \in \mathcal{N}(r), q \neq r, \tag{10}$$

where A_{qr}^n represents the conditional probability that the transition from q to r is admissible, that is, that site r accepts the particle from site q, at time step n. With these probabilities of transition, eq. (5) becomes

$$P_r^{n+1} - P_r^n = \frac{1}{4}\left(\sum_{\substack{q \in \mathcal{N}(r) \\ q \neq r}} P_q^n A_{qr}^n - P_r^n \sum_{\substack{q \in \mathcal{N}(r) \\ q \neq r}} A_{rq}^n \right) + \tfrac{1}{4} P_r^n \sum_{\substack{q \in \mathcal{N}(r) \\ q \neq r}} P_q^n \left(A_{rq}^n - A_{qr}^n \right). \tag{11}$$

If the automaton rule for the synchronous implementation were to give rise to a probability function A such that everywhere on the lattice $A_{qr}^n = A_{rq}^n$, then the term in large parentheses would be diffusive with a diffusion coefficient represented by A (see section 3), and the second term would vanish, reproducing the finite difference form of Fick's equation.

To illustrate the disparity between Fick's diffusion and this synchronous model, we consider in detail the one-dimensional case in which particles move left or right with probability $\frac{1}{2}$, respecting the exclusion principle. The one-dimensional neighborhood of site x_j consists of sites x_{j-1}, x_j and x_{j+1}. The transition probabilities W_{ij} corresponding to (10) are the following:

$$W_{ij}^n = \tfrac{1}{2}(1 - P_j^n) A_{ij}^n, \quad i = j-1, j+1, \tag{12}$$

where

$$A_{j-1\,j}^n = 1 - \tfrac{1}{4} P_{j+1}^n,$$

$$A_{j+1\,j}^n = 1 - \tfrac{1}{4} P_{j-1}^n.$$

The dynamics are still Markovian and eq. (5) describes the probability evolution, which takes the following form with the transition probabilities of (12):

$$P_j^{n+1} - P_j^n = \tfrac{1}{2}\left(P_{j+1}^n - 2P_j^n + P_{j-1}^n \right) + E_j^n, \tag{13}$$

where

$$E_j^n = \tfrac{1}{8}\left[P_{j-2}^n (1 - P_{j-1}^n) P_j^n - 2 P_{j-1}^n (1 - P_j^n) P_{j+1}^n + P_j^n (1 - P_{j+1}^n) P_{j+2}^n \right].$$

Eq. (13) is the one-dimensional finite difference form of the diffusion equation (2) with

$$D = \Delta x^2 / 2\,\Delta t$$

plus the "correction" term E_j^n, which, for a uniform population (P constant), is zero. The effect of this extra term on the simulation is illustrated in an example of one-dimensional diffusion at equilibrium, fig. 1. The diffusion problem

$$\frac{\partial u}{\partial t} = \frac{1}{2} \frac{\partial^2 u}{\partial x^2} \quad \text{with } u(0,t) = 0,\ u(1,t) = 1 \tag{14}$$

has an equilibrium solution $u(x,t) = x$ on the unit interval (law of Fourier). Constraining the automaton dynamics to one dimension, we use the synchronous model to conduct 100 simultaneous simulations on the two-dimensional lattice to study this problem. The particles are initially distributed so as to correspond to the equilibrium solution of problem (14), the dashed line in fig. 1. The steady state of the model appears in fig. 1: the smooth curve is the fixed point of the numerical iteration defined by (13), with $P(x,0) = x$, and the "experimental" data are the measurements of

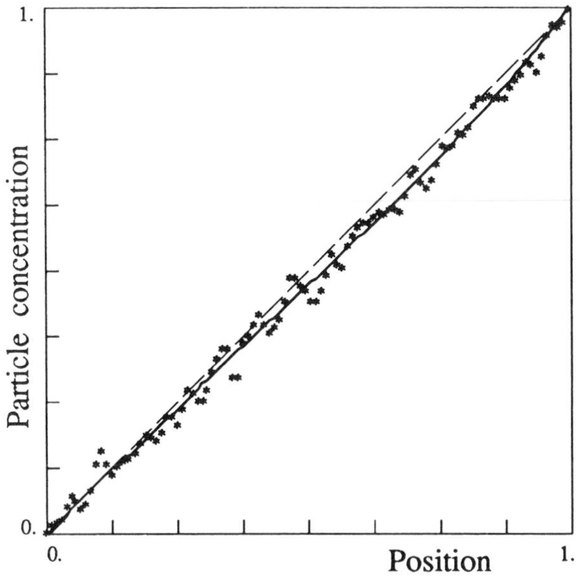

Fig. 1. Equilibrium in the non-synchronized automaton model. The straight line is the steady-state solution of problem (14) (law of Fourier). The solid curve is the expected equilibrium of this synchronous automaton, the fixed point of the numerical iteration defined by eq. (13). The asterisks indicate the automaton concentration measurements at equilibrium.

the particle distribution averaged over the 100 automaton experiments, at time step 20000. Equilibrium for the synchronous system is not the straight line equilibrium solution of (14), nor does it approach the straight line as the discretization is made increasingly fine. The effect of the competition for free sites is to introduce a form of hard-core interactions and to slightly inhibit the diffusion.

Now, consider the synchronous model in which the decisions governing the random walks of the particles are not independent, but are linked such that the particles move in synchronization. In this case, competition between particles for the free sites does not arise and consequently the probabilities A^n_{qr} described above are identically 1. In a one-dimensional model, the particles, in unison, choose a direction of movement, either to the right or to the left, with equal probability $\frac{1}{2}$. The exclusion principle described above is respected. Thus during the time interval from t_n to t_{n+1}, a particle is prohibited from moving to a site which is already occupied at moment t_n, inhibiting the simple translation of neighboring particles. The master equations are of the following form (the discretized diffusion equation):

$$P_j^{n+1} - P_j^n = \tfrac{1}{2}\left(P_{j+1}^n - 2P_j^n + P_{j-1}^n\right). \qquad (15)$$

Implementing this algorithm for problem (14) on 100 independent systems (on the 100×100 lattice, with 1D dynamics on each row) and averaging over the ensemble of the experiments, gives the results shown in fig. 2a, with the corresponding configuration "snapshot" of fig. 2b.

In a two-dimensional model, the algorithm is the following: choose either vertical or horizontal motion, each with probability $\frac{1}{2}$, thus constraining the movement of the particles either to columns or to rows. If the chosen movement is vertical (horizontal), then on each column (row) choose between moving up or down (left or right), with equal probability $\frac{1}{2}$. The resulting master equation has just the correct form of the two-dimensional discretized diffusion equation:

$$P_r^{n+1} - P_r^n = \frac{1}{4}\left(\sum_{\substack{q \in \mathcal{N}(r)\\ q \neq r}} P_q^n - 4P_r^n\right). \qquad (16)$$

Results for this parallel automaton model, motivated by problem (14), are depicted in fig. 3. In fig. 3a we see the one-dimensional representation of the particle concentration of a single two-dimensional system. The ensemble average is approximated by the column-by-column average on the configuration, depicted in fig. 3b.

We shall refer to this model as the synchronized model, given that the particles' movements are synchronized as well as synchronous.

4. Non-linear diffusion

Eq. (2) is an example of Fick's law of diffusion in the special case of a constant coefficient D. Non-uniform diffusion, as in an inhomogeneous

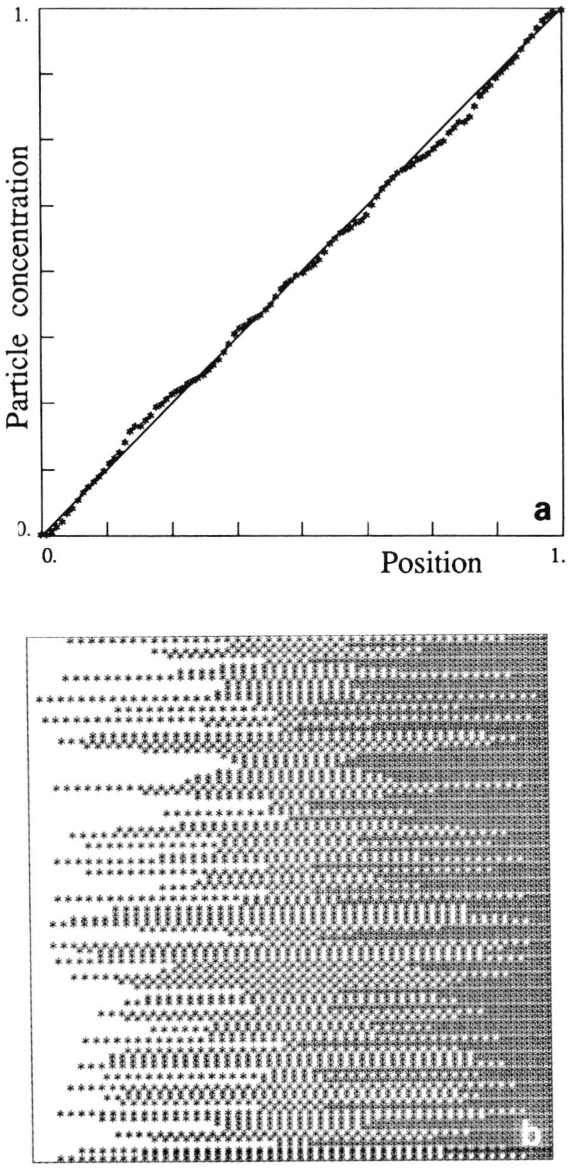

Fig. 2. Equilibrium in the 1D synchronized automaton model (law of Fourier). The straight line is the expected equilibrium for this automaton. In (a) the asterisks indicate the average over the ensemble of the systems at time step 20 000. (b) shows the corresponding configuration "snapshot".

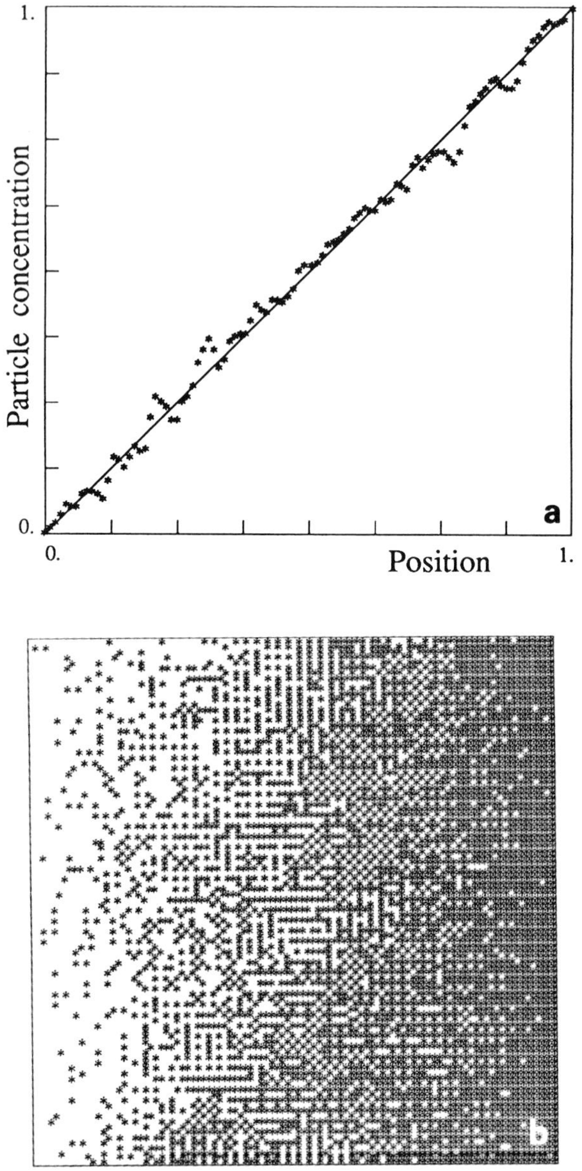

Fig. 3. Equilibrium in the 2D synchronized automaton model (law of Fourier). The straight line represents the automaton equilibrium along the x-axis. In (a) the asterisks indicate the one-dimensional representation of the particle concentration in a single two-dimensional system. The particle concentration is approximated by the column-by-column average on the configuration "snapshot" shown in (b).

porous medium, can also be modelled by Fick's equation:

$$\frac{\partial u}{\partial t} = \operatorname{div}(\mathscr{D} \operatorname{grad} u) \tag{17}$$

in which the coefficient $\mathscr{D}$ is a positive function of position or of the solution u itself. The effects of the porosity are described by an effective diffusion coefficient which captures the phenomenology of the system and governs the model accordingly. The coefficient can be supplied either from theoretical considerations or from experimental observations, and when it is a function of the solution u, eq. (17) is non-linear. Examples of such modelling are furnished in studies of drying and reaction in concrete [4, 5].

In our simulation of non-linear diffusion, we do not attempt to recreate the microscopic details which can influence diffusive behavior. Simply assuming the diffusion coefficient given, we use it to influence the displacement probabilities in the automaton in such a way as to reproduce the effect of the coefficient and the corresponding behavior of eq. (17).

The idea is the following. Suppose that in a one-dimensional model, where originally a particle targeted one of the neighboring sites with probability $\frac{1}{2}$, it now chooses its displacement direction with probability $\frac{1}{2}D$, where D is a given function evaluated on the lattice. D_j^n denotes the value of D at site x_j at moment t_n, and we define D_{ij}^n as the arithmetic average of D_i^n and D_j^n, $D_{ij}^n = \frac{1}{2}(D_i^n + D_j^n)$. Thus we define D on the lattice edges, and a particle at site x_i moves in the direction of neighboring site x_j with a displacement probability of $\frac{1}{2}D_{ij}^n$. Notice that $0 \le D \le 1$ must hold since we use D as a probability. D is constant on lattice edges, $D_{ij}^n = D_{ji}^n$, and since the dynamics are those of the synchronized model described above, the transition probabilities are the following:

$$W_{ij}^n = \tfrac{1}{2}D_{ij}^n(1 - P_j^n), \quad i = j-1, j+1. \tag{18}$$

As before, eq. (5) describes the evolution of the resulting probability distribution at a site x_j and, with the probabilities of (18), takes the following form:

$$P_j^{n+1} - P_j^n = \frac{1}{2}\left(\sum_{\substack{i \in \mathscr{N}(j) \\ i \neq j}} P_i^n D_{ij}^n - P_j^n \sum_{\substack{i \in \mathscr{N}(j) \\ i \neq j}} D_{ij}^n \right). \tag{19}$$

Eq. (19) is a finite difference formulation of eq. (17) in one dimension, with

$$\mathscr{D} = (\Delta x^2 / 2\,\Delta t) D.$$

Thus for a given function $\mathscr{D}$ governing the diffusive behavior in eq. (17), we can implement it, to within a scaling constant, and reproduce the same behavior in the automaton. When $\mathscr{D}$ is a function of the solution u, D must be evaluated as a function of the occupation probability P.

Fig. 4 shows a comparison between synchronized automaton results and a finite element solution [5] for the non-linear diffusion equation (17) in one dimension, with the coefficient $\mathscr{D}$ a function of u as shown. The one-dimensional automaton dynamics are once again implemented in 100 independent systems in such a way as to impose a fixed boundary condition along the border $x = 1$ and a zero flux condition along the other border at $x = 0$. The smooth curves show the finite element solution of the one-dimensional equation at the indicated times, and the experimental points are the automaton concentrations (the ensemble average), measured at the same moment of simulated time.

Since D is a function of the occupation probability P, at each site x_j and time t_n, we must approximate this probability by a particle concentration measured in the automaton. One possibility is to approximate P by the average occupation number in a neighborhood of each site. While this respects the strictly local nature of a cellular automaton, the approximation of the solution and thus of the diffusion coefficient D may be deficient. A more reliable possibility, which we have

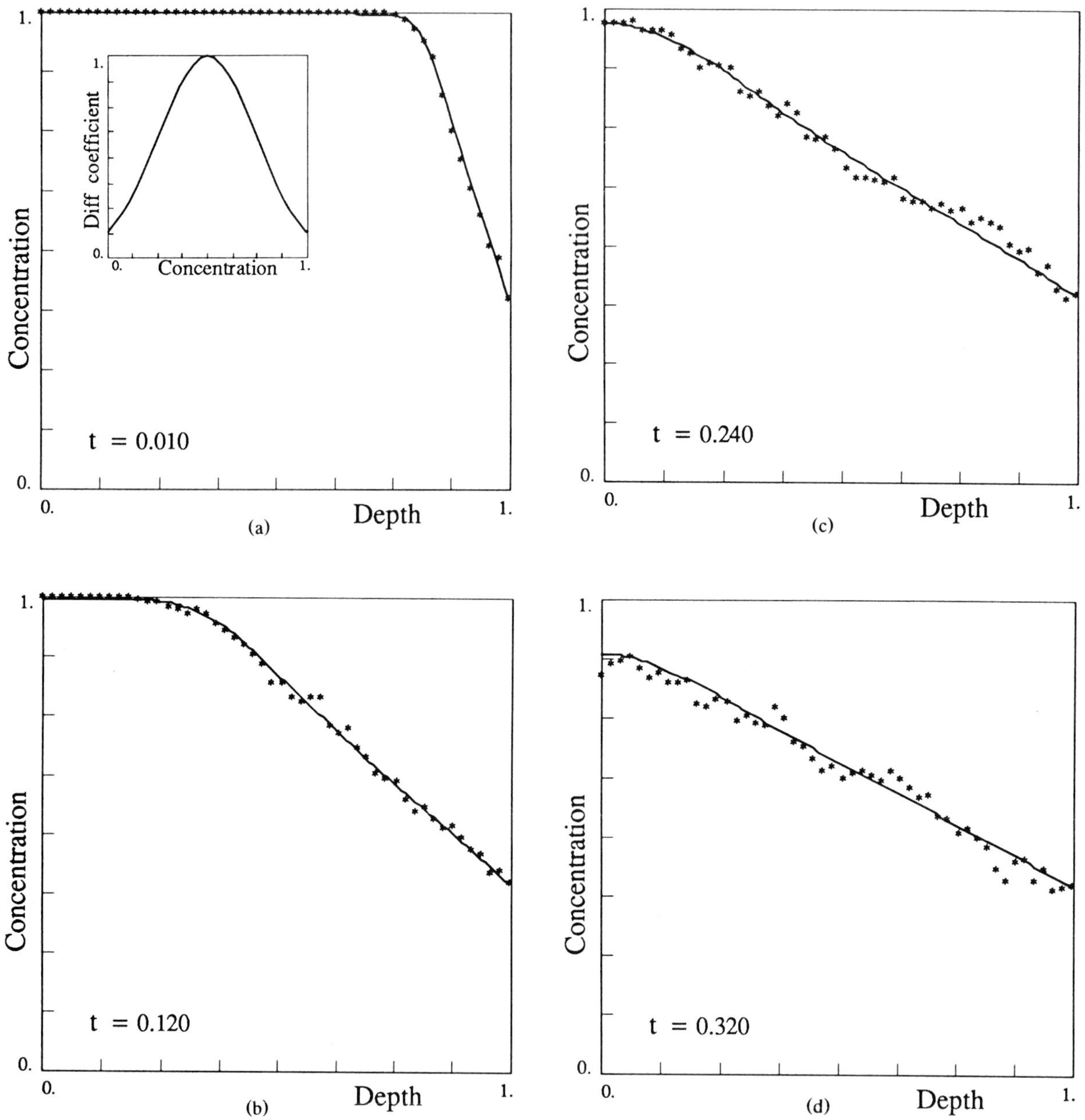

Fig. 4. The comparison between synchronized automaton results and a finite element solution for the one-dimensional non-linear diffusion equation (17), with coefficient $\mathscr{D}(u)$ as shown. The smooth curves show the finite element solution of the one-dimensional equation at the indicated times, and the experimental points are the automaton concentrations (the ensemble average), measured at the same moment of simulated time.

chosen for this presentation, is to use ensemble averages over equivalent systems to approximate the probability P. This obliges that the global statistics of the ensemble be collected and injected into the model at each time step, at the cost of the local character of the dynamics.

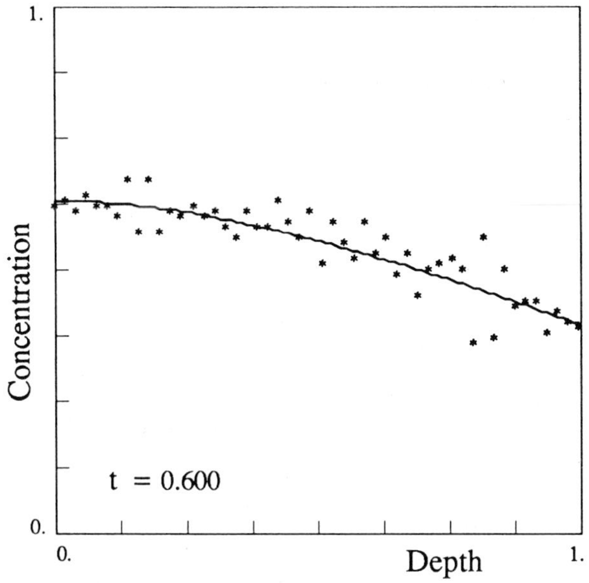

(e)

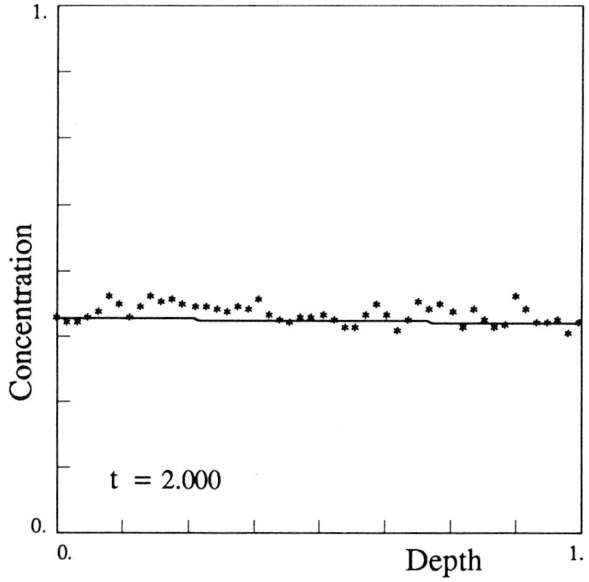

(f)

Fig. 4. Continued.

Such a comparison as that of fig. 4 necessitates that we introduce units into the heretofore dimensionless automaton. The explicit finite difference form of eq. (17) for arbitrary $\mathscr{D}$ can be written as follows:

$$u_j^{n+1} - u_j^n = \frac{k\,\Delta t}{\Delta x^2}\left(\sum_{\substack{i\in\mathscr{N}(j)\\ i\neq j}} u_i^n D_{ij}^n - u_j^n \sum_{\substack{i\in\mathscr{N}(j)\\ i\neq j}} D_{ij}^n\right), \tag{20}$$

where k is the normalization constant so that $\mathscr{D} = kD$, $0 \le D \le 1$. If Δt is given by the following relation

$$\Delta t = \Delta x^2/2k \tag{21}$$

and D is the function implemented in the simulation, then eq. (20) corresponds exactly to the balance equation (19) of the automaton. This time interval Δt represents the real time elapsed during one configuration update in the automaton.

5. Conclusions

The automaton model presented above simulates the solution of Fick's equation, providing a stochastic algorithm for solving approximately this partial differential equation. More generally, whenever a given partial differential equation can be fitted with an automaton model whose master equations adequately approximate it, we can use the automaton dynamics to simulate the solution. The validity of this approach for a given problem depends not on the stochastic nature of the system studied but on our ability to formulate the problem so that it can be solved by making use of random numbers in a discrete system. These models constitute a class of massively parallel algorithms, whose potential becomes more tantalizing with the advent of parallei and dedicated machines.

There is a large category of recalcitrant problems, typically simulations of nonstationary complex physical phenomena, for which automaton models hold promise. One such example is found in the study of the carbonation of concrete, a problem of reaction–diffusion in a porous medium [6]. Cellular automata represent discrete idealizations of physical systems in which the state of each site, permitted a finite number of admissible values, evolves according to a set of rules. The direct and exact implementation of such a model dispenses with the need to define an associated numerical solution scheme, and the task of constructing or modifying a simulation is accordingly facilitated. Moreover, since the results of the simulation are not filtered through an accompanying numerical scheme but are the direct image of the model, the problem of evaluating the validity of the model is straightforward.

Acknowledgement

One of the authors (E.B.) was supported by the Swiss National Science Foundation.

References

[1] F.A.L. Dullien, Porous Media – Fluid Transport and Pore Structure (Academic Press, New York, 1979).
[2] W. Feller, An Introduction to Probability Theory and its Applications (Wiley, New York, 1964).
[3] J.L. Lions, Cours d'analyse numérique, Département de Mathématiques Appliquées, Éditions de l'Ecole Polytechnique, Palaiseau, France (1984).
[4] Z.P. Bazant and L.J. Najjar, Cement Concrete Res. 1 (1971) 461.
[5] L.M. Brieger and F.H. Wittmann, in: Proceedings of the International Colloquium Materials Science and Restoration, Esslingen (1986) p. 635.
[6] L.M. Brieger and E. Bonomi, J. Comput. Phys., to appear.

Physica D 47 (1991) 169–188
North-Holland

MULTISPECIES 2D LATTICE GAS WITH ENERGY LEVELS: DIFFUSIVE PROPERTIES

D. BERNARDIN, O.E. SERO-GUILLAUME and C.H. SUN
LEMTA UA CNRS 875, 24-30 Rue Lionnois, B.P. 3137, Nancy Cedex, France

Received 15 February 1990
Revised manuscript received 22 May 1990

We consider two particular applications of a multispecies, multispeed lattice gas with energy levels. In the first one, we study the mass diffusion properties of a model where the collisions preserved the partial masses. In the second one, we are looking at heat diffusion for a model where the total mass, momentum and energy are the only conserved quantities. The diffusion equations are derived by the Chapman–Enskog method and some numerical simulations are presented.

1. Introduction

In order to simulate two-dimensional diffusion processes with lattice gases we have proposed in ref. [1] to introduce particles with different speed and masses, which generalise earlier works on lattice gases with coloured particles (see for example refs. [2, 3]). These models can also yield a true energy equation, but the possible nontrivial collisions are limited by the microscopic conservation of kinetic energy. To remove this restriction we have introduced in ref. [4] what we have called energy levels, which are massless particles but without velocity contrary to the "photons" [5]. The advantage is, as for the models with rest particles (see for example refs. [6, 7]), that only a minimum of supplementary bits is introduced. In this paper we complete the studies of refs. [1, 4] and present a simulation of a mass diffusion process without convection and of a thermal relaxation experiment for a model with temperature. In sections 2 and 3 we describe the models and give the general expression of equilibrium populations. In section 4 we detailed these equilibrium populations for the two gases that we have simulated, while in section 5 we briefly show how these models can also account for simple chemical kinetics. Finally, in section 6 we give the expressions of the diffusivities obtained from a Chapman–Enskog procedure and we present the results of the simulations.

2. Description of the models

The hexagonal lattice, denoted by $\mathscr{R}$, is generated by the two vectors $\boldsymbol{e}_1$ and $\boldsymbol{e}_2$, the length of the mesh is Δx. A particle of kind ν, $\nu = 0,\dots,p-1$ is characterised by its mass m_ν and its speed V_ν, it has six possible velocities: $\boldsymbol{c}_{j\nu} = r^j(V_\nu)$, where r is the rotation of angle $\pi/3$. We impose that V_ν takes one of the two following forms: $V_\nu = p_\nu \boldsymbol{e}_1/\Delta t$ or $V_\nu = p_\nu(\boldsymbol{e}_1+\boldsymbol{e}_2)/\Delta t$, which insures that the set of velocities is invariant under the symmetry group of the lattice; p_ν is an integer and Δt is the time step. At each node of the lattice we also associate a scalar ϕ which has the dimension of an energy and takes one of the n

0167-2789/91/$03.50 © 1991 – Elsevier Science Publishers B.V. (North-Holland)

values: $\phi_1 < \phi_2 < \ldots < \phi_n$. The state of a node is described by a vector $\boldsymbol{X}$ of $\mathbf{E} = \{0,1\}^{6p+n} \cap \{\Sigma_i X_{6p+i} = 1\}$, where $X_{j+6\nu}$ stands for the presence of a particle of kind ν with velocity $\boldsymbol{c}_{j\nu}$, the X_{6p+i} stand for the energy levels. For $n = p = 1$ one recovers the FHP-I model, a minimum model for mass diffusion corresponds to $p = 2$, $n = 1$ and a realistic model with thermodynamics is obtained with $p = n = 2$.

A configuration of the medium is an element ω of the set $\Omega = \mathbf{E}^{\mathscr{R}}$, where $\omega(\alpha)$ is the state of node α. The evolution rules are the usual rules for lattice gases: after a collisional stage, particles propagate in the direction of their velocities, but the energies ϕ_i are not convected. At each node collisions preserve, at least the total mass $\Sigma_{j,\nu} X_{j+6\nu} m_\nu$, the total momentum $\Sigma_{j,\nu} X_{j+6\nu} m_\nu \boldsymbol{c}_{j\nu}$, the total energy $\Sigma_{j,\nu} X_{j+6\nu} m_\nu V_\nu^2/2 + \Sigma_i X_{6p+i}\phi_i$ and eventually other physical quantities such as partial masses or number of atoms, depending on the macroscopic model we want to obtain.

We assume that the collisional process is independent of the evolution of the automaton, and we shall let $\zeta(\boldsymbol{X} \to \boldsymbol{Y})$ be the local transition probability from a state $\boldsymbol{X}$ to a state $\boldsymbol{Y}$ and let $\zeta(\omega \to \omega') = \Pi_{\mathscr{R}} \zeta(\omega(\alpha) \to \omega'(\alpha))$ be the global transition probability from a configuration ω to a configuration ω'. We shall assume that the transition probabilities are invariant under the symmetry group of the lattice and obey a semi-detailed balance, i.e.

$$\sum_{\boldsymbol{X}} \zeta(\boldsymbol{X} \to \boldsymbol{Y}) = \sum_{\boldsymbol{Y}} \zeta(\boldsymbol{X} \to \boldsymbol{Y}). \tag{1}$$

If we want to simulate a chemical process, one can introduce the following description. We consider that the gas is composed of chemical species, and for simplicity we assume that a chemical species is represented by particles of only one kind ν, which we shall call S_ν. Thus there are p different chemical species, and at a given node there are no more than six particles of a given species. Each species is composed of atoms $A^1, A^2, \ldots, A^a$ and we note:

$$S_\nu = A^1_{k_{\nu 1}} A^2_{k_{\nu 2}} \ldots A^a_{k_{\nu a}}.$$

We shall suppose that $a \le p$ and that the rank of $[k_{\nu i}]$ is a. We then consider that there are $p - a$ independent reactions given by a stoichiometric matrix $[\lambda_{\nu j}]$ for $\nu = 0, \ldots, p-1$ and $j = 1, \ldots, p-a$, such that we have

$$\sum_\nu \lambda_{\nu j} k_{\nu i} = 0 \quad \forall i, j.$$

We then assume that the collisions preserve the number of each kind of atom: $\Sigma_{j,\nu} X_{j+6\nu} k_{\nu i}$ and thus the total mass is not an independent conserved quantity.

We recover a diffusion model [1] if we take $a = p$ and $[k_{\nu i}] = \mathbf{Id}$. The energy levels become stationary and independent from the dynamics of the particles. A thermal nonreactive model [4] is also obtained by setting $a = 1$ and $k_{\nu i} = 1$ for each ν.

A collisional invariant is a vector $\boldsymbol{I}$ of $\mathbf{R}^{6p+n}$ which satisfies

$$\forall \boldsymbol{X}, \boldsymbol{Y} \in \mathbf{E}^2 \colon \zeta(\boldsymbol{X} \to \boldsymbol{Y}) \sum_j I_j (Y_j - X_j) = 0.$$

The set $\mathscr{I}$ of all collisional invariants is a linear subset of $\mathbf{R}^{6p+n}$. We shall assume that the invariants associated with the conservation of atoms, momentum and energy form a basis of $\mathscr{I}$. In fact there is a supplementary invariant associated with the relation $\Sigma_i X_{6p+i} = 1$ which expresses that ϕ_i takes only one of its n possible values.

3. Statistical description and equilibrium distribution

We are interested in one-parameter families ($\wp(t)$) of probability distributions on Ω which obey the so-called Liouville equation:

$$\wp(t+\Delta t)(s(\omega)) = \sum_{\omega' \in \Omega} \wp(t)(\omega')\,\zeta(\omega' \to \omega), \tag{2}$$

which is identical to that proposed in ref. [8] for monocomponent gases. The shift operator s is defined for all α and ω by

$$[s(\omega)]_{j+6\nu}(\alpha) = \omega_{j+6\nu}(\alpha - \Delta t\, \boldsymbol{c}_{j\nu}),$$

$$[s(\omega)]_{6p+i}(\alpha) = \omega_{6p+i}(\alpha).$$

We then introduce the mean populations per kind and direction:

$$N_{j\nu}(\alpha, t) = \sum_{\omega \in \Omega} \wp(t)(\omega)\,\omega_{j+6\nu}(\alpha), \tag{3}$$

and the probabilities per energy level:

$$E_i(\alpha, t) = \sum_{\omega \in \Omega} \wp(t)(\omega)\,\omega_{6p+i}(\alpha). \tag{4}$$

In the following we shall call $\boldsymbol{N}$ the vector of $\mathbf{R}^{6p+n}$ whose components are the $N_{j\nu}$ and the E_i. We shall use the following mean quantities: $\rho = \Sigma_{j,\nu} N_{j\nu} m_\nu$, $\rho_\nu = \Sigma_j N_{j\nu} m_\nu$, $n_i = \Sigma_{j,\nu} N_{j\nu} k_{\nu i}$, $\rho \boldsymbol{v} = \Sigma_{j,\nu} N_{j\nu} m_\nu \boldsymbol{c}_{j\nu}$, $e = \Sigma_{j,\nu} N_{j\nu} m_\nu V_\nu^2/2 + \Sigma_i E_i \phi_i$, and $u = e - \frac{1}{2}\rho \boldsymbol{v}^2$, which are respectively the total and partial densities of mass, the densities of atoms, of momentum, and of total and internal energy. By definition a natural state variable of the system associated with the invariant $\boldsymbol{I}$ will be the mean quantity: $\Sigma_{j,\nu} N_{j\nu} I_{j+6\nu} + \Sigma_i E_i I_{6p+i}$. Thus $\rho, n_i, \rho\boldsymbol{v}, e$ are always state variables but not necessarily independent.

The $N_{j\nu}$ and the E_i obey the following evolution equations:

$$N_{j\nu}(\alpha + \Delta t\, \boldsymbol{c}_{j\nu}, t + \Delta t) - N_{j\nu}(\alpha, t) = \Delta_{j+6\nu}(\alpha, t),$$

$$E_i(\alpha, t + \Delta t) - E_i(\alpha, t) = \Delta_{6p+i}(\alpha, t), \tag{5}$$

where

$$\Delta_k = \sum_{\omega \in \Omega} \wp(t)(\omega)\,\delta_k(\omega(\alpha))$$

with

$$\delta_k(X) = \sum_{(X',Y)\in \mathbf{E}^2} \left(\zeta(X' \to Y)\,(Y_k - X'_k) \prod_1^{6p} X_i^{X'_i}(1-X_i)^{1-X'_i} \prod_1^{n} X_{6p+i}^{X'_{6p+i}} \right).$$

Let us note that if $\wp$ a completely factorised distribution on Ω, i.e.

$$\wp(t)(\omega) = \prod_{\mathscr{R}} \left(\prod_{j,\nu} N_{j\nu}^{\omega_{j+6\nu}(\alpha)}(1-N_{j\nu})^{1-\omega_{j+6\nu}(\alpha)} \prod_1^n E_i^{\omega_{6p+i}(\alpha)} \right), \tag{6}$$

where the $N_{j\nu}$ and the E_i depend on α and t, then the r.h.s. of eq. (5) is simply given by $\delta(\boldsymbol{N}(\alpha, t))$.

Eq. (2) admits stationary and spatially uniform factorised solutions characterised by the

Proposition 1. Let $\boldsymbol{N} = (N_{10}, \ldots, N_{j\nu}, \ldots, E_1, \ldots, E_n)$ be a given vector with $0 < N_{j\nu} < 1$ and $0 < E_i < 1$ and $\Sigma_i E_i = 1$. The following statements are equivalent:

(i) $\wp$, given by (6) with $\boldsymbol{N}(\alpha, t) = \boldsymbol{N}$ for all α, t, is a solution of the Liouville equation,

(ii) $\delta(\boldsymbol{N}) = 0$,

(iii) $\log(\overline{\boldsymbol{N}}) \in \mathscr{I}$,

(iv) $\langle \log(\overline{\boldsymbol{N}}), \delta(\boldsymbol{N}) \rangle = 0$,

where $\overline{\boldsymbol{N}}$ is the vector of $\mathbf{R}^{6p+n}$ whose components are given by

$$\overline{N}_{j+6\nu} = \frac{N_{j\nu}}{1 - N_{j\nu}}, \quad \overline{N}_{6p+i} = E_i \quad \text{for } j = 1, \ldots 6, \quad \nu = 0, \ldots, p-1, \quad i = 1, \ldots, n.$$

A proof of this proposition can be deduced from ref. [9] with only slight modifications.

If one introduces the entropy $S = -\Sigma_\omega \wp(\omega) \log(\wp(\omega))$, one can show that among all the distributions on Ω, those which maximize S with the state variables given as constraints are factorised and satisfy $\log(\overline{\boldsymbol{N}}(\alpha)) \in \mathscr{I}$. Thus the stationary and uniform solutions of (2) appear as global equilibrium distributions for the system. From point (iii) of the proposition one deduces that those global equilibrium solutions are in general given by

$$N_{j\nu} = \frac{1}{1 + \exp\left(m_\nu \boldsymbol{b} \cdot \boldsymbol{c}_{j\nu} + \beta m_\nu V_\nu^2/2 - \mu_\nu\right)},$$
$$E_i = \frac{\exp(-\beta\phi_i)}{\Sigma_j \exp(-\beta\phi_j)}. \tag{7}$$

The $N_{j\nu}$ obey as usual a Fermi–Dirac distribution, due to the exclusion principle. The coefficients $\beta, \mu_\nu, \boldsymbol{b}$ are constants fixed by the values of the state variables. The μ_ν are not independent, they are related by the $p - a$ equations:

$$\sum_\nu \lambda_{\nu j} \mu_\nu = 0, \quad j = 1, \ldots, p - a,$$

where the $\lambda_{\nu j}$ are the stoichiometric coefficients of the jth reaction. This last relation expresses the classical mass action law at equilibrium.

At equilibrium, the entropy density of the system is given by

$$s = -\left(\sum_{j,\nu} N_{j\nu}\log(N_{j\nu}) + (1-N_{j\nu})\log(1-N_{j\nu}) + \sum_i E_i \log(E_i)\right) \tag{8}$$

and we have

$$\mathrm{d}s = \beta\,\mathrm{d}e + \boldsymbol{b}\cdot\mathrm{d}(\rho\boldsymbol{v}) - \sum_\nu \mu_\nu\,\mathrm{d}(\rho_\nu/m_\nu)$$

or equivalently

$$\mathrm{d}s = \beta\,\mathrm{d}u + \left[\boldsymbol{b}\cdot\mathrm{d}(\rho\boldsymbol{v}) + \tfrac{1}{2}\beta\,\mathrm{d}(\rho\boldsymbol{v}^2)\right] - \sum_\nu \mu_\nu\,\mathrm{d}(\rho_\nu/m_\nu).$$

Then in contrast to a classical system, the entropy is not invariant under Galilean transformations since, as we shall see later, $\boldsymbol{b}$ is not equal to $-\beta\boldsymbol{v}$. But one can prove that $\boldsymbol{b}=\boldsymbol{0}$ if and only if $\boldsymbol{v}=\boldsymbol{0}$, and one obtains for static equilibria the following differential relation:

$$\mathrm{d}s = \beta\,\mathrm{d}u - \sum_\nu \mu_\nu\,\mathrm{d}(\rho_\nu/m_\nu). \tag{9}$$

Thus at zero mean velocity, by analogy with classical thermodynamics one can identify $1/\beta$ as the thermodynamic temperature of the system and μ_ν/β as its chemical potentials. Moreover there is thermal equilibrium between the particles and the nodes. Let us then point out that in this statistical description the thermodynamic pressure is not defined since the volume is not a natural state variable for the system: A variation of the density corresponds to a change of the number of particles.

4. Two particular equilibrium distributions

4.1. Mass diffusion models

We are interested here in models where $a=p$ and $[k_{\nu i}]=\mathbf{Id}$. Thus the conservation of atoms reduces to the conservation of partial masses, and the conservation of the total energy to that of each energy level. The independent state variables are now ρ_ν and $\rho\boldsymbol{v}$. In this case, the coefficient β then disappears. Relation (7) reduces to

$$N_{j\nu} = \frac{1}{1+\exp(m_\nu \boldsymbol{b}\cdot\boldsymbol{c}_{j\nu} - \mu_\nu)},$$

$$E_i = E_{i0},$$

where E_{i0} are fixed, independent of $N_{j\nu}$. In the neighbourhood of $\boldsymbol{v}=\boldsymbol{0}$, the mean populations are implicit functions of the ρ_ν and $\rho\boldsymbol{v}$. A Taylor expansion in $\rho\boldsymbol{v}$ is, up to order 2:

$$N_{j\nu} = \frac{\rho_\nu}{6m_\nu} + \alpha_\nu\langle\rho\boldsymbol{v}, \boldsymbol{c}_{j\nu}\rangle + \beta_\nu\left(2\boldsymbol{c}_{j\nu}\boldsymbol{c}_{j\nu} - V_\nu^2\mathbf{Id}\right):\rho\boldsymbol{v}\,\rho\boldsymbol{v} + \mathcal{O}(\rho\boldsymbol{v}^3)$$

with

$$\alpha_\nu = \frac{\rho_\nu(6m_\nu - \rho_\nu)}{3m_\nu \Sigma_\mu \rho_\mu(6m_\mu - \rho_\mu)V_\mu^2},$$

$$\beta_\nu = \frac{\rho_\nu(6m_\nu - \rho_\nu)(3m_\nu - \rho_\nu)}{3m_\nu\left[\Sigma_\mu \rho_\mu(6m_\mu - \rho_\mu)V_\mu^2\right]^2}.$$

4.2. Thermal non-reactive models

In order to obtain thermal properties without chemical reactions we set all the masses equal to m, and we choose $a = 1$ and $k_{\nu i} = 1$ for all ν. The conservation of atoms then reduces to the conservation of the total mass, the independent variables are now $\rho, e, \rho\boldsymbol{v}$. The μ_ν are all equal, the equilibrium distribution is then given by (7) with $\mu_\nu = \mu$ for each ν.

The set $(\rho, e, \rho\boldsymbol{v})$ is not a usual set of macroscopic variables since e is the total energy. We then choose to express β, μ and $\boldsymbol{b}$ as functions of $\rho, u, \rho\boldsymbol{v}$ where $u = e - \frac{1}{2}\rho v^2$ is the internal energy density for the system; they are indeed implicit analytic functions of $(\rho, u, \rho\boldsymbol{v})$ at least in a neighbourhood of $(\rho_0, u_0, 0)$ for any $\rho_0 \neq 0$ and $u_0 \neq 0$. They can then be expanded in powers of $\rho\boldsymbol{v}$. Up to order 3 one obtains

$$\beta(\rho, u, \rho\boldsymbol{v}) = 1/T + \beta_2(\rho, u)\, \rho\boldsymbol{v}\cdot\rho\boldsymbol{v} + \mathcal{O}(\rho\boldsymbol{v}^4),$$

$$\mu(\rho, u, \rho\boldsymbol{v}) = \mu_0 + \mu_2(\rho, u)\, \rho\boldsymbol{v}\cdot\rho\boldsymbol{v} + \mathcal{O}(\rho\boldsymbol{v}^4),$$

$$\boldsymbol{b}(\rho, u, \rho\boldsymbol{v}) = -b_1\rho\boldsymbol{v} + b_3(\rho, u)(\rho\boldsymbol{v}\cdot\rho\boldsymbol{v})\, \rho\boldsymbol{v} + \mathcal{O}(\rho\boldsymbol{v}^5).$$

Because of relation (9) and in contrast to ref. [4], following Molvig et al. [5], we set T to be the thermodynamic temperature of the system and $\mu_0 T$ its chemical potential. In practice that does not modify the main results in ref. [4] since they are given for low mean velocities. Let us note that with these definitions T and μ_0 only depend on ρ and u as in classical thermodynamics.

The coefficients β_2, μ_2, b_1 and b_3 are implicit functions of ρ, u through T and μ_0. Their expressions are given in the appendix. One can verify that $b_1\rho T \neq 1$, and so $\boldsymbol{b}$ is not equal to $-\beta\boldsymbol{v}$ as said before. A Taylor expansion of the mean populations in $\rho\boldsymbol{v}$ gives, up to order 3:

$$\begin{aligned} N_{j\nu} &= \frac{\rho_{\nu 0}}{6m} + \alpha_\nu\langle \rho\boldsymbol{v}\cdot\boldsymbol{c}_{j\nu}\rangle + \left(2\theta_\nu \boldsymbol{c}_{j\nu}\boldsymbol{c}_{j\nu} - \gamma_\nu V_\nu^2 \mathbf{Id}\right) : \rho\boldsymbol{v}\,\rho\boldsymbol{v} \\ &\quad + \left(2\zeta_\nu \boldsymbol{c}_{j\nu}\boldsymbol{c}_{j\nu} - \delta_\nu V_\nu^2 \mathbf{Id}\right)\boldsymbol{c}_{j\nu} : \rho\boldsymbol{v}\,\rho\boldsymbol{v}\,\rho\boldsymbol{v} + \mathcal{O}(\rho\boldsymbol{v}^4), \\ E_i &= E_{i0}(T)\left[1 + (\bar{\phi} - \phi_i)\beta_2 \rho\boldsymbol{v}\cdot\rho\boldsymbol{v}\right] + \mathcal{O}(\rho\boldsymbol{v}^4). \end{aligned} \tag{10}$$

We have used the following notations:

$$\frac{\rho_{\nu 0}}{6m} = N_{j\nu}(\rho, u, \boldsymbol{0}) = d_{\nu 0} = \frac{1}{1 + \exp\left(mV_\nu^2/2T - \mu_0\right)},$$

$$E_{i0}(T) = E_{i0}(\rho, u, \boldsymbol{0}) = \frac{\exp(-\phi_i/T)}{\Sigma_j \exp(-\phi_j/T)}. \tag{11}$$

$\overline{\phi}$ is the mean energy level: $\Sigma_i E_{i0}(T)\,\phi_i$. All the coefficients are functions of ρ, u through T and μ_0; they are given in the appendix. For static equilibria one obtains:

$$du = \kappa\, d\rho + C\, dT, \tag{12}$$

where

$$C = \frac{3m^2 \Sigma_{\mu,\nu} d_{\nu 0} d_{\mu 0}(1-d_{\mu 0})(1-d_{\nu 0})\left(V_\mu^2 - V_\nu^2\right)^2}{4T^2 \Sigma_\mu d_{\mu 0}(1-d_{\mu 0})} + \frac{\overline{\phi^2} - (\overline{\phi})^2}{T^2},$$

$$\kappa = \frac{\Sigma_\mu d_{\mu 0}(1-d_{\mu 0}) V_\mu^2}{2\Sigma_\mu d_{\mu 0}(1-d_{\mu 0})}.$$

The term $\overline{\phi^2} - (\overline{\phi})^2$ in C is due to the presence of massless particles and thus we do not want to discuss nor define the specific heat for the whole system.

5. The Boltzmann equation

The Boltzmann hypothesis for lattice gases [8] assumes that particles which enter a collision have no prior correlations. Thus eq. (5) yields the so-called Boltzmann equation for the lattice gases:

$$\begin{aligned} &N_{j\nu}(\alpha + \Delta t\, \boldsymbol{c}_{j\nu}, t + \Delta t) - N_{j\nu}(\alpha, t) = \delta_{j+6\nu}(\boldsymbol{N}(\alpha, t)), \\ &E_i(\alpha, t + \Delta t) - E_i(\alpha, t) = \delta_{6p+i}(\boldsymbol{N}(\alpha, t)). \end{aligned} \tag{13}$$

The r.h.s. of this equation only depends on N and is obtained by replacing in (5) averages of products by products of averages.

Following the demonstration of Hénon [8] one obtains a global H-theorem (or a local one for a spatially uniform case) for the function $H = -\Sigma_{\alpha \in \mathscr{R}} s(\alpha, t)$ with s given by (8) for N the solution of the finite difference equation (13). This insures that the second law of thermodynamics is valid for these lattice gases in the Boltzmann approximation. The main argument of the proof is semi-detailed balance.

In the following we shall study the dissipative properties of these models from a Chapman–Enskog expansion of the solutions of (13). Now, we shall briefly show how this equation can take into account simple chemical kinetics.

Let us write, as usual, a given reaction as

$$\sum_\nu \lambda^1_{\nu j} \mathrm{S}_\nu \underset{2}{\overset{1}{\rightleftharpoons}} \sum_\nu \lambda^2_{\nu j} \mathrm{S}_\nu$$

with $0 \le \lambda^1_{\nu j}$ and $0 \le \lambda^2_{\nu j}$ and $\lambda^2_{\nu j} - \lambda^1_{\nu j} = \lambda_{\nu j}$. For simplicity, we shall assume there is no more than one reaction per collision, with the input state exactly given by the stoichiometry. We can then divide the set of all pairs $(\boldsymbol{X}, \boldsymbol{Y})$ which have a non-zero transition probability $\zeta(\boldsymbol{X} \to \boldsymbol{Y})$ in the $r+1$ classes C_j, where for $j \le r$ each C_j contains the set of all collisions $(\boldsymbol{X} \to \boldsymbol{Y})$ associated with the jth reaction, and C_{r+1} contains all the non-reactive collisions. These classes are disjoint and we shall note C^1_j and C^2_j the elements of C_j which respectively correspond to the jth reaction in direction 1 or 2. We are then interested in spatially uniform and isotropic solutions of (13), and so we define N_ν as the numerical density of species ν. With these conventions, one obtains the following production rates:

$$\frac{1}{\Delta t}[N_\nu(t+\Delta t) - N_\nu(t)] = \sum_{j=1}^{r} \lambda_{\nu j}\left(K_j^1 \prod_\mu N_\mu^{\lambda_{\mu j}^1}(6-N_\mu)^{6-\lambda_{\mu j}^1} - K_j^2 \prod_\mu N_\mu^{\lambda_{\mu j}^2}(6-N_\mu)^{6-\lambda_{\mu j}^2}\right),$$

where the coefficients K_j^1 and K_j^2 are given by

$$K_j^1 = \sum_{(X,Y)\in C_j^1} \left(\zeta(X\to Y)\prod_1^n E_i^{X_{6p+i}}\right)\frac{1}{6^{6p}\,\Delta t},$$

$$K_j^2 = \sum_{(X,Y)\in C_j^2} \left(\zeta(X\to Y)\prod_1^n E_i^{X_{6p+i}}\right)\frac{1}{6^{6p}\,\Delta t}.$$

These can be interpreted as the kinetic coefficients of the reactions, and in a first approximation near equilibrium, they only depend on the temperature. The presence of terms like $(6-N_\mu)$ is due to the exclusion principle, but at low densities one recovers the usual kinetic laws where the orders are equal to the stoichiometric coefficients. We discuss in more detail these chemical models in a report [10].

6. Macroscopic properties

6.1. The Chapman–Enskog expansion

We shall define $\boldsymbol{y}$ as the set of macroscopic state variables associated with a given basis of $\mathscr{I}$ (for example $\boldsymbol{y}=(\rho, e, \rho\boldsymbol{v})$ for the models of section 3.2). We are looking for slowly varying phenomena and we choose $\Delta t = \varepsilon\tau$, where τ is a reference time scale and ε a typical small parameter. We assume that $\Delta t\, \boldsymbol{c}_{j\nu} = \mathscr{O}(\varepsilon)$ and we then Taylor expand eq. (13). Let us note that this choice is not unique, it depends on the characteristic scales which one wants to observe. We are then looking for a solution to eq. (13) as an asymptotic expansion of the form:

$$\boldsymbol{N} = \sum_n \varepsilon^n \boldsymbol{N}^{(n)}, \qquad \frac{\partial \boldsymbol{y}}{\partial t} = \sum_n \varepsilon^n \boldsymbol{F}^{(n)}, \tag{14}$$

where the functions $\boldsymbol{N}^{(n)}$ and $\boldsymbol{F}^{(n)}$ only depend on $\boldsymbol{y}$ and its successive gradients up to order n and $n+1$, respectively. One finds that $\delta(\boldsymbol{N}^{(0)}) = \boldsymbol{0}$. Thus $\boldsymbol{N}^{(0)}$ is a local equilibrium distribution and only depends on $\boldsymbol{y}$. As usual in kinetic theory we assume that the perturbations of $\boldsymbol{N}$ do not change the values of the state variables $\boldsymbol{y}$ which are evaluated for $\boldsymbol{N}^{(0)}$. Equations for the macroscopic behaviour of the gas to order n are obtained by truncating the expansion of $\partial \boldsymbol{y}/\partial t$ at order n and assuming that the calculation of $\boldsymbol{N}^{(n+1)}$ for the next order is possible. One then obtains $\boldsymbol{N}^{(1)}$ as the solution of the linear system:

$$\mathbf{₵B}^{-1}(\boldsymbol{N}^{(1)}) = \boldsymbol{U} \tag{15}$$

where $\mathbb{C}$ is a linear negative Fredholm operator in $\mathbf{R}^{6p+n}$ called the linear collision operator. Its kernel is $\mathscr{I}$. $\boldsymbol{U}$ is the vector of the first-order terms in the Taylor expansion of (13) and $\mathbf{B}$ is the diagonal matrix:

$$B_{j+6\nu,j+6\nu} = N_{j\nu}^{(0)}\left(1 - N_{j\nu}^{(0)}\right), \quad j = 1,\ldots,6, \quad \nu = 0,\ldots,p-1,$$
$$B_{ii} = E_i^{(0)}, \qquad i = 6p+1,\ldots,6p+n.$$

The general expression for $\mathbb{C}$ is the following:

$$\left[\mathbb{C}(V)\right]_j = \prod_{j,\nu}\left(1 - N_{j\nu}^{(0)}\right) \sum_{(X,Y)\in\mathbf{E}^2} \zeta(X \to Y)(Y_j - X_j)\langle X, V\rangle \prod_{j,\nu} \overline{N}_{j\nu}^{(0)X_{j+6\nu}} \prod_i E_i^{(0)X_{6p+i}} \tag{16}$$

and $\mathbb{C}, \mathbf{B}, \boldsymbol{U}$, only depend on $\boldsymbol{N}^{(0)}$, $\boldsymbol{y}$ and $\nabla \boldsymbol{y}$.

6.2. Mass diffusion properties

6.2.1. Theoretical results

We study here the conservation equations for partial masses for the models of section 3.1. The detailed calculations can be found in ref. [1]. We shall only give the main results. In order to obtain diffusion properties we perform the Chapman–Enskog expansion up to order one. The conservation of the total mass yields the continuity equation:

$$\frac{\partial \rho}{\partial t} + \operatorname{div}(\rho \boldsymbol{v}) = 0$$

and because of the symmetry properties of $\mathbb{C}$, $\boldsymbol{N}^{(1)}$ take the general form:

$$N_{j\nu}^{(1)} = \sum_\mu \frac{\tau}{3m_\nu V_\nu^2}\alpha^{\nu\mu}\left[\tfrac{1}{2}\nabla\left(\rho_\mu V_\mu^2\right) - 3m_\mu \alpha_\mu V_\mu^2 \nabla p^0\right]\boldsymbol{c}_{j\nu} + \tau\psi^\nu\left(2\boldsymbol{c}_{j\nu}\boldsymbol{c}_{j\nu} - V_\nu^2\mathbf{Id}\right): \nabla\rho\boldsymbol{v}$$
$$+ \sum_\mu \tau\beta^{\nu\mu}\left(2\boldsymbol{c}_{j\nu}\boldsymbol{c}_{j\nu} - V_\nu^2\mathbf{Id}\right): \rho\boldsymbol{v}\,\nabla\rho_\mu + \mathscr{O}\left(\boldsymbol{v}^3, \boldsymbol{v}^2\,\nabla\boldsymbol{v}\right),$$

where $p^0 = \sum_\nu \frac{1}{2}\rho_\nu V_\nu^2$ is the hydrostatic pressure of the gas. The coefficients $\alpha^{\nu\mu}$, $\beta^{\nu\mu}$, and ψ^ν are functions of the ρ_μ and are obtained by inverting (15) for $\boldsymbol{v} = \boldsymbol{0}$. Since $\boldsymbol{N}^{(1)}$ belongs to $\mathscr{I}^\perp$, they obey linear relations that we omit here. We shall define c_ν as the concentrations and we obtain the following equations for the conservation of partial masses:

$$\rho\left(\frac{\partial c_\nu}{\partial t} + \nabla c_\nu \boldsymbol{v}\right) = -\operatorname{div}(\boldsymbol{I}_\nu), \tag{17}$$

where the flux vectors $\boldsymbol{I}_\nu$ are given by

$$\boldsymbol{I}_\nu = \rho\left(\sum_{\mu\neq\nu}\left[D^{\nu\mu}\,\nabla c_\mu + B^{\nu\mu}\,\nabla(\rho^{-1}) + \left(c_\mu - 3m_\mu\alpha_\mu V_\mu^2\right)\boldsymbol{v}\right]\right) + \mathscr{O}\left(\boldsymbol{v}^3, \varepsilon\boldsymbol{v}^3, \varepsilon\boldsymbol{v}^2\,\nabla\boldsymbol{v}\right).$$

The diffusion coefficients $D^{\nu\mu}$ and $B^{\nu\mu}$ are functions of the ρ_μ (cf. ref. [1]) and for low mean velocities and density gradients one recovers the usual diffusion equations. In the case of a two-component mixture we have

$$D = D^{10} = D^{01} = \frac{2\alpha - 1}{4} \frac{\varepsilon t V_0^2 V_1^2 [\rho_0(6m_0 - \rho_0) + \rho_1(6m_1 - \rho_1)]}{\rho_0(6m_0 - \rho_0)V_0^2 + \rho_1(6m_1 - \rho_1)V_1^2},$$

$$B = -B^{10} = B^{01} = \frac{2\alpha - 1}{4} \frac{\varepsilon t V_0^2 V_1^2 \rho_0 \rho_1 (6m_1 - \rho_1 - 6m_0 + \rho_0)}{\rho_0(6m_0 - \rho_0)V_0^2 + \rho_1(6m_1 - \rho_1)V_1^2},$$

where α is the common value of α^{10} and α^{01}. Thus with the above assumptions, we obtain (for $c = c_1 = 1 - c_0$) up to leading order, Fick's diffusion equation:

$$\rho \frac{\partial c}{\partial t} = \operatorname{div}(\rho D \nabla c). \tag{18}$$

Let us note that the factor α in D arises from the collisions. One can prove from the negativity of the linear collision operator that it is always positive (see ref. [1]), the remaining part which is negative comes from the discretisation of the l.h.s. of (13).

6.2.2. Diffusion simulations

We have simulated one-dimensional diffusion of a binary mixture in a rectangular box of length L (see plate I). We shall use the following choice of masses and velocities: $m_0 = 3m$, $\boldsymbol{V}_0 = V\boldsymbol{e}_1/\Delta x$, $m_1 = m$, $\boldsymbol{V}_1 = V(\boldsymbol{e}_1 + \boldsymbol{e}_2)/\Delta x$, $\Delta t V = \Delta x$ and a 512×256 lattice or a 1024×512 lattice. It is natural to set the reference time τ introduced before to the order of the relaxation time L^2/D. However we have $D = \varepsilon \tau V^2 D_0$ where D_0 is a dimensionless function of the transition probabilities and of the numerical densities of species. Thus in order to obtain reasonable relaxation time, D_0 should be of unit order. Under these conditions, a dimensional analysis of eq. (17) shows that for $\boldsymbol{v} = \mathscr{O}(\varepsilon)$ and for $\nabla(\rho/24m) \ll \nabla c$, one can expect to simulate (18) with this lattice gas. To obtain suitable values of D_0 we have been led to choose a model where we take into account all the possible nontrivial collisions. We have noticed that the more we take into account nontrivial collisions, the lower are the values of D_0 and its variations with the densities. To illustrate this dependence we plot in fig. 1 D_0 versus d_0 and d_1 for the model with all possible collisions and for the model with only binary mixing collisions exposed in ref. [1] ($d_0 = \rho_0/6m_0$ and $d_1 = \rho_1/6m_1$ are the species numerical densities per link).

Let us now briefly describe this extended model. We can divide $\mathbf{E}$ in disjoint subsets where all the elements of a given subset have the same partial masses and the same momentum. There are 823 nontrivial subsets, i.e. with more than one element, and 688 trivial subsets. For trivial subsets, we set $\zeta(\boldsymbol{X} \to \boldsymbol{X}) = 1$, and for nontrivial subsets we set $\zeta(\boldsymbol{X} \to \boldsymbol{X}) = 0$ and $\zeta(\boldsymbol{X} \to \boldsymbol{Y}) = 1/(n-1)$ for $\boldsymbol{X} \neq \boldsymbol{Y}$, where n is the number of elements of the subset or equivalently the number of possible output states for a given input state. The maximum value of n is 18. One can verify that semi-detailed balance and the invariance requirements are satisfied. We have then for α the following expression:

$$\alpha = \left[\sum_{i,j=0}^{6} \left(\frac{f_{ij}^0}{d_0(1-d_0)} + \frac{f_{ij}^1}{d_1(1-d_1)} \right) \phi_{ij} \right]^{-1},$$

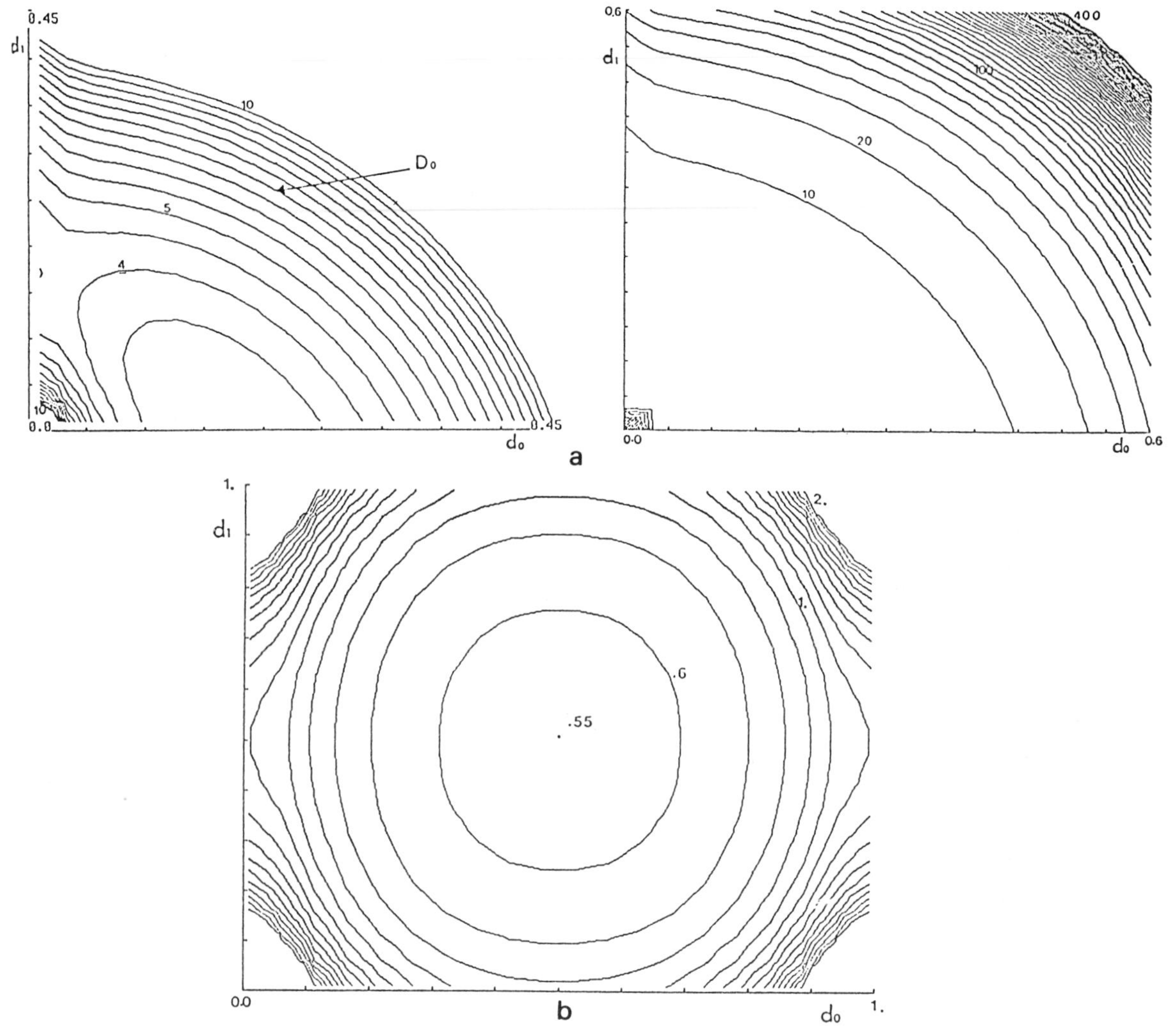

Fig. 1. D_0 versus d_0 and d_1 for the model of ref. [1] with only binary mixing collisions (a) and for the model with all possible collisions (b).

where

$$\phi_{ij} = (1-d_1)^{6-i} d_1^i d_0^j (1-d_0)^{6-j},$$

the f_{ij}^1 are given in table 1, and we have the relation: $f_{ij}^0 = \frac{1}{3} f_{ij}^1$. D is then given by the expression

$$D = \frac{\varepsilon \tau V^2}{4} \frac{9d_0(1-d_0) + d_1(1-d_1)}{3d_0(1-d_0) + d_1(1-d_1)} (2\alpha - 1). \tag{19}$$

The lattice is a closed rectangular box with pure specular reflections on the boundaries. It is divided in three parts as shown in scheme 1 with $L_1 = \frac{1}{4} L$. Initially each part of the box is filled with a spatially

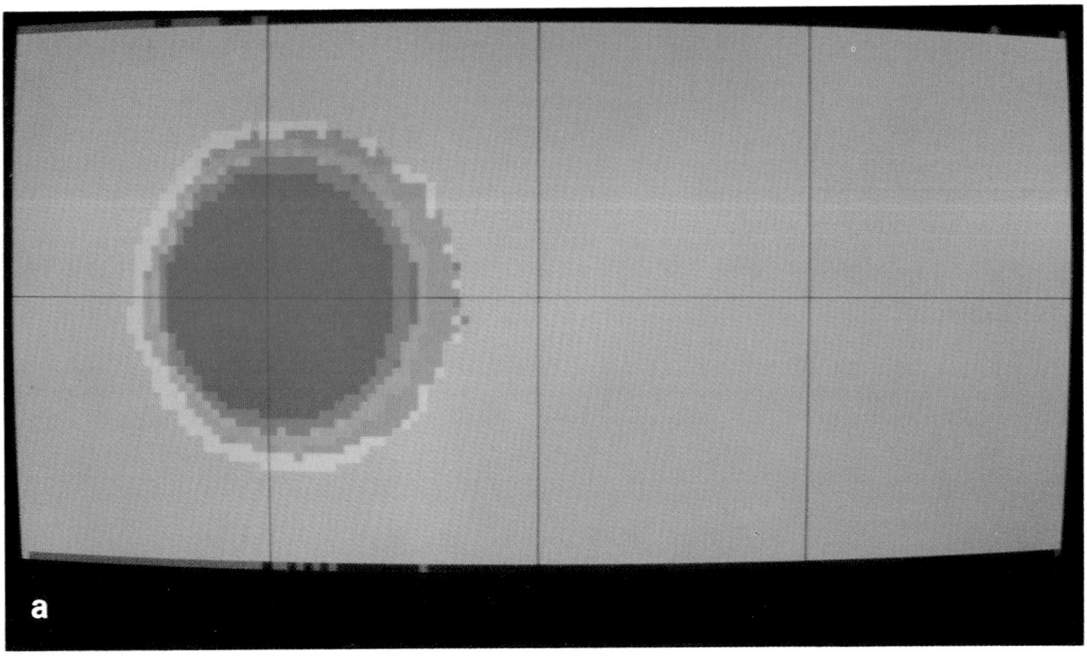

a

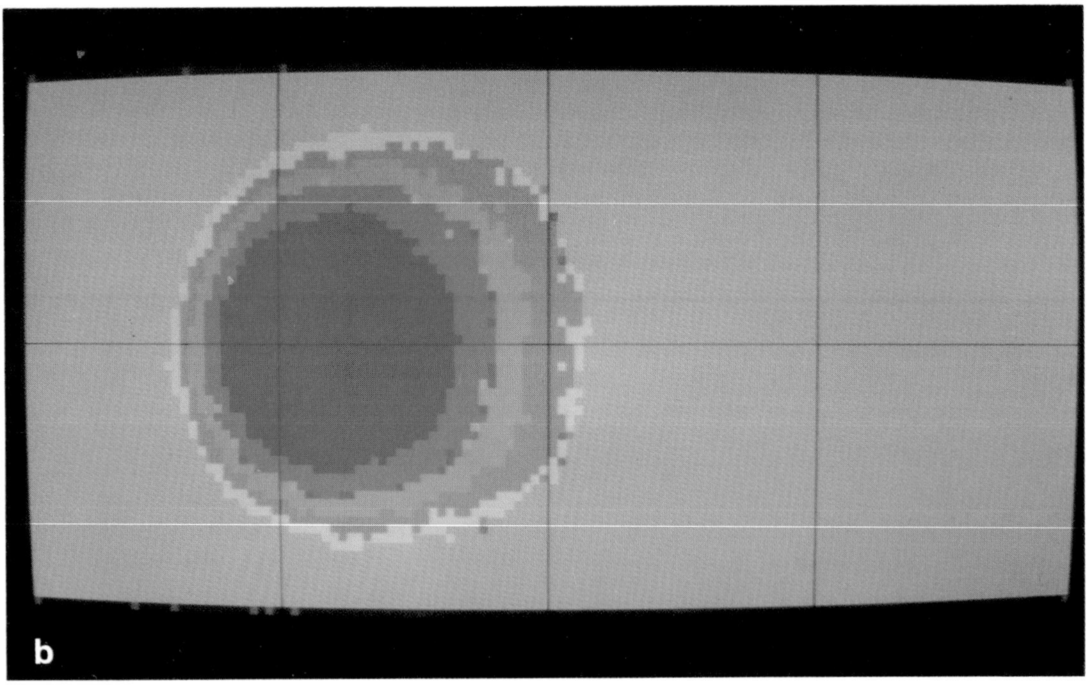

b

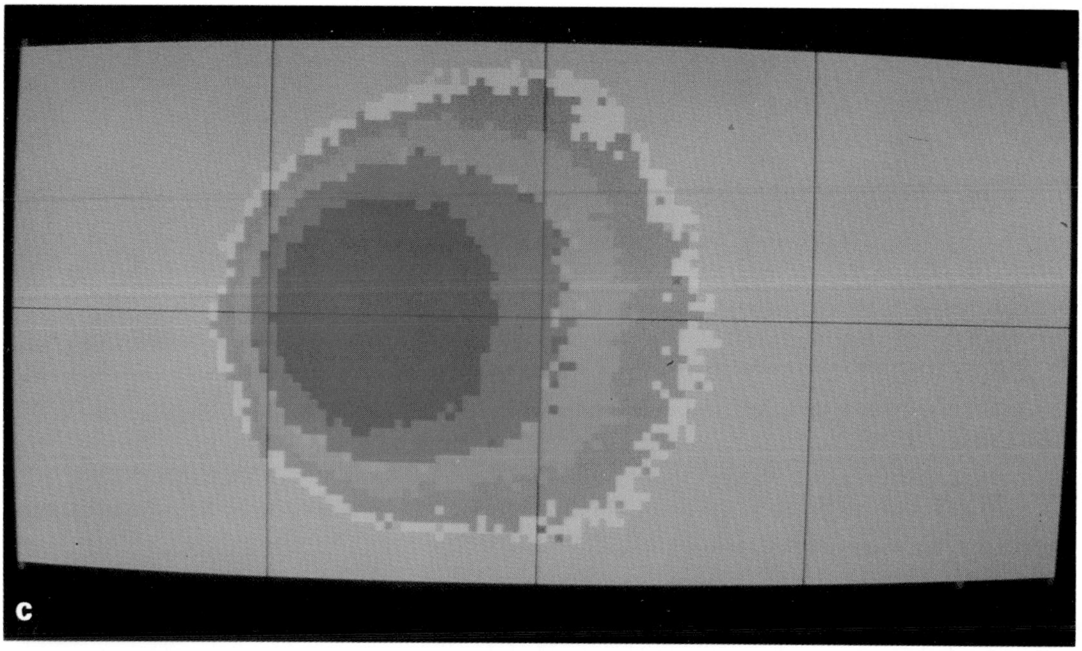

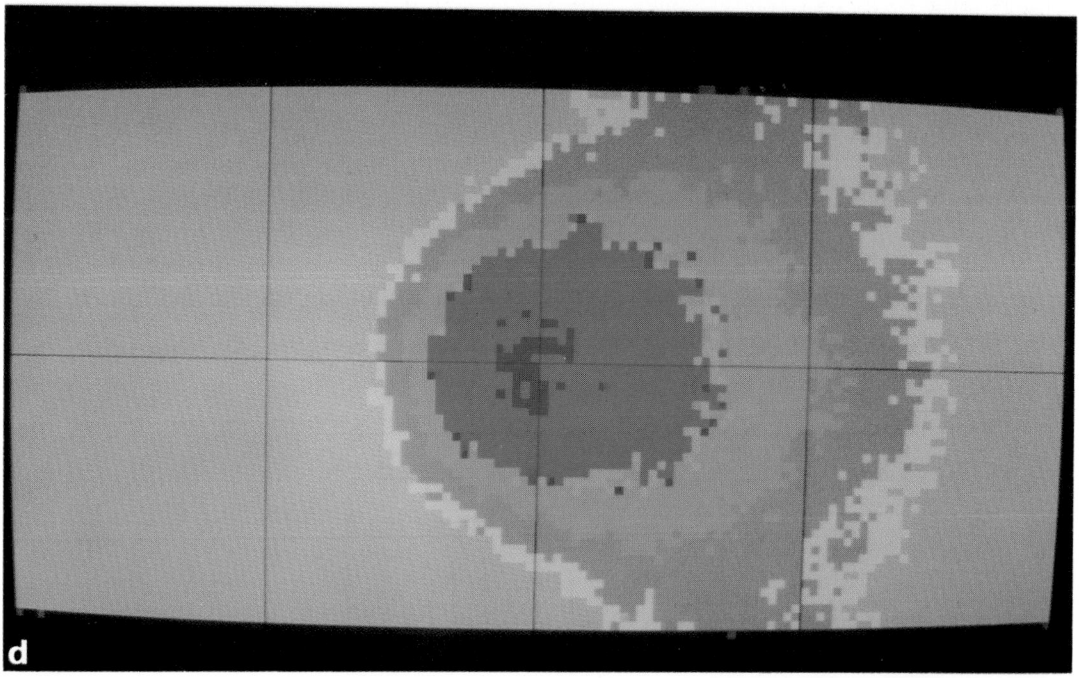

Plate I. The simulation is run on a 512×256 hexagonal lattice with the "extended" model described in section 6.2.2. At time $t=0$ we inject a bubble of radius R ($=256/6$ nodes), at rest, filled with blue particles of species 0 ($d_0=2/3$, $d_1=0$) in a uniform flow of green particles of species 1 ($d_0=0$, $d_1=1/2$). The mean velocity $v=0.2V$ is along the horizontal axis, from the left to the right. The Reynolds number, computed with the diameter of the bubble, is Re=29.25 (the dynamic viscosity of the green particles is $1.75m\ \Delta t\ V^2$). One clearly observes the growth of the trail in front of the bubble. (a)–(d) respectively correspond to 100, 300, 500 and 900 iteration steps. The darkness of the blue decreases with the concentration of species 0.

Table 1

$f^1_{i,j}$	0	1	2	3	4	5	6
0	0	0	0	0	0	0	0
1	0	6	387/20	122/5	387/20	6	0
2	0	63/4	1093/21	273/4	1093/21	63/4	0
3	0	86/5	5351/85	5264/55	5351/85	86/5	0
4	0	63/4	1093/21	273/4	1093/21	63/4	0
5	0	6	387/20	122/5	387/20	6	0
6	0	0	0	0	0	0	0

uniform equilibrium distribution with zero mean velocity. The total density ρ is uniform, the partial densities in part I and III are ρ_1^0 and $\rho_0^0 + \Delta\rho_0^0$ and in part II: ρ_0^0 and $\rho_1^0 + \Delta\rho_1^0$, with $\Delta\rho_i^0 \ll \rho$. Moreover the $\Delta\rho_i^0$ are chosen such that $\Delta p^0 = 0$. The spatial averages are done over strips of 16×256 nodes for the small lattice or 32×512 for the other, and over 500 iteration times. We have also averaged over 16×16 and 32×32 cells and verified that there is no significant variation along the y axis.

Thus we have compared the results of the simulations with the concentration fields $c(x,t)$ obtained by solving the one-dimensional diffusion equation:

$$\frac{\partial c}{\partial t} = D\frac{\partial^2 c}{\partial t^2} + \frac{1}{\rho}\frac{\partial \rho D}{\partial x}\frac{\partial c}{\partial x}, \qquad \frac{\partial c}{\partial x}(t,0) = \frac{\partial c}{\partial x}(t,L) = 0$$

with an explicit finite difference scheme where D is computed using relation (19).

There is good agreement between the evolutions of the two fields up to the stationary equilibrium state for a large range of initial densities as illustrated in figs. 2, 3, 4 and 5. We have then verified that the mean velocity and the variations of the total density remain small. In fact, they remain in the range of the noise and thus, in these experimental conditions, there is no unphysical convection as described, for example, in ref. [11].

These simulations show that the Boltzmann approximation leads to a good evaluation of the diffusion coefficient, at least in a binary mixture. We refer to ref. [10] for more complete numerical results and for a discussion of the time and space steps in the FD scheme.

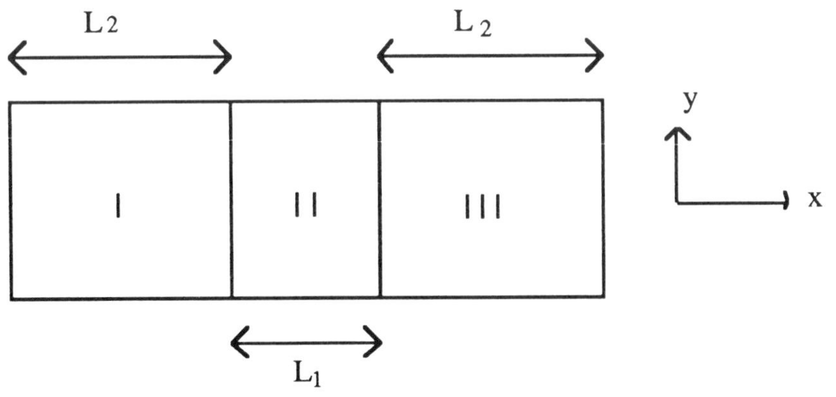

Scheme 1.

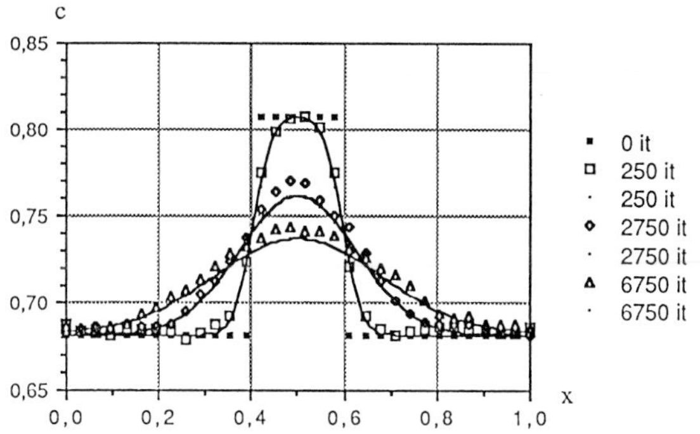

Fig. 2. $c(x,t)$ versus time on a 512×256 lattice. The continuous lines are obtained by the FD method; $d_0 = d_1 = 3,5/6$; $\Delta d_0 = -\Delta d_1 = -1/6$.

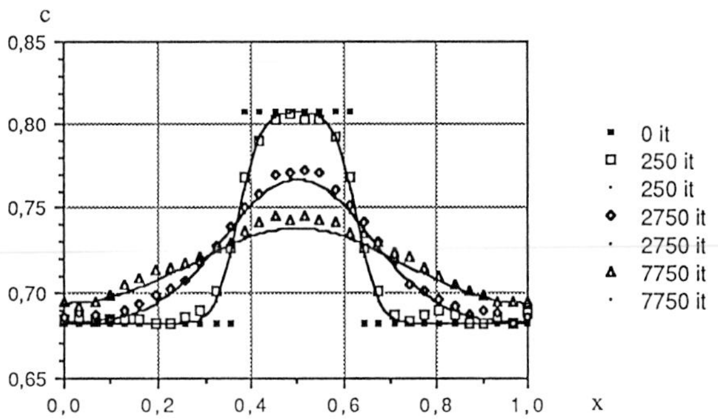

Fig. 3. $c(x,t)$ versus time on a 512×256 lattice. The continuous lines are obtained by the FD method; $d_0 = d_1 = 3, 5/15$; $\Delta d_0 = -\Delta d_1 = -1/15$.

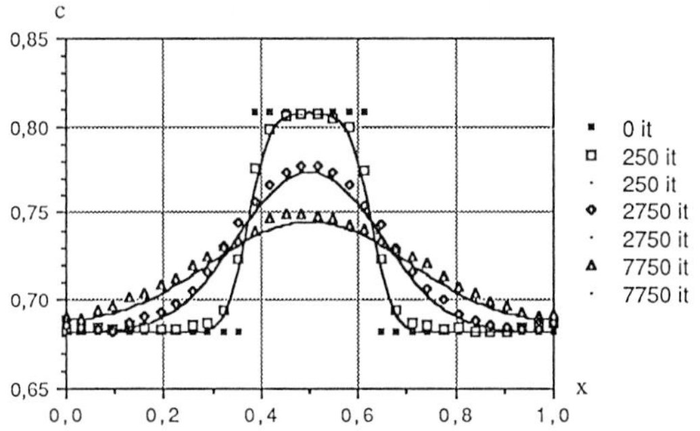

Fig. 4. $c(x,t)$ versus time on a 512×256 lattice. The continuous lines are obtained by the FD method; $d_0 = d_1 = 4, 9/6$; $\Delta d_0 = -\Delta d_1 = -1, 4/6$.

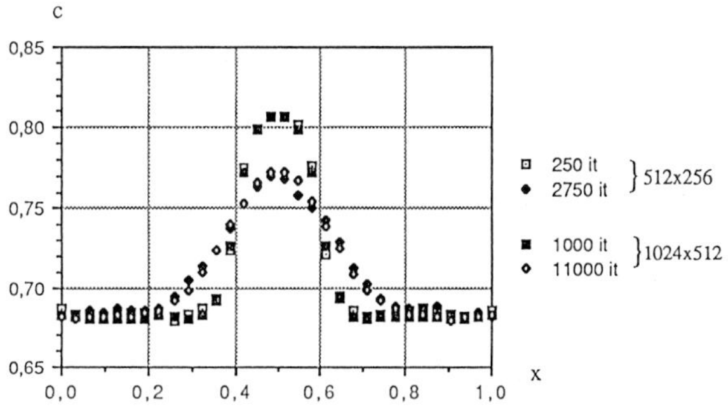

Fig. 5. Comparison between the concentration fields for a 512×256 lattice and a 1024×512 lattice for the same initial conditions as in fig. 2.

6.3. Thermal conduction properties

6.3.1. Theoretical results

We now study macroscopic properties of models described in section 3.2. The detailed calculations can be found in ref. [4]. The conservation equations for mass, momentum and energy to first order in the Chapman–Enskog expansion are:

$$\frac{\partial \rho}{\partial t} + \operatorname{div}(\rho \boldsymbol{v}) = 0, \tag{20}$$

$$\rho\left(\frac{\partial \boldsymbol{v}}{\partial t} + \nabla(\boldsymbol{v}) \cdot \boldsymbol{v}\right) = \operatorname{div}(\boldsymbol{\Pi}^0 + \varepsilon \boldsymbol{\Pi}^1), \tag{21}$$

$$\frac{\partial u}{\partial t} + \operatorname{div}(u\boldsymbol{v}) = (\boldsymbol{\Pi}^0 + \varepsilon \boldsymbol{\Pi}^1) : \nabla \boldsymbol{v} - \operatorname{div}(\boldsymbol{q}^0 + \varepsilon \boldsymbol{q}^1). \tag{22}$$

The stress tensor $\mathbf{\Pi} = \mathbf{\Pi}^0 + \varepsilon \mathbf{\Pi}^1$ and the vector $\boldsymbol{q} = \boldsymbol{q}^0 + \varepsilon \boldsymbol{q}^1$ are given by

$$\mathbf{\Pi}^0 = -\left(\sum_{j,\nu} m N_{j\nu}^{(0)} \boldsymbol{c}_{j\nu} \boldsymbol{c}_{j\nu} - \rho \boldsymbol{v}\boldsymbol{v} \right), \tag{23}$$

$$\mathbf{\Pi}^1 = -\left[\sum_{j,\nu} m N_{j\nu}^{(1)} \boldsymbol{c}_{j\nu} \boldsymbol{c}_{j\nu} + \frac{\tau}{2} \frac{\mathrm{d}}{\mathrm{d}y} \left(\sum_{j,\nu} m N_{j\nu}^{(0)} \boldsymbol{c}_{j\nu} \boldsymbol{c}_{j\nu} \right) (F^{(0)}) + \frac{\tau}{2} \operatorname{div}\left(\sum_{j,\nu} m N_{j\nu}^{(0)} \boldsymbol{c}_{j\nu} \boldsymbol{c}_{j\nu} \boldsymbol{c}_{j\nu} \right) \right], \tag{24}$$

$$\boldsymbol{q}^0 = \sum_{j,\nu} m \frac{V_\nu^2}{2} N_{j\nu}^{(0)} \boldsymbol{c}_{j\nu} + ({}^t\mathbf{\Pi}^0 - e\,\mathbf{Id}) \cdot \boldsymbol{v}, \tag{25}$$

$$\boldsymbol{q}^1 = \sum_{j,\nu} m \frac{V_\nu^2}{2} N_{j\nu}^{(1)} \boldsymbol{c}_{j\nu} + {}^t\mathbf{\Pi}^1 \cdot \boldsymbol{v} + \frac{\tau}{2} \frac{\mathrm{d}}{\mathrm{d}y} \left(\sum_{j,\nu} m \frac{V_\nu^2}{2} N_{j\nu}^{(0)} \boldsymbol{c}_{j\nu} \right) (F^{(0)}) + \frac{\tau}{2} \operatorname{div}\left(\sum_{j,\nu} m \frac{V_\nu^2}{2} N_{j\nu}^{(0)} \boldsymbol{c}_{j\nu} \boldsymbol{c}_{j\nu} \right). \tag{26}$$

$\mathbf{\Pi}^0$ and $\boldsymbol{q}^0$ are the zero order flux in the Chapman–Enskog expansion. The factor of $\tau/2$ arises from the discretisation. Relations (20), (21) and (22) have been written in their usual form in continuum mechanics, but there are some important remarks. Firstly (21) and (22) are not Gallilean invariant and $\mathbf{\Pi}$ does not satisfy the material objectivity principle [12]. This problem is well known for lattice gases (see for example refs. [2, 8, 11]), but it leads to an important problem with the equation of energy. In fact, since the system is not Galilean invariant, the external power supply has no reason to take the usual form: $\operatorname{div}({}^t\mathbf{\Pi} \cdot \boldsymbol{v})$. Thus the definition of the heat flux vector is ambiguous. Nevertheless, identifying formally eq. (22) with the classical one for continuous media, one could say that $\boldsymbol{q}$ is the heat flux vector as in ref. [7].

Because of the symmetry properties of the linear collisional operator, we have

$$N_{j\nu}^{(1)} = \frac{\tau}{3m} c^\nu\, \nabla(\rho) \cdot \boldsymbol{c}_{j\nu} + \frac{\tau}{3mV_\nu^2} k^\nu\, \nabla(u) \cdot \boldsymbol{c}_{j\nu} + \frac{\tau}{3mV_\nu^2} \left(2 f^u \boldsymbol{c}_{j\nu} \boldsymbol{c}_{j\nu} - g^\nu V_\nu^2 \mathbf{Id} \right) : \nabla \rho \boldsymbol{v} + \mathcal{O}(\rho \boldsymbol{v}),$$

$$E_i^{(1)} = \frac{\tau}{m} h^i \operatorname{div}(\rho \boldsymbol{v}) + \mathcal{O}(\rho \boldsymbol{v}),$$

where the non-dimensional coefficients c^ν, k^ν, f^u, g^ν and h^i are functions of ρ and u and are obtained by inverting relation (15) for $\boldsymbol{v} = \boldsymbol{0}$. Since $N^{(1)}$ belongs to $\mathcal{I}^\perp$, they obey linear relations that we omit here. We now focus our attention on the heat conduction properties of the gas. At low mean velocities and low velocity gradients we obtain, up to leading order, the following heat diffusion equation:

$$\frac{\partial u}{\partial t} = \operatorname{div}[D_{\text{th}}(\nabla u - \kappa \nabla \rho)],$$

where the thermal diffusivity of the model is given by

$$D_{\text{th}} = \frac{\varepsilon\tau}{2} \left[-\sum_\nu k^\nu V_\nu^2 - \frac{1}{4} \sum_\nu \frac{\partial \rho_{\nu 0}}{\partial u} V_\nu^2 \left(V_\nu^2 - 3m \sum_\mu \alpha_\mu V_\mu^4 \right) \right].$$

Its collisional part, $\frac{1}{2}\varepsilon\tau\,(-\Sigma_\nu k^\nu V_\nu^2)$ is positive while its discrete part is negative. Since u is a function of ρ and T corresponding to the pseudo-equilibrium distribution $N^{(0)}$. Using relation (12) one obtains the heat conductivity of the medium:

$$\lambda = CD_{\text{th}}.$$

In the case of two sets of velocities (that is $\nu = 0, 1$) we have $k^0 = -k^1$ and the collisional part of λ is given by

$$\lambda_{\text{coll}} = -\frac{9\varepsilon\tau m^2\left(V_0^2 - V_1^2\right)^2}{2T^2\Sigma_{i=1,\,j=1}^{12,\,12}\Phi_{ij}X_i\cdot X_j}$$

the $X_{j+6\nu}$, $j = 1,\ldots,6$, $\nu = 0, 1$ are the vectors $[(-1)^\nu/d_\nu(1-d_\nu)V_\nu^2]c_{j\nu}$, where as previously the $d_\nu = \rho_{\nu 0}/6m$ are the species numerical densities per link. The discrete part is given by

$$\lambda_{\text{dis}} = -\frac{3\varepsilon\tau m^2 d_0(1-d_0)d_1(1-d_1)V_0^2V_1^2\left(V_0^2 - V_1^2\right)^2}{8T^2\left[d_0(1-d_0)V_0^2 + d_1(1-d_1)V_1^2\right]}.$$

For these models with two (and up to four) different set of velocities one can obtain analytical expressions for λ as function of ρ and T as shown in ref. [4]. In this last reference we have in particular studied a model with only binary collisions for the following choice: $m_0 = m_1 = m$, $V_0 = Ve_1/\Delta x$, $V_1 = V(e_1 + e_2)/\Delta x$, $\phi_0 = 0$, $\phi_1 = 2mV^2$. We have noticed that the medium tends to be a heat superconductor for $d = \rho/12m$ greater than $1/2$ and when T goes to zero.

6.3.2. Simulations

In order to illustrate the thermodynamics of these models we propose a numerical one-dimensional simulation similar to that for mass diffusion, but we impose periodic boundary conditions. We use the model described in ref. [4]. At initial time, each part of the box is filled with an equilibrium distribution at zero mean velocity and with the same total mass density ρ. Parts I and III (which are contiguous) are set at temperature T_1 while part II is set at temperature T_2. We have plotted in figs. 6, 7 and 8 the evolution of the mean entropy density (obtained as the ratio of the total entropy divided by the number of cells) of this closed system versus time for different sizes of the box but with the same initial conditions. $N_{j\nu}$ and E_i are computed over 16×16 cells.

One observes that the entropy grows to reach a maximum value, as expected. We have then determined from relation (11) the equilibrium distribution which corresponds to the equilibrium state of the system, obtained assuming that the total mass and the total energy are the same as at the initial time (for the initial data of the figures we have $T \cong 1.22\ \mathrm{mV}^2$). We have simulated this equilibrium state on the three lattices and measured the entropy, the averages being always taken over 16×16 cells. We have found for the smaller up to the larger lattice: $s = 5.690 \pm 0.002$, 5.691 ± 0.001, 5.693 ± 0.001. These values fit very well the limits obtained for long time in the relaxation experiments.

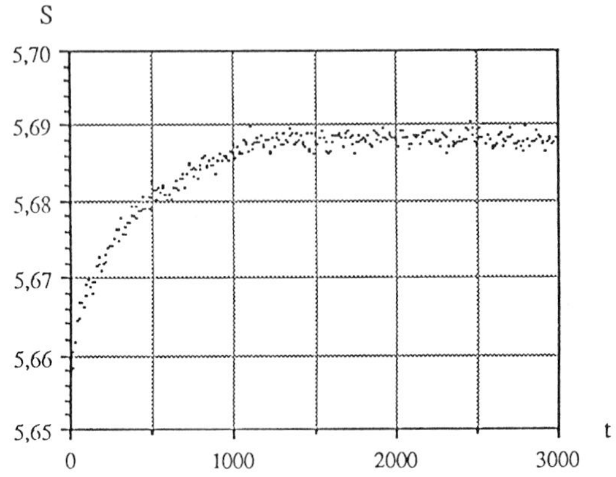

Fig. 6. s versus time for a 256×128 lattice. The initial conditions are: $T_1 = 1\ mV^2$, $T_2 = 2.5\ mV^2$, $\rho = 2m$.

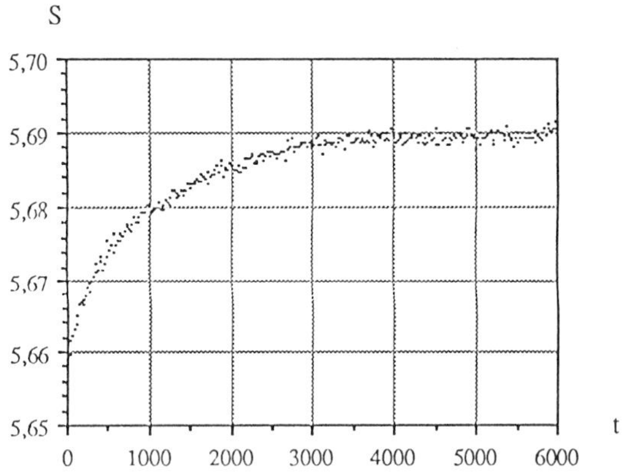

Fig. 7. s versus time for a 384×192 lattice. The initial conditions are the same as in fig. 6.

Finally, one observes that the equilibrium time of the system grows approximately as the square of the length of the box as we expected for a one-dimensional simulation. Moreover, the statistical fluctuations decrease as the size of the system increases.

7. Conclusion

The models presented here allow several kinds of macroscopic behaviour depending on the collisional invariants, and a large class of simulations is possible. For example, in a forthcoming paper we will present simulations of thermal conduction with Dirichlet and Neumann conditions at the boundary. However, let us note that if the extension to three-dimensional models is theoretically easy, in practice it is not so simple since it leads to a much larger state space.

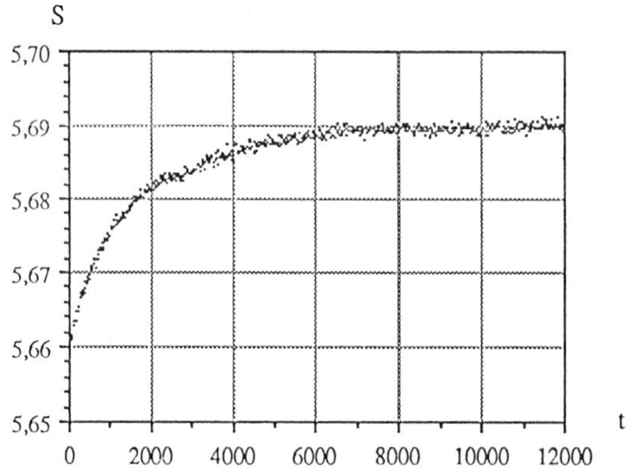

Fig. 8. s versus time for a 512×256 lattice. The initial conditions are the same as in fig. 6.

Appendix

1. Expressions of b_1, b_3, β_2, μ_2 for the "thermal" model. The notations are these of section 3.2. We have

$$\rho = 6m \sum_\nu d_{\nu 0}, \qquad \rho \boldsymbol{v} = 6m \sum_\nu d_{\nu 0} \boldsymbol{c}_{j\nu},$$

$$u = 3m \sum_\nu d_{\nu 0} V_\nu^2 + \sum_i E_{i0}\phi_i - \tfrac{1}{2}\rho v^2.$$

One then obtains

$$b_1 = -\frac{1}{3m^2 \Sigma_\nu d_{\nu 0}(1-d_{\nu 0})V_\nu^2},$$

$$\beta_2 = -\frac{1}{2\rho C T^2} - \frac{1}{12 m C T^2 \Sigma_\nu d_{\nu 0}(1-d_{\nu 0})V_\nu^2}\left(\frac{\Sigma_\nu d_{\nu 0}(1-d_{\nu 0})(1-2d_{\nu 0})V_\nu^2}{\Sigma_\nu d_{\nu 0}(1-d_{\nu 0})} - \frac{\Sigma_\nu d_{\nu 0}(1-d_{\nu 0})(1-2d_{\nu 0})V_\nu^4}{\Sigma_\nu d_{\nu 0}(1-d_{\nu 0})V_\nu^2}\right),$$

$$\mu_2 = \frac{m}{2\Sigma_\nu d_{\nu 0}(1-d_{\nu 0})}\left(-\frac{m b_1^2}{2}\sum_\nu d_{\nu 0}(1-d_{\nu 0})(1-2d_{\nu 0})V_\nu^2 + \beta_2 \sum_\nu d_{\nu 0}(1-d_{\nu 0})V_\nu^2\right),$$

$$b_3 = \frac{b_1}{\Sigma_\nu d_{\nu 0}(1-d_{\nu 0})V_\nu^2}\left(\frac{m^2 b_1^2}{8}\sum_\nu d_{\nu 0}(1-d_{\nu 0})[6d_{\nu 0}(1-d_{\nu 0})-1] - \frac{m\beta_2}{2}\sum_\nu d_{\nu 0}(1-d_{\nu 0})(1-2d_{\nu 0})V_\nu^4 + \mu_2 \sum_\nu d_{\nu 0}(1-d_{\nu 0})(1-2d_{\nu 0})V_\nu^2\right).$$

2. Expressions of the coefficients in the expansion of the $N_{j\nu}$ and of the E_i.

$$\alpha_\nu = \frac{d_{\nu 0}(1-d_{\nu 0})}{3m\Sigma_\mu d_{\mu 0}(1-d_{\mu 0})V_\mu^2},$$

$$\theta_\nu = \frac{d_{\nu 0}(1-d_{\nu 0})(1-2d_{\nu 0})}{\left[6m\Sigma_\mu d_{\mu 0}(1-d_{\mu 0})V_\mu^2\right]^2},$$

$$\gamma_\nu = \frac{d_{\nu 0}(1-d_{\nu 0})}{V_\nu^2 \Sigma_\mu d_{\mu 0}(1-d_{\mu 0})}\left(\frac{m\beta_2}{2}\sum_\mu d_{\mu 0}(1-d_{\mu 0})\left(V_\nu^2 - V_\mu^2\right) + \sum_\mu \theta_\mu V_\mu^2\right),$$

$$\zeta_\nu = \frac{d_{\nu 0}(1-d_{\nu 0})[6d_{\nu 0}(1-d_{\nu 0})-1]}{3\left[6m\Sigma_\mu d_{\mu 0}(1-d_{\mu 0})V_\mu^2\right]^3},$$

$$\delta_\nu = m\left[\gamma_\nu(1-2d_{\nu 0})V_\nu^2 b_1 - d_{\nu 0}(1-d_{\nu 0})b_3\right].$$

References

[1] D. Bernardin and O.E. Sero-Guillaume, Lattice gas mixture models for mass diffusion, Eur. J. Mech. B/Fluids 9(1) (1989) 21–46.

[2] D. D'Humières, P. Lallemand and G. Searby, Numerical experiments on lattice gases: Mixtures and gallilean invariance, Complex Systems 1 (1987) 633–647.

[3] H. Chen and W.H. Matthaeus, Cellular automaton formulation of passive scalar dynamics, Phys. Fluids 30 (1987) 1235–1237.

[4] O.E. Sero-Guillaume and D. Bernardin, A lattice gas model for heat transfer and chemical reactions, Eur. J. Mech. B/Fluids 9(2) (1990) 177–196.

[5] K. Molvig, P. Donis, R. Miller, J. Myczkowsky and G. Vichniac, Thermodynamics of multi-species lattice gases, in: Cellular Automata and Modelling of Complex Physical Systems (Springer, Berlin, 1989).

[6] D. D'Humières and P. Lallemand, Numerical simulations of hydrodynamics with lattice gas automata in two dimensions, Complex Systems 1 (1987) 599–632.

[7] S. Chen, M. Lee, K.H. Zhao and G. Doolen, A lattice gas model with temperature, Physica D 37 (1989) 42–59.

[8] U. Frisch, D. D'Humières, B. Hasslacher, P. Lallemand, Y. Pomeau and J.P. Rivet, Lattice gas hydrodynamics in two and three dimensions, Complex Systems 1 (1987) 649–707.

[9] R. Gatignol, Théorie Cinétique des Gaz à Répartitions Discrète de Vitesses, Lectures Notes in Physics (Springer, Berlin, 1975).

[10] C.H. Sun, D. Bernardin and O.E. Sero-Guillaume, Internal Report, Lemta (1989).

[11] V. Zehnlé and G. Searby, Lattice gas experiments on a non-exothermic diffusion flame in a vortex field, J. Phys. 50 (1989) 1083–1097.

[12] W. Noll, A mathematical theory of the mechanical behavior of continuous media, Arch. Ration. Mech. Anal. 2 (1958) 197–226.

CHAPTER 5

BASIC RELATIONS AND LONG-TIME CORRELATIONS

Physica D 47 (1991) 191–197
North-Holland

TAGGED PARTICLE DIFFUSION IN 3D LATTICE GAS CELLULAR AUTOMATA

M.A. VAN DER HOEF and D. FRENKEL
FOM Institute for Atomic and Molecular Physics, Kruislaan 407, 1098 SJ Amsterdam, The Netherlands

Received 2 January 1990

We report simulations of tagged particle diffusion in three-dimensional lattice gas cellular automata (LGCA). In particular we looked at the decay of the velocity autocorrelation function (VACF) using a new technique that is about a million times more efficient than the conventional techniques. For longer times the simulations clearly show the algebraic $t^{-3/2}$ tail of the VACF. We compare the observed long-time tail with the predictions of mode-coupling theory. In three dimensions, the amplitude of this tail is found to agree within the (small) statistical error with these predictions.

1. Introduction

The velocity autocorrelation function (VACF) is a function of fundamental importance in atomic fluids. For instance, its time integral is equal to the diffusion constant D. In the Boltzmann approximation (collisions are uncorrelated) it is easy to demonstrate that the VACF should decay exponentially. However, for long times the breakdown of the Boltzmann approximation will become apparent and the decay will be much slower than exponential due to memory effects in the fluid. This behavior was first demonstrated by Alder and Wainwright [1] in a molecular dynamics simulation. Their findings indicated that for times much longer than the mean free time t_0, the VACF decays algebraically with an exponent $d/2$ where d is the dimensionality. In 1971 the algebraic decay was described theoretically by Ernst, Hauge and van Leeuwen [2]. In their analysis it is assumed that the long-time tail is the consequence of coupling between particle diffusion and shear modes in the fluid. More recent simulations of hard-core fluids in both two and three dimensions confirmed the existence of the algebraic tail [3, 4, 21]. However, it proved very difficult to determine the VACF with enough accuracy to find agreement with the theoretical prediction of the amplitude within reasonable small errors.

The idea to use lattice gases as a simulation model to calculate the VACF of a tagged particle is not new. The first attempts were by Boon and Noullez [5] for the FHP model, and by Binder and d'Humières [6] for 2D lattice Lorentz gas systems. Unfortunately, it turned out that for these calculations the statistics of lattice gases was worse than that of continuous systems. Hence the presence of a long-time tail was hidden in the statistical noise and a spectral analysis was required to demonstrate that the simulation data are, in fact, compatible with the presence of an algebraic tail of approximately the expected amplitude [7].

We recently developed a new method to calculate the VACF in a lattice gas simulation much more efficiently. The corresponding gain in computer time proved to be about a factor 10^6 for the 2D FHP model and a factor 10^9 for the lattice Lorentz gas model. The philosophy and theoretical details of this method are being published elsewhere [8, 9]. In this paper we will only show the results of such calculations for the three-dimensional FCHC model.

This paper consists of two parts. First we will very briefly go over some theoretical aspects of the model. In the second part we will give the simulation results.

2. Theoretical aspects

2.1. LGCA models

Let us briefly summarize the essentials of lattice gas cellular automata, in order to establish the notation. For more details the reader is referred to, for instance, refs. [10–13]. Our notations will be as close as possible to the notations in the first papers by Frisch et al., and is as follows:

Particles can have possible velocities $\boldsymbol{c}_i$, $i \in \{1, 2, \ldots, b\}$, $|\boldsymbol{c}_i| = c$. The time evolution of the LGCA consists of two steps:

(1) Propagation: All particles move in one time step Δt from their initial lattice position $\boldsymbol{r}$ to a new position $\boldsymbol{r}' = \boldsymbol{r} + \boldsymbol{c}_i \, \Delta t$. For convenience we choose $\Delta t = 1$.

(2) Collision: The particles at all lattice nodes undergo a collision that conserves the total number of particles and the total momentum at each node. The collision rules may or may not be deterministic.

The state of the automaton at time(-point) t is completely given by $s_i(\boldsymbol{r}, t)$, which is equal to 1 (0) if a particle is present (absent) on node $\boldsymbol{r}$ with velocity $\boldsymbol{c}_i$. The density is then defined as $d = \sum_i \langle s_i(\boldsymbol{r}, t) \rangle / b$, which is the average number of particles per link. The lattice used in our simulation is the face-centered hyper-cubic (FCHC) lattice [11, 13]. In this lattice there are 24 possible velocities, so a collision would require a 2^{24}-word lookup table, which requires a very large shared memory [14]. In the algorithm used in the present paper the 24-bit state is split into two 12-bit substates [15], which requires only a small 12-bit lookup table. This splitting can be done in 6 different ways, one of which is chosen randomly at every collision. The parameter $\mathrm{Re}_*^{\max}$ measuring the effectiveness of the collision rules is about 2, similar to Hénon's isometric rules [16]. Much higher values for $\mathrm{Re}_*^{\max}$ can be achieved [14, 17]. However, for the present simulation a high value of $\mathrm{Re}_*^{\max}$ is not necessary, hence we employ the simple 12-bit rules. An explicit expression for the kinematic viscosity and the diffusion coefficient that follows from the Boltzmann–Enskog equation for this 3D model is given in the next section.

2.2. Diffusion coefficient and viscosity of the 3D LGCA

As mentioned before, the diffusion coefficient is equal to the time integral of the VACF. The time-discrete equivalent to this so-called Green–Kubo relation has proved to apply to LGCA models [18] and reads (with time step $\Delta t = 1$):

$$D = \tfrac{1}{2} \langle v_x^2(0) \rangle + \sum_{n=1}^{\infty} \langle v_x(0) \, v_x(n) \rangle \,. \qquad (1)$$

In the Boltzmann–Enskog approximation this may be written as

$$D_0 = \tfrac{1}{2} \langle v_x^2(0) \rangle + \langle v_x^2(0) \rangle \frac{Z_{\mathrm{N}}(1)}{1 - Z_{\mathrm{N}}(1)} \qquad (2)$$

with $Z_{\mathrm{N}}(1)$ the normalized correlation function after one time step. In the 3D lattice gas this function is equal to

$$Z_{\mathrm{N}}(1) = \sum_s d^{p-1} (1-d)^{24-p} \sum_{i=1}^{24} \frac{s_i c_{ix}}{\sum_j s_j} \,,$$

where the summation is over all configurations with $s_1 = 1$, with the convention that $c_{1x} = 1$. The summation can be carried out over all states with the same number of particles p separately,

Table 1
Coefficients C_p and A_p determining the kinematic viscosity and the Enskog diffusion coefficient, respectively, for the FCHC model used in this paper. p is the number of particles.

p	C_p	A_p	p	C_p	A_p
1	0	1	13	630920	49742
2	4	11	14	465708	35530
3	78	77	15	280756	21318
4	718	385	16	137952	10659
5	4170	1463	17	54828	4389
6	17296	4389	18	17296	1463
7	54828	10659	19	4170	385
8	137952	21318	20	718	77
9	280756	35530	21	78	11
10	465708	49742	22	4	1
11	630920	58786	23	0	0
12	698148	58786	24	0	0

yielding

$$Z_N(1) = \sum_{p=1}^{24} A_p d^{p-1}(1-d)^{24-p}, \tag{3}$$

where the coefficients A_p are given in table 1. The initial value $\langle v_x^2(0)\rangle$ of the VACF is 0.5 for this FCHC model. An explicit formula for the kinematic viscosity in the Boltzmann approximation has been derived by Hénon [19]. For our collision rules this formula can be written in the following way:

$$\nu_0 = \frac{1}{4}\left(\sum_{p=1}^{24} C_p d^{p-1}(1-d)^{23-p}\right)^{-1} - \frac{1}{6}. \tag{4}$$

The coefficients C_p are given in table 1.

2.3. Mode-coupling theory for LGCA

The formula for the first-order long-time corrections to the VACF as derived by Ernst, Hauge and van Leeuwen for continuous 3D systems is

$$\langle v_x(0)\, v_x(t)\rangle = \frac{2}{3}\frac{\langle v_x^2(0)\rangle}{\rho[4\pi(D_0+\nu_0)t]^{3/2}}, \tag{5}$$

where ρ, ν_0 and D_0 are the particle density, the diffusion constant and the kinematic viscosity, respectively, both in the Boltzmann–Enskog approximation. The extension of this formula to lattice gases is to some extent straightforward, but there are a few subtle points which are peculiar to the discrete system, like the Fermi exclusion principle, which give rise to non-trivial modifications. The exact derivation is given by Ernst [20]. See also ref. [9]. Here we will only give the final result for the three-dimensional system:

$$\langle v_x(0)\, v_x(t)\rangle = \frac{2}{3}\frac{N_0(1-\rho/b)\langle v_x^2(0)\rangle}{\rho[4\pi(D_0+\nu_0)t]^{3/2}} = \frac{d_0^*\langle v_x^2(0)\rangle}{t^{3/2}} = \frac{d_0}{t^{D/2}}, \tag{6}$$

where d_0^* is the normalized tail coefficient. The density ρ is defined as the average number of particles per site. The values of the parameters b, $\langle v_x^2(0)\rangle$ and N_0 formula (16) are respectively 24, $\frac{1}{2}$, 1 for the FCHC model.

2.4. Moment-propagation method

We will now briefly discuss the new technique to calculate the VACF of a tagged particle in a lattice gas simulation.

In the techniques used thus far, a single tagged particle is followed along its classical trajectory. An estimate of $\langle v_x(0)\, v_x(t)\rangle$ is then obtained as an average of $v_x(0)\, v_x(t)$ over different time origins in this trajectory, different particles and different initial conditions. To obtain reasonable statistics with these "brute force" methods one needs lengthy simulations on large systems.

In the moment-propagation method we make use of the fact that we can choose our stochastic collision rules such that in a collision with the tagged particle involved any outgoing particle is equally likely to be the tagged particle. This results in a large number of possible paths for the tagged particle, each associated with a certain probability. It can be shown [8, 9] that because of the lattice symmetry the calculation of the VACF as an average over all these paths can be done very efficiently. Secondly we can make use of the fact that the particles are identical, and use every particle in the lattice as a starting particle in one simulation. We do not need to identify the contribution of each individual particle, as long as we make a correction for the different initial velocities of the different particles. The result is that in one run over t time steps in a system with N particles, the averaging is over approximately $N\rho^t$ different events, where ρ is the average number of particles per site. Although the events are not totally uncorrelated, the number $N\rho^t$ is for times $t > 10$ of such magnitude that the gain in accuracy is enormous compared to the computational effort and, most importantly, the accuracy increases with increasing t, which makes the method extremely useful for calculating the VACF for longer times. In our simulations using this new technique we observed statistical errors of the order of 10^{-6} in both two and three dimensions. Compared with the only known simulation results for two-dimensional LGCA using conventional techniques [5, 6], this corresponds to a gain of 10^6 in computer time. Using this method we should not only be able to observe long-time tails, but furthermore be able to give a quantitative estimate of the amplitude of the long-time tails with such an accuracy that a comparison with mode-coupling theory is possible.

3. Simulation results

3.1. General aspects

The simulations were carried out on systems of up to $60 \times 60 \times 60$ lattice points. In all cases correlations were only computed for time intervals less than the shortest time in which any particle could cross the periodic box. This in contrast to corresponding simulations of long-time tails in atomic fluids [21] where time intervals up to five times the acoustic wave traversal time had to be used. In the present simulation the VACF is calculated for different densities varying from $d = 0.05$ to $d = 0.90$, where d is defined as the average number of particles per link per node ($d = \rho/b$). In order to estimate the statistical error of the VACF five to ten independent simulations per density were performed. All calculations were performed on a NEC-SX2 supercomputer.

3.2. Results and discussion

As a test we looked first at the VACF after one time step. Hydrodynamic effects are not yet possible and the Boltzmann value as derived in section 2.2 (formula (3)) should be observed in the simulation. In table 2 we show both the theoretical value and the simulation result for the VACF after one time step for different densities. We see that there is agreement almost within the error and there is no systematic deviation. We will first look in detail at the results for one density $d = 0.1$. In fig. 1 we show the normalized VACF for this density. The crosses are the simulation data, the squares are the error, which is of the order of 10^{-6}. For shorter times we see that the decay is

Table 2
Normalized velocity autocorrelation function after one time step for different densities d. The theoretical value is computed from the formula (3) of section 2.2.

d	$\langle v_x(0)\, v_x(1)\rangle_N$		
	theory	simulation	error
0.05	0.5721835	0.5720450	0.27E-03
0.10	0.3566233	0.3565930	0.16E-03
0.15	0.2405122	0.2407160	0.63E-04
0.20	0.1728864	0.1727750	0.16E-03
0.25	0.1302603	0.1303010	0.48E-04
0.30	0.1014215	0.1015020	0.17E-03
0.35	0.0807413	0.0808057	0.49E-04
0.40	0.0652169	0.0651721	0.45E-04
0.45	0.0531400	0.0531490	0.20E-04
0.50	0.0434783	0.0435255	0.86E-04
0.55	0.0355731	0.0355466	0.54E-04
0.60	0.0289855	0.0290297	0.21E-04
0.65	0.0234114	0.0234080	0.19E-04
0.70	0.0186335	0.0186572	0.14E-04
0.75	0.0144928	0.0145033	0.68E-05
0.80	0.0108696	0.0108432	0.10E-04
0.90	0.0048309	0.0048343	0.27E-05

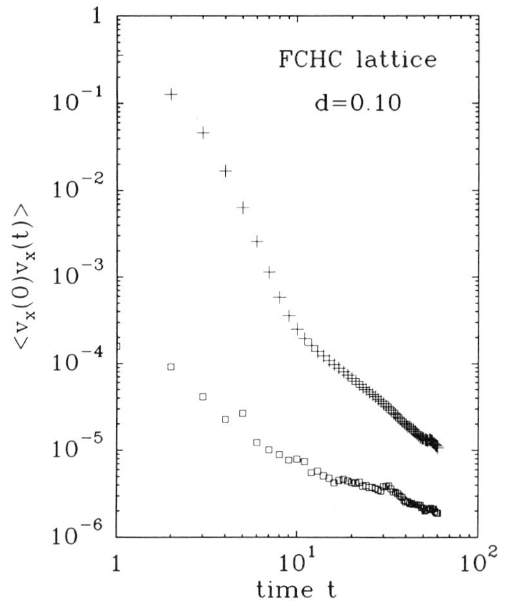

Fig. 1. Log–log plot of the normalized velocity autocorrelation function of a three-dimensional lattice gas cellular automaton on a FCHC lattice for density $d = 0.10$. The time is in units of the CA time step $\Delta t(=1)$. The crosses are the simulation data, the squares are the estimated error. Note that the error decreases with increasing t to a value of order 10^{-6}.

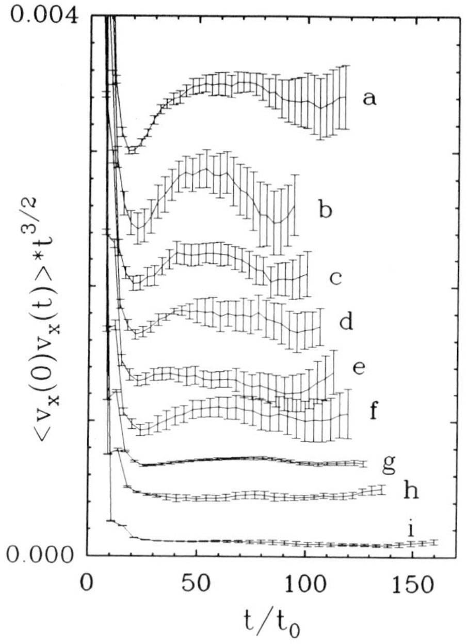

Fig. 2. Velocity autocorrelation function multiplied by $t^{3/2}$ for different densities: (a) $d = 0.45$, (b) $d = 0.5$, (c) $d = 0.55$, (d) $d = 0.6$, (e) $d = 0.65$, (f) $d = 0.7$, (g) $d = 0.75$, (h) $d = 0.8$, (i) $d = 0.9$. The time is in units of mean free time t_0.

approximately exponential, and for longer times we clearly observe the algebraic decay, which appears in the log–log plot as a straight line. The slope of the line has about the correct value of the exponent, $-3/2$. Note that only the fact that the error is decreasing with increasing time allows us to observe long-time tails. A more convenient way to present the results is by multiplying the VACF with $t^{3/2}$. This is shown in fig. 2 for a range of densities. If the decay is algebraic with the exponent $-3/2$ these functions should reach a constant value in the limit $t \gg t_0$. This behavior is indeed observed. These plateau values can be identified with the amplitude of the tail and can be compared with the prediction of mode-coupling theory. In fig. 3 we show these amplitudes (squares) together with the theoretical prediction (solid line). We see that there is agreement with the predicted amplitude for all densities. We wish to stress that this agreement is achieved with no

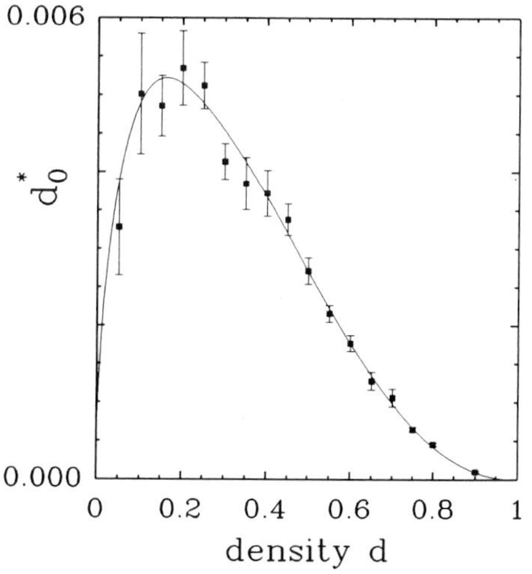

Fig. 3. Normalized tail coefficient d_0^* for the three-dimensional system as a function of the density (see eq. (6)). The solid line is the prediction of mode-coupling theory, the points are the simulation data. The density is in units of average number of particles per link.

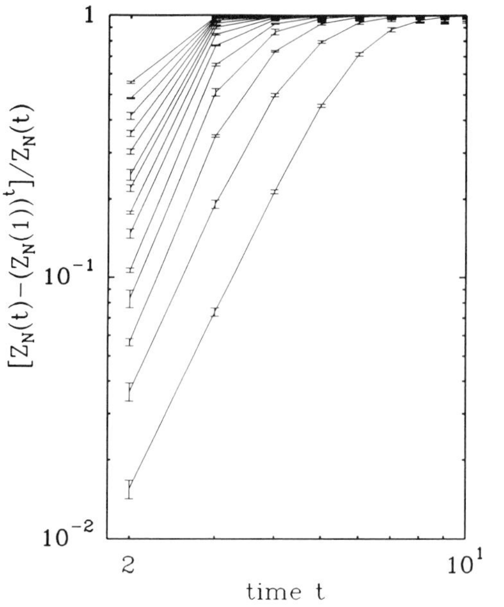

Fig. 4. The relative deviation of the observed VACF $Z_N(t)$ from the exponential decaying VACF $[Z_N(1)]^t$ for different densities ranging from $d = 0.15$ (lowest curve) to $d = 0.8$ (highest curve). The change in d from one curve to the next is 0.05. Time is in units of the CA time step Δt.

Table 3
Comparison between the Boltzmann–Enskog diffusion coefficient (theoretical value) and the "true" diffusion coefficient calculated from the simulations, which incorporates the hydrodynamic contributions. d is the density.

d	Diffusion coefficient		
	Boltzmann–Enskog	simulation ("true")	error
0.05	0.9187252	0.9206111	0.18E-05
0.10	0.5271497	0.5305632	0.19E-05
0.15	0.4083385	0.4120427	0.12E-05
0.20	0.3545119	0.3578774	0.24E-05
0.25	0.3248846	0.3280692	0.20E-05
0.30	0.3064344	0.3091144	0.15E-05
0.35	0.2939165	0.2963107	0.25E-05
0.40	0.2848834	0.2870077	0.26E-05
0.45	0.2780612	0.2800275	0.93E-06
0.50	0.2727273	0.2743430	0.14E-05
0.55	0.2684426	0.2697141	0.10E-05
0.60	0.2649254	0.2660255	0.84E-06
0.65	0.2619863	0.2628251	0.10E-05
0.70	0.2594937	0.2601514	0.11E-05
0.75	0.2573529	0.2578591	0.73E-07
0.80	0.2554945	0.2557927	0.22E-06
0.90	0.2524272	0.2525252	0.12E-06

adjustable parameter. In comparison with theory we note that the hydrodynamic correction to the VACF is only predicted by mode-coupling theory for the long-time regime. Our data are, however, accurate enough to give a meaningful prediction of the deviation from exponential decay for shorter times. In fig. 4 we plotted the relative difference between the exponentially decaying VACF and the observed VACF for a number of densities. The deviation proved to have approximately the same density dependence as the long-time amplitude. Note that the deviation from Boltzmann at $t = 2$ is due to correlations that are caused by the periodicity of only one lattice spacing in the fourth dimension in the FCHC model. Finally we look at the influence of the hydrodynamic corrections on the transport coefficient, that is, the deviation of the observed, "true" diffusion coefficient from the Boltzmann–Enskog value. In table 3 we show the "true" diffusion coefficient, which we calculated as the time sum of the observed VACF (see ref. [18]), and the Boltzmann–Enskog value as exploited in section 2.2, both for a range of densities. As we see, the "true" diffusion constant is at most one percent larger than the Boltzmann–Enskog value. This justifies the assumption that we could use the Boltzmann–Enskog coefficient D_0 in eq. (13) instead of the "true" diffusion coefficient, with respect to the observed error in the tail amplitude. Similarly the difference between ν and ν_0 proved to be very small [22]. Another factor that might influence the mode-coupling prediction is the presence of spurious invariants. It is known that many LGCA models have spurious "unphysical" conserved quantities that are related to the discrete nature of the model and may couple to the hydrodynamic modes [23]. However, none of the known invariants in the FCHC model has the right symmetry to affect the results of the lowest-order mode-coupling theory (eq. (16)).

In summary, we used a new technique that allowed us to calculate the velocity autocorrelation function in a simulation with sufficient accuracy to give a quantitative estimate of the amplitude of the hydrodynamic long-time tail. In this way we could test the prediction of mode-coupling theory for the density dependence of this amplitude. In three dimensions the comparison indicated that, for all densities studied, the coefficient of the algebraic long-time tail is predicted by mode-coupling theory within the (small) statistical error. This finding strongly supports the validity of the basic assumptions of mode-coupling theory for 3D atomic fluids in general.

Acknowledgements

We would like to thank Tony Ladd, Matthieu Ernst and Uriel Frisch for stimulating and useful discussions and for their interest in our work. This work is part of the research program of the Stichting voor Fundamenteel Onderzoek der Materie (Foundation for Fundamental Research on Matter) and was made possible by financial support from the Nederlandse Organisatie voor Zuiver Wetenschappelijk Onderzoek (Netherlands Organization for the Advancement of Research). Computer time on the NEC-SX2 at NLR was made available through a grant by the NFS (Nationaal Fonds Supercomputers).

References

[1] B.J. Alder and T.E. Wainwright, Phys. Rev. A 1 (1970) 18.
[2] M.H. Ernst, E.H. Hauge and J.M.J. van Leeuwen, Phys. Rev. A 4 (1971) 2055.
[3] D. Levesque and W.T. Ashurst, Phys. Rev. Lett. 33 (1974) 277.
[4] J.J. Erpenbeck and W.W. Wood, Phys. Rev. A 26 (1982) 1648.
[5] J.-P. Boon and A. Noullez in: Proceedings of Workshop on Discrete Kinetic Theory, Lattice Gas Dynamics and Foundations of Hydrodynamics, ed. R. Monaco (World Scientific, Singapore, 1989).
[6] P.M. Binder and D. d'Humières, Los Alamos preprint LA-UR-1341 (1988); in Proceedings of Workshop on Discrete Kinetic Theory, Lattice Gas Dynamics and Foundations of Hydrodynamics, ed. R. Monaco (World Scientific, Singapore 1989).
[7] A. Noullez and J.P. Boon, Physica D 47 (1991) 212, these Proceedings.
[8] D. Frenkel and M.H. Ernst, Phys. Rev. Lett. 63 (1989) 2165.
[9] M.A. van der Hoef and D. Frenkel, Phys. Rev. A 41 (1990) 4277.
[10] U. Frisch, B. Hasslacher and Y. Pomeau, Phys. Rev. Lett. 56 (1986) 1505.
[11] U. Frisch, D. d'Humières, B. Hasslacher, P. Lallemand, Y. Pomeau and J.-P. Rivet, Complex Systems 1 (1987) 649.
[12] D. d'Humières and P. Lallemand, Complex Systems 1 (1987) 599.
[13] D. d'Humières, P. Lallemand and U. Frisch, Europhys. Lett. 2 (1986) 291.
[14] J.-P. Rivet, M. Hénon, U. Frisch and D. d'Humières, Europhys. Lett. 7 (1988) 231.
[15] A.J.C. Ladd and D. Frenkel, in: Proceedings of Workshop on Cellular Automata and Modelling of Complex Physical Systems, ed. P. Manneville (Springer, Berlin, 1989).
[16] M. Hénon, Complex Systems 1 (1987) 475.
[17] J.A. Somers and P.C. Rem, in: Proceedings of Workshop on Cellular Automata and Modelling of Complex Physical Systems, ed. P. Manneville (Springer, Berlin, 1989).
[18] M.H. Ernst and J.W. Dufty, Phys. Lett. A 138 (1989) 391; J.W. Dufty and M.H. Ernst, J. Chem. Phys. 93 (1989) 7015.
[19] M. Hénon, Complex Systems 1 (1987) 763.
[20] M.H. Ernst, Physica D 47 (1991) 198, these Proceedings.
[21] J.J. Erpenbeck and W.W. Wood, Phys. Rev. A. 32 (1985) 412.
[22] A.J.C. Ladd, private communication.
[23] G. Zanetti, Phys. Rev. A 40 (1989) 1539.

Physica D 47 (1991) 198–211
North-Holland

MODE-COUPLING THEORY AND TAILS IN CA FLUIDS

M.H. ERNST
Institute for Theoretical Physics, University of Utrecht, Utrecht, The Netherlands

Received 20 December 1989

After a summary of important effects of mode-coupling theory in continuous fluids, the theory is extended to CA fluids and the long-time tails calculated for the velocity correlation function and for the stress–stress correlation function, including the spurious contributions from the staggered momentum densities. At finite densities the Fermi statistics strongly suppresses the long-time singularities arising from two fluid modes, but not those from tagged particle diffusion.

1. Introduction

The velocity correlation function and other current correlation functions in equilibrium fluids have long-time behavior with power law decay. In computer simulations of hard disks one finds [1] that the velocity autocorrelation function (VACF) of a tagged particle, $\langle \boldsymbol{v}(t) \cdot \boldsymbol{v}(0) \rangle$ has a *positive* long-time tail, proportional to $t^{-d/2}$, where d is the dimensionality of the system. It is caused by the following mechanism. The initial momentum of the tagged particle is transferred in part to the surrounding fluid particles. This sets up a hydrodynamic vorticity flow around the tagged particle, which in turn transfers some of its momentum back to the tagged particle ("kick in the back"), thus yielding a positive correlation between $\boldsymbol{v}(0)$ and $\boldsymbol{v}(t)$. This correlation extends over hydrodynamic time scales. In a Lorentz gas [2] however $\langle \boldsymbol{v}(t) \cdot \boldsymbol{v}(0) \rangle \sim -t^{-1-d/2}$, i.e. after a long time t the moving particle returns to its point of origin on the average moving in a direction opposite to its initial velocity, instead of parallel as in the fluid case. The VACF is *negative* at large times and the correlations are weaker than in the fluid.

A quantitative calculation of long time tails for general current–current correlation functions in continuous fluids can be made using mode-coupling theory [3–8] or hard-sphere kinetic theory [9]. It was found that all current–current correlation functions have positive long-time tails $\sim t^{-d/2}$. The mode-coupling theory has been used in ref. [3, 6] to calculate the dominant and subleading tails of time correlation functions and to derive non-analytic dispersion relations for the relaxation rates of hydrodynamic excitations. The theoretical results have been confirmed in part in extensive computer simulations on hard-sphere systems [1].

There are a number of interesting implications of long-time tails in continuous fluids. The goal of this paper is to investigate to what extent these results are relevant for *lattice gas cellular automata* (LGCA). They will be briefly reviewed before any quantitative details on LGCA will be discussed.

(i) *Non-existence of two-dimensional linear hydrodynamics*. This is the most dramatic consequence. As the self-diffusion and Navier–Stokes transport coefficients are given as time integrals (or time sums in LGCA) over slowly decaying tails $\sim t^{-1}$, these integrals are diverging as $\ln(t)$ and the transport coefficients do not exist. In one-dimensional systems this slow decay would imply an even stronger divergence of the transport coefficients. It is an interesting open problem to determine the form of the macroscopic fluid dynamic equations in these low-dimensional systems. Only above the crossover dimension ($d = 2$) the standard Navier–Stokes equations apply.

(ii) *t^{-1} tails with bare transport coefficients*. The general structure of the long-time tails, say for the VACF, $\langle \boldsymbol{v}(t) \cdot \boldsymbol{v}(0) \rangle \sim [(D + \nu)t]^{-d/2}$, involve the transport coefficients themselves, where D is the coefficient of self-diffusion and ν the kinematic viscosity. In two dimensions this result is inconsistent for asymptotically large times, because D and ν are diverging logarithmically. However, in computer simulations a large time regime has been observed, up to a few hundred mean free times between collisions [1, 10], where t^{-1} behavior is observed. This behavior is well described by mode-coupling theory if the Navier–Stokes transport coefficients $D, \nu, \ldots$ are replaced by their *bare*, or *short-time* or Boltzmann–Enskog values $D^0, \nu^0, \ldots$. These bare transport coefficients can be calculated from the Boltzmann–Enskog equation, which is exact in the limit of short times.

(iii) *Non-analytic frequency dependence* of the transport coefficients. The long-time tails $\sim t^{-d/2}$ imply *small-s singularities* in the Laplace transform $\Phi(s)$ of the current–current correlation functions $\varphi(t)$, namely

$$\begin{aligned} \Phi(s) &\sim \ln s && (d=2), \\ \Phi(s) &\sim \Phi_0 + a\sqrt{s} && (d=3), \\ \Phi(s) &\sim \Phi_0 + z\Phi_1 + \ldots + as^{d/2-1}/\Gamma(-d/2) && (d>3). \end{aligned} \tag{1}$$

(iv) *Non-analytic dispersion relations*. The elementary long-wavelength excitations in a continuous fluid are the hydrodynamic modes, $\psi^{\mu}(\boldsymbol{k}, t)$, where $\boldsymbol{k}$ is the wave vector of the excitation. According to the Landau–Placzek theory their time correlation functions or *hydrodynamic propagators*, $G_{\mu}(\boldsymbol{k}, t)$, decay as

$$G_{\mu}(\boldsymbol{k}, t) = \langle \psi^{\mu}(\boldsymbol{k}, t) | \psi^{\mu}(\boldsymbol{k}) \rangle \simeq \mathrm{e}^{-t\omega_{\mu}(\boldsymbol{k})}, \tag{2}$$

where $\omega_{\mu}(\boldsymbol{k})$ are the hydrodynamic frequencies or relaxation rates. The modes are linear combinations of the Fourier transforms of the fluctuations in number, energy, longitudinal and transverse momentum densities [6]. They are normalized such that $G_{\mu}(\boldsymbol{k}, 0) = 1$. The inner product is defined as $\langle a|b \rangle = V^{-1}\langle ab^* \rangle$. The standard Navier–Stokes equations yield analytic dispersion relations

$$\omega_{\mu}(k) \simeq \mathrm{i}\, c_{\mu} k + \Gamma_{\mu} k^2 + \ldots, \tag{3}$$

for small wave numbers. Here c_{μ} is the sound velocity for a propagating mode and $c_{\mu} = 0$ for diffusive modes such as heat and shear modes. However, the mode-coupling theory yields the following asymptotic expansion at small k:

$$\begin{aligned} \omega_{\mu}(k) &\simeq \mathrm{i}\, c_{\mu} k + \gamma_{\mu} k^2 \ln k + \ldots && (d=2), \\ \omega_{\mu}(k) &\simeq \mathrm{i}\, c_{\mu} k + \gamma_{\mu} k^2 + (\ldots) k^{5/2} + (\ldots) k^{7/4} + \ldots && (d=3). \end{aligned} \tag{4}$$

There is partial experimental confirmation of these theoretical predictions in neutron scattering experiments.

(v) *Higher-order tails*. The higher-order mode-coupling corrections to the frequencies in (3) also induce corrections to long-time tails [5–8]. One finds for the VACF,

$$\frac{\langle \boldsymbol{v}(t)\cdot\boldsymbol{v}(0)\rangle}{\langle \boldsymbol{v}\cdot\boldsymbol{v}\rangle} \simeq \frac{d_0}{t} + \frac{d_1 \ln(t/\tau_0)}{t} + \ldots \quad (d=2),$$

$$\simeq \frac{d_0}{t^{3/2}} + \frac{d_1}{t^{7/4}} + \ldots \quad (d=3). \tag{5}$$

Similar results apply to the current–current correlation functions with explicit expressions for the coefficients d_l $(l = 0, 1, \ldots)$ given in ref. [7]. In the two-dimensional case these coefficients involve bare transport coefficients. The $t^{-1}\ln(t)$ "correction" to the t^{-1} tail only sets in after a few hundred mean free times. It is a reflection of the non-existence of the linear transport coefficient in two-dimensional systems. The sign of d_1 is *negative*, signalling already that in a fully self-consistent theory the "superlong" time behavior would actually decay faster than given by (5).

(vi) *Self-consistent mode-coupling theory* [8]. In item (ii) it was noted that the mode coupling theory for the VACF yields

$$D(t) \sim \int_0^t \mathrm{d}\tau\, \varphi(\tau), \qquad \varphi(t) \sim [(D+\nu)t]^{-d/2} \tag{6}$$

leading to inconsistencies for $d \leq 2$. Alternatively, one may consider (6) and similar time integrals for the Navier–Stokes transport coefficients as a self-consistent set of equations defining $D(t)$ and $\nu(t)$. The solutions of these equations supposedly provide the "superlong" time tails of the correlation functions. To illustrate the structure of the solutions we drop $\nu(t)$ in (6). This yields the differential equation

$$\mathrm{d}D/\mathrm{d}t \sim (Dt)^{-d/2} \quad (t \to \infty). \tag{7}$$

Its solution in two dimensions is

$$D(t) \sim \sqrt{\ln t}\,, \quad \varphi(t) \sim (t\sqrt{\ln t})^{-1} \quad (t \to \infty). \tag{8}$$

The same time dependence applies to the Navier–Stokes transport coefficients in two dimensions. Indeed, the decay is (marginally) faster at superlong times. This implies for the Laplace transform $\Phi(s)$ of $\varphi(t)$

$$\Phi(s) \sim \sqrt{\ln s} \quad (s \to 0). \tag{9}$$

In the one-dimensional case the self-consistent treatment would yield the superlong time behavior or singular small-s behavior,

$$D(t) \sim t^{1/3}, \qquad \varphi(t) \sim t^{-2/3} \quad (t \to \infty), \qquad \Phi(s) \sim s^{-1/3} \quad (s \to 0). \tag{10}$$

In the context of lattice gas cellular automata fluids one has considered one-dimensional fluid-type models with momentum conservation [14]. Sound waves are the only two hydrodynamic modes of such systems and the only possible sources for long-time behavior in the mode-coupling theory. The

Green–Kubo formula for the longitudinal viscosity or sound damping constant $\Gamma(s)$ contains the time correlation function $\varphi_\nu(t)$ of the microscopic stress.

If we assume that the self-consistent mode-coupling theory applies, the long-time tail of $\varphi_\nu(t) \sim t^{-2/3}$ yields a non-analytic contribution to the sound damping constant and $\Gamma(s) \sim s^{-1/3}$ for small s. On the basis of these estimates one may expect that the small-k form of the sound dispersion relation has the asymptotic solution

$$\omega_+(k) \simeq \mathrm{i}\, c_0 k + (\ldots) k^{5/3} \quad (k \to 0). \tag{11}$$

A qualitative confirmation of this crude estimate is found in ref. [14], where computer simulations show that the damping or real part of (11), divided by k^2, increases anomalously at small k values. The above self-consistency argument predicts a small-k increase as $k^{-1/3}$.

For LGCA the mode coupling theory has first been formulated by Kadanoff et al. [15]. They have investigated the consequences of the small-s divergence in the two-dimensional Navier–Stokes transport coefficient by non-equilibrium simulation of Poisseuille flow using the FHP model. Their analysis of the results using mode-coupling theory for the long-time tails leads to good agreement. Many simulations of long-time tails have been performed [10–12, 16, 17]. Only the simulations of Frenkel and van der Hoef have sufficient accuracy to allow an analysis in terms of mode-coupling theory.

After this review the lattice gas models will be defined and the notation established. A lattice gas or CA fluid is defined as a collection of N unlabeled indistinguishable point particles, located at the sites $\boldsymbol{r} = r_1\boldsymbol{e}_1 + r_2\boldsymbol{e}_2 + \ldots$ of a d-dimensional Bravais lattice, containing V sites and having periodic boundary conditions, where $\{\boldsymbol{e}_i\}$ with $i = 1, 2, \ldots, d$ is the set of basis vectors spanning the unit cell. The total volume of the unit cell and system are v_0 and $v_0 V$, respectively. For example, on the square lattice $\boldsymbol{e}_1 = (1,0)$, $\boldsymbol{e}_2 = (0,1)$ and $v_0 = 1$ and on the triangular lattice $\boldsymbol{e}_1 = (1,0)$, $\boldsymbol{e}_2 = \frac{1}{2}(1, \sqrt{3})$ and $v_0 = \frac{1}{2}\sqrt{3}$. At each site there exists a set of b allowed velocity states ($i = 0, 1, \ldots, b-1$), connecting the site with its nearest neighbors, next-nearest neighbors, etc. The model is also referred to as a b-bit model. Rest particles, having $|\boldsymbol{c}| = 0$, are also allowed in certain models. In each state, specified by $\{\boldsymbol{c}_i, \boldsymbol{r}\}$, at most one particle is allowed. So the particles obey the *Fermi exclusion rule*. The state of the whole system is described by the set of occupation numbers $n(\cdot) = \{n(\boldsymbol{c}_i, \boldsymbol{r})\}$. If a state specified by $\{\boldsymbol{c}_i, \boldsymbol{r}\}$ is occupied or empty, $n(\boldsymbol{c}_i, \boldsymbol{r})$ equals 1 or 0, respectively. The total phase space contains bV one-particle states.

The time evolution of the system is governed by propagation and collisions that conserve the total number of particles N and the total momentum $\boldsymbol{P}$ at each site, at least in fluid-type models that propagate sound waves. Energy is either trivially conserved or not conserved at all. The dynamics can be expressed formally by the time evolution equation, where the collision term depends in a non-linear way on the occupation numbers $n(\boldsymbol{c}_i, \boldsymbol{r}; t)$ at site $\boldsymbol{r}$ with $i = 0, 1, 2, \ldots, b-1$. Beyond this general form no further restrictions are required on the collision term.

When attempting to compute the VACF of a particle in a LGCA, one is immediately confronted with a conceptual problem. As all lattice gas particles are indistinguishable, the VACF of a particle is ill defined. As soon as a particle has collided it is no longer possible to tell which of the outgoing particles is the original one whose VACF we are attempting to compute. To avoid this problem, the particle under consideration must be labeled differently from the rest (say, a "red" one in a sea of "whites") and collision rules for the "red" and "white" particles must be specified separately. This can and has been done in many different ways [10–12, 16, 17].

Frenkel and collaborators have defined the following tagged particle dynamics, described in refs. [10, 11]. The collision rules for a tagged particle with different untagged particles result in the occupation of the same output states as in the case of collisions between untagged particles. And, most importantly, the tagged particle *has equal probability to be in any of the occupied output states*. Hence, for the tagged particle the collision rules are stochastic, although to a "color-blind" observer, the rules appear deterministic. Note also that even in "collisions" that have the same input and output states the tagged particle may still change its state. The essential feature of these rules is that the average velocity of a tagged particle after a collision at site $\boldsymbol{r}$, depends only on the (colorless) state at that site. The details of tagged particle dynamics does not enter in the mode coupling theory, except through the value of the diffusion coefficient (see section 3).

The plan of this paper is as follows. In section 2 a more intuitive derivation for the long time tails of the VACF is presented, which is a straightforward generalization of the concepts for continuous fluids [3]. In section 3 the more formal mode-coupling theory of ref. [4], adapted to LGCA in ref. [18], is applied to the stress–stress correlation function in the two-dimensional FHP model. The coefficient of the t^{-1} tail not only involves the usual hydrodynamic modes, but also the staggered momentum modes. The suppression of the long-time tails in CA fluids through the Fermi statistics is explained and illustrated by computer simulations for the time-dependent longitudinal viscosity in a one-dimensional CA fluid.

2. Velocity correlations in CA fluids

A typical example of equilibrium time correlation functions is the velocity autocorrelation function,

$$\langle v_x(t)\, v_x(0)\rangle = \sum_{rc} \sum_{r_0 c_0} c_{0x} c_x \langle n^*(\boldsymbol{c}_0, \boldsymbol{r}_0; 0)\, n^*(\boldsymbol{c}, \boldsymbol{r}; t)\rangle, \tag{12}$$

where the velocity of the tagged particle can be expressed in occupation numbers as

$$\boldsymbol{v} = \sum_{rc} \boldsymbol{c} n^*(\boldsymbol{c}, \boldsymbol{r}; t). \tag{13}$$

Here $n^*(\boldsymbol{c}, \boldsymbol{r}; t)$ and $n(\boldsymbol{c}, \boldsymbol{r}; t)$ are respectively the occupation numbers of a tagged and a fluid particle. The value $n^* = 1$ or $n = 1$ indicates that the occupied state contains respectively a tagged or a fluid particle.

To calculate its long-time behavior I follow the mode coupling theory of ref. [3] and introduce the special non-equilibrium ensemble average $\langle \dots \rangle_s$,

$$\langle A(\boldsymbol{c}, \boldsymbol{r}; t)\rangle_s = \langle A(\boldsymbol{c}, \boldsymbol{r}; t)\, n^*(\boldsymbol{c}_0, \boldsymbol{r}_0; 0)\rangle / \langle n^*(\boldsymbol{c}_0, \boldsymbol{r}_0; 0)\rangle. \tag{14}$$

The velocity correlation function becomes then

$$\langle v_x(t)\, v_x(0)\rangle = (bV)^{-1} \sum_{rc} \sum_{r_0 c_0} c_x c_{0x} \langle n^*(\boldsymbol{c}, \boldsymbol{r}; t)\rangle_s. \tag{15}$$

Its initial value is

$$\langle v_x^2(0)\rangle = c_0^2 = (bd)^{-1} \sum_c |\boldsymbol{c}|^2, \tag{16}$$

where c_0 is the speed of sound. We assume that the local tagged particle current density can be approximated for large space and time scales by the coarse grained expression

$$\sum_{rc} \boldsymbol{c}\langle n^*(\boldsymbol{c},\boldsymbol{r};t)\rangle_s \simeq \sum_{r} \boldsymbol{u}(\boldsymbol{r},t)\, P(\boldsymbol{r},t) = V^{-1}\sum_{q} \boldsymbol{u}(-\boldsymbol{q},t)\, P(\boldsymbol{q},t). \tag{17}$$

This assumption implies that the velocity $\boldsymbol{c}$ of the tagged particle at site $\boldsymbol{r}$ is equal to the local fluid velocity $\boldsymbol{u}(\boldsymbol{r},t)$. Then $\Sigma_c\langle n^*(\boldsymbol{c},\boldsymbol{r},t)\rangle$ reduces to the probability $P(\boldsymbol{r},t)$ to find the tagged particle at site $\boldsymbol{r}$ in the special non-equilibrium ensemble (14). Similarly $\boldsymbol{u}(\boldsymbol{r},t)$ is the local fluid velocity in that ensemble

$$P(\boldsymbol{r},t) = \sum_{c} \langle n^*(\boldsymbol{c},\boldsymbol{r};t)\rangle_s, \qquad \rho \boldsymbol{u}(\boldsymbol{r},t) = \sum_{c} \boldsymbol{c}\langle n(\boldsymbol{c},\boldsymbol{r};t)\rangle_s. \tag{18}$$

Note that the first equation refers to the tagged particle (n^*) and the second to a fluid particle (n). Furthermore $\rho = bf$ is the fraction of occupied sites in the equilibrium state and the basic averages are

$$\langle n\rangle = f = N/bV, \qquad \langle n^*\rangle = f/N = 1/bV. \tag{19}$$

Here $1/N$ is the fraction of tagged particles, V is the number of sites and b the number of velocity states per site. The total volume $v_0 V$ is a V-tuple of the volume of the unit cell v_0. Below we also need the fluctuation formulae

$$\begin{aligned}
\langle n^*(\boldsymbol{c}\boldsymbol{r})\, n^*(\boldsymbol{c}_0\boldsymbol{r}_0)\rangle &= \langle n^*\rangle^2(1-\delta_{cc_0}\delta_{rr_0}) + \langle n^*\rangle \delta_{cc_0}\delta_{rr_0} \\
&\simeq (bV)^{-1}\delta_{cc_0}\delta_{rr_0} + \mathcal{O}(V^{-2}), \\
\langle n(\boldsymbol{c}\boldsymbol{r})\, n^*(\boldsymbol{c}_0\boldsymbol{r}_0)\rangle &= (bV)^{-1}f + (bV)^{-1}(1-f)\delta_{cc_0}\delta_{rr_0}.
\end{aligned} \tag{20}$$

To determine the long-time behavior of (17) we use the solution of the diffusion equation for the tagged particle, $P(\boldsymbol{q},t) = \exp(-Dq^2 t)\, P(\boldsymbol{q},0)$, and decompose $\boldsymbol{u}(\boldsymbol{q},t)$ in hydrodynamic modes

$$\boldsymbol{u}(\boldsymbol{q},t) = \hat{q}u_\ell(\boldsymbol{q},t) + \boldsymbol{u}_\perp(\boldsymbol{q},t), \tag{21}$$

where $u_\ell = \hat{q}\cdot\boldsymbol{u}$ is the longitudinal component of $\boldsymbol{u}(\boldsymbol{q},t)$ and $\boldsymbol{u}_\perp$ are the remaining transverse components. The component $u_\ell(\boldsymbol{q},t)$ will decay as a sound mode and will finally contribute terms proportional to t^{-2} for long times in two dimensions [3]. The transverse components decay as $d-1$ degenerate shear modes

$$u_{\perp\alpha}(\boldsymbol{q},t) = \mathrm{e}^{-\nu q^2 t}(\delta_{\alpha\beta} - \hat{q}_\alpha\hat{q}_\beta)u_\beta(\boldsymbol{q},0). \tag{22}$$

In the two-dimensional case the transport coefficients D and ν have their (short-time) Enskog values. The initial values of the slow densities can be calculated from eqs. (19), (18) with the result

$$P(\boldsymbol{r},0) = \delta_{rr_0}, \qquad \rho u_\alpha(\boldsymbol{r},0) = (1-f)c_{0\alpha}\delta_{rr_0}. \tag{23}$$

The extra factor $1-f$ in (23) guarantees that the state occupied by the tagged particle does not contain a fluid particle. Inserting these results into (17) yields

$$\begin{aligned}\sum_{rc} c_\alpha \langle n^*(\boldsymbol{c},\boldsymbol{r};t)\rangle_s &\simeq (\rho V)^{-1}(1-f)\sum_{\boldsymbol{q}}(\delta_{\alpha\beta}-\hat{q}_\alpha\hat{q}_\beta)c_{0\beta}\,\mathrm{e}^{-(\nu+D)q^2t} \\ &\simeq v_0 c_{0\alpha}(bf)^{-1}(1-f)\int\frac{\mathrm{d}\boldsymbol{q}}{(2\pi)^d}\,\mathrm{e}^{-(\nu+D)q^2t} \\ &= v_0 c_{0\alpha}(bf)^{-1}(1-f)d^{-1}(d-1)[4\pi(\nu+D)t]^{-d/2}. \qquad (24)\end{aligned}$$

In the transition from a Fourier sum to an integral we also changed from the oblique coordinate system with $\boldsymbol{q}=q_1\boldsymbol{e}_1^*+q_2\boldsymbol{e}_2^*+\dots$ to Cartesian coordinates $\boldsymbol{q}=\{q_x,q_y,\dots\}$. Consequently $\mathrm{d}q_1\,\mathrm{d}q_2\dots = v_0\,\mathrm{d}q_x\,\mathrm{d}q_y\dots = v_0\,\mathrm{d}\boldsymbol{q}$, where the volume of the unit cell v_0 is the Jacobian of this transformation. This can be directly verified from the definition of the biorthonormal sets of basis vectors $\{\boldsymbol{e}_i, i=1,2,\dots,d\}$ and their reciprocals $\{\boldsymbol{e}_j^*=1,2,\dots,d\}$ with $\boldsymbol{e}_j^*\cdot\boldsymbol{e}_i=\delta_{ij}$. We recall that $v_0=1$ for square and cubic lattices and $v_0=\frac{1}{2}\sqrt{3}$ for the triangular lattice.

Combining eqs. (24) and (15) yields the VACF for LGCA at long times:

$$\begin{aligned}\langle v_x(t)\,v_x(0)\rangle &\simeq d^{-1}(d-1)(bf)^{-1}(1-f)v_0c_0^2[4\pi(\nu+D)t]^{-d/2} \\ &\equiv d_0^* c_0^2 t^{-d/2}, \qquad (25)\end{aligned}$$

where c_0 is the sound velocity. The coefficient of the long-time tail in the normalized VACF, $C(t)=\langle v_x(t)\,v_x(0)\rangle/\langle v_x^2(0)\rangle$ is the same as in the continuous fluid, apart from the extra factor $(1-f)$, discussed below (23). Here $\rho/v_0=bf/v_0$ is the number density per unit volume or surface area.

In LGCA there exist unphysical or spurious conservation laws [15, 18], leading to the slow staggered momentum modes. These modes can in principle give rise to spurious components in the long-time tails, as will be demonstrated in the next section for the stress–stress correlation function. However, the staggered modes do not couple to the tagged particle current $\Sigma_c \boldsymbol{c}n^*(\boldsymbol{c},\boldsymbol{r};t)$, and the coefficient $d_0^*=(1-f)d_0$ of the long-time tail has almost the same form as in the continuous fluid, apart from the correction factor $(1-f)$ due to Fermi statistics.

To investigate the long-time tail in LGCA and to test the mode-coupling theory, computer simulations [10–12] have been performed on the two-dimensional FHP-III model and on the three-dimensional FCHC model. The measured amplitude are compared with the predictions from mode-coupling theory in the accompanying paper by van der Hoef [12], where Boltzmann–Enskog values ν_0 and D_0 for the transport coefficients have been used. The kinematic viscosity for the FHP-III model was calculated by d'Humières and Lallemand [20] and the self-diffusion coefficient for the tagged particle, with the collision rules specified above, was calculated by Frenkel and the author [10] with the result

$$\begin{aligned}\nu_0 &= 28f(1-f)/[1-\tfrac{8}{7}f(1-f)]-\tfrac{1}{8}, \\ D_0 &= \tfrac{3}{7}(1/\lambda_0-\tfrac{1}{2}), \\ \lambda_0 &= 7\sum_{m=0}^{5}\frac{1}{m+2}\binom{5}{m}f^{m+1}(1-f)^{5-m} \qquad (26)\end{aligned}$$

with $f=\frac{1}{7}\rho$. Similar calculations have been carried out in refs. [11, 12] for the three-dimensional FCHC model.

The simulation results of Frenkel and collaborators constitute by far the most accurate verification to date of the mode-coupling prediction for hydrodynamic long-time tails in two- and three-dimensional systems (compare ref. [1]). In two dimensions there is a small but systematic deviation of a few percent between theory and simulations in the density range $0 < f < 0.5$, which remains to be explained.

In the two-dimensional lattice gases, in the long-time intervals studied, one would expect to observe the $t^{-1}\ln(t)$ correction to the t^{-1} tail. However, the coefficient d_1^* for the LGCA has not been calculated. The value of d_1/d_0, calculated for the hard-disk system should be easily observable in the simulations of ref. [10]. However, there is no evidence for a faster than t^{-1} decay at any density. It is not known to what extent the Fermi statistics of the fluid particle affects the numerical value of the ratio d_1^*/d_0^*. In general, the Fermi statistics strongly suppresses the magnitude of the long-time singularities in LGCA, as I will show in the next section. In three-dimensional LGCA the coefficient d_1 of the $t^{-7/4}$ tail has neither been measured nor calculated.

3. Stress–stress correlations in CA fluids

In this section a more formal and general presentation of mode-coupling theory is given, that includes cases of spurious slow modes, the staggered momentum densities. No specific dimension, lattice structure or CA dynamics is implied, however. It can be applied directly to the time correlation functions in the Green–Kubo relations, as derived in ref. [19].

We use the mode-coupling theory for LGCA of ref. [18], where the long-time behavior of the current–current correlation function is given by

$$\langle \hat{J}_\alpha(\boldsymbol{k},t)|\hat{J}_\beta(\boldsymbol{k})\rangle \to (2V)^{-1} \sum_{\boldsymbol{q},\mu,\nu} M^{\mu\nu}_{\alpha\beta}(\boldsymbol{q},\boldsymbol{k}-\boldsymbol{q})\, G_\mu(\boldsymbol{q},t)\, G_\nu(\boldsymbol{k}-\boldsymbol{q},t). \tag{27}$$

Here $G_\mu(\boldsymbol{k},t)$ are the hydrodynamic propagators, defined in (2). For $d>2$ they decay as pure exponentials in the long-wavelength limit with $G_\mu(\boldsymbol{k},t)=\exp[-t\omega_\mu(k)]$. For dimensionality $d \le 2$ this need not be true. The two-mode strength factors are given by

$$M^{\mu\nu}_{\alpha\beta}(\boldsymbol{q},\boldsymbol{l}) = \langle \hat{J}_\alpha(\boldsymbol{k})|\psi^\mu(\boldsymbol{q})\,\psi^\nu(\boldsymbol{l})\rangle\langle \psi^\mu(\boldsymbol{q})\,\psi^\nu(\boldsymbol{l})|\hat{J}_\beta(\boldsymbol{k})\rangle. \tag{28}$$

The hydrodynamic eigenvalues in (3) presume the existence of the small-s limit of the frequency-dependent damping constants or transport coefficients $\Lambda(s)$. This is always possible to assure for finite geometry, but in two-dimensional systems the limit may not exist in the large system size limit. The source of this difficulty lies in the assumption that the hydrodynamic eigenvalues are analytic in the wavevector k. More generally, the dispersion relations for the slow excitations have the form (4) or equivalently

$$\omega_\mu(k) = \mathrm{i}\, c_\mu k + \Gamma_\mu(k)\, k^2, \tag{29}$$

where $\Gamma_\mu(k)$ is in general a non-analytic function of k,

$$\Gamma_\mu(k) = \Gamma^0_\mu + \tfrac{1}{2}\sum_{t=0}^{\infty} \mathrm{e}^{\omega_\mu(k)t} \sum_{\boldsymbol{q},\rho,\nu} M^{\rho\nu}_{\mu\mu}(\boldsymbol{q},\boldsymbol{k}-\boldsymbol{q})\, G_\rho(\boldsymbol{q},t)\, G_\nu(\boldsymbol{k}-\boldsymbol{q},t) \tag{30}$$

and Γ_μ^0 is the bare transport coefficient arising from the faster excitations. With the above results the hydrodynamic description for the long-time behavior of time correlation functions has also been extended to LGCA. The reader is referred to ref. [18] for the explicit forms of the microscopic modes ψ_k^μ, appropriate for lattice gases, and for the hydrodynamic relaxation rates expressed in Green–Kubo expressions for the Navier–Stokes transport coefficients $\Lambda_\mu(s)$.

As a direct application of the mode-coupling formula (27) we quote results for the long-time tail in the stress–stress correlation function, obtained by Naitoh et al. [21],

$$\varphi_\eta(t) = \langle \hat{J}_{xy}(t) | \hat{J}_{xy} \rangle = V^{-1} \langle \hat{J}_{xy}(t) \hat{J}_{xy} \rangle \tag{31}$$

with microscopic stress

$$\hat{J}_{xy}(t) = \sum_{rc} c_x c_y [n(c, r; t) - f] \,. \tag{32}$$

The result applies only to LGCA with an isotropic viscosity tensor. For a calculation of the two mode strength factors one needs fluctuation formulae, similar to those in (20). From here on the calculations are very similar to those for the continuous fluid [4]. The contributions from shear and sound modes read for dimensionality $d \geq 2$

$$\varphi_\eta^{\eta\eta}(t) \simeq A_{\eta\eta} \left(\frac{1}{8\pi\nu t} \right)^{d/2} , \tag{33}$$

$$\varphi_\eta^{+-}(t) \simeq A_{+-} \left(\frac{1}{8\pi \Gamma t} \right)^{d/2} , \tag{34}$$

where ν is the kinematic viscosity and Γ the sound mode damping constant, defined in (3). The sound velocity c_0 is defined in (16) and the constants A have a density dependence

$$A \sim (1 - 2f)^2 / f(1 - f) \tag{35}$$

The long-time tail $\sim t^{-d/2}$ in the stress–stress correlation function has the same general form as in continuous fluids [4]. The shear viscosity is related to the time integral of this correlation function. Consequently the *viscosity will diverge logarithmically* in two dimensions. However, these formulae are expected to be still applicable in the two-dimensional case for intermediate time intervals up to a few hundred collision times. Then the actual transport coefficients ν and Γ can be replaced by bare or Enskog transport coefficients ν^0 and Γ^0. This expectation is fully confirmed for the special case of the velocity correlation function [10, 12].

For the two-dimensional FHP model there exist three *spurious slow modes*, the staggered momentum densities, which also have to be included in the two mode terms. The calculations are somewhat more involved [21] and yield for this model

$$\varphi_\eta^{\theta\theta}(t) \simeq A_{\theta\theta} \frac{1}{\pi t \sqrt{D_1^0 (D_1^0 + D_2^0)}} , \tag{36}$$

where D_1^0 and D_2^0 are the bare diffusivities of the staggered modes [18, 22]. It is of interest to point out that the Fermi statistics strongly suppresses the long-time tails of *fluid-type* current correlation functions as opposed to the tagged particle velocity correlation function in section 2. This occurs around the half-filled lattice ($f = 1/2$), because all long-time tails in eqs. (33)–(36) are proportional to $(1-2f)^2$. For instance at $f = 0.4$ the prefactor is only 4% of its low-density value. The source of this suppression are the mode-coupling amplitudes in (28). They are proportional to $\langle(\delta n)^3\rangle = f(1-f)(1-2f)$ as a consequence of the Fermi statistics.

One-dimensional fluid-type models frequently exhibit pathological features, due to the strong dynamical correlations between particles moving on a line. However, as already discussed in (11), it appears that mode-coupling theory accounts, at least in a qualitative manner, for the anomalous damping of sound waves in a one-dimensional CA fluids [14]. It seems therefore worthwhile to further test the robustness of mode-coupling concepts in one-dimensional systems. So we consider a one-dimensional fluid-type cellular automaton, with five velocities per site ($c = 0, \pm 1, \pm 2$; $b = 5$), introduced by d'Humières et al. [14], with possible collisions

$$\begin{aligned} &(-1) + (+1) \Leftrightarrow (+2) + (-2), \\ &(0) + (-1) \Leftrightarrow (+1) + (-2), \\ &(0) + (+1) \Leftrightarrow (-1) + (+2). \end{aligned} \tag{37}$$

One may consider two versions of the model: the *self-dual version, where the collisions occur irrespective of the presence of the fifth particle* with respective velocities (0), (-2) or $(+2)$ at the same site, and the *non-self-dual version, where the collisions only occur in absence of a fifth particle*. The model has two collisional invariants ($a_n = 1$, $a_t = c$) and $c_0^2 = \frac{1}{5}\sum_c c^2 = 2$. The frequency-dependent longitudinal viscosity is given by

$$\nu(s) = \frac{1}{\chi} \sum_{t=0}^{\infty} e^{-st} \varphi_\nu(t) - \tfrac{1}{2}\varphi_\nu(0), \tag{38}$$

where the susceptibility $\chi_\ell = 10f(1-f)$ and the correlation function of longitudinal stresses is

$$\varphi_\nu(t) = \langle \hat{J}(t) | \hat{J} \rangle \tag{39}$$

with a subtracted current

$$\hat{J}(t) = \sum_{rc} (c^2 - 2)[n(\boldsymbol{c}, \boldsymbol{r}; t) - f]. \tag{40}$$

If Navier–Stokes transport coefficients would exist, the limit $s \to 0$ of (38) would yield the transport coefficient. However, the Green–Kubo expression (38) in one dimension is expected to diverge strongly as $s \to 0$ due to long memory effects.

Use of the mode-coupling results (27)–(30), interpreted as a self-consistent definition for $\nu(s)$, gives the equation

$$\nu(s) = \nu_0 + M/2\chi\sqrt{s\nu(s)} \tag{41}$$

with the amplitude

$$M = (\tfrac{7}{10})^2(1-2f)^2. \tag{42}$$

The small-s solution of this equation shows that the longitudinal viscosity grows as

$$\nu(s) \simeq (M/2\chi)^{2/3} s^{-1/3} \quad (s \to 0). \tag{43}$$

In order to test these predictions de Smet et al. [23] have performed computer simulations on this one-dimensional fluid and measured the time-dependent longitudinal viscosity, defined as

$$\tilde{\nu}(t) = \frac{1}{\chi} \sum_{\tau=0}^{t} \varphi_\nu(\tau) - \tfrac{1}{2}\varphi_\nu(0). \tag{44}$$

Its long-time behavior follows from the small-s behavior in (43) with the help of Tauberian theorems with the result

$$\tilde{\nu}(t) \simeq (M/2\chi)^{2/3} t^{1/3} / \Gamma(4/3) \quad (t \to \infty). \tag{45}$$

If only the uncorrelated collisions of the Boltzmann approximation would be taken into account, one would find that $\tilde{\nu}(t)$ approaches for long times the bare viscosity

$$\begin{aligned} \nu^0 &= 3/10f(1-f) - \tfrac{7}{10} \quad \text{self-dual}, \\ &= 3/10f(1-f)^2 - \tfrac{7}{10} \quad \text{non-self-dual}. \end{aligned} \tag{46}$$

Fig. 1 shows preliminary results [23] for the collisional part $\tilde{\nu}_c(t) = \tilde{\nu}(t) + \frac{7}{10}$ of the longitudinal viscosity in this one-dimensional CA fluid at reduced densities $f = 0.1$ and $f = 0.5$. The low-density simulations cover only a relatively short time interval of about $100f = 10$ mean free times. Therefore one does not expect that $\tilde{\nu}(t)$ has reached its asymptotic tail $\sim t^{1/3}$. Nevertheless the measured values tend to keep increasing with t, in qualitative agreement with (45).

On the other hand, around the half-filled lattice ($f = \frac{1}{2}$) the mode-coupling contributions are substantially reduced by a factor $M^{2/3} \sim (1-2f)^{4/3}$ on account of (45). The measured $\tilde{\nu}(t)$ values in this density range seem to reach a plateau value in a few mean free times, which is still in qualitative agreement with the Boltzmann result (46). For instance, the measured values of $\tilde{\nu}_c(100)$ after hundred time steps in the non-self-dual model are 2.27 (2.04), 2.46 (2.40), 5.04 (4.76) at reduced densities $f = 0.3, 0.5, 0.7$, respectively. The numbers inside parentheses are the Boltzmann values ν_c^0 of the collisional part in (46).

4. Conclusions

Lattice gas cellular automata may be considered as bona fide, although extremely simplified, statistical mechanical models of non-equilibrium systems with many degrees of freedom. In particular, they are able to model many aspects of hydrodynamics and transport properties of fluids. They seem therefore most suitable to tackle and test some fundamental problems in nonequilibrium statistical mechanics and kinetic theory of fluids, such as the long-time tails of current–current correlation functions.

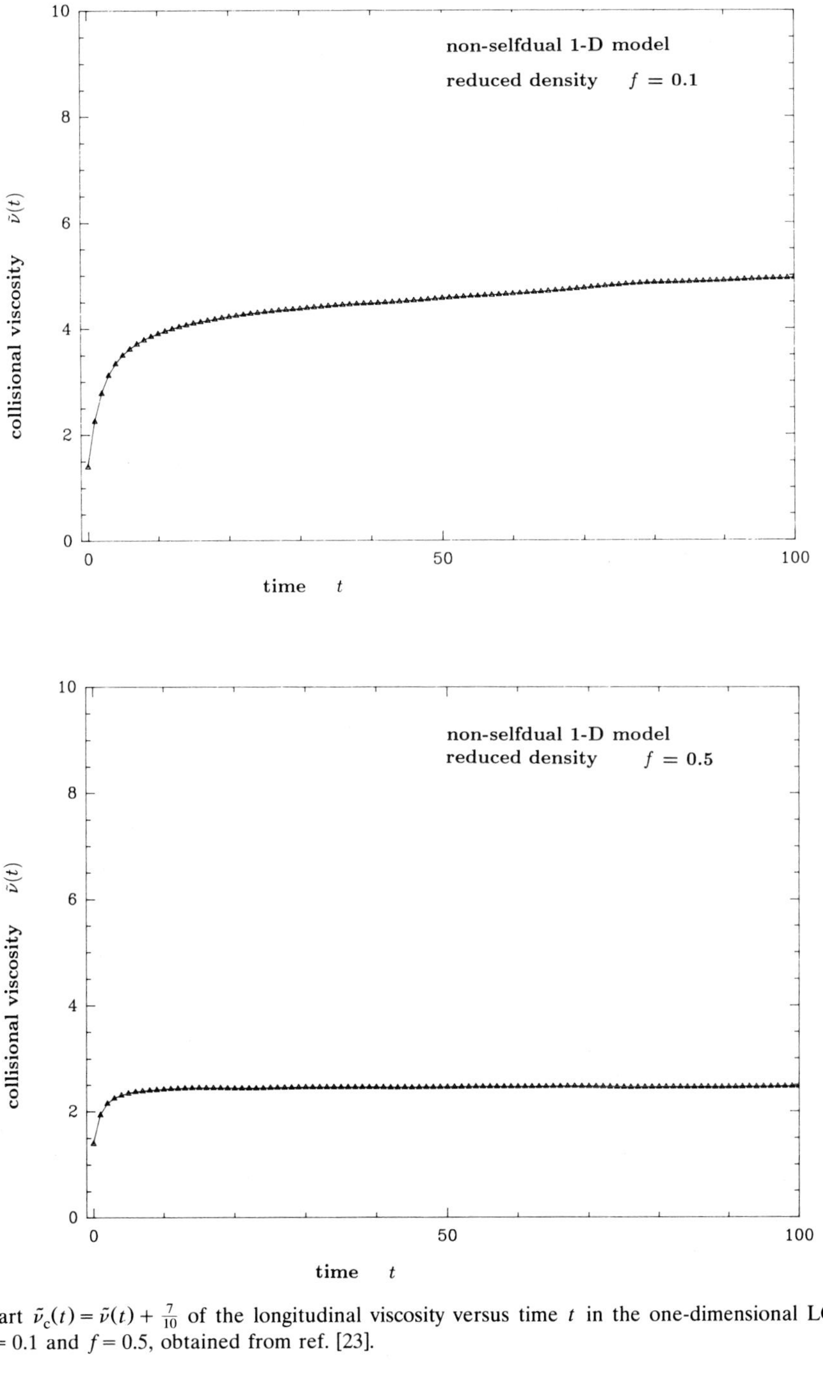

Fig. 1. Collisional part $\tilde{\nu}_c(t) = \tilde{\nu}(t) + \frac{7}{10}$ of the longitudinal viscosity versus time t in the one-dimensional LGCA of eq. (37) at reduced densities $f = 0.1$ and $f = 0.5$, obtained from ref. [23].

The fundamental problems on the existence of long-time tails in continuous fluids has been reviewed in section 1 and the mode-coupling theory has been extended to lattice gases in section 2 and 3. In fluid-type LGCA that admit sound waves there exists a *positive* long-time tail of the form $At^{-d/2}$, where d is the dimensionality. The constant of proportionality A is calculated in section 2 for several typical cases. In the work of Frenkel and collaborators [10–12] these results are compared with computer simulations. The simulation results on the velocity autocorrelation function of a tagged particle in the two-dimensional FHP fluid [10] and in the three-dimensional FCHC fluid [11, 12] are in excellent agreement with the mode-coupling results, presented here. We also discussed some simulation results of a one-dimensional CA fluid, shown in fig. 1, mainly to emphasize the point that the Fermi statistics strongly suppresses the hydrodynamic long-time singularities and also to show the robustness of the mode-coupling theory, even for the low-dimensional systems where the Navier–Stokes equations no longer exist.

The long-time tail indicates that the transport coefficients do not exist in dimensions less than or equal to 2. Therefore it is expected that Navier–Stokes hydrodynamics has transport properties that depend intrinsically on the boundary conditions, or alternatively that the hydrodynamic equations have a more complex structure. This makes the study of linearized hydrodynamics, by theory or simulation, particularly interesting for the low-dimensional LGCA.

Acknowledgements

It is a pleasure to acknowledge stimulating discussions with D. Frenkel, M. van der Hoef and J.W. Dufty. I am grateful to G. Zanetti for making a remark about a factor of $\frac{1}{2}\sqrt{3}$, which removed the last disagreement between the two-dimensional results for the VACF in the FHP model, obtained from mode-coupling theory and computer simulations. I also want to thank CNLS at Los Alamos National Laboratory for their hospitality and support during the summer of 1989, when part of this work has been done.

References

[1] B.J. Alder and T.E. Wainwright, Phys. Rev. A 1 (1970) 18;
J.J. Erpenbeck and W.W. Wood, Phys. Rev. A 26 (1982) 1648; Phys. Rev. A 32 (1985) 412.
[2] E.H. Hauge, in: Transport processes, eds. G. Kirczenow and J. Marro, Lecture Notes in Physics, No. 31 (Springer, Berlin, 1974) p. 338.
[3] M.H. Ernst, E.H. Hauge and J.M.J. van Leeuwen, Phys. Rev. A 4 (1971) 2055.
[4] M.H. Ernst, E.H. Hauge and J.M.J. van Leeuwen, J. Stat. Phys. 15 (1976) 7.
[5] Y. Pomeau, Phys. Rev. A 5 (1973) 2569.
[6] M.H. Ernst and J.R. Dorfman, J. Stat. Phys. 12 (1976) 311.
[7] I.M. de Schepper and M.H. Ernst, Physica A 87 (1977) 35;
I.M. de Schepper, H. van Beijeren and M.H. Ernst, Physica 75 (1974) 1.
[8] Y. Pomeau and P. Resibois, Phys. Rep. 19 (1975) 64.
[9] J.R. Dorfman and E.G.D. Cohen, Phys. Rev. A 6 (1972) 776; A 12 (1975) 292.
[10] D. Frenkel and M.H. Ernst, Phys. Rev. Lett 63 (1989) 2165.
[11] M. van der Hoef and D. Frenkel, Phys. Rev. A 41 (1990) 4277
[12] M. van der Hoef and D. Frenkel, Physica D 47 (1991) 191–197, these Proceedings.
[13] J. Bosse, W. Götze and M. Lücke, Phys. Rev. A 20 (1979) 1603;
I.M. de Schepper, P. Verkerk, A.A. van Well and L.A. de Graaf, Phys. Lett. A 104 (1984) 29;
P. Verkerk, J. Westerweel, U. Bafile, L.A. de Graaf, W. Monfrooij and I.M. de Schepper, Phys. Rev. A 40 (1989) 2860.

[14] D. d'Humières, P. Lallemand and Y.H. Qian, Compt. Rend. Acad. Sci. Paris II 308 (1988) 585; preprint ENS (1989).
[15] L.P. Kadanoff, G. McNamara and G. Zanetti, Phys. Rev. A 40 (1990) 4527.
[16] P.M. Binder, D. d'Humières and L. Poujol, in: Proceedings of Workshop on Discrete Kinetic Theory, Lattice Gas Dynamics and Foundations of Hydrodynamics, ed. R. Monaco (World Scientific, Singapore, 1989) p. 38.
[17] J.P. Boon and A. Noullez, in: Proceedings of Workshop on Discrete Kinetic Theory, Lattice Gas Dynamics and Foundations of Hydrodynamics, ed. R. Monaco (World Scientific, Singapore, 1989) p. 400.
[18] M.H. Ernst and J.W. Dufty, J. Stat. Phys. 58 (1990) 57.
[19] M.H. Ernst and J.W. Dufty, Phys. Lett. A 138 (1989) 391.
[20] D. d'Humières and P. Lallemand, in: Lattice Gas Methods for Partial Differential Equations, ed. G.D. Doolen (Addison–Wesley, Reading, MA, 1989) p. 297.
[21] T. Naitoh, M.H. Ernst and J.W. Dufty, Phys. Rev. A to be published.
[22] G. Zanetti, Phys. Rev. A 40 (1989) 1539.
[23] J. de Smet, D. Frenkel and M.H. Ernst, Internal Report, University of Utrecht (November 1989).

Physica D 47 (1991) 212–215
North-Holland

LONG-TIME CORRELATIONS IN A 2D LATTICE GAS

Alain NOULLEZ and Jean-Pierre BOON
Physique Non-linéaire et Mécanique Statistique, Faculté des Sciences, C.P. 231, Université Libre de Bruxelles, 1050 Bruxelles, Belgium

Received 15 February 1990

We present the results of CA simulations of a moderately dense lattice gas using an extended version of the FHP model for colored automata. By tracking a tagged particle, we construct its velocity autocorrelation function and we show that the corresponding power spectrum exhibits a low-frequency contribution characteristic of the long-time power law behavior ($\approx t^{-1}$, in 2D).

The persistence of long-time correlations in fluid systems and the ensuing consequences in 2D and 3D hydrodynamics have been of considerable interest in non-equilibrium statistical mechanics [1]. From the theoretical viewpoint, this long-time effect is important because its existence is in contradiction of the Boltzmann–Enskog transport theory. In physical terms, momentum transfer from a tagged particle to the surrounding fluid sets up a vortical back-flow acting on the particle and this feedback process is a manifestation of the violation of the molecular chaos hypothesis. Such a back-flow effect has been known to exist at the macroscopic level since the mid-nineteen century [2]; however the stimulating impact in modern physics was triggered, about twenty years ago, by computer simulations and theoretical work showing the consequences of hydrodynamic long-time correlations at the molecular scale. The stimulation came from simulation results for the velocity autocorrelation function (VACF) of hard spheres [3] confirmed subsequently by further computer experiments on a Lennard-Jones type fluid [4].

The now conventionally called *long-time tail* (LTT) effect met a recent regain of interest through the development of lattice gas automata (LGA) methods [5] which provide a new approach to fluid dynamical problems and use powerful (yet simple) computational techniques for simulating systems with a large number of colliding particles. It is worth mentioning that LGAs are well suited for the investigation of long-time tail effects. Indeed, the amplitude of the effect decreases with viscosity (responsible for the particle slowing down) and diffusion coefficient, and globally increases with density (by emphasis of the hydrodynamic back-flow inducing retardation of the decay of correlations) [6]; the lattice gas exhibits a conveniently accessible range of moderate densities where the kinematic viscosity and the diffusion coefficient are optimally minimized [7].

We performed LGA simulations using a two-color fourteen-bit extension of the FHP-IV two-dimensional seven-bit model. The color-blind collision rules optimize the number of non-transparent collisions to maximize second-order momentum transfer, thereby minimizing the kinematic viscosity. Correspondingly, the value of the diffusion coefficient is reduced by optimizing the number of collisions with "colored" momentum transfer. This LGA model is referred to as "limited diffusion model" and has been described previously [7]. Preliminary simulations of tagged particle diffusion using this model were reported elsewhere [8]. Recent computations were performed with system sizes up to 480×384 at a density $d = 0.3$, and the tagged particle velocities were recorded up to 2^{18} steps. The velocity autocorrelation function $\Psi(t)$ was constructed as described in ref. [8].

We define the discrete Fourier transform of $\Psi(t)$

as

$$\tilde{\Psi}(\omega) = \sum_{t=-\infty}^{\infty} \Psi(t) \exp(-i\omega t). \qquad (1)$$

Note that the diffusion coefficient D_s (including what is generally called the "propagative" contribution) is given by the zero-frequency limit of (1), i.e.

$$D_s = \frac{1}{2D} \lim_{\omega \to 0} \tilde{\Psi}(\omega) , \qquad (2)$$

where D denotes the space dimension (here $D = 2$). Since the number of data points is finite and $\Psi(t)$ is known only at integer times, (1) must be rewritten as

$$\tilde{\Psi}\left(\frac{\pi k}{M\tau}\right) = \sum_{j=-M}^{M} \Psi(j\tau) \exp\left(-\frac{i\pi jk}{M}\right),$$
$$k = 0, \ldots, M, \qquad (3)$$

where M is the number of points used in the computation of $\tilde{\Psi}(\omega)$ and τ is the propagation time (taken here as unit of time, so that $\tau = 1$). Corresponding to the finite time resolution of $\Psi(t)$, there is an upper limit to the frequency range over which $\tilde{\Psi}(\omega)$ is known (Shannon sampling theorem). For a signal sampled at successive discrete times with increment $\Delta t = \tau$, this maximum ($\omega_{\max} = \pi/\tau$) corresponds to the frequency $(2\tau)^{-1}$. In order to extend the frequency range beyond $\omega_{\max}$, and to reduce aliasing from the higher frequencies, one defines $\Psi(t)$ at non-integer times by linear interpolation between data points, i.e.

$$\Psi(t + \delta) = \Psi(t) + \delta[\Psi(t+1) - \Psi(t)],$$
$$0 \leq \delta \leq 1. \qquad (4)$$

This linear interpolation of the VACF follows from the fact that in LGA, the velocity is constant between collision steps and changes abruptly at integer times. We used the interpolation method to obtain an eight-fold increase in resolution on $\Psi(t)$, so that we could compute $\tilde{\Psi}(\omega)$ up to $\omega = 8\pi/\tau$ corresponding to $f = 4/\tau$.

If in (3), the sum extends to the number N of data points of the VACF, that is to the number of velocity data points, the relative error on the power spectrum will be 100 %, whatever the number of data points N and their own relative accuracy ε; therefore, we must take $M \ll N$ to reduce the variance of the estimator (3) of $\tilde{\Psi}(\omega)$. We chose $M = 128$, because at 128τ, the VACF has decreased well below the noise level ($t = 128$ corresponds to ≈ 300 times the characteristic time of $\Psi(t)$ for $d = 0.3$), so that the noise signal is not included in the summation (3). To obtain a smooth decay to zero of the noise included in the VACF, we used a Hamming window in the time domain, i.e. we replaced $\Psi(t)$ by

$$\Psi_s(t) = \Psi(t)\,(0.54 + 0.46 \cos(\pi t/M), \qquad t \leq M,$$
$$= 0, \qquad t > M. \qquad (5)$$

Fig. 1a shows the VACF at various densities and fig. 2 shows the corresponding power spectrum $\tilde{\Psi}(f)/2D$ for the density $d = 0.3$. Note that the value at zero frequency $\tilde{\Psi}(0)/2D = 0.223$ is in agreement with the value of D_s obtained by random walk experiments performed with the present model at density $d = 0.3$ [7]. Our main result is the evidence of a peak centered at zero frequency in the power spectrum which corresponds to the t^{-1} power law of the 2D velocity long-time correlations. As discussed above, the manifestation of the LTT is enhanced in the optimal density range. Indeed, we could not detect its manifestation in simulations performed at $d = 0.05$ and $d = 0.1$ [8]. At these low densities, the VACF has a monotonic decay (fig. 1a) and its corresponding power spectrum is fit to very good accuracy with a single narrow Lorentzian centered at zero frequency, making it difficult to detect the presence of a low-amplitude zero-frequency narrow peak. For $d \geq 0.27$, the VACF acquires an oscillatory decay behavior as a function of time (*cage* effect); see fig. 1a. Consequently (without the LTT contribution) the power spectrum consists of a symmetrical double Lorentzian, with a dip at zero frequency. As a result, the spectral contribution of the long-time correlations is then most clearly evidenced since it manifests as a central peak (see fig. 2).

It should be noted that, even at these moderate densities, the exponentially decaying oscillat-

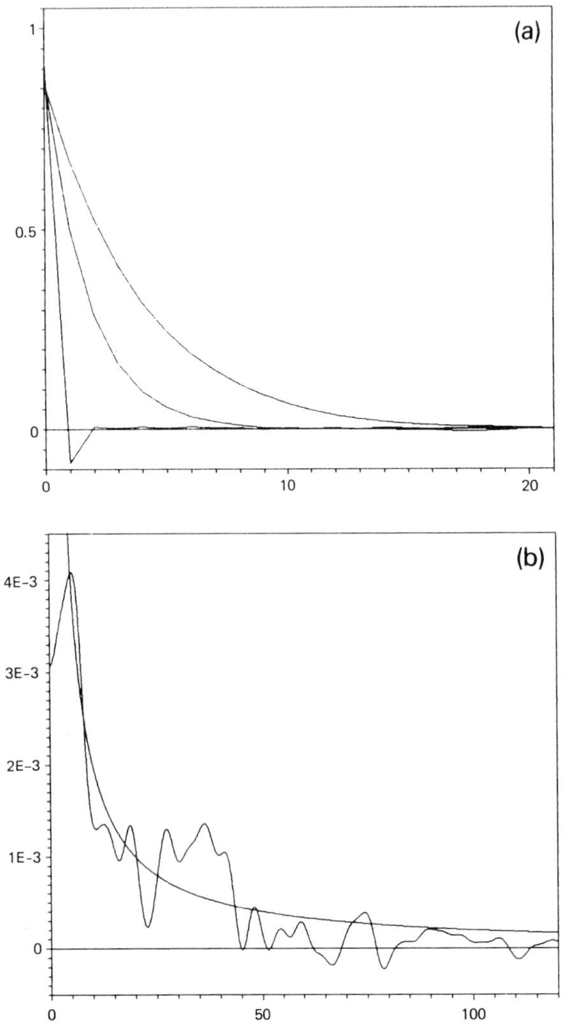

Fig. 1. The velocity autocorrelation function $\Psi(t)$ as a function of time (in lattice units). (a) VACF for increasing densities (from top to bottom: $d = 0.05$, $d = 0.1$, $d = 0.3$); (b) the long-time contribution obtained by subtracting the exponentially decaying Boltzmann function from the full VACF at $d = 0.3$; the solid curve is a t^{-1} fit.

ing part of the VACF can be predicted and fit to excellent accuracy (better than 0.5 %) by a Boltzmann calculation [9]. So we subtract the power spectrum of the Boltzmann part of $\Psi(t)$ from the full $\tilde{\Psi}(\omega)$ and inverse Fourier transform the difference, to obtain the long-time contribution to $\Psi(t)$. The result is shown in fig. 1b along with a fit $\Psi_{\mathrm{LT}}(t) = d_0/t$ with $d_0 = 2 \times 10^{-2}$. The time representation of the LTT is quite noisy and it

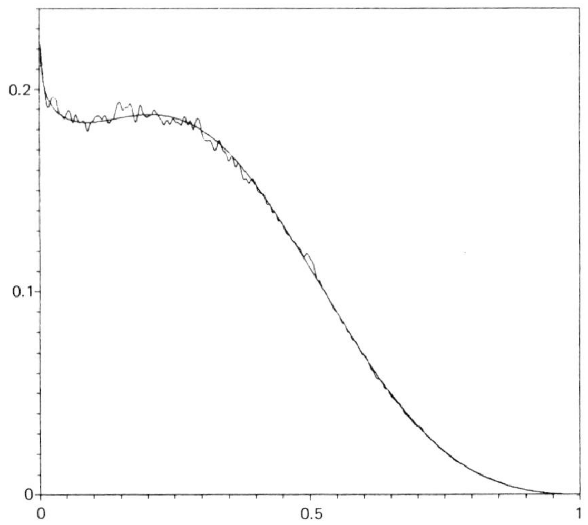

Fig. 2. The power spectrum of the VACF at $d = 0.3$ shown in fig. 1: $\tilde{\Psi}(f)/2D$ (unit of $f = \tau^{-1}$). The solid curve is a fit computed by combining the Boltzmann shifted Lorentzian and the Fourier transform of $\Psi_{\mathrm{LT}}(t)$, eq. (6).

is difficult to assess quantitatively the functional form of the long-time contribution or its amplitude. We then fit the long-time part of the VACF by

$$\begin{aligned} \Psi_{\mathrm{LT}}(t) &= at^2 \exp(-t/\tau), \qquad t \leq 4\tau, \\ &= d_0/t \qquad\qquad\qquad t \geq 4\tau, \end{aligned} \tag{6}$$

where a is such that the two expressions match at $t = 4\tau$. (This particular form has been chosen such that the LTT starts smoothly from zero amplitude at time 0 and so can be integrated.)

The Fourier transform of the combination of this fit with the Lorentzian Boltzmann part of $\tilde{\Psi}(\omega)$ is shown to give an excellent fit to the measured power spectrum (fig. 2). The error estimate for $\tilde{\Psi}(\omega)$ as computed from the simulation data is at most $\sqrt{2M+1}$ times the uncertainty on the VACF. The latter is $\approx 1.25 \times 10^{-3}$, which, with $M = 128$, yields for the error on $\tilde{\Psi}(\omega)/2D$ a value $\approx 5 \times 10^{-3}$, whereas the amplitude of the peak corresponding to the LTT is $\approx 3.7 \times 10^{-2}$, thus well above noise level. (It should be noted that this amplitude is determined by the long-time tail amplitude d_0, but also by the number M of data points used in the computation and so cannot be directly used to estimate the value of d_0; for an infinite number of points M, this peak should become infinitely narrow and of infinite amplitude.)

In conclusion, we have shown that the long-time persistence of velocity correlations is evidenced in a 2D lattice gas at moderate densities. This result corroborates the conclusions drawn from other LGA direct [10] or indirect [11] measurements of long-time correlations.

Acknowledgements

We benefited from stimulating discussions with D. d'Humières and D. Frenkel. J.-P.B. acknowledges support by the Fonds National de la Recherche Scientifique (FNRS, Belgium) and A.N. has benefited from support by the Belgium PAI program. This work was performed under EEC contract (grant SC1-0212).

References

[1] Y. Pomeau and P. Résibois, Phys. Rep. 19 (1975) 63.

[2] L.D. Landau and E.M. Lifschitz, Fluid Mechanics (Addison-Wesley, Reading, MA, 1959) ch. 2;
J.-P. Boon and S. Yip, Molecular Hydrodynamics (McGraw-Hill, New York, 1980) ch. 3.

[3] B.J. Alder and T.E. Wainwright Phys. Rev. A 1 (1970) 18.

[4] D. Levesque and W.T. Ashurst, Phys. Rev. Lett. 33 (1974) 977.

[5] G. Doolen, ed., Lattice Gas Methods for Partial Differential Equations (Addison-Wesley, Reading, MA, 1990).

[6] M.H. Ernst, in: Liquids, Freezing, and the Glass Transition, eds. D. Levesque, J.P. Hansen, and J. Zinn-Justin (Elsevier Science Publishers, Amsterdam, 1990).

[7] D. d'Humières, P. Lallemand, J.-P. Boon, D. Dab, and A. Noullez, in: Chaos and Complexity, eds. R. Livi, S. Ruffo, S. Ciliberto and M. Buiatti (World Scientific, Singapore, 1988) pp. 278–301.

[8] J.-P. Boon and A. Noullez, in: Discrete Kinematic Theory, Lattice Gas Dynamics and Foundations of Hydrodynamics, ed. R. Monaco (World Scientific, Singapore, 1989) pp. 400–408.

[9] A. Noullez, Ph.D. Thesis, University of Brussels (1990).

[10] D. Frenkel and M.H. Ernst, Phys. Rev. Lett. 63 (1989) 2165;
M.A. Van der Hoef and D. Frenkel, Tagged particle diffusion in 3D lattice gas cellular automata, Physica D 47 (1991) 191–197; these Proceedings.

[11] L.P. Kadanoff, G.R. McNamara and G. Zanetti, Phys. Rev. A 40 (1989) 4527;
G.R. McNamara, Diffusion in a lattice gas automaton, Europhys. Lett. 12 (1990) 329.

CHAPTER 6

LATTICE BOLTZMANN

Physica D 47 (1991) 219–230
North-Holland

THE LATTICE BOLTZMANN EQUATION: A NEW TOOL FOR COMPUTATIONAL FLUID-DYNAMICS

Sauro SUCCI[a], Roberto BENZI[b] and Francisco HIGUERA[c]
[a]*IBM European Center for Scientific and Engineering Computing, Via Giorgione 159, 00147 Rome, Italy*
[b]*Università di Roma "Tor Vergata", Via Orazio Raimondo, 00173 Rome, Italy*
[c]*School of Aeronautics, Universidad Politecnica de Madrid, Pza Cardenal Cisneros 3, 28040 Madrid, Spain*

Received 20 October 1989

We present a series of applications which demonstrate that the lattice Boltzmann equation is an adequate computational tool to address problems spanning a wide spectrum of fluid regimes, ranging from laminar to fully turbulent flows in two and three dimensions.

1. Introduction

Recently, a new computational technique, based on the resolution of a Boltzmann equation in a discrete lattice with appropriate symmetries, has been proposed as an alternative tool to investigate problems in the domain of two- and three-dimensional fluid dynamics [1, 2]. The lattice Boltzmann equation (LBE) is essentially the kinetic equation resulting from ensemble-averaging of the discrete dynamics of the Frisch–Hasslacher–Pomeau (FHP) cellular automaton (and its generalizations) [3] supplemented with the assumption of molecular chaos. The idea is that instead of solving the Boolean equations which govern the microdynamical evolution of the cellular automaton (the analogue of Hamilton equations in classical mechanics) one only aims to follow the evolution of the mean populations living in the lattice. A definite advantage of the LBE formulation is the suppression of the large amount of noise which usually plagues cellular automata (CA) simulations. In addition, a quasi-linear version of LBE allows a straightforward implementation of three-dimensional cases which is known to be highly problematic in the Boolean version.

The price to pay for that is twofold: on the theoretical side one loses the possibility of looking at many-body correlations that are by definition discarded in the Boltzmann approach. On the practical side, since the mean populations are real-valued non-negative quantities, the resulting simulations rely upon floating-point algebra, thus losing the important property of CA models to be exactly solvable on a digital computer (this problem however can probably be circumvented by using a fixed point representation of the mean populations).

Whether the advantages outweigh the disadvantages is a question which cannot be answered in general; what can be stated is that, with the present state of the art, purely hydrodynamic situations are handled much more economically by the LB rather than by the CA method over virtually the whole range of Reynolds numbers attainable on present-day computers. Obviously such a conclusion only pertains to general purpose computing environments (i.e. computers with no hardware optimization for Boolean algebra) and any extrapolation to different situations is by no means justified.

2. The lattice Boltzmann equation

Lattice gas hydrodynamics (LGH) is based on a special class of cellular automata whose dynamics is governed by local rules based on mass and momentum conservation. These automata can be viewed as a set of pseudo-particles which are constrained to move with a unit speed and a unit mass along the links of a regular lattice, in such a way that the state of the automaton is entirely specified in terms of a set of Boolean variables n_{ij} which take the value one or zero according to whether the jth site holds a particle moving along the ith link or not. The interaction between the pseudo-particles is governed by a set of collision rules which, site by site, mimic the momentum transfer occurring between molecules in a real fluid. In addition, an exclusion principle holds, by which the simultaneous presence of two or more particles with the same speed at the same spatial location is forbidden.

These simple prescriptions allow one to construct the microdynamical equations which govern the evolution of the Boolean field $n_i(\boldsymbol{x}, t)$. These take the form

$$T_i n_i \equiv n_i(\boldsymbol{x}_j + \boldsymbol{c}_i, t+1) - n_i(\boldsymbol{x}_j, t) = \Omega_i(n), \quad (1)$$

where $\boldsymbol{c}_i$ $(i = 1, \dots, b)$ is the set of unit vectors connecting a given site of the lattice to its b neighbors and the term Ω_i represents the change in the ith occupation number due to the collisional interaction. This can be written as

$$\Omega_i = \sum_{ss'} (s_i' - s_i) A_{ss'} \prod_{j=1}^{b} n_j^{s_j} (1 - n_j)^{1-s_j}, \quad (2)$$

where $A_{ss'}$ is the matrix element mediating the collision transforming the input state s into s', s and s' being Boolean strings of b bits.

Owing to mass and momentum conservation, the collision term fulfills the following relations:

$$\sum_{i=1}^{b} \Omega_i = 0, \quad \sum_{i=1}^{b} \Omega_i \boldsymbol{c}_i = \boldsymbol{0}. \quad (3)$$

Starting from the microdynamical equations (1) and taking the appropriate limits of small Mach and small Knudsen numbers [5], one ends up with a set of macroscopic equations for the density and the velocity of the automaton fluid which, as a consequence of the aforementioned conservation laws, take the same structure of the Navier–Stokes equations.

As a matter of fact, in order to obtain exactly the Navier–Stokes equations, the lattice has to be symmetric enough to guarantee the isotropy of fourth-order tensors. This is the case for the hexagonal FHP lattice in two dimensions and for the face-centered hypercube (FCHC) in four dimensions. In addition, the lack of translational symmetry of the discrete lattice becomes manifest through the appearance of a density-dependent factor $g(d)$ in front of the advective term of the Navier–Stokes equations. This anomaly is, however, easily removed by a simple rescaling of the time variable $t \to t/g$.

Eq. (1) is very appealing from the computational viewpoint in that it lends itself to massive vector and parallel computing. In addition, all the operations can be performed in Boolean logic so that the resulting simulations are automatically freed from the round-off and truncation errors, which commonly affect standard floating-point simulations. This simplicity has, however, a price; first, the fact that to recover smooth hydrodynamics a large amount of automata are needed (the well-known problem of noise common to any particle method) and second, the implementation of the collision operator Ω_i in three dimensions. In two dimensions, where $b = 6$ (hexagonal lattice), Ω_i can be expressed in closed form as a combination of the appropriate elementary Boolean operators. In three dimensions there is no lattice for single-speed particles possessing the symmetries required to ensure isotropy at a macroscopic level and the remedy is to use a four-dimensional face-centered hypercube (FCHC), which is subsequently projected back in three dimensions, resulting in multiple (two) speeds. In the FCHC lattice each node of the lattice is coupled to 24 neighbors. Owing to this large number of neighbors, no closed Boolean expression for the collision operator can be found so that Ω_i has to be coded via a pre-constructed lookup table mapping the input state s onto the output state s'. Since these states are represented by strings of 24 bits, this means working with a 48 Mbyte large lookup table which has to be accessed randomly, thus rendering the corresponding algorithm very memory intensive even on a large mainframe [6].

The lattice Boltzmann equation is a possible way out of these problems (even though other solutions start to be available [7]). In the Boltzmann approach the occupation numbers n_i are replaced by corresponding mean populations $N_i = \langle n_i \rangle$, where the angular brackets denote ensemble averaging. This means that instead of following each pseudo-particle in detail, one is only interested in the story of an average particle. By supplementing this averaging procedure with some further statistical assumptions (molecular chaos), one derives a lattice Boltzmann equation which has exactly the same form as eq. (1) with the occupation numbers n_i replaced by the mean populations N_i. This yields

$$T_i N_i = \Omega_i(N). \tag{4}$$

The advantage of eq. (4) over eq. (1) is that the functions N_i (which are no longer Boolean but real variables) are now smooth functions of space and time because the single-particle fluctuations have been averaged out by the very definition of the mean populations N_i. Another practical advantage is that 3D hydrodynamics can be simulated without using the huge lookup table required by the Boolean version, as we will detail in the sequel. On the other hand, it is also clear that eq. (4) contains much less physical information than eq. (1): in particular all the physics related to particle correlations is lost.

In summary, the successful application of LBE hinges on the fact that in many practical situations the information lost in the averaging procedure does not significantly affect the large-scale hydrodynamic behavior of the automaton fluid.

3. LBE with enhanced collisions

The LBE is a nonlinear finite difference equation (even though it does not result from the discretization of any partial differential equation!). An immediate practical problem associated with its numerical resolution concerns the manipulation of polynomials of order b, which can give rise to severe round-off errors in three dimensions ($b = 24$). A partial alleviation of this problem is obtained by linearizing the collision operator Ω_i around the uniform steady state $N_i = d$, d being the density per link of the discrete fluid. This leads to the following difference equation

$$T_i N_i = \sum_{j=1}^{b} \Omega_{ij}\left(N_j - N_j^{\text{equil}}\right), \tag{5}$$

where $\Omega_{ij} \equiv \partial \Omega_i / \partial N_j$ at $N_i = d$, is given by

$$\Omega_{ij} = \sum_{ss'} (s'_i - s_i) d^{p-1} (1-d)^{b-p-1} A_{ss'} (s'_j - s_j), \tag{6}$$

p being the number of particles involved in the collision and N_j^{equil} the (Fermi–Dirac or Bose–Einstein) equilibrium distribution expanded to the second order in the local velocity field u. By taking advantage of the symmetries of the problem, it can be shown that, despite its apparent linearity, eq. (5) accounts for second-order terms in the expansion of the collision operator as a function of the local flow speed u. Since the validity of the LGH method is in any case limited to low-speed (i.e. low Mach numbers) regimes, it follows that eq. (5) is indeed compatible with the aim of a large-scale description of the automaton fluid.

Very recently, we proposed a new class of lattice Boltzmann equations which can be used to achieve maximum efficiency regardless of the collision rules [8]. For this purpose, let us consider the matrix element Ω_{ij} representing the change due to collisions of the ith population about the reference state as induced by a unit change in the jth population. Due to the isotropy of the reference state, Ω_{ij} only depends on the angle between directions i and j, which has only a limited set of values for a given lattice. So, for the 2D triangular lattice the values of the angles are $0°, 60°, 120°, 180°$ while for the 4D FCHC lattice one has 0°, 60°, 90°, 120°, and 180°. Accordingly, the number of possibly different elements Ω_{ij} is four for the 6-particle FHP and five for 24-particle FCHC models, respectively.

The number of independent matrix elements is further decreased by the conditions of conservation of mass and momentum, requiring that $D+1$ eigenvalues of the collision operator be zero (Goldstone modes) with the corresponding null subspace being spanned by the $D+1$ vectors,

$$V_0 = (1_i), \quad V_\alpha = (c_{i\alpha}),$$
$$i = 1, \ldots, b, \quad \alpha = 1, \ldots, D. \tag{7}$$

For FCHC models these conditions lead to the relations

$$a_0 + 8a_{60} + 6a_{90} + 8a_{120} + a_{180} = 0,$$
$$a_0 + 4a_{60} - 4a_{120} - a_{180} = 0, \tag{8}$$

where a_θ are the matrix elements linking pairs of directions that make an angle θ.

The eigenvalues and eigenvectors of these matrices are easily computed by exploiting the fact that Ω_{ij} is a circulant matrix [9]. For 24-particle FCHC models, the non-zero eigenvalues are

$$\lambda = a_0 - 2a_{90} + a_{180},$$
$$\sigma = \tfrac{3}{2}(a_0 - a_{180}),$$
$$\tau = \tfrac{3}{2}(a_0 + 6a_{90} + a_{180}), \tag{9}$$

with multiplicities 9, 8 and 2. In both cases the eigenvectors are independent of the values of the matrix elements. In particular, the eigenvectors associated with the eigenvalue λ are the $\frac{1}{2}D(D+1) - 1$ linearly independent elements of the set

$$V_{\alpha\beta} = Q_{i\alpha\beta} \equiv c_{i\alpha}c_{i\beta} - (c^2/D)\delta_{\alpha\beta}.$$

We are now in a position to appreciate the impressive reduction of computational complexity one achieves in passing from the Boolean dynamics, eq. (1), to the quasi-linear LBE, eq. (4). In the former case, one has to employ the huge (48 Mbyte) matrix $A_{ss'}$ while with LBE, one only needs the $b \times b$ matrix Ω_{ij}, which, as we have just seen, can be expressed in terms of only three parameters, namely its non-zero eigenvalues λ, σ, τ. These can be chosen at will, with the only proviso that they be negative. Also, the eigenvalue λ can be tuned to minimize the viscosity according to relation

$$\nu = -\tfrac{1}{3}(\tfrac{1}{2} + 1/\lambda). \tag{10}$$

On account of this, all the simulations presented in the sequel have been performed on the basis of eq. (4), with either the "classical" FCHC collision operator, eq. (6), or its "enhanced" version, eqs. (8)–(10).

4. Applications

4.1. Application 1: Three-dimensional flows in complex geometries

One of the most advocated merits of the LG method, besides the fact of being ideal candidate for massive vector and parallel computing, is its flexibility with respect to complex geometries. This is easily understood by recalling that, even though based on a very stylized cellular automaton dynamics, the LG method is basically a particle tracking technique, and as such, it allows one to treat the intricacies associated with complex boundary conditions in simple terms of particle reflections and bounces at appropriate spatial locations flagged as wall sites. These properties point to the LG method as to an excellent candidate to address a long-standing problem in the physics of random media, namely the calculation of transport coefficients in terms of the medium's microscopic geometry. Pioneering work along these lines has been developed by Rothman [10], who applied the six-bit FHP Boolean automaton to the study of two-dimensional flows in porous media.

Very recently, we have been able to extend Rothman's work to three-dimensional cases [11]. The simulation of the random medium was performed on a small resolution (32^3) cubic lattice at a density per link $d = 0.328$ with the standard FCHC scheme. The complex medium was modelled as a random sequence of elementary blocks of four units in size distributed in such a way that no free pore with a cross section smaller than 4×4 lattice units was allowed. This size was determined by a preliminary series of tests aiming to ascertain under which parameter conditions LBE is able to quantitatively reproduce a Poiseuille flow in a square channel. On the surfaces of the blocks, as well as on the lateral walls, no-slip boundary conditions were imposed, i.e. particles impinging along a given direction were reflected back into the fluid along the opposite direction. Along the streaming direction (x) periodic conditions were applied.

The first goal of this investigation was to test the validity of Darcy's law

$$U = -(k/\mu)G, \qquad (11)$$

k being the permeability of the medium and μ the dynamic viscosity. This was done by varying the total pressure gradient $G \equiv \partial(p - \rho f x)/\partial x$ via the input parameter f, which represents the force per unit mass imparted to the fluid.

The results are indicated in fig. 1, where the numerical flux Q is displayed as a function of G for three decreasing values of the porosity $\Phi = 0.635, 0.5, 0.375$. From this figure, we see that Darcy's law is well fulfilled for values of G up to $G = 1.4 \times 10^{-3}$ corresponding to an effective speed $u \equiv Q/H^2$ in the range of a few milli-lattice units per step. According to the theory of lattice gas, the Reynolds number can be computed as $\mathrm{Re} = R^{\times} HM$, where $R^{\times} = 7.57$ is a dimensionless coefficient specific of the FCHC scheme and $M = \sqrt{2}u$ is the Mach number of the flow. According to this formula, the data reported in fig. 1 correspond to a maximum Reynolds number of order one as is appropriate for laminar flows. No significant deviations from Darcy's law have been observed for values of the Reynolds number up to about 5.

For the present simulation ($H = 32$ and $\mu = 24 d\nu \approx 0.25$) the permeability k can be directly deduced by fig. 1 as

$$k = \mu Q/GH^2 \approx 2.5 \times 10^{-4} Q/G. \qquad (12)$$

The above formula yields $k \approx 0.5, 0.18, 0.07$ for $\Phi = 0.635, 0.5, 0.375$, respectively. In the absence of a specifically designed companion experiment, the assessment of these values is necessarily restricted to a qualitative level; in fact, analytical estimates based on quick visual inspection of the flow pattern, so helpful in two dimensions, become much less sensible in our case due to the genuinely three-dimensional nature of the flow. The latter can be appreciated by looking at figs. 2 and 3, in which we show two cross sections of the flow in the plane $z = 18$ as a function of x, y and in the plane $x = 18$ as a function of y, z. While in fig. 2 there is evidence of a systematic drift along the x direction, as imposed by the pressure gradient, fig. 3 clearly shows that the fluid wraps

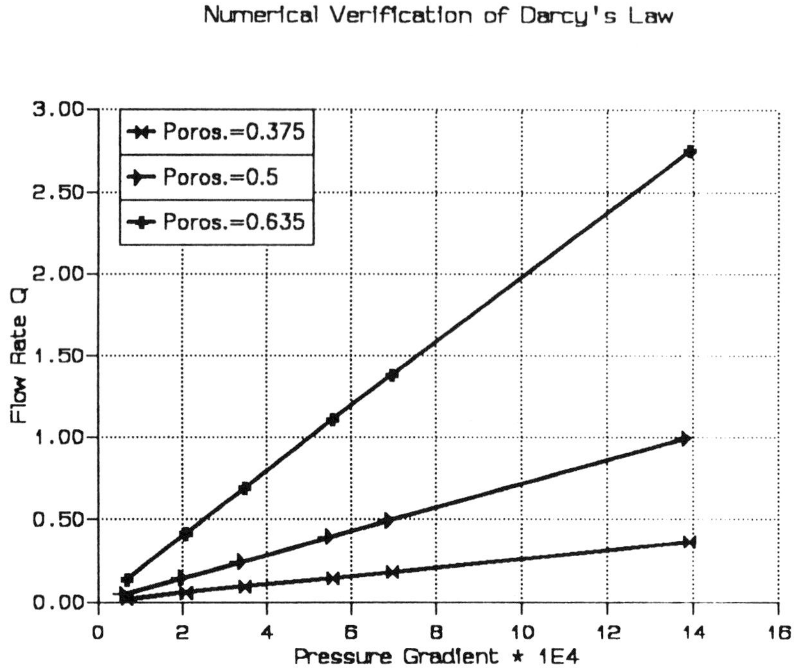

Fig. 1. The volumetric flow Q as a function of the pressure gradient for three different values of the porosity $\Phi = 0.635, 0.5, 0.375$.

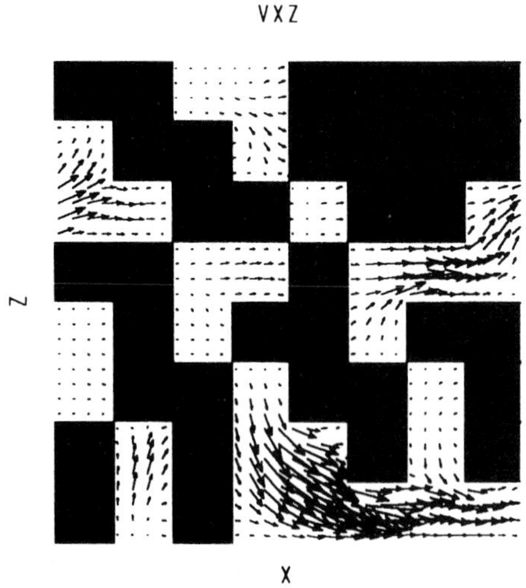

Fig. 2. Flow through the top-view section $z = 18$ for $\Phi = 0.5$. Note the systematic drift along the x direction imposed by the pressure gradient.

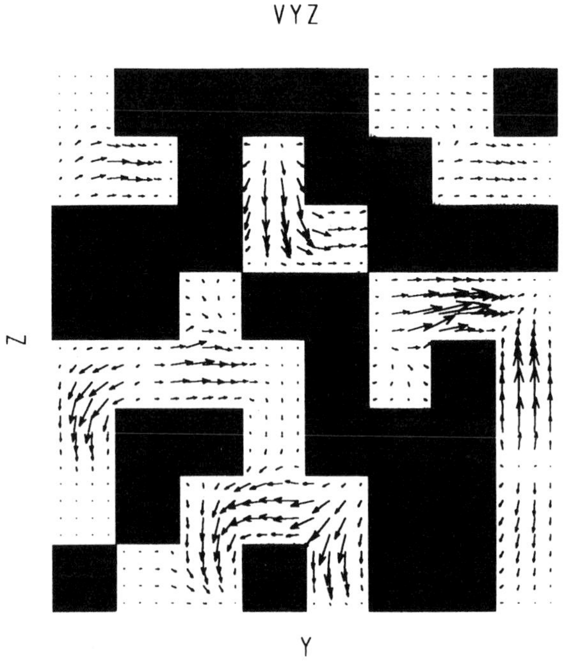

Fig. 3. Flow through the longitudinal section $x = 18$. Note that, contrarily to the case of fig. 2, the flow has no systematic drift in this plane but it wanders erratically thus developing a fully three-dimensional structure.

around the solid obstacles along any of the three directions available to the motion.

A semi-qualitative comparison with theoretical results based on simple pictures of shrinking-tube and grain-consolidation models, showed that the numerical simulations yield a permeability which agrees with analytical estimates within about a factor two.

Apart from the quantitative aspects, the main result of this study is the demonstration of the adherence of the LB model to Darcy's law for a three-dimensional flow through a complex medium. The simulation or real rocks is likely to require mesh sizes of the order of 128 cubed, with a corresponding increase of almost two orders of magnitude in computing power. Given the fact that, on a IBM 3090/VF, the LBE scheme is processed at a rate of about 0.1 Msites/s (with a single processor), running, say, a thousand steps on a 128^3 lattice will take approximately 10 h CPU time. This figure can be easily brought down to a couple of hours by running on a six-headed 3090 multiprocessor. Hence, the problem is already feasible on a present-day supercomputer even though, for parametric studies, a significant improvement of these processing speeds is certainly desirable.

4.2. Application 2: Bifurcations of a two-dimensional Poiseuille flow

It has recently been shown that by exciting obstacle-induced shear-layer instabilities it is possible to promote fluid turbulence in a Poiseuille flow well below the critical Reynolds number of the same flow in a free channel [12]. Apart from the practical motivation of minimizing the cost of mechanical power versus thermal transport, the study of these instabilities provides an interesting scenario of a transition from laminar to turbulent regimes at moderate values of the Reynolds number, typically of the order of hundred. We have studied this problem by considering a plane channel containing a periodic array of identical plates. The plates, one fifth of the channel width ($H =$ 181), are placed normal to the flow stream and spaced 6.6 times the channel half-width. This spacing is about three times the wavelength of the channel which needs to be excited in order to destabilize the least stable mode of the Poiseuille flow. The Reynolds number of the system is defined as $\mathrm{Re} = UH/2\nu$, where H is the width of the channel and U is the maximum velocity of a Poiseuille flow leading to the same flux as the actual flow.

The laminar-turbulent transition can be described within the framework of Landau's theory, with the amplitude of the primary bifurcated mode $A = \rho\, \mathrm{e}^{\mathrm{i}\theta}$ playing the role of the order parameter. The dynamics of the transition is described by Landau's equations

$$\dot{\rho} = S_r \rho - L_r \rho^3,$$
$$\dot{\theta} = S_i - L_i \rho^2, \tag{13}$$

where S and L are two complex coefficients which characterize the transition. In the vicinity of the critical point ($S_r = 0$), they can be taken in the form $S_r = \alpha r$ and $S_i = S_0 + \beta r$, where $r \equiv (\mathrm{Re} - \mathrm{Re_c})/\mathrm{Re_c}$ is the reduced Reynolds number. Above the critical threshold ($r > 0$), the steady-state solution of eqs. (13) reads

$$\rho = \sqrt{S_r/L_r} \propto r^{0.5},$$
$$\omega \equiv \dot{\theta} = S_0 + (\beta - \alpha L_i/L_r) r, \tag{14}$$

which shows the typical mean-field theory critical exponent 0.5.

We found that the critical Reynolds number $\mathrm{Re_c}$ is about 86, i.e. much smaller than the critical Reynolds number for the same flow in the absence of any obstacle. For $\mathrm{Re} \geq \mathrm{Re_c}$ the flow exhibits a stable periodic limit cycle with frequency $\omega = 0.38$ (measured in units of $2U/H$) in close agreement with the results given in ref. [12]. Streamlines of a typical flow configuration for $\mathrm{Re} \approx 100$ are shown in fig. 4. Beyond $\mathrm{Re} > 90$ a secondary bifurcation of much smaller amplitude and higher frequency was observed, which indi-

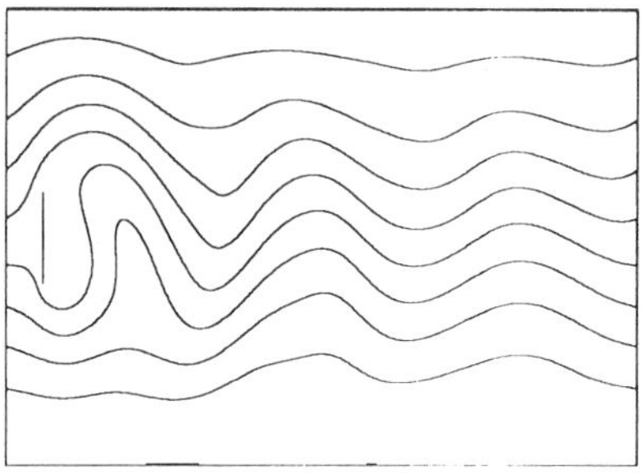

Fig. 4. Streamlines of the unstable Poiseuille flow at Re ≈ 100.

cates the onset of fully developed turbulence not describable by eqs. (13).

The $r^{0.5}$ scaling of the primary bifurcated mode was roughly recovered (see fig. 5) even though large fluctuations were observed. On the contrary, the linear correction of ω to ω_c was found to be too small to be detected. A complete characterization of the transition, which requires the numerical determination of the coefficients appearing in the equations is still under investigation.

4.3. Application 3: Fully developed two-dimensional forced turbulence

We have already noticed that the enhanced version of LBE offers the possibility of entering zero-viscosity fluid regimes by simply tuning the eigenvalues of the collision matrix Ω_{ij} (see eq. (10)). Thus, Reynolds numbers as high as allowed by the finite size of the grid can be accessed. In this context, homogeneous turbulence provides a particularly severe test in the sense that all dy-

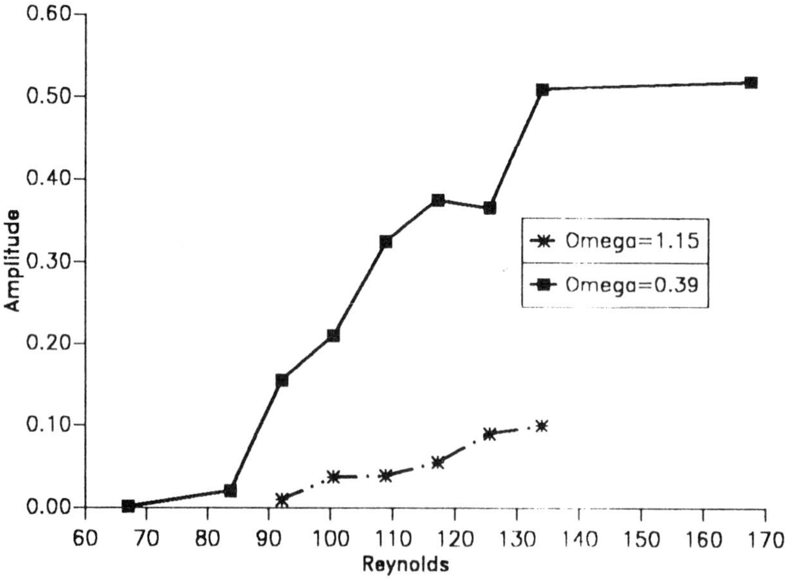

Fig. 5. The amplitude of the primary and secondary bifurcated modes of the Poiseuille flow as a function of the Reynolds number.

namical scales ranging from the macroscopic scale L to the dissipative scale $l = L\,\mathrm{Re}^{-0.5}$ are effectively excited. It is therefore of primary interest to test the ability of LBE to reproduce the basic physics of fully turbulent flows and possibly estimate its computational efficiency with respect to other conventional techniques. For this purpose, we have studied a two-dimensional flow in a square box of size 2π with periodic boundary conditions and a periodic long-wave forcing of the type $F = F_0 \sin(2\pi py/L)$ along the x axis on the wavenumber $p = 4$.

As a first instance we have compared the physical results obtained by running the LBE and a pseudo-spectral code in a 64^2 grid, i.e. at a moderately low resolution. It is worth mentioning that this comparison is particularly severe since the spectral methods are probably the most well-established numerical technique for this kind of problems.

The spectral viscosity was fixed at the minimum value compatible with the grid resolution, that is to say $\nu \equiv U\Delta x$. The spectral time step was $\Delta t = 0.01$, corresponding to 1/382 LBE units after the appropriate scaling between spectral and lattice units (subscripts sp and B refer to spectral and Boltzmann, respectively):

$$L_{\mathrm{sp}} = (2\pi/N)L_{\mathrm{B}}, \quad U_{\mathrm{sp}} = 25U_{\mathrm{B}},$$

where N is the number of grid points/dimension in the Boltzmann simulation and the factor 25 between speeds stems from the requirement that the maximum speed produced by the force F corresponds to $U_{\mathrm{B}} = 0.2$ (complete details are given in ref. [13]).

The total energy and enstrophy for both LB and spectral simulations are shown in fig. 6 as a function of time. From this figure we see that indeed the two simulations yield quite similar results, even though the intermittent spikes of enstrophy are somewhat higher in the LB model.

To get a more refined idea of the agreement between the two models, we have increased the resolution by a factor two and repeated the experiment with the same values of the other parameters. The energy spectrum corresponding to LBE and spectral simulations after about 150 physical time units is presented in fig. 7, from which a remarkable agreement is apparent. The

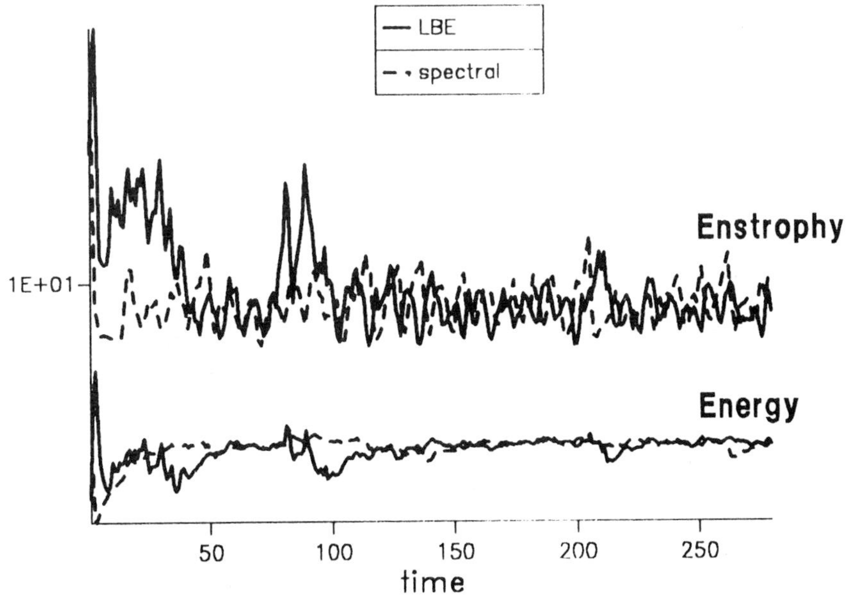

Fig. 6. Time evolution of the total energy and enstrophy for the spectral and LBE models on a 64^2 grid.

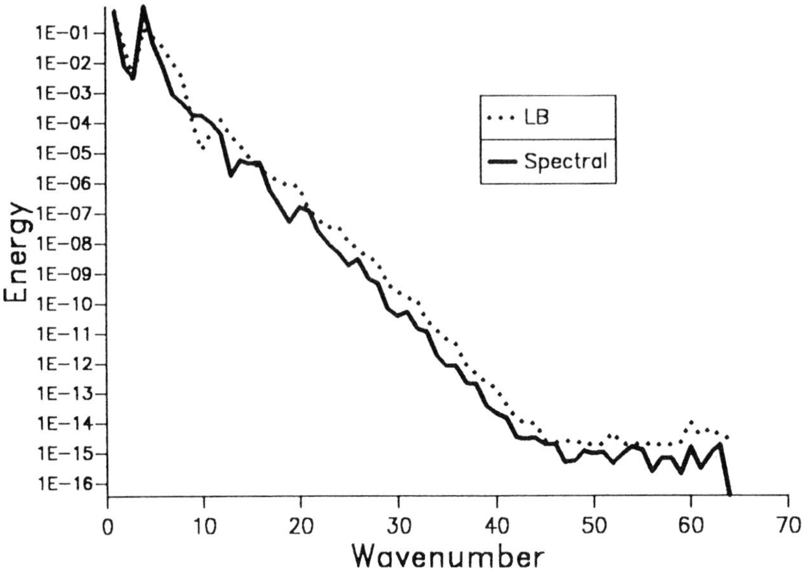

Fig. 7. Energy spectra of the spectral and LBE simulations on a 128^2 grid.

flat high-wavenumber region of these spectra is due to an excess of resolution with respect to the actual Reynolds number of the simulation: the remarkable point is that the flat region begins at the same wavenumber for both spectra.

However, the hardest challenge for LBE is to test whether it is able to reproduce the statistical properties of two-dimensional fully developed turbulence, namely the enstrophy inertial range at high k and the energy inertial range at low k. For this purpose we have performed a series of high-resolution experiments on a 512^2 grid. A typical series of five nearly steady energy spectra is shown in fig. 8, from which we see that the predicted k^{-3} law is not obtained and steeper spectra with slope ≈ 4–5 are observed instead. This discrepancy may tentatively be attributed to the absence in the LBE model of any sort of superviscosity effects, i.e. higher-order dissipative terms of the form Δ^p with $p > 1$ that allow coherent structures to survive down to small scales which would otherwise be overdamped by normal viscosity ($p = 1$). Indeed, the absence of small-scale structures is easily seen by looking at an instantaneous vorticity map (see fig. 9). This figure highlights the presence of strong large-scale agglomerates which are the result of the inverse energy cascade producing energy pile-up at low wavenumbers. This energy pile-up is easily removed in pseudo-spectral simulations by adding an inverse Laplacian term [14]. The inclusion of both superviscosity and inverse dissipation is quite straightforward in a spectral method, since it reduces to the trivial algebraic replacement $\Delta \to -k^2$. Unfortunately, there is nothing equivalently simple in a lattice approach.

Having appreciated the physical picture, we can proceed to examine the computational efficiency of the methods. On a single processor of the IBM 3090 vector multiprocessor, for a 128^2 gird, the LBE required 60 ms/step and the spectral code (no dealiasing and a two-step time-marching scheme) 300 ms/step. Each step of the spectral method requires two units of computational work (see below) to advance 0.01 physical units, as opposed to LBE, which requires one unit of work to advance 1/382 physical units. According to these data, the conclusion is that the LBE method supersedes the spectral method by a factor 5/2.

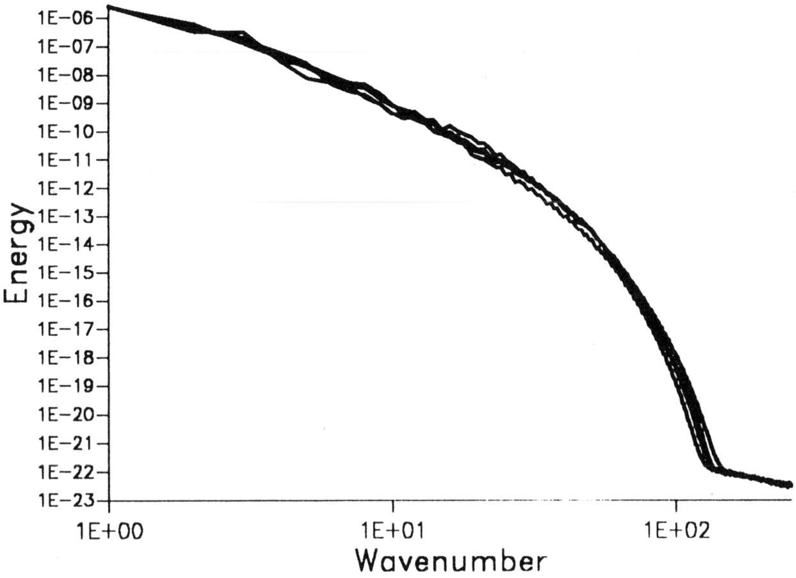

Fig. 8. Energy spectra of the LBE simulation on a 512^2 grid.

Apart from this specific example, a much more far-reaching consideration is suggested by inspection of the number of floating-point operations required to complete a unit of computational work. For LBE we have

$$N_{\mathrm{fpo}} \sim 150 N^2,$$

while for the spectral method

$$N_{\mathrm{fpo}} \sim (25 \log_2 N) N^2.$$

For the present case, $N = 128$, the number of operations is practically the same for the two methods; however, the important observation is that for LBE the factor in front of the N^2 term is a constant as opposed to the logarithmic dependence exhibited by the spectral method [15].

As a result, even though this difference is shadowed by the limited problem size allowed by present-day computers, it is easy to predict that this favourable scaling will become more and more apparent as the future developments of computer technology will push the frontier of large-scale computing further ahead.

Fig. 9. Instantaneous vorticity map for a 512^2 LBE simulation.

We may summarize these results by stating that the LBE yields indeed a faithful picture of two-dimensional incompressible turbulence and performs at least at the same rate as the best conventional technique known so far for homogeneous isotropic flows. However, at the present stage of the theory, it does not lend itself to the same kind of numerical tricks which are readily available in a spectral context to increase the effective resolution.

Apart from any specific consideration related to its actual performance, the main merit of LBE remains basically its ideal amenability to massively vector and parallel computing.

5. Conclusions

We have presented a variety of fluid-dynamics applications ranging from laminar to fully turbulent flows. Quantitative analysis and comparisons with other well-established numerical techniques seem to indicate that, even though a large margin of improvement is still left, LBE provides a new viable tool for two- and three-dimensional computational fluid dynamics.

Acknowledgement

R.B. and F.H. wish to thank the support of IBM European Center for Scientific and Engineering Computing where all the numerical simulations were performed.

References

[1] F.J. Higuera, in: Proceedings of Workshop on Discrete Kinetic Theory, Lattice Gas Dynamics and Foundations of Hydrodynamics, ed. R. Monaco (World Scientific, Singapore, 1989) p. 162.
[2] G. McNamara and G. Zanetti, Phys. Rev. Lett. 61 (1988) 2332.
[3] U. Frisch, B. Hasslacher and Y. Pomeau, Phys. Rev. Lett. 56 (1986) 1505.
[4] F. Higuera and S. Succi, Europhys. Lett. 8 (1989) 517.
[5] U. Frisch, D. d'Humieres, B. Hasslacher, P. Lallemand, Y. Pomeau and J.P. Rivet, Complex Systems 1 (1987) 649.
[6] J.P. Rivet, Ph.D. Thesis, Université de Nice (June 1988), and references therein.
[7] P. Rem and J. Somers, Proceedings of a Workshop on Cellular Automata and Modeling of Complex Physical Systems, Les-Houches (1989), Springer Proc. Phys. 46 (1989) 161.
[8] F. Higuera, S. Succi and R. Benzi, Europhys. Lett. 9 (1989) 345.
[9] S. Wolfram, J. Stat. Phys. 45 (1986) 471.
[10] D. Rothman, Geophysics 53 (1988) 509.
[11] S. Succi, E. Foti and F. Higuera, Europhys. Lett. 10 (1989) 433.
[12] G.E. Karniadakis, B.B. Mikic and A.T. Patera, J. Fluid Mech. 192 (1988) 365.
[13] R. Benzi and S. Succi, J. Phys. A 23 (1990) L1.
[14] B. Legras, P. Santangelo and R. Benzi, Europhys. Lett. 5 (1988) 37.
[15] G.S. Patterson and S. Orszag, Phys. Fluids 14 (1971) 2538.

Physica D 47 (1991) 231–232
North-Holland

RELATION BETWEEN THE LATTICE BOLTZMANN EQUATION AND THE NAVIER–STOKES EQUATIONS

U. FRISCH
CNRS, Observatoire de Nice, B.P. 139, 06003 Nice Cedex, France

Received 29 December 1989

It is shown that the lattice gas Boltzmann equation may be rewritten in a form which brings out its close relation with the Navier–Stokes equations. Various consequences are pointed out.

It has been recently proposed to use the lattice Boltzmann approximation as an alternative to Boolean lattice gas simulations [1–4]. In this approach, the mean populations N_i (real numbers) are used and their evolution is computed exactly from the lattice Boltzmann equation. One advantage is the absence of the noise inherent to Monte Carlo methods. The drawbacks have been clearly discussed in refs. [1,2], so that we need not come back to them.

Given that one then works with a limited number (say, 6 to 24) of real fields, defined on a space–time grid, the question arises if there is any *direct* relation to the finite-difference discretized Navier–Stokes equations.

The present investigation has been carried out so far only in a restricted framework: the Boltzmann approximation in differential form with a six-velocity two-dimensional model of the FHP-I type [5,7]. Generalizations of the present approach to FCHC-based models in two and three dimensions have been worked out recently [6]. In the following, notation is taken from ref. [7].

We start from the Boltzmann equation in discrete form

$$\partial_t N_i + \partial_\alpha c_{i\alpha} N_i = \Delta_i, \tag{1}$$

where Δ_i is the collision term. We decompose the N_i's according to a suitably chosen orthogonal basis, such that the first components correspond to the hydrodynamical fields and that the other ones should display as much invariance as possible:

$$N_i = \tfrac{1}{6}\rho + \tfrac{1}{3}\rho c_{i\alpha} u_\alpha + Q_{i\alpha\beta} S_{\alpha\beta} + (-1)^i \sigma. \tag{2}$$

The tensor $S_{\alpha\beta}$ is symmetric and traceless (two independent components). We multiply (1) successively by 1, $c_{i\alpha}$, $Q_{i\alpha\beta}$, $(-1)^i$ and we sum over i. We then expand the collision term, treating N_i as a perturbation of $\frac{1}{6}\rho$; this is justified at low Mach numbers. The first order of the expansion is expressible from the linearized collision matrix A_{ij}. It is necessary to go to second order in $\boldsymbol{u}$ to obtain the quadratic terms in the Navier–Stokes equations and their coefficient $g(\rho)$ [1,8]. Higher than quadratic terms are irrelevant.

Using the invariance properties, we obtain the following system (irrelevant terms omitted):

$$\partial_t \rho + \partial_\alpha(\rho u_\alpha) = 0, \tag{3}$$

$$\partial_t(\rho u_\alpha) + \tfrac{1}{2}\partial_\alpha \rho + \tfrac{3}{2}\partial_\beta S_{\alpha\beta} = 0, \tag{4}$$

$$\tfrac{3}{2}\partial_t S_{\alpha\beta} + \tfrac{1}{4}[\partial_\alpha(\rho u_\beta) + \partial_\beta(\rho u_\alpha) - \partial_\gamma(\rho u_\gamma)\delta_{\alpha\beta}] - \mu(u_\alpha u_\beta - \tfrac{1}{2}u^2\delta_{\alpha\beta}) + R_{\alpha\beta\gamma}\partial_\gamma \sigma = -\lambda S_{\alpha\beta}, \tag{5}$$

$$6\partial_t \sigma + R_{\alpha\beta\gamma}\partial_\gamma S_{\alpha\beta} = -\kappa\sigma. \tag{6}$$

The positive coefficients λ, μ, and κ depend on the density and on the collision rules. They are not given explicitly here. The third-order tensor $R_{\alpha\beta\gamma}$ is given by

$$R_{\alpha\beta\gamma} = \sum_i c_{i\alpha} c_{i\beta} c_{i\gamma} (-1)^i. \tag{7}$$

It is $\mathcal{G}$-invariant but not isotropic.

In the usual hydrodynamical regime, where the variables ρ and $\boldsymbol{u}$ vary slowly, the variables $S_{\alpha\beta}$ and σ are *slaved* because of the presence of negative eigenvalues (proportional to $-\lambda$ and $-\kappa$) in the linearized collision operator. $S_{\alpha\beta}$ is (up to a factor) the stress tensor, that is the momentum flux tensor. It is slaved to a combination of the rate-of-strain and of the nonlinear flux. In the slaved regime, we exactly recover the equations of hydrodynamics (with the nonlinear term modified as usual by a coefficient $g(\rho) = \frac{3}{2}\mu/\lambda\rho$). An implication is that the rate-of-strain tensor (symmetrical part of the velocity gradient) can be obtained directly from the mean populations N_i (via the tensor $S_{\alpha\beta}$) without taking any space-derivatives; this remark does unfortunately not extend to the vorticity (antisymmetrical part). As for the field σ, it is irrelevant in the hydrodynamical limit. Its (tensor) generalization to the FCHC case (pseudo-4D) can nevertheless become relevant for models with *negative viscosities* [9]: it is then necessary to include spatial derivatives up to fourth order in the hydrodynamical equations; also, it may not be legitimate to truncate the expansion of the collision term.

The slaving of $S_{\alpha\beta}$, that is the possibility to neglect $\partial_t S_{\alpha\beta}$, will hold so much more so as the eigenvalue λ (roughly, the inverse collision viscosity) is larger. Thus, efforts to minimize the (collision) viscosity are also likely to improve the validity of the hydrodynamic approximation. With this view-point, the Boltzmann equation (1) can also be regarded as an approximation to the Navier–Stokes equations. This approximation is *hyperbolic* and introduces a slight *delay* (of order $1/\lambda$) in the slaving of the stress tensor.

We may now consider the *lattice* Boltzmann equations as a finite difference approximation to (1) and, thereby to the Navier–Stokes equations. The lattice Boltzmann equations read:

$$N_i(t_\star + 1, \boldsymbol{r}_\star + \boldsymbol{c}_i) - N_i(t_\star, \boldsymbol{r}_\star) = \Delta_i. \tag{8}$$

It is known that existence–uniqueness–regularity problems are very hard at the Navier–Stokes level (in three dimensions), and even harder at the Boltzmann level. However, at the lattice Boltzmann level, we have a simple polynomial iteration which poses no existence–uniqueness–regularity problems. Furthermore, the H theorem (Hénon's version; see appendix F of ref. [7]) ensures "good" behaviour, even for large times, for those models satisfying semi-detailed balance.

We may also consider the system (3)–(6) from the view-point of "large eddy simulations", that is for situations where the Reynolds number is too large for the resolution retained. Eq. (5) provides us then with an equation for the sub-grid-scale stresses.

We have thus shown that the lattice Boltzmann technique may be regarded as a new finite-difference technique for the Navier–Stokes equations having the property of unconditional stability.

One final remark: Grad [10] has shown that the Navier–Stokes equations can be obtained from the full (non-discrete) Boltzmann equation by an expansion procedure involving Hermite polynomials in the microscopic velocity. Our approach to the lattice Boltzmann equation can be viewed as a Grad-type expansion, but one which terminates after finitely many terms, because of the discreteness of velocity space.

We are grateful to D. d'Humières and Y. Pomeau for useful remarks.

References

[1] F.J. Higuera and J. Jiménez, Europhys. Lett. 9 (1989) 663.

[2] G.R. Mc Namara and G. Zanetti, Phys. Rev. Lett. 61 (1988) 2332.

[3] F.J. Higuera, Lattice gas simulation based on the Boltzmann equation, in: Discrete Kinetic Theory, Lattice Gas Dynamics and Foundations of Hydrodynamics, Torino, September 20–24, ed. R. Monaco (World Scientific, Singapore, 1989) pp. 162–177.

[4] F.J. Higuera and S. Succi, Europhys. Lett. 8 (1989) 517.

[5] U. Frisch, B. Hasslacher and Y. Pomeau, Phys. Rev. Lett. 56 (1986) 1505.

[6] M. Vergassola, R. Benzi and S. Succi, Europhys. Lett. 13 (1990) 411.

[7] U. Frisch, D. d'Humières, B. Hasslacher, P. Lallemand, Y. Pomeau and J.P. Rivet, Complex Systems 1 (1987) 632.

[8] B. Dubrulle, Complex Systems 2 (1988) 577.

[9] B. Dubrulle, U. Frisch, M. Hénon and J.P. Rivet, Low viscosity lattice gases, J. Stat. Phys. 59 (1990) 1187.

[10] H. Grad, Commun. Pure Appl. Math. 2 (1949) 331; Phys. Fluids 6 (1963) 147.

Physica D 47 (1991) 233–240
North-Holland

BOUNDARIES IN LATTICE GAS FLOWS

Paul LAVALLÉE[a], Jean Pierre BOON[b] and Alain NOULLEZ[b]
[a]*Département de Physique, Université du Québec à Montréal, CP 8888, Montreal, Canada, H3C 3P8*
[b]*Faculté des Sciences, Université Libre de Bruxelles, CP 231, 1050 Bruxelles, Belgique*

Received 15 January 1990

A one-dimensional lattice gas model is used to study the interaction of fluid flows with solid boundaries. Various interaction mechanisms are examined. Lattice Boltzmann simulations show that bounce-back reflection is not the only interaction that yields "no-slip" boundary conditions (zero velocity at a fixed wall) and that Knudsen-type interaction is also appropriate.

1. Introduction

A large class of problems in fluid mechanics involves solid boundaries. Standard analysis requires that the velocity of the fluid be zero at a fixed boundary, in agreement with experimental observation at the hydrodynamic level: this is known as the "no-slip" condition. This condition is satisfactory for simulation of the Navier–Stokes equations with traditional computational methods, but obviously it does not shed any light on the mechanisms that effectively reduce the velocity to zero on the wall. On the other hand, there are very few experiments that explore the interaction of fluids and solid boundaries at the molecular level. In this respect, lattice gas methods prove useful because even though lattice gas automata are constructed as fictitious microworlds, their operational algorithms must include the proper basic microscopic mechanisms in order to produce correct hydrodynamic behavior. Therefore the lattice gas approach to the boundary flow problem requires a model interaction mechanism with solid boundaries. We present a simple lattice gas model, for the study of flows near walls; the model is easily implemented on a small computer and is free of the usual Monte Carlo noise inherent to a description at the microscopic level, so that computations run fast.

2. The lattice gas model

We consider the six-particle FHP model [1]. This automaton runs on a triangular lattice with hexagonal symmetry. All particles have unit mass, move with unit velocity in one of the six possible directions of the lattice (fig. 1a) and are subject to an exclusion principle (at a given node, no two particles occupy the same link at a given time); thus, the maximum number of particles per node is six. When two particles meet at a node, they collide at integer times. There are essentially two collision rules, one involving head-on binary collisions and one involving three particles converging symmetrically towards a node (fig. 1b). The presence of three-particle collisions is essential in order to remove a non-physical conservation law [2] so that the only conserved quantities are mass and momentum. All

0167-2789/91/$03.50 © 1991 – Elsevier Science Publishers B.V. (North-Holland)

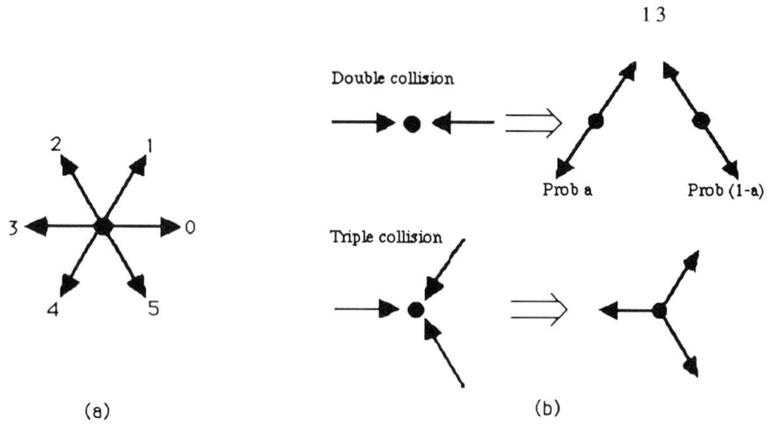

Fig. 1. (a) Indices for velocity orientations on lattice nodes; (b) collision rules.

other collisions leave the particle configuration unchanged, and so do not contribute to momentum transfer. After the collision, particles move one lattice link length per unit time in the direction of their velocity. These are the simplest rules that produce correct hydrodynamic behavior at the macroscopic level. Specific collision rules are used to model interactions of particles with solid boundaries; for instance, particles are either specularly reflected or bounced back as shown in fig. 2. Collision rules can be conveniently written as a collision table. In the simple six-particle FHP model, the collision table has 64 input states (the number of possible input configurations is 2^b, where b is the number of sites per node).

The equations governing the evolution of the gas can be written by inspection. The position of a lattice node is denoted by $\boldsymbol{r}$. The occupancy variable n_i (direction $i = 0, 1, 2, \ldots, 5$) is either 0 or 1 (because of the exclusion principle) and n_i is a function of position $\boldsymbol{r}$ on the lattice and of time t. The evolution of the automaton in space and time proceeds by sequences of two consecutive events: propagation followed by collision. During propagation, all particles move one lattice unit in the direction of their velocity $\boldsymbol{c}_i$, and during collision, all nodes are updated, i.e., particles are reorganized according to the collision rules. Note that after a large number of updates, it does not matter whether collision or propagation occurred first [2].

In establishing the equations governing the evolution of the lattice gas obeying the simple FHP collision rules (fig. 1b) given above one must account for the fact that there are two possible output configurations for binary head-on collisions which are assigned complementary probabilities a and $1 - a$. The value $a = \frac{1}{2}$ is most common, that is on the average every other collision leads to one of the possible configurations. The microdynamical equations are [2]

$$n_i(t+1, \boldsymbol{r} + \boldsymbol{c}_i) = n_i(t, \boldsymbol{r}) + \Delta_i(n), \quad i = 0, 1, \ldots, 5, \tag{1}$$

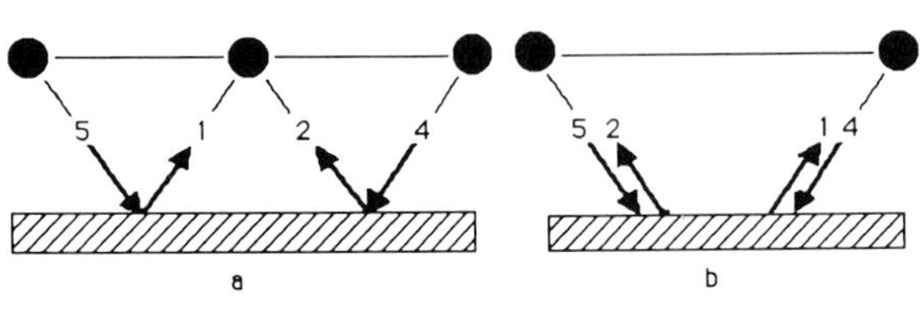

Fig. 2. Collision rules for wall interactions. (a) Specular reflection; (b) bounce-back reflection.

where $n_i(t, \boldsymbol{r})$ is the propagation term and $\Delta_i(n)$ is the collision term:

$$\Delta_i(n) = a n_{i+1} n_{i+4} \bar{n}_i \bar{n}_{i+2} \bar{n}_{i+3} \bar{n}_{i+5} + (1-a) n_{i+2} n_{i+5} \bar{n}_i \bar{n}_{i+1} \bar{n}_{i+3} \bar{n}_{i+4} - n_i n_{i+3} \bar{n}_{i+1} \bar{n}_{i+2} \bar{n}_{i+4} \bar{n}_{i+5}$$
$$+ n_{i+1} n_{i+3} n_{i+5} \bar{n}_i \bar{n}_{i+2} \bar{n}_{i+4} - n_i n_{i+2} n_{i+4} \bar{n}_{i+1} \bar{n}_{i+3} \bar{n}_{i+5}, \tag{2}$$

where $\bar{n}_i = 1 - n_i$.

The first line on the right-hand side of (2) refers to collisions involving two particles and the second line to those involving three particles; terms with a plus (minus) sign represent populating (depopulating) contributions to $n_i(t+1, \boldsymbol{r}+\boldsymbol{c}_i)$. A lattice gas obeying the microdynamical equations

(1) has an equilibrium state, and the equilibrium distribution is of the Fermi–Dirac type, resulting from the exclusion principle [2];

(2) evolves according to the Navier–Stokes equations in the incompressible hydrodynamic limit.

The set of eqs. (2) defines the cellular automaton to be implemented for simulation purposes. These equations are well suited for efficient implementation since all variables are Boolean. All particles on the lattice are first propagated in the direction of their velocity; then each node is updated according to the collision rules. This operation is most conveniently performed by means of a lookup table, which yields directly the output configuration resulting from a given input configuration. All variables are thus updated by "Boolean processors" of the cellular automaton type.

As appealing as this method may seem, it may not be the optimal computation technique for investigating boundary flow mechanisms, because it is by essence inherently noisy. Therefore to obtain a velocity profile with the required degree of accuracy one needs long times for the average velocities to emerge from the noise. So it appears more efficient to use averaged quantities N_i for the occupancy rather than the Boolean microscopic quantities n_i. This is accomplished by standard of statistical mechanical methods.

One defines a probability distribution P that assigns each microstate a probability of occurrence in phase space Γ of all possible assignments $\{n_i = \{n_i(\boldsymbol{r})\}, \boldsymbol{r} \in \text{lattice}\}$ of the Boolean field $n_i(\boldsymbol{r})$ [2]. Averaged quantities can then be defined; in particular, macroscopic probabilities of occupancy N_i are given by

$$N_i(r_*) = \langle n_i(r_*) \rangle = \sum_{\{m_j\} \in \Gamma} n_i(r_*)\, P(\{m_j\}). \tag{3}$$

The averaging process defined above applied to the microdynamical equations (1) yields

$$N_i(t+1, \boldsymbol{r}+\boldsymbol{c}_i) = N_i(t, \boldsymbol{r}) + \Delta_i(n), \tag{4}$$

where

$$\Delta_i(n) = a\langle n_{i+1} n_{i+4} \bar{n}_i \bar{n}_{i+2} \bar{n}_{i+3} \bar{n}_{i+5} \rangle + (1-a)\langle n_{i+2} n_{i+5} \bar{n}_i \bar{n}_{i+1} \bar{n}_{i+3} \bar{n}_{i+4} \rangle$$
$$- \langle n_i n_{i+3} \bar{n}_{i+1} \bar{n}_{i+2} \bar{n}_{i+4} \bar{n}_{i+5} \rangle + \langle n_{i+1} n_{i+3} n_{i+5} \bar{n}_i \bar{n}_{i+2} \bar{n}_{i+4} \rangle$$
$$- \langle n_i n_{i+2} n_{i+4} \bar{n}_{i+1} \bar{n}_{i+3} \bar{n}_{i+5} \rangle. \tag{5}$$

With the factorization hypothesis (Boltzmann approximation) applied to eqs. (5), eq. (4) becomes the lattice Boltzmann equation where the collision term (for the FHP model) reads:

$$\begin{aligned}\Delta_i(N) = {} & aN_{i+1}N_{i+4}(1-N_i)(1-N_{i+2})(1-N_{i+3})(1-N_{i+5}) \\ & +(1-a)N_{i+2}N_{i+5}(1-N_i)(1-N_{i+1})(1-N_{i+3})(1-N_{i+4}) \\ & -N_iN_{i+3}(1-N_{i+1})(1-N_{i+2})(1-N_{i+4})(1-N_{i+5}) \\ & +N_{i+1}N_{i+3}N_{i+5}(1-N_i)(1-N_{i+2})(1-N_{i+4}) \\ & -N_iN_{i+2}N_{i+4}(1-N_{i+1})(1-N_{i+3})(1-N_{i+5}).\end{aligned} \tag{6}$$

The local observables, such as the density ρ and the velocity $\boldsymbol{U}$ are obtained as

$$\rho = \sum_i N_i, \qquad \rho\boldsymbol{U} = \sum_i N_i\boldsymbol{c}_i. \tag{7}$$

For the type of problem considered here, computation from the lattice equation is quite straightforward; the lattice Boltzmann method is then particularly fast and convenient, as compared to microscopic simulations and to conventional numerical techniques.

3. Flow near a flat wall

We consider the interaction of a moving fluid with an infinitely long flat boundary; however, because of translational invariance, it suffices to consider a short section of the boundary (actually, a single column of nodes perpendicular to the wall). The system is initialized as a fluid with density ρ (or link density $d=\rho/6$) at rest near the wall. At time $t=0$, the fluid is made to flow instantaneously with velocity U_0 and the bulk of the fluid is in the stationary state corresponding to density ρ and velocity $(U_0, 0)$ (fig. 3).

After the first time step, only the layer adjacent to the wall will experience the influence of the boundary and the velocities at all nodes in that first layer will have decreased by an equal amount ΔU. All subsequent layers (labeled second layer, third layer,...) remain at the free flow velocity U_0. After the second time step, not only has the first layer been further slowed down by the wall, but now the velocity gradient imposed between the first and third layers has slowed down the second layer. Because of the initial conditions in the simulation, all the nodes of the second layer are slowed down by an equal amount. Again all layers beyond the second one remain at velocity U_0.

After the nth time step, friction effects from the wall affect all layers up to the nth layer. All velocities within a given layer will be reduced by an equal amount and all nodes above the nth layer have velocity U_0. Thus, the velocity profile of the gas is translationally invariant: for flat boundaries of infinite extent, we have a one-

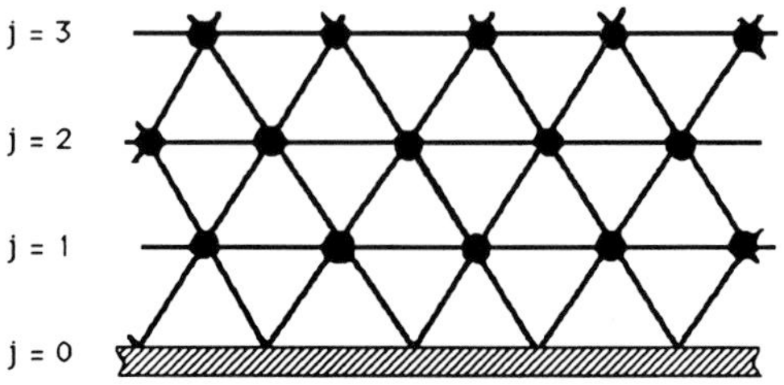

Fig. 3. Configuration of the lattice near the wall (layer indices).

dimensional problem. Such a problem can be easily implemented. Each new computation step adds a new layer to the simulation, while all layers beyond are at the stationary velocity U_o.

We define the boundary layer thickness δ as the distance from the wall where the velocity is $0.99U_o$. δ increases with time at a much slower rate than the rate at which the number of layers involved in the computation increases.

Because of translational invariance, there is no velocity gradient along the wall and the Navier–Stokes equation for the flow considered reduces to (in the incompressible limit):

$$\frac{\partial u}{\partial t} = \nu \frac{\partial^2 u}{\partial y^2}, \tag{8}$$

where ν is the kinematic viscosity, u is the velocity in the x direction along the wall and y denotes the axis orthogonal to the wall. If the velocity on the wall is zero, the boundary conditions are $u(0,t)=0$, and $u(\infty,t)=U_o$, and the solution to (8) reads [3]

$$\begin{aligned} u(y,t) &= \frac{U_o}{\sqrt{\pi\nu t}} \int_0^y \exp\left(\frac{-y'^2}{4\nu t}\right) \mathrm{d}y' \\ &= U_o \operatorname{erf}\left(\frac{y}{\sqrt{4\nu t}}\right). \end{aligned} \tag{9}$$

Two important properties emerge from this solution:

(1) Velocity profiles have a space-dependent scaling property;

(2) The boundary layer thickness δ increases as the square root of time.

Note that when the free flow velocity is constant, time can be converted into distance along the wall ($x = U_o t$) and the variation of δ with respect to t can be equally expressed in terms of x. In classical laminar boundary theory applied to flow along a semi-infinite plane [3] (Blasius), δ increases as $x^{1/2}$, where x is the horizontal distance from the leading edge of the wall. Since U_o is constant, we can interpret the successive velocity profiles as profiles at increasing distance along the wall whose edge is located at $x=0$ (i.e. at $t=0$).

4. Lattice Boltzmann simulations

Simulations with zero velocity boundary condition obtained by the bounce-back rule on the wall (fig. 2b) match very well the theoretical prediction, eq. (9), for the velocity profile (fig. 4) and for the growth of the boundary layer thickness (fig. 5). The velocity profiles are also quite close to the classical Blasius profile (fig. 4).

When the assumption of zero velocity on the wall is relaxed, one no longer has an error-function-type velocity profile. In this respect, we now specifically address the question of the relationship between the velocity profile and the particle–wall interaction. Bounce-back reflection with solid bodies (fig. 2b) is commonly used in lattice gas simulations; however, there are good reasons to examine other types of reflections. One reason stems from consideration of experiments at

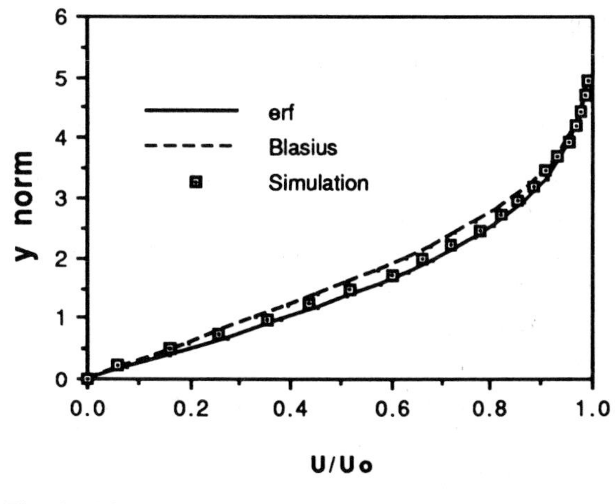

Fig. 4. Velocity profile for flow along a fixed wall. Theoretical error function, Blasius profile, and lattice gas simulation data after 150 time steps (particle density $d=0.1833$, free flow velocity $U_o=2.72$, bounce-back reflection coefficient $r=1$). y norm is the normalized space coordinate measuring the distance from the wall. y norm $=4.99j/j_o$ where j_o is the interpolated row index value corresponding to $U=0.99U_o$. The factors 4.99 and 0.99 are standard in boundary layer theory [3].

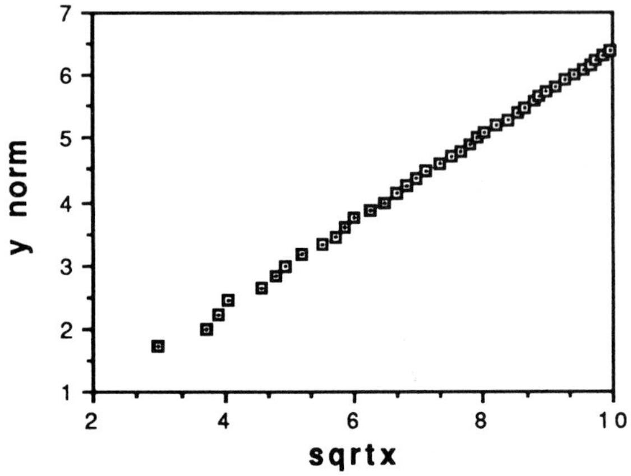

Fig. 5. Boundary layer growth with distance (up to 41 distance units). Same conditions as in fig. 4. y norm is defined in the caption of fig. 4. The distance from the leading edge is tU^*. Since the boundary layer thickness is defined at $U^* = 0.99U_0$, we define x by converting directly t into distance (ignoring the constant factor U^*).

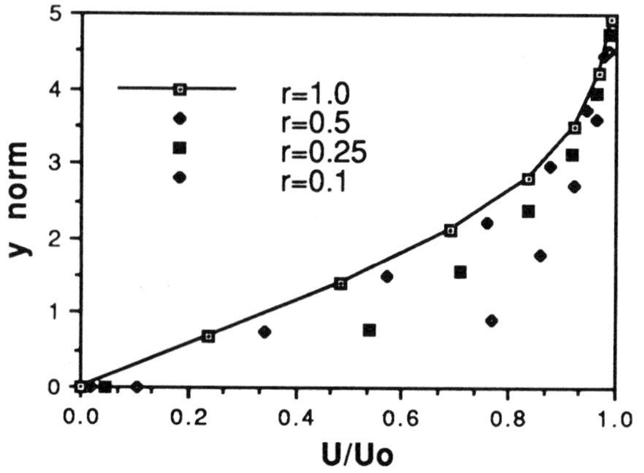

Fig. 6. Velocity profiles for various bounce-back reflection ratios. (gas density: $d = 0.233$; free flow velocity: $U_0 = 0.500$). The data shown were obtained after 20 time steps.

the molecular level. In his book published in 1934, Knudsen [4] described an experiment where molecules are directed towards a wall at a fixed angle of incidence; he observed that the molecules are randomly scattered in all directions. In a lattice gas, this would correspond to a combination of specular and bounce-back reflections in equal proportions. A second reason is that, with purely deterministic interactions with the wall, the Boltzmann assumption of no correlation between particles prior to collision is not valid; note, however, that the hypothesis is correct in a statistical sense for 50% combination of bounce-back and specular reflections.

In a lattice gas, zero velocity at the boundary is obtained (in a strict sense) only with pure bounce-back reflection. If bounce-back reflection is present (in any proportion), more fluid particles will be reflected in the backward direction than the forward direction. This results in a friction effect that slows down the fluid near the wall until its velocity vanishes; in other words, the velocity on the wall should go as close to zero as we want, if we wait long enough.

Our model is particularly well suited to verify this assertion. We define a reflection coefficient r as the ratio of bounce-back to specular reflections in the interactions with the wall: $r = 1$ corresponds to pure bounce-back reflection and $r = 0$ to pure specular reflection. Fig. 6 shows the velocity profiles obtained for different values of r, all other parameters (computation time, particle density and free flow velocity) being kept constant. As expected, velocity on the wall is found to be zero for $r = 1$, and increases progressively to U_0 as r decreases to zero. However, for long computation times (with all other parameters kept constant) (fig. 7), the velocity on the wall decreases and goes arbitrarily close to zero.

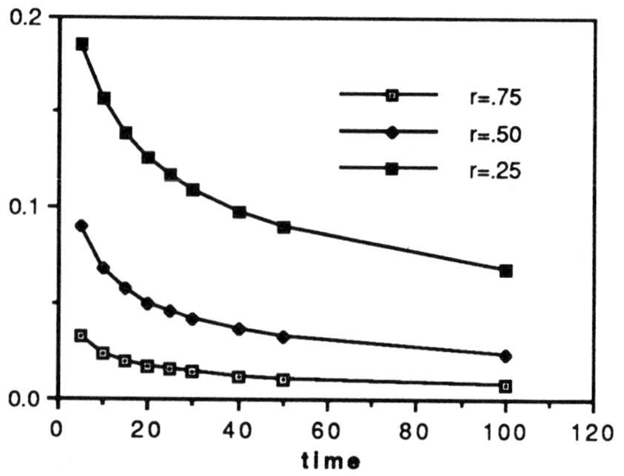

Fig. 7. Velocity at the boundary as a function of time for various coefficient values.

In order to support the claim that zero velocity on the wall can be obtained in the long-time limit, fig. 8 shows the computation time required to obtain the velocity value $0.1U_o$ on the wall as a function of r (all other parameters being kept constant). We find that the velocity on the wall can be expected to go close to zero very rapidly for $r > 0.4$ (60% specular reflection proportion or less); for $0.1 > r > 0.35$, the computation time increases exponentially with $(1 - r)$.

The validation of lattice gas modelling of physical phenomena requires knowledge of these phenomena at the hydrodynamic level, but also, to some extent, at the molecular level. Conversely, lattice gas simulations can be useful to providing information as to what simple models can do and how simplified microdynamics can provide correct hydrodynamics. Boundary interactions are a good example of such a situation. In a real gas there are no "layers" and molecules do not move in discrete time steps. However, if we imagine layers of thickness dy parallel to the wall, and if we assume that all molecules in each layer have the same average velocity, then a random redistribution of molecules colliding with a wall would not give a zero velocity stricto sensu; to first approximation the redistribution would result in a velocity on the wall with a value equal to one half the velocity in the first layer. Of course, as $dy \to 0$, this value would go infinitesimally close to zero. However, even without going to this limit, the velocity on the wall is expected to get closer to zero in the course of time because of viscosity effects. This suggests that $r = 0.5$ might be an alternate choice to $r = 1$ for boundary condition.

The situation discussed above is different from Couette channel flow (one side of the channel moves with velocity U_o) where the flow effectively reaches a steady state. With $r = 1$ on the fixed wall, theory predicts a steady-state velocity profile increasing linearly from zero on the wall at rest to U_o on the moving wall. For the semi-infinite Poiseuille system, the profile never reaches a steady state and the boundary layer grows indefinitely. Lattice gas simulations of Couette flow show the expected linear profile. The velocity on the fixed wall is zero for a reflection coefficient

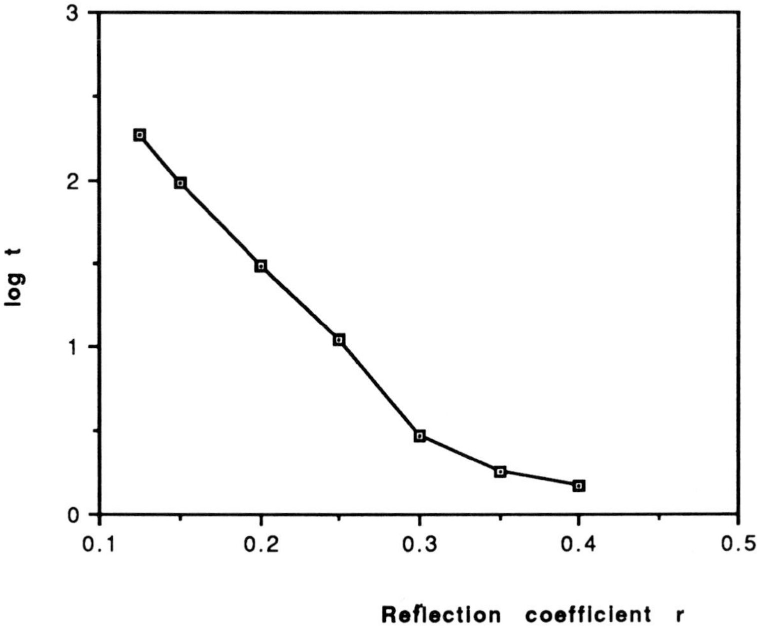

Fig. 8. Logarithm of the number of computation steps versus reflection coefficient ($d = 0.25$, $U_o = 0.4$).

$r = 1$. However, for $r < 1$, the velocity on the fixed wall reaches a non-zero steady-state value which increases as specular reflection becomes more important; indeed, one should not expect zero velocity on the boundary, since along the wall momentum transferred from the adjacent layers cannot be transferred back to the fluid at each time step and consequently there is a constant velocity on the wall. This effect is expected to be negligible when the width of the channel is large compared to the mean free path.

5. Conclusion

We have shown that in a lattice gas flow along a fixed solid boundary, pure bounce-back reflection is not the only type of interaction that yields no-slip condition at the boundary. Zero velocity on the wall can be reached when bounce-back reflection is combined with specular reflection in any proportion. However, the time required to obtain no-slip condition increases considerably when the reflection coefficient becomes smaller than 0.4. These results validate the use of the Knudsen interaction ($r = 0.5$) model which is in agreement with actual experimental observation.

Acknowledgements

We would like to thank Dominique d'Humières, Geoffrey Searby and Fernand Hayot for helpful discussions. A.N. has benefited from a PAI grant. JPB acknowledges support from the "Fonds National de al Recherche Scientifique" (FNRS, Belgium). This work was supported by European Community grant ST2J-0190. Part of this work was done under PAFAC grant Z217-L128.

References

[1] U. Frisch, B. Hasslacher and Y. Pomeau, Lattice gas automata for the Navier–Stokes equation, Phys. Rev. Lett. 56 (1986) 1505.

[2] U. Frisch, D. d'Humières, B. Hasslacher, P. Lallemand, Y. Pomeau and J.P. Rivet, Lattice gas hydrodynamics in two and three dimensions, Complex Systems 1 (1987) 648.

[3] G.K. Batchelor, An Introduction to Fluid Dynamics (Cambridge Univ. Press, Cambridge, 1967).

[4] M. Knudsen, The Kinetic Theory of Gases (Methuen Monographs, London, 1934).

Physica D 47 (1991) 241–259
North-Holland

A KNUDSEN LAYER THEORY FOR LATTICE GASES

R. CORNUBERT
BERTIN et Cie, 59, rue Pierre-Curie, B.P. 3, 78373 Plaisir Cedex 05, France

D. D'HUMIÈRES
CNRS and Université Pierre et Marie Curie, Laboratoire de Physique Statistique de l'ENS, 24, rue Lhomond, 75231 Paris Cedex 05, France

and

D. LEVERMORE
Department of Mathematics, University of Arizona, Tucson, AZ 85721, USA

Received 15 April 1990

A Knudsen layer theory is presented for lattice gases with arbitrary boundary conditions. Analytical results are obtained for two special orientations; these exhibit anisotropic Knudsen layers provided suitable conditions are satisfied. However, the standard boundary conditions used in previous simulations are shown to be isotropic, the bulk steady state extending everywhere in the gas. This theory allows a more accurate localization of the obstacle with respect to the lattice nodes. These results are in good agreement with the numerical simulations.

1. Introduction

During the last few years, several models of gas particles moving on a regular lattice have been used to simulate fluid mechanics. These models, called lattice gases, have shown that very simple cellular automata are able to produce complicated hydrodynamical behavior such as two- or three-dimensional flows, sometimes including chemical reactions or multiphase flows [1]. [#1] While their basic theory is now well understood in the context of periodic boundary conditions [3,4], the problem of more general boundary conditions has not yet been solved and only heuristic solutions have been used for the simulations: "bounce-back" obstacles or "wind tunnel" [5]. These solutions have produced satisfactory results for fixed obstacles but early attempts to move obstacles in the gas have shown that more care should be exercised to avoid unexpected effects [6]. Lattice gases have also been used to model flows in porous media. While the first simulations used wide pores [7], more effective models have led to much smaller pores [8]. In this case, quantitative comparisons with experiments and other simulations require a better localization of the walls and a better understanding of their effects on the flow.

The modification of the bulk equilibrium by a wall leads to a standard problem of kinetic theory, referred to as the Kramers problem [9]. It can be shown that near a wall there exists a thin layer with a thickness on the order of the mean free path in which the steady state is different from the bulk one. This layer, called the kinetic boundary or Knudsen layer, depends on the details of the scattering rules on the wall and, in general, cannot be fully described analytically. The effects of the Knudsen layer are important for rarefied gases and high-speed flows (when both the hydrodynamical boundary and Knudsen

[#1] For the earliest papers see ref. [2]. An extended bibliography can also be found at the end of these Proceedings.

layers have the same magnitude). In most applications, they can only be approximated by slip conditions (non-zero velocity on the wall) and temperature jump at the wall for the thermal effects.

In lattice gases, due to the discrete velocity set and the simple scattering rules, the Kramers problem can be completely solved analytically for the linearized Boltzmann approximation. Within this framework, we present 1D steady solutions for the six-bit FHP models [1] in geometries where the boundary is either parallel or perpendicular to a link of the lattice. In the parallel case, the local equilibrium, already given by the other approaches, is found unchanged by the boundary. However, when the boundary is perpendicular to a link of the lattice, new solutions appear that are exponentially damped in the vicinity of the boundaries. This result shows the existence of anisotropic Knudsen layers that must be avoided when isotropic conditions are required at the boundaries. From this, it can be shown that the standard bounce-back or specular reflection conditions are "good", in the sense that they do not change the local equilibria. However, the effective position of the obstacles with the bounce-back condition depends on the boundary orientation, the effective wall being slightly displaced from the nodes where the boundary conditions are implemented. While this theory was initially intended for the original Boolean implementation of lattice gases, it can be immediately applied to lattice Boltzmann models where the Boolean variables are replaced by real numbers [10,11]. In a similar spirit there is the work of Gatignol [12] where the effects of boundary conditions were studied for discrete velocity gases; anisotropies were also found in this case.

The second section will be devoted to the presentation of the bulk kinetic theory for lattice gases using a formalism very close to the one used in refs. [3,13]. In section 3 the computation will be explicitly carried out for the two simplest geometries of the FHP models and then applied to the standard boundary conditions. Numerical results will be presented in section 4.

2. Kinetic theory of lattice gases

2.1. Lattice gas description

Lattice gas automata describe the motion of particles on a regular D-dimensional lattice $\mathcal{L} = \{\vec{r}_\star\}$, all the particles being on the nodes of the lattice at discrete time steps $t_\star$. Each time step is split into a *propagation step* and a *collision step*. During the propagation step, the particles hop from one node of the lattice $\vec{r}_\star$ to one of its neighbors $\vec{r}_\star + \vec{c}_i$ with an increment chosen from a finite set $\{\vec{c}_i\}$ of velocities. This set has the same local symmetry group $\mathcal{G}$ as the lattice. During the collision step, all the particles sitting at a given node are shuffled according to rules specific to the model so as not to change the conserved quantities of the physical world (mass, momentum, energy, species, etc.). In order to strictly enforce these conservation laws with finite precision arithmetic, an exclusion principle is added: the number of particles at a given node with a given velocity is not greater than the maximum number that can be represented (1 for Boolean arithmetic).

When this exclusion principle is applied in the Boolean case (at most one particle per velocity at a node), the system can be described at time $t_\star$ by b Boolean fields $\{n_i(\vec{r}_\star, t_\star)\}$, $\vec{r}_\star \in \mathcal{L}$, $i \in \{0, \ldots, b-1\}$, where b is the number of possible velocities. These fields allow each node to be described by a b-vector $\boldsymbol{n}(\vec{r}_\star, t_\star)$ representing its internal state just before the collision step. In this paper, arrows will denote vectors in the lattice space and bold face vectors in the state space; their components will be indexed by greek and latin letters, respectively.

The collision process is described by the probability $A(\boldsymbol{s} \to \boldsymbol{s}')$ that the input state $\boldsymbol{s}$ is changed to the output state $\boldsymbol{s}'$. These probabilities fulfill the following constraints:

(i) the normalization

$$\sum_{\boldsymbol{s}'} A(\boldsymbol{s} \to \boldsymbol{s}') = 1, \quad \forall \boldsymbol{s}; \tag{2.1}$$

(ii) some conservation laws, there exist p b-vectors $\boldsymbol{u}_a$ such that

$$A(\boldsymbol{s} \to \boldsymbol{s}')(\boldsymbol{s}' - \boldsymbol{s}) \cdot \boldsymbol{u}_a = 0\,, \quad \forall a, \boldsymbol{s}, \boldsymbol{s}'\,; \tag{2.2}$$

(iii) the same local symmetry group as the lattice

$$A(g(\boldsymbol{s}) \to g(\boldsymbol{s}')) = A(\boldsymbol{s} \to \boldsymbol{s}')\,, \quad \forall g \in \mathcal{G}\,, \forall \boldsymbol{s}, \boldsymbol{s}'\,; \tag{2.3}$$

(iv) semi-detailed balance

$$\sum_{\boldsymbol{s}} A(\boldsymbol{s} \to \boldsymbol{s}') = 1\,, \quad \forall \boldsymbol{s}'\,. \tag{2.4}$$

The time evolution of the gas is given by

$$n_i(\vec{r}_\star + \vec{c}_i, t_\star + 1) = n_i(\vec{r}_\star, t_\star) + \Delta_i\left(\boldsymbol{n}(\vec{r}_\star, t_\star)\right)\,, \tag{2.5}$$

where the collision terms Δ_i are polynomials of the n_j:

$$\Delta_i = \sum_{\boldsymbol{s},\boldsymbol{s}'} (s_i' - s_i)\, a(\boldsymbol{s} \to \boldsymbol{s}') \prod_j n_j^{s_j} (1 - n_j)^{1-s_j}\,, \tag{2.6}$$

where $a(\boldsymbol{s} \to \boldsymbol{s}')$ is a Boolean variable, set to 1 with a probability $A(\boldsymbol{s} \to \boldsymbol{s}')$ and to 0 with a probability $1 - A(\boldsymbol{s} \to \boldsymbol{s}')$. #2, #3

2.2. Boltzmann approximation

While the Boolean variables n_i describe the microscopic behavior of the lattice gas, the "macroscopic" behavior is described through their averaged values (over space, time or ensemble) N_i. The time evolution of the N_i is given by the averaging of eq. (2.5):

$$N_i(\vec{r}_\star + \vec{c}_i, t_\star + 1) = N_i(\vec{r}_\star, t_\star) + C_i(\vec{r}_\star, t_\star)\,, \tag{2.7}$$

where $C_i(\vec{r}_\star, t_\star) = \langle \Delta_i\left(\boldsymbol{n}(\vec{r}_\star, t_\star)\right)\rangle$ is the "macroscopic" collision term. Since space and time correlations are present, the $C_i(\vec{r}_\star, t_\star)$ depend not only on $N_i(\vec{r}_\star, t_\star)$ but also on the neighboring and past quantities.

In the Boltzmann approximation, all the correlations between sites are neglected so the C_i are replaced by the polynomials Δ_i evaluated at N_j:

$$C_i = \sum_{\boldsymbol{s},\boldsymbol{s}'} (s_i' - s_i)\, A(\boldsymbol{s} \to \boldsymbol{s}') \prod_j N_j^{s_j} (1 - N_j)^{1-s_j}\,. \tag{2.8}$$

Until now, no physical meaning was given to the conserved quantities $\{\boldsymbol{u}_a\}$. In what follows, models are considered with only mass and momentum conservation for particles having unit mass and one speed c: $\boldsymbol{u}_0 = \boldsymbol{1}$ and $\boldsymbol{u}_\alpha = \boldsymbol{c}_\alpha$ ($\alpha \in \{1, \ldots, D\}$), where $\boldsymbol{u}_0 = \{1, \ldots, 1\}$ and $\boldsymbol{c}_\alpha$ is the b-vector built by the components of the b velocities in direction α. It is useful to introduce the following b-set of traceless symmetric second-order tensors $(Q_{i\alpha\beta})$ defined by

$$Q_{i\alpha\beta} = c_{i\alpha} c_{i\beta} - (c^2/D)\delta_{\alpha\beta}\,, \tag{2.9}$$

and the corresponding b-vectors $\boldsymbol{Q}_{\alpha\beta}$. #4 In the case where $\boldsymbol{Q}_{\alpha\beta}$ are nonzero for all α and β, it is easy to check that $\boldsymbol{1}$, $\boldsymbol{c}_\alpha$, and $\boldsymbol{Q}_{\alpha\beta}$ ($\alpha \in \{1, \ldots, D\}$, $1 \le \beta \le \alpha$, and $\beta \ne D$) are $D(D+3)/2$ pairwise orthogonal

#2 We use the convention $0^0 = 1$.

#3 For practical implementation, $a(\boldsymbol{s} \to \boldsymbol{s}')$ is obtained from some Boolean fields describing the possible values of $A(\boldsymbol{s} \to \boldsymbol{s}')$.

#4 There are at most $(D-1)(D+2)/2$ such independent vectors.

vectors. Along with $b - D(D+3)/2$ other ones, they can be used to define a new basis for the b-vectors. As pointed out by Frisch [15], the components of the macroscopic state vectors $\boldsymbol{N}$ in this basis are directly related to the density ρ, the momentum $\vec{j}$, and the stress tensor $(S_{\alpha\beta})$, by

$$\rho = \boldsymbol{N} \cdot \boldsymbol{1} = \sum_i N_i \,, \tag{2.10}$$

$$j_\alpha = \boldsymbol{N} \cdot \boldsymbol{c}_\alpha = \sum_i c_{i\alpha} N_i \,, \tag{2.11}$$

$$S_{\alpha\beta} = \boldsymbol{N} \cdot \boldsymbol{Q}_{\alpha\beta} = \sum_i Q_{i\alpha\beta} N_i \,. \tag{2.12}$$

From eqs. (2.7) and (2.8) and using the semi-detailed balance condition, it is possible to show that a homogeneous state evolves toward an equilibrium described by a Fermi–Dirac distribution determined from the conserved quantities ρ and $\vec{j}$. For $\vec{j} = 0$, the equilibrium distribution is $\boldsymbol{N} = f\boldsymbol{1}$, where $f = \rho/b$.

A further simplification occurs in the linear case:

$$N_i(\vec{r}_\star + \vec{c}_i, t_\star + 1) = N_i(\vec{r}_\star, t_\star) - (\mathsf{A}\boldsymbol{N}(\vec{r}_\star, t_\star))_i \,, \tag{2.13}$$

where $-\mathsf{A} = -(A_{ij})$ is the collision operator [#5] linearized around the uniform density $\rho = bf$:

$$A_{ij} = \tfrac{1}{2} \sum_{\boldsymbol{s},\boldsymbol{s}'} (s_i' - s_i)(s_j' - s_j)\, A(\boldsymbol{s} \to \boldsymbol{s}')\, f^{q-1}(1-f)^{b-q-1} \,, \tag{2.14}$$

where $q = \sum_i s_i$.

The conservation laws lead directly to

$$\mathsf{A} \cdot \boldsymbol{1} = \mathsf{A} \cdot \boldsymbol{c}_\alpha = 0 \,. \tag{2.15}$$

For one speed lattice gases with sufficient symmetry, the $\boldsymbol{Q}_{\alpha\beta}$ are eigenvectors of A:

$$\mathsf{A} \cdot \boldsymbol{Q}_{\alpha\beta} = \lambda_B \boldsymbol{Q}_{\alpha\beta} \,, \quad \forall \alpha\beta \,, \tag{2.16}$$

where the positive eigenvalue λ_B depends on the details of the collision rules.

Thus, the state of the nodes can be expressed in two bases. In the first one, the b-vector $\boldsymbol{N}$ is expressed by its projections on the b directions $\vec{c}_i$: $\boldsymbol{N} = \{N_i\}$. This basis will be called "direction basis", and is the most suitable to express the streaming operator: $N_i(\vec{r}_\star + \vec{c}_i, t_\star + 1) - N_i(\vec{r}_\star, t_\star)$. The second useful basis, called "physical basis", is built by the vectors $\boldsymbol{1}$, $\boldsymbol{c}_\alpha$, $\boldsymbol{Q}_{\alpha\beta}$, along with the $b - D(D+3)/2$ other eigenvectors of A, and is the most suitable to represent the linearized collision operator. When needed, a prime ($\boldsymbol{N} = \{N_i'\}$) will denote the components of a b-vector in the second basis, after normalization of the eigenvectors of A.

3. Knudsen layers for the six-bit FHP gases

3.1. Bulk equilibrium

Henceforth we consider Boolean particles moving on a bounded triangular lattice with constant velocities chosen from the set $\{\vec{c}_i\}$, where

$$\vec{c}_i = (\cos(i\pi/3), \sin(i\pi/3)) \,, \qquad i = 0, \ldots, 5 \,. \tag{3.1}$$

[#5] The minus sign is introduced in order to get a nonnegative definite matrix A.

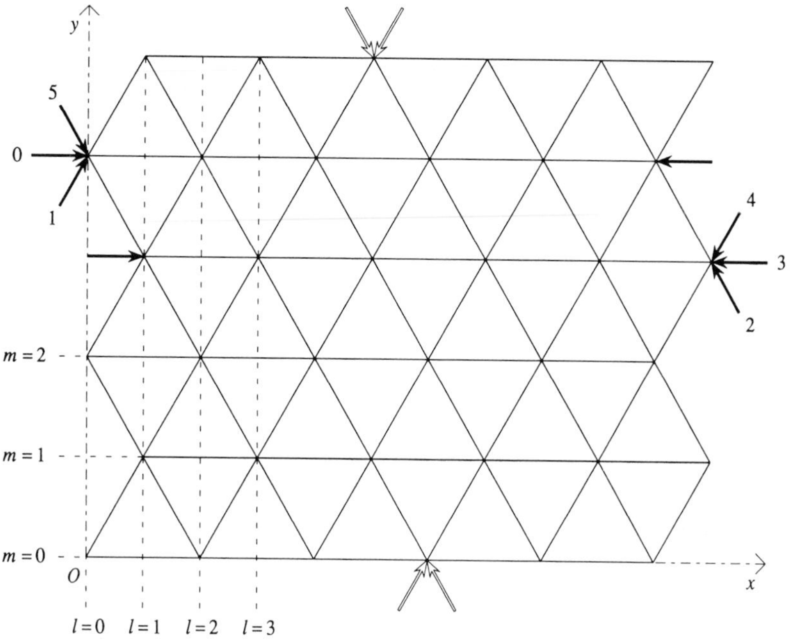

Fig. 1. Schematic of the triangular lattice, open (filled) arrows represents the cut links for the parallel (perpendicular) orientation of the boundary.

The lattice sites are represented by ordered pairs of integers (l,m) such that $l+m$ is even; $l = 2x_\star$ and $m = (2/\sqrt{3})y_\star$, where $0 \le x_\star \le L_x$ and $0 \le y_\star \le L_y$. Two kinds of boundary conditions are considered. In the first case (parallel case) the lattice is periodically wrapped in the x direction and there are two propagation directions cut on each horizontal side of the lattice (see fig. 1). In the second case (perpendicular case) the lattice is periodically wrapped in the y direction and there are three propagation directions cut on each vertical side of the lattice; the way these directions are cut depending on the parity of the horizontal lines.

The vectors $\mathbf{1}$, $\boldsymbol{c}_x$, $\boldsymbol{c}_y$, $\boldsymbol{Q}_{xx}$, and $\boldsymbol{Q}_{xy}$ are five pairwise orthogonal vectors. [#6] The vector $\boldsymbol{p} = \{(-1)^i\}$ is the sixth vector required to build an orthogonal basis and the corresponding projection of the state vector $\boldsymbol{N}$ is denoted $\sigma = \boldsymbol{N}\cdot\boldsymbol{p}$. The normalized vectors $\{\boldsymbol{e}_0,\ldots,\boldsymbol{e}_5\}$ composing the physical basis are therefore $\boldsymbol{e}_0 = (1/\sqrt{6})\mathbf{1}$, $\boldsymbol{e}_1 = (1/\sqrt{3})\boldsymbol{c}_x$, $\boldsymbol{e}_2 = (1/\sqrt{3})\boldsymbol{c}_y$, $\boldsymbol{e}_3 = (2/\sqrt{3})\boldsymbol{Q}_{xx}$, $\boldsymbol{e}_4 = (2/\sqrt{3})\boldsymbol{Q}_{xy}$, and $\boldsymbol{e}_5 = (1/\sqrt{6})\boldsymbol{p}$. From the symmetry properties assumed for the general model it can be checked that $\mathsf{A}\boldsymbol{p} = \lambda_T\boldsymbol{p}$, and the linearized collision operator is diagonal in the physical basis, with diagonal elements $\{0,0,0,\lambda_B,\lambda_B,\lambda_T\}$.

The most general model for this six-velocity gas is defined by the probabilities assigned to the different collisional possibilities conserving number of particle and momentum.

(i) Head-on collision (see fig. 2a): two particles with velocities $\vec{c}_i$ and $\vec{c}_{i+3}$ stay unchanged with probability $1-p_2$ or give with probability $p_2/2$ one of the following two configurations: particles with velocities $\vec{c}_{i+1}$ and $\vec{c}_{i+4}$ or velocities $\vec{c}_{i+2}$ and $\vec{c}_{i+5}$.

(ii) Symmetric three-body collision: three particles with velocities $\vec{c}_i$, $\vec{c}_{i+2}$, and $\vec{c}_{i+4}$ stay unchanged with probability $1-p_{3\mathrm{s}}$ or give particles with velocities $\vec{c}_{i+1}$, $\vec{c}_{i+3}$ and $\vec{c}_{i+4}$ with probability $p_{3\mathrm{s}}$.

(iii) Asymmetric three-body collision: three particles with velocities $\vec{c}_i$, $\vec{c}_{i+3}$, and $\vec{c}_j$, $j\in\{i+2,i+5\}$ ($j\in\{i+1,i+4\}$), stay unchanged with probability $1-p_{3\mathrm{n}}$ or give particles with velocities $\vec{c}_{i+1}$, $\vec{c}_{i+4}$ and $\vec{c}_j$ ($\vec{c}_{i+2}$ and $\vec{c}_{i+5}$ and $\vec{c}_j$), with probability $p_{3\mathrm{n}}$.

[#6] In the two-dimensional case, $\alpha = 1$ and 2 will be replaced by indices x and y.

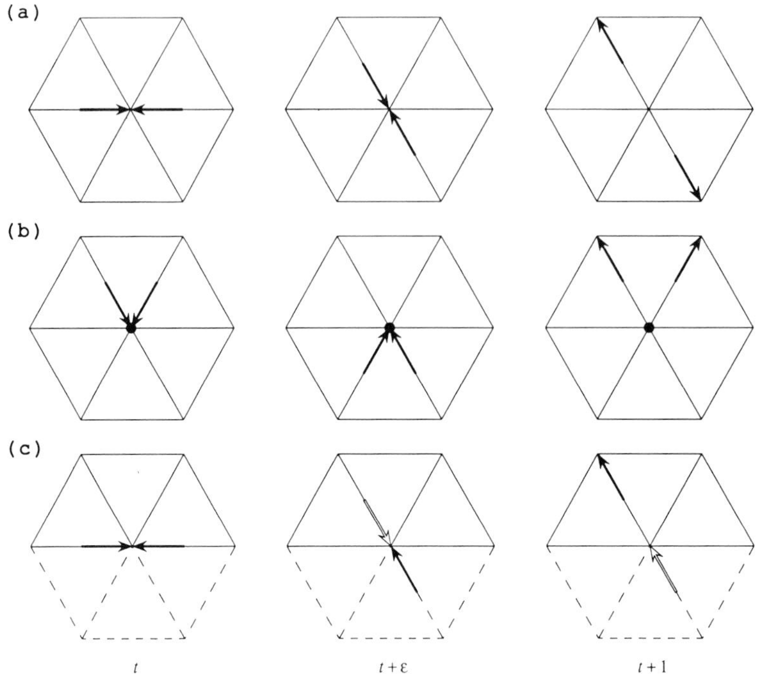

Fig. 2. Time evolution of the FHP lattice gas, at time t, time $t+\varepsilon$ just after the collision step, and time $t+1$ just after the propagation step, for: (a) head-on collision on a node in the bulk; (b) bounce-back condition on a node $\vec{r}_{w\star}$ (the • is the node boundary); (c) bounce-back condition and head-on collision on a node $\vec{r}_{g\star}$ (link boundary, $i_g = 0$), the dashed lines represent the link outside the gas, the outlined arrows show the particle with a different propagation rule.

(iv) Four-body collision: four particles with velocities $\vec{c}_i$, $\vec{c}_{i+1}$, $\vec{c}_{i+3}$, and $\vec{c}_{i+4}$ stay unchanged with probability $1-p_4$ or give with probability $p_4/2$ one of the following two configurations: particles with velocities $\vec{c}_{i+1}$, $\vec{c}_{i+2}$, $\vec{c}_{i+4}$, and $\vec{c}_{i+5}$, or velocities $\vec{c}_i$, $\vec{c}_{i+2}$, $\vec{c}_{i+3}$, and $\vec{c}_{i+5}$.

The values of λ_B and λ_T are related to p_2, p_{3s}, p_{3n}, p_4, and the reduced density $f=\rho/6$, by

$$\lambda_B = 3p_2\, f(1-f)^3 + 12p_{3n}\, f^2(1-f)^2 + 3p_4\, f^3(1-f)\,, \tag{3.2}$$

$$\lambda_T = 6p_{3s}\, f^2(1-f)^2\,. \tag{3.3}$$

For the original FHP model (FHP-I) $p_2 = p_{3s} = 1$ and $p_{3n} = p_4 = 0$, thus

$$\lambda_B = 3f(1-f)^3 \quad \text{and} \quad \lambda_T = 6f^2(1-f)^2\,. \tag{3.4}$$

The maximum values of λ_B and λ_T are obtained for $p_2 = p_{3s} = p_{3n} = p_4 = 1$ and are given by

$$\lambda_B = 3f(1-f)[1+2f(1-f)] \quad \text{and} \quad \lambda_T = 6f^2(1-f)^2\,. \tag{3.5}$$

Looking for solutions with a linear spatial dependence: $\boldsymbol{N}(x,y) = \boldsymbol{v} + x\boldsymbol{v}_x + y\boldsymbol{v}_y$, the following general solution is obtained:

$$\begin{aligned}\boldsymbol{N}(x,y) &= \tfrac{1}{6}\rho\mathbf{1} + \tfrac{1}{3}(j_{0x} + y\,\partial_y j_x)\boldsymbol{c}_x + \tfrac{1}{3}(j_{0y} + x\,\partial_x j_y)\boldsymbol{c}_y \\ &\quad - (1/3\lambda_B)(\partial_x j_x - \partial_y j_y)(\partial_y j_x + \partial_x j_y)\boldsymbol{Q}_{xy}\,.\end{aligned} \tag{3.6}$$

This formula, already found in refs. [3,13], relates the steady non-uniform equilibrium distribution to the physical parameters and their gradients. In addition, it can be used to determine the injection rates that must be applied on the cut links to achieve a given equilibrium everywhere in the lattice.

3.2. Knudsen layer

3.2.1. General orientation

In the previous section injection rates were obtained that produce a steady equilibrium with a linear spatial dependence in the momentum. If there are more degrees of freedom than the five parameters appearing in eq. (3.6), these injection rates can be altered to produce some deviation from the above equilibrium. The goal of this section is to state the problem and to solve it for simple orientations of the boundaries. Since eq. (2.7) is a linear difference equation that is translation invariant over the lattice, it is natural to seek steady solutions with the exponential space dependence

$$\boldsymbol{N}(x, y) = \eta^{(2/\sqrt{3})y} \xi^{2x} \boldsymbol{v} . \tag{3.7}$$

For these solutions, the streaming operator is linear and given in the direction basis by

$$(\mathsf{S}_{ij}) = \begin{pmatrix} \xi^2 - 1 & & & & & \\ & \eta\xi - 1 & & & \mathbf{0} & \\ & & \dfrac{\eta - \xi}{\xi} & & & \\ & & & \dfrac{1 - \xi^2}{\xi^2} & & \\ & \mathbf{0} & & & \dfrac{1 - \eta\xi}{\eta\xi} & \\ & & & & & \dfrac{\xi - \eta}{\eta} \end{pmatrix} . \tag{3.8}$$

Since the vector $\boldsymbol{v}$ solves $(\mathsf{S} + \mathsf{A})\boldsymbol{v} = 0$, the solubility condition $\det(\mathsf{S} + \mathsf{A}) = 0$ must be satisfied. When S is invertible ($\xi \neq \pm 1$, η, or $1/\eta$), the determinant of $\mathsf{S} + \mathsf{A}$ can be written

$$\det(\mathsf{S} + \mathsf{A}) = \det(\mathsf{S}) \det(\mathsf{I} + \mathsf{S}^{-1}\mathsf{A}) , \tag{3.9}$$

where I is the identity operator. Since A is a diagonal matrix with three zero diagonal elements in the physical basis, it is natural to use this representation to evaluate eq. (3.9). The matrix $\mathsf{I} + \mathsf{S}^{-1}\mathsf{A}$ then becomes

$$\mathsf{I} + \mathsf{S}^{-1}\mathsf{A} = \begin{pmatrix} 1 & 0 & 0 & 0 & 0 & \dfrac{\lambda_T}{6}(d_a - d_{aa}) \\ 0 & 1 & 0 & \dfrac{\lambda_B}{12}(4d_a - d_{aa}) & \dfrac{\lambda_B}{4\sqrt{3}} d_{sa} & 0 \\ 0 & 0 & 1 & -\dfrac{\lambda_B}{4\sqrt{3}} d_{sa} & \dfrac{\lambda_B}{4} d_{aa} & 0 \\ 0 & 0 & 0 & 1 - \dfrac{\lambda_B}{2} & 0 & \dfrac{\lambda_T}{6\sqrt{2}}(2d_a + d_{aa}) \\ 0 & 0 & 0 & 0 & 1 - \dfrac{\lambda_B}{2} & -\dfrac{\lambda_T}{2\sqrt{6}} d_{sa} \\ 0 & 0 & 0 & \dfrac{\lambda_B}{6\sqrt{2}}(2d_a + d_{aa}) & -\dfrac{\lambda_B}{2\sqrt{6}} d_{sa} & 1 - \dfrac{\lambda_T}{2} \end{pmatrix} , \tag{3.10}$$

where $d_a = \hat{\xi}\tilde{\xi}/(\tilde{\xi}^2 - 4)$, $d_{aa} = 2\hat{\xi}/(\tilde{\xi} - \tilde{\eta})$, $d_{sa} = -2\hat{\eta}/(\tilde{\xi} - \tilde{\eta})$, and

$$\hat{\eta} = \eta - \frac{1}{\eta} , \qquad \tilde{\eta} = \eta + \frac{1}{\eta} , \qquad \hat{\xi} = \xi - \frac{1}{\xi} , \qquad \tilde{\xi} = \xi + \frac{1}{\xi} .$$

The determinant of S + A is given by

$$\det(\mathsf{S}+\mathsf{A}) = -(1-\tfrac{1}{2}\lambda_B)^2(1-\tfrac{1}{2}\lambda_T)\{(\tilde{\xi}^2-4)(\tilde{\xi}-\tilde{\eta})^2 - \tfrac{1}{4}a^2[(4+\tilde{\eta}\tilde{\xi}-2\tilde{\xi}^2)^2 + 3(\tilde{\eta}^2-4)(\tilde{\xi}^2-4)]\}, \tag{3.11}$$

where

$$a^2 = \frac{8\lambda_B\lambda_T}{9(2-\lambda_B)(2-\lambda_T)}, \quad \text{with} \quad a > 0. \tag{3.12}$$

3.2.2. Parallel orientation

The simplest orientation for the boundaries corresponds to the situation where the lattice is cut along a line parallel to a velocity direction (this case will be called the parallel case), cutting two links per node. For the chosen lattice orientation, one of these cases occurs for a cut parallel to the x direction. The solutions invariant by translation along this direction correspond to $\xi = 1$ in eq. (3.7) ($\hat{\xi} = 0$). The matrix S is not invertible, but the determinant of S + A, expressed in the physical basis, is the product of two 3×3 ones. The solubility condition is given by

$$\det(\mathsf{S}+\mathsf{A}) = \tfrac{1}{9}\lambda_B\lambda_T(2-\lambda_T)(\eta-1)^4/\eta^2 = 0\,. \tag{3.13}$$

The solution is the fourth-order root $\eta = 1$. The four corresponding modes are density, momenta along the x and y directions, and the gradient along the y direction of the momentum in the x direction. The equilibrium distribution is related to the injection parameters by eq. (3.6).

For a finite slice of the lattice $0 \le y_\star \le L_y$, the equilibrium distribution is determined by the four parameters defining the average distribution on the two incoming links per node on each side of the slab: $p_1 = N_1(0)$, $p_2 = N_2(0)$, $p_4 = N_4(L_y)$, and $p_5 = N_5(L_y)$. Using eq. (3.6) and $\partial_x j_y = 0$, it follows that

$$\rho = \tfrac{3}{2}(p_1+p_2+p_4+p_5)\,, \tag{3.14}$$

$$j_{0x} + \tfrac{1}{2}L_y\partial_y j_x = j_x(\tfrac{1}{2}L_y) = \tfrac{3}{2}(p_1-p_2-p_4+p_5)\,, \tag{3.15}$$

$$j_{0y} = \tfrac{1}{2}\sqrt{3}(p_1+p_2-p_4-p_5)\,, \tag{3.16}$$

$$\partial_y j_x = \frac{3\lambda_B}{\lambda_B L_y + \sqrt{3}}(-p_1+p_2-p_4+p_5)\,. \tag{3.17}$$

3.2.3. Perpendicular orientation

The next simplest boundary condition is when the lattice is cut along the bisector of two consecutive velocity directions (this case will be called perpendicular case). One possible orientation is the y direction, for which the translation invariant solutions correspond to $\eta = 1$ ($\hat{\eta} = 0$). For this case the lines with different parity of $y_\star$ are not equivalent, on the left edge there are three links cut for the even values of $y_\star$ and only one for the odd values, while the role of even and odd values are exchanged on the right side. Thus, by an elementary count of the incoming links, the solution has eight degrees of freedom. In addition to the four modes defining the bulk equilibrium, four boundary modes are expected, two for each side of the slab. The solubility condition is given by

$$\det(\mathsf{S}+\mathsf{A}) = -(2-\lambda_B)^2(2-\lambda_T)\frac{(\xi-1)^4}{8\xi^4}[(\xi^2-1)^2 - a^2(1+\xi+\xi^2)^2] = 0\,. \tag{3.18}$$

The following bounds can be obtained from eqs. (3.5):

$$0 < \lambda_B \le \tfrac{9}{8}\,, \qquad 0 < \lambda_T \le \tfrac{3}{8}\,, \qquad 0 < a^2 \le \tfrac{24}{91}\,. \tag{3.19}$$

In addition of the four roots $\xi = 1$ there are four more roots, given by

$$\xi_{\varepsilon_1\varepsilon_2} = \frac{\varepsilon_1 a + \varepsilon_2\sqrt{4-3a^2}}{2(1-\varepsilon_1 a)}, \tag{3.20}$$

where $\varepsilon_1 = \pm 1$ and $\varepsilon_2 = \pm 1$ and

$$\xi_{+-} < -1 < \xi_{--} < 0 < \xi_{-+} < 1 < \xi_{++} \,. \tag{3.21}$$

It can be easily checked that replacing ε_1 with $-\varepsilon_1$ changes ξ to $1/\xi$, the corresponding pair of modes reflects the existence of identical solutions on each side of the slab, an exponential decreasing mode on the left side becoming an exponential increasing mode on the right side. For a slice $0 \le x \le L_x$, with $L_x \gg 1/|2\log(|\xi_{\varepsilon_1\varepsilon_2}|)|$, only one pair of modes is relevant on each side: $\varepsilon_1 = -1$ on the left and $\varepsilon_1 = 1$ on the right. These two modes are: an exponential one for $\varepsilon_2 = 1$ and an oscillating exponential for $\varepsilon_2 = -1$. The latter mode reflects the existence of different equilibria on the odd and even horizontal lines of the lattice corresponding to odd or even $y_\star$.

The modes associated to the values $\xi_{\varepsilon_1\varepsilon_2}$ are the solutions of

$$(\mathsf{S}_{\varepsilon_1\varepsilon_2} + \mathsf{A})\boldsymbol{v}_{\varepsilon_1\varepsilon_2} = 0\,, \tag{3.22}$$

and are given, up to multiplicative constants, by

$$\boldsymbol{v}_{\varepsilon_1\varepsilon_2} = \frac{\varepsilon_2\lambda_T\sqrt{4-3a^2}}{6a}\mathbf{1} + \frac{\varepsilon_1}{4}(2-\lambda_T)\left(2-\varepsilon_2\sqrt{4-3a^2}\right)\boldsymbol{c}_x - \frac{8\lambda_T}{3a(2-\lambda_B)}\boldsymbol{Q}_{xx} + \varepsilon_1\boldsymbol{p}\,. \tag{3.23}$$

A very surprising feature of these modes is the component along the $\boldsymbol{c}_x$ *b*-vector or a nonzero local value of the momentum in the x direction. A naïve argument would lead to the conclusion there exists a nonuniform steady solution that does not satisfy the mass conservation. Indeed, this apparent paradox disappears if the mass flux is measured across a line at $x_\star + \frac{1}{4}$, taking some care about the order of the collision and propagation steps that is used to derive eq. (2.13). The mass flux across the line at $x_\star + \frac{1}{4}$ is given by the sum of densities in direction 0, 1, and 5 at nodes $x_\star + \frac{1}{2}$ and direction 0 at nodes $x_\star + 1$, minus the sum of densities in direction 2, 3, and 4 at nodes $x_\star$ and direction 3 at nodes $x_\star - \frac{1}{2}$. Let $\boldsymbol{w}_{\varepsilon_1\varepsilon_2}$ be defined as $\{\xi_{\varepsilon_1\varepsilon_2}(\xi_{\varepsilon_1\varepsilon_2}+1), \xi_{\varepsilon_1\varepsilon_2}, -1, -(\xi_{\varepsilon_1\varepsilon_2}+1)/\xi_{\varepsilon_1\varepsilon_2}, -1, \xi_{\varepsilon_1\varepsilon_2}\}$ in the direction basis, the mass flux is given by $\boldsymbol{w}_{\varepsilon_1\varepsilon_2}\cdot\boldsymbol{v}_{\varepsilon_1\varepsilon_2}$. After some tedious algebra it can be verified that $\boldsymbol{w}_{\varepsilon_1\varepsilon_2}\cdot\boldsymbol{v}_{\varepsilon_1\varepsilon_2} = 0$.

The total contribution of the exponential modes is given by

$$\boldsymbol{N}(x) = \sum_{\varepsilon_1\,\varepsilon_2} \kappa_{\varepsilon_1\varepsilon_2}\boldsymbol{v}_{\varepsilon_1\varepsilon_2}\xi_{\varepsilon_1\varepsilon_2}^{2x-(1+\varepsilon_1)L_x}\,, \tag{3.24}$$

where the term $-(1+\varepsilon_1)L_x$ is added in order to have equal contributions of the parameters $\kappa_{\varepsilon_1\varepsilon_2}$ at both edges of the lattice slice.

3.3. Applications

Most flow simulations use boundary conditions to implement solid walls. There exist two main conditions: "bounce-back" and "specular reflection"; in the first case the particles reverse their velocity on the obstacle, in the second case they are reflected on the wall like a light ray. Obviously, a mixed case can be considered with probability p of bounce-back and $1-p$ of specular reflection. There are also two ways to implement these laws: at the level of the nodes [5] or at the level of the links [16].

In the node implementation, the bounce back condition is independent of the orientation of the wall and can be defined by a set $\{\vec{r}_{\mathrm{w}\star}\}$ of nodes that are outside the gas. The evolution equation (2.5) is replaced on $\{\vec{r}_{\mathrm{w}\star}\}$ by (see fig. 2b)

$$n_i(\vec{r}_{\mathrm{w}\star} + \vec{c}_i, t_\star + 1) = n_{i+3}(\vec{r}_{\mathrm{w}\star}, t_\star)\,, \tag{3.25}$$

where the indices are computed modulo six. The specular reflection can be implemented by bit exchanges for the parallel or the perpendicular orientations only. The walls must be defined as a set of triplets $\{(\vec{r}_{w\star}, \|, i_w)\}$, resp. $\{(\vec{r}_{w\star}, \perp, i_w)\}$, where $\vec{r}_{w\star}$ is the location of the node, and the wall is parallel (resp. pependicular) to the direction $\vec{c}_{i_w}$, ($i_w \in \{0,1,2\}$). The evolution equation (2.5) is replaced on $\{\vec{r}_{w\star}\}$ by

$$n_i(\vec{r}_{w\star} + \vec{c}_i, t_\star + 1) = n_i(\vec{r}_{w\star}, t_\star), \qquad \text{for } i = i_w\,,\ i_w + 3\,, \tag{3.26}$$

$$n_i(\vec{r}_{w\star} + \vec{c}_i, t_\star + 1) = n_{2i_w - i}(\vec{r}_{w\star}, t_\star), \qquad \forall i \neq i_w\,,\ i_w + 3\,, \tag{3.27}$$

in the parallel case and by

$$n_i(\vec{r}_{w\star} + \vec{c}_i, t_\star + 1) = n_{3+2i_w - i}(\vec{r}_{w\star}, t_\star), \qquad \forall i\,, \tag{3.28}$$

in the perpendicular case.

To stress the difference between the two implementations, the index 'w' used above for the nodes is replaced by a 'g' for the links. In the latter implementation both bounce-back and specular reflection depend on the orientation of the wall and must be defined by the same set of triplets as above. In that case the nodes $\{\vec{r}_{g\star}\}$ are inside the gas, i_g is in $\{0,\ldots,5\}$, and the direction of $\vec{c}_{i_g}$ defines where the gas is with respect to the wall.

For the parallel orientation, the directions $i_g + i$, $i = 0$ to 3, point into the gas, and their evolution equations (2.5) are unchanged. The directions $i_g - 1$ and $i_g - 2$ point outwards and the corresponding propagation rules are modified to keep the particles inside the gas:

$$n_{i_g - i + 3}(\vec{r}_{g\star}, t_\star + 1) = n_{i_g - i}(\vec{r}_{g\star}, t_\star) + \Delta_{i_g - i}(\boldsymbol{n}(\vec{r}_{g\star}, t_\star))\,, \qquad i = 1,\ 2 \tag{3.29}$$

for the bounce-back condition (see fig. 2c) and

$$n_{i_g + i}(\vec{r}_{g\star}, t_\star + 1) = n_{i_g - i}(\vec{r}_{g\star}, t_\star) + \Delta_{i_g - i}(\boldsymbol{n}(\vec{r}_{g\star}, t_\star))\,, \qquad i = 1,\ 2 \tag{3.30}$$

for the specular reflection condition.

For the perpendicular orientation, the directions $i_g + i$, $i = -1$ to 1, point inwards, and their evolution equations are unchanged. The directions $i_g + i$, $i = 2$–4, point outwards and, as above, their evolution equations are

$$n_{i_g + i - 3}(\vec{r}_{g\star}, t_\star + 1) = n_{i_g + i}(\vec{r}_{g\star}, t_\star) + \Delta_{i_g + i}(\boldsymbol{n}(\vec{r}_{g\star}, t_\star))\,, \qquad i = 2,\ 3,\ 4 \tag{3.31}$$

for the bounce-back, and

$$n_{i_g - i - 3}(\vec{r}_{g\star}, t_\star + 1) = n_{i_g + i}(\vec{r}_{g\star}, t_\star) + \Delta_{i_g + i}(\boldsymbol{n}(\vec{r}_{g\star}, t_\star))\,, \qquad i = 2,\ 3,\ 4 \tag{3.32}$$

for the specular reflection.

3.3.1. Parallel orientation

In this section, let us consider a bulk gas, invariant in the x direction and defined for $0 \leq y_\star$ ($i_w = i_g = 0$, $y_{w\star} = -\frac{1}{2}\sqrt{3}$, or $y_{g\star} = 0$, with the notations of the previous section). For a mixture of bounce-back and specular reflection with probability p and $1 - p$, respectively, omitting the $x_\star$ dependence and taking the average in the Boltzmann approximation, eqs. (3.25) and (3.26) yield

$$\begin{aligned} N_1(0, t+1) &= pN_4(-\tfrac{1}{2}\sqrt{3}, t) + (1-p)N_5(-\tfrac{1}{2}\sqrt{3}, t)\,, \\ N_2(0, t+1) &= pN_5(-\tfrac{1}{2}\sqrt{3}, t) + (1-p)N_4(-\tfrac{1}{2}\sqrt{3}, t)\,, \end{aligned} \tag{3.33}$$

for the node implementation, while eqs. (3.29) and (3.30) lead to

$$\begin{aligned} N_1(0,t+1) &= pN_4(-\tfrac{1}{2}\sqrt{3},t+1)+(1-p)N_5(-\tfrac{1}{2}\sqrt{3},t+1)\,, \\ N_2(0,t+1) &= pN_5(-\tfrac{1}{2}\sqrt{3},t+1)+(1-p)N_4(-\tfrac{1}{2}\sqrt{3},t+1)\,, \end{aligned} \tag{3.34}$$

for the link implementation. Here the $N_i(y_\star, t)$ at $y_\star = -\frac{1}{2}\sqrt{3}$ are those obtained for the bulk gas at this location (compare the right-hand terms of eqs. (2.6) and (3.29) to (3.32)). Comparing these two sets of equations, it is easy to see that the node implementation introduces a time delay in the closure equations at the boundary. Thus, a difference can be expected between the two implementations when considering staggered invariants [14], while they are the same for the steady solutions.

The steady state produced by these boundary conditions is obtained by the values of the free parameters in eqs. (3.6) that satisfy the relations (3.33) (or(3.34)). By translation invariance in the direction x, it is obvious that $\partial_x j_y = 0$. The sum and difference of the closure relations yield

$$\begin{aligned} N_1(0)+N_2(0) &= N_4(-\tfrac{1}{2}\sqrt{3})+N_5(-\tfrac{1}{2}\sqrt{3})\,. \\ N_1(0)-N_2(0) &= (1-2p)[N_5(-\tfrac{1}{2}\sqrt{3})-N_4(-\tfrac{1}{2}\sqrt{3})]\,. \end{aligned} \tag{3.35}$$

Using (3.6), the first equation implies $\rho + \sqrt{3}j_{0y} = \rho - \sqrt{3}j_{0y}$, or $j_{0y} = 0$, and there is no mass flux across the wall, as expected. The second equation gives

$$j_{0x} - \frac{\sqrt{3}}{2\lambda_B}\partial_y j_x = (1-2p)\left(j_{0x} + \frac{\sqrt{3}(1-\lambda_B)}{2\lambda_B}\partial_y j_x\right), \tag{3.36}$$

or

$$2\,p\,j_{0x} = \frac{\sqrt{3}}{2\lambda_B}[p\lambda_B + (1-p)(2-\lambda_B)]\partial_y j_x\,. \tag{3.37}$$

If $p = 0$, $\partial_y j_x = 0$, then the specular reflection enforces a zero gradient at the wall and corresponds to free-slip condition. If $p \neq 0$, eq. (3.37) becomes

$$j_{0x} = \sqrt{3}\left[\frac{1}{4} + \frac{1-p}{2p}\left(\frac{1}{\lambda_B} - \frac{1}{2}\right)\right]\partial_y j_x = \sqrt{3}\left(\frac{1}{4} + 2\nu\frac{1-p}{p}\right)\partial_y j_x\,, \tag{3.38}$$

where $\nu = (2-\lambda_B)/8\lambda_B$ is the kinematic viscosity [3]. Using eq. (3.6) and this value of j_{0x}, the horizontal momentum j_x is zero for

$$y_0 = -\sqrt{3}\left(\frac{1}{4} + 2\nu\frac{1-p}{p}\right)\,. \tag{3.39}$$

Thus the nonspecular conditions correspond to the no-slip condition for an imaginary wall located at y_0. For the bounce-back condition $p = 1$ and the wall is located at $y_0 = -\frac{1}{4}\sqrt{3}$, halfway between the last horizontal line in the gas and the first one outside.

3.3.2. Perpendicular orientation

In this section, let us consider a bulk gas, invariant in the y direction and defined for $0 \le x_\star$ ($i_\mathrm{w} = i_\mathrm{g} = 0$, $y_{\mathrm{w}\star} = -\frac{1}{2}$ and -1, or $y_{\mathrm{g}\star} = 0$ and $\frac{1}{2}$). In both implementations, the closure relations are the same for the steady states:

$$\begin{aligned} N_0(0) &= N_3(-1)\,, \\ N_0(\tfrac{1}{2}) &= N_3(-\tfrac{1}{2})\,, \\ N_1(0) &= pN_4(-\tfrac{1}{2})+(1-p)N_2(-\tfrac{1}{2})\,, \\ N_5(0) &= pN_2(-\tfrac{1}{2})+(1-p)N_4(-\tfrac{1}{2})\,. \end{aligned} \tag{3.40}$$

The steady state produced by these boundary conditions is obtained by the values of the free parameters in eqs.(3.6) and (3.24) that satisfy relations (3.40). This linear problem can be split in two steps. First find the solution of eqs. (3.6) and (3.40) ($\kappa_{\varepsilon_1\varepsilon_2} = 0$). Then check for the existence of solutions with nonzero values of the $\kappa_{\varepsilon_1\varepsilon_2}$.

As in the previous section, the sum and the difference of the last two relations of (3.40) lead to

$$\begin{aligned} N_1(0) + N_5(0) &= N_2(-1/2) + N_4(-1/2)\,, \\ N_1(0) - N_5(0) &= (1-2p)[N_2(-1/2) - N_4(-1/2)]\,. \end{aligned} \tag{3.41}$$

For $\kappa_{\varepsilon_1\varepsilon_2} = 0$, these two relations give $\partial_y j_x = 0$, $j_{0x} = 0$, and

$$j_{0y} - \frac{1}{2\lambda_B}\partial_x j_y = (1-2p)\left(j_{0y} + \frac{1-\lambda_B}{2\lambda_B}\partial_x j_y\right). \tag{3.42}$$

If $p = 0$, $\partial_x j_y = 0$, the specular reflection enforces a zero gradient at the wall or a free-slip condition. If $p \neq 0$ then

$$j_{0y} = \left(\frac{1}{4} + 2\nu\frac{1-p}{p}\right)\partial_x j_y \tag{3.43}$$

and the vertical momentum j_y is zero for

$$x_0 = -\left(\frac{1}{4} + 2\nu\frac{1-p}{p}\right). \tag{3.44}$$

Thus the nonspecular conditions correspond to the no-slip condition for an imaginary wall located at x_0 ($x_0 = -\frac{1}{4}$, for the bounce-back condition). Note that for a given p the position of the effective wall is not the same for the parallel and perpendicular orientations (a factor $\sqrt{3}$ in eq. (3.39).) This effect should be taken into account for small obstacles.

To compute the effects of the exponential modes, the lattice will be assumed to extend to infinity on the right. Thus, the contributions of the modes for $\varepsilon_1 = 1$ is zero. Using eq. (3.24) and the first two relations of (3.40), the components κ_{-+} and κ_{--} of the exponential modes must satisfy

$$\begin{aligned} \kappa_{-+}v_{0-+} + \kappa_{--}v_{0--} &= \frac{\kappa_{-+}v_{3-+}}{\xi^2_{-+}} + \frac{\kappa_{--}v_{3--}}{\xi^2_{--}}\,, \\ \kappa_{-+}v_{0-+}\xi_{-+} + \kappa_{--}v_{0--}\xi_{--} &= \frac{\kappa_{-+}v_{3-+}}{\xi_{-+}} + \frac{\kappa_{--}v_{3--}}{\xi_{--}}\,, \end{aligned} \tag{3.45}$$

or

$$\begin{aligned} \kappa_{-+}\left(v_{0-+}\xi_{-+} - \frac{v_{3-+}}{\xi_{-+}}\right)\frac{1}{\xi_{-+}} + \kappa_{--}\left(v_{0--}\xi_{--} - \frac{v_{3--}}{\xi_{--}}\right)\frac{1}{\xi_{--}} &= 0\,, \\ \kappa_{-+}\left(v_{0-+}\xi_{-+} - \frac{v_{3-+}}{\xi_{-+}}\right) + \kappa_{--}\left(v_{0--}\xi_{--} - \frac{v_{3--}}{\xi_{--}}\right) &= 0\,. \end{aligned} \tag{3.46}$$

The determinant of this system is given by

$$\det = \left(\frac{1}{\xi_{-+}} - \frac{1}{\xi_{--}}\right)\left(v_{0-+}\xi_{-+} - \frac{v_{3-+}}{\xi_{-+}}\right)\left(v_{0--}\xi_{--} - \frac{v_{3--}}{\xi_{--}}\right). \tag{3.47}$$

Some algebra leads to

$$v_{0\varepsilon_1\varepsilon_2}\xi_{\varepsilon_1\varepsilon_2} - \frac{v_{3\varepsilon_1\varepsilon_2}}{\xi_{\varepsilon_1\varepsilon_2}} = -\varepsilon_1(2-\lambda_T)(1-\varepsilon_2\sqrt{4-3a^2}), \tag{3.48}$$

and

$$\det = -3\left(\frac{1}{\xi_{-+}} - \frac{1}{\xi_{--}}\right)(1-a^2)(2-\lambda_T)^2 \neq 0; \tag{3.49}$$

thus, the determinant of (3.46) is nonzero and the solution of this system is $\kappa_{-+} = \kappa_{--} = 0$.

Then, the mixing of bounce-back and specular reflection does not build any Knudsen layer near the wall, at least for the two orientations studied here. While this result remains to be generalized, it can be conjectured that it holds for more general orientations and that the boundary conditions used in the previous simulations do not build any spurious anisotropic boundary layer.

4. Numerical experiments

The theoretical results obtained in the previous section were compared to those obtained by numerical simulations. We used the numerical scheme described in ref. [5] in which 64 bits are packed in one computer word and the collision rules are implemented by bitwise logical operations. For computer efficiency, the inner loop (I variable, $1 \leq \mathtt{I} \leq \mathtt{NXA}$) of the program sweeps the lattice in the x direction perpendicular to the cut edges of the lattice, while the bit packing is done in the y direction parallel to the edges (J variable, $1 \leq \mathtt{J} \leq \mathtt{NYA}$). Then, the orientation of the lattice is the one used in this paper in the perpendicular case, while it is rotated counter-clockwise 90° in the parallel case (the horizontal and vertical extensions of the lattice are: $L_x = \frac{1}{2}\sqrt{3}(\mathtt{NXA}-1)$ and $L_y = (64 * \mathtt{NYA} - 1)$ for the parallel orientation, and $L_x = (\mathtt{NXA}-1)$ and $L_y = \frac{1}{2}\sqrt{3}(64 * \mathtt{NYA} - 1)$ for the perpendicular one).

The propagation step takes advantage of the packing to simultaneously move 64 particles. To minimize the number of logical shifts, the bit packing was not done for contiguous nodes but by splitting the lattice in 64 strips and using the bit n of the word $\mathtt{A}_i(\mathtt{I},\mathtt{J})$ for the particle moving in direction i at node $\{\mathtt{I}, \mathtt{J}+n*\mathtt{NYA}\}$ ($n \in \{0,\ldots,63\}$). With this packing, the propagation is done by writing

$$\mathtt{A}_i(\mathtt{I},\mathtt{J}) = \mathtt{A}_i(\mathtt{I} + \mathtt{IncX}(i,\mathtt{I},\mathtt{J}), \mathtt{J} + \mathtt{IncY}(i,\mathtt{I},\mathtt{J}))$$

and only one circular shift is needed in order to implement the periodic boundary condition in the y direction for J=1 or J=NYA. $\mathtt{IncX}(i,\mathtt{I},\mathtt{J})$ and $\mathtt{IncY}(i,\mathtt{I},\mathtt{J})$ are the horizontal and vertical displacements in direction i that implement the mapping of the hexagonal lattice on a square array, using the trick given in ref. [5]. In the parallel case, $\mathtt{IncX}(i,\mathtt{I},\mathtt{J}) = 2c_{iy}/\sqrt{3}$, and $\mathtt{IncY}(i,\mathtt{I},\mathtt{J}) = \{-1,0,1,1,1,0\}$ (resp. $\{-1,-1,0,1,0,-1\}$) for I odd (resp. even) and $i \in \{0,\ldots,5\}$. In the perpendicular case, $\mathtt{IncX}(i,\mathtt{I},\mathtt{J}) = \{-1,0,1,1,1,0\}$ (resp. $\{-1,-1,0,1,0,-1\}$) for J odd (resp. even) and $i \in \{0,\ldots,5\}$, and $\mathtt{IncY}(i,\mathtt{I},\mathtt{J}) = -2c_{iy}/\sqrt{3}$.

4.1. Modes

The first set of simulations was a study of the hydrodynamical and Knudsen modes. An "injection" step was added after each propagation one. In the parallel case, particles were injected along directions 4 and 5 on the left edge of the lattice and along directions 1 and 2 on the right edge. In the perpendicular case, particles were injected along directions 0, 1, and 5 on the even lines and direction 0 on the odd lines of the left edge, and along direction 3 on the even lines and directions 2, 3, and 4 on the odd lines of the right edge. The probabilities of injection along the different links are determined by eqs. (3.6) and (3.24), and the Boolean particles are set by comparison of a random number and the above probabilities (1 if the random number is smaller, 0 otherwise). The initial configuration is also set by the same method. To

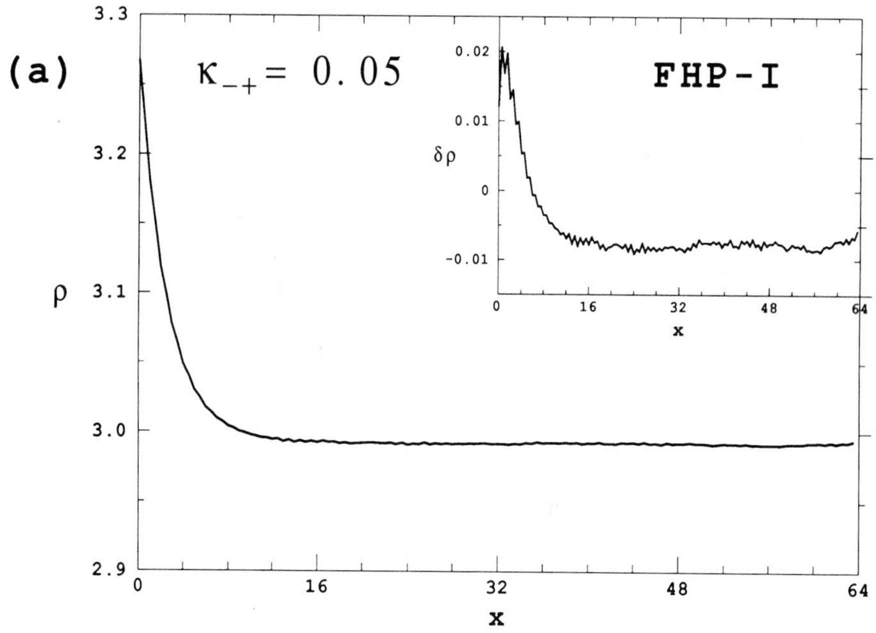

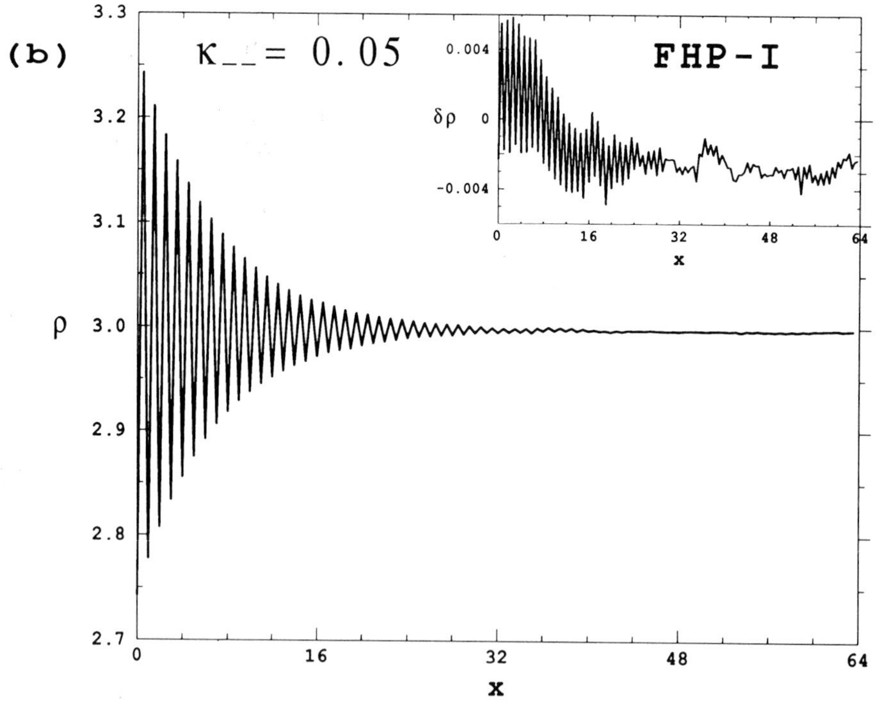

Fig. 3. Exponential (a) and oscillating exponential (b) modes on the left edge of the lattice. The inserts show the difference between the measured values and those computed from the theory.

decrease the statistical noise, the Boolean fields are averaged both along the y direction and over several time steps (typically 4000).

Just as the theory predicted, the simulations in the parallel case did not show any strange effects at the boundaries, so we focused our main interest on the perpendicular case. The results obtained for a uniform density $\rho = 3$ far from the edge, $\kappa_{-+} = 0.05$ (resp. $\kappa_{--} = 0.05$), and all the other parameters (momentum, gradient, and $\kappa_{\varepsilon_1\varepsilon_2}$) set to zero, are shown in figs. 3a (resp. fig. 3b) by displaying the local density $\rho(x)$ as a function of x. The inserts show the difference between the actual measurements and the theoretical values for the chosen parameters. The results agree with these values within ten percent.

There are several explanations for this discrepancy. First, the injection rules implement the chosen distribution on average only. For a finite simulation time, the actual values of the injection parameters differ slightly from the expected ones. This phenomenon changes not only the effective density in the simulation but also the excitation level of the boundary layer modes. While the injection rates are picked to produce a pure Knudsen mode, the actual ones may produce a combination of a dominant mode plus a small component of the other ones. Another correction is due to the neglected nonlinear terms in the Boltzmann equation. This effect was kept minimal by the choice of $\rho = 3$.

4.2. Lattice Boltzmann calculations

To get a better understanding of the possible corrections, the statistical noise can be removed by the direct simulation of the Boltzmann equation (2.7). Since all the results given here are derived from the Boltzmann approximation, they remain valid if we simulate directly the Boltzmann equation with the same collision term C_i as in eq. (2.5). This macroscopic equation uses real numbers and the noise due to the Boolean particles disappears. The results of these simulations are given in figs. 4a and 4b for the same parameters values as those in the previous section.

The inserts of figs. 3a and 4a show that the nonlinear corrections are present for the chosen amplitude of the exponential mode and have roughly the same magnitude for the Boolean and the Boltzmann models. For the Boolean simulations, the noise and the corrections to the actual values of the parameters (bulk density for instance) are much smaller than the nonlinear corrections for this mode.

The insert of fig. 4b shows that, for the chosen amplitude, the nonlinear corrections are an order of magnitude smaller for the oscillating mode than for the exponential one. A quantitative explanation would require a nonlinear analysis of the problem that is left for future work. The surprising feature is the magnitude of the noise for the corresponding Boolean simulations; it is larger than the nonlinear corrections by at least a factor of two. While we do not have yet an explanation for this effect, we suspect that it is linked to the choice of the possible output state for two-body collisions that depends on the parity of the horizontal lines. This asssumption is supported by preliminary studies of the averages on odd and even time steps that show strong spatial oscillations depending on the choice between the two possible output configurations. This effect and its link with the spurious invariants are still under study.

Using the Boltzmann simulations, the amplitude of the nonlinear effects was measured as a function of $\kappa_{\varepsilon_1\varepsilon_2}$ and found to fit a quadratic law. A qualitative derivation of the space dependence of the nonlinear terms shows that the dominant term is $\xi^{4x}_{\varepsilon_1\varepsilon_2}$. Since this term has no reason to be a factor of the b-vector $\boldsymbol{v}_{\varepsilon_1\varepsilon_2}$, this leads to second-order corrections to $\kappa_{\varepsilon_1\varepsilon_2}$, $\kappa_{\varepsilon_1\bar{\varepsilon}_2}$ (where $\bar{\varepsilon} = -\varepsilon$), ρ, and j_x, that consequently changes the values of $\kappa_{\bar{\varepsilon}_1\varepsilon_2}$ and $\kappa_{\bar{\varepsilon}_1\bar{\varepsilon}_2}$ (a small component of the Knudsen modes on the opposite side). The inserts of figs. 4 are the superposition of all these contributions. This derivation can be continued for higher-order corrections. For $|\kappa_{\varepsilon_1\varepsilon_2}| \leq 10^{-3}$ and using this expansion up to the fourth order terms, it is possible to adjust the amplitude of the different exponentials in order to get a residue smaller than the numerical noise for the double precision computation.

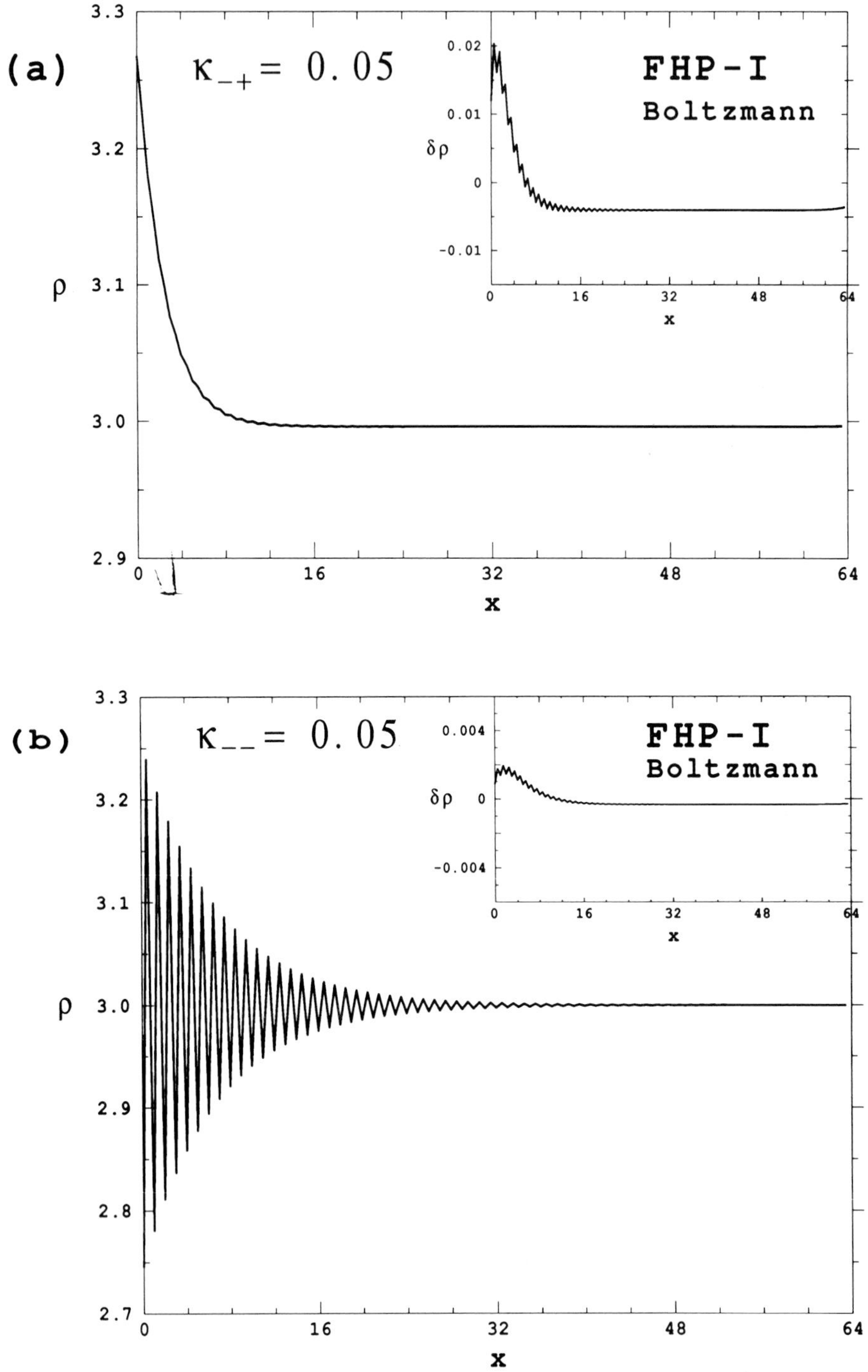

Fig. 4. Same modes as in fig. 3 for the Boltzmann equation.

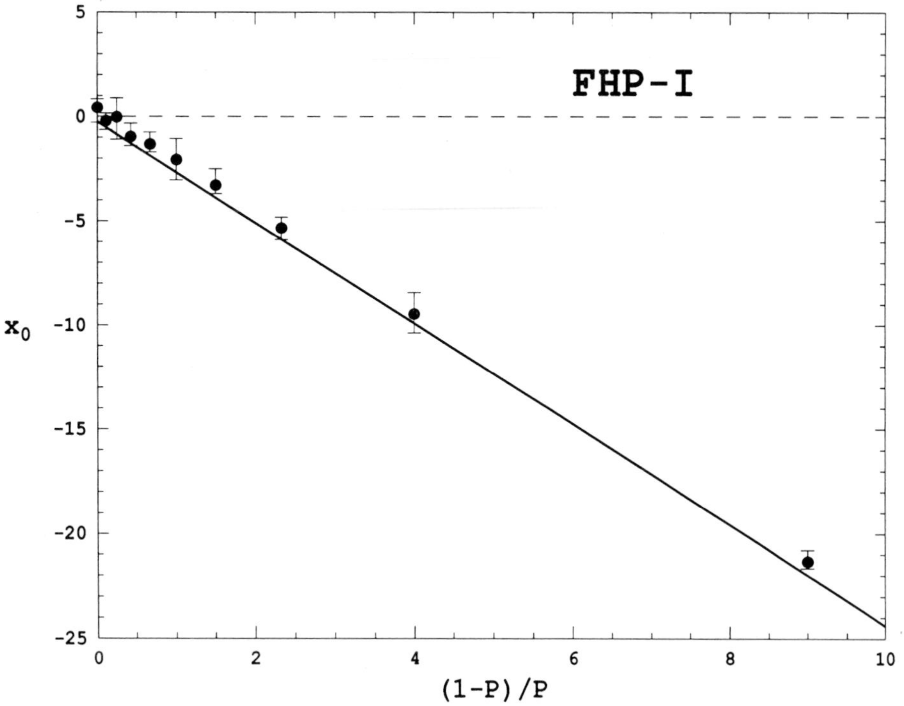

Fig. 5. Position of the zero velocity line as a function of the probability p of bounce-back condition, for the node implementation of the boundary and a gas extension $0 \leq x_*$ (above the dashed line). • indicates the average value over six runs, and the error bars correspond to the extreme values. The solid line is given by eq. (3.44).

4.3. Boundary conditions

We have also studied different implementations of solid walls for the perpendicular orientation by varying the probability p of bounce-back versus specular reflection. For a given p, the velocity has been extrapolated to locate the line with zero velocity and the found position compared to the one given by eq. (3.44), as shown in fig. 5 for the node implementation of the boundary. The error bars are of the order of the node spacing, making fine comparisons between the simulations and the theory quite difficult.

Again using the Boltzmann equation to remove the statistical noise, excellent agreement was found between the theory and the Boltzmann simulations. Nonlinear corrections proportional to the square of the gradient near the wall were observed. These corrections also appear on the opposite edge of the lattice. Consequently the linear dependence of the momentum must be fit close to the wall to get the correct position. This latter effect may explain the systematic bias of the data in the Boolean case.

During this study of the Boltzmann equation, small differences were found between the node and the link implementations. These appear linked to the differing behavior of the spurious invariant associated with j_y [14] in the two implementations. In the second case the quadratic law is true even for very small values of the gradient (up to 10^{-6}), while in the first case the corrections remain constant ($\approx 5 \times 10^{-6}$) as soon as the gradient is smaller than 10^{-5}. A more complete analysis of these phenomena is included in our subsequent work [17].

5. Conclusions

Using the linear Boltzmann approximation, a Knudsen layer theory was built and solved analytically for the six-bit models with one-dimensional space dependence for boundary orientations parallel and perpendicular to the links of the lattice. Knudsen layers are shown to be anisotropic. Standard boundary conditions suppress them, but do implement effective walls slightly off the nodes of the lattice in a way that depends on the wall orientation. We have also found a very good agreement between this theory and the simulations.

This work can be continued in several directions:

(i) Time dependent solutions, the study of which is almost finished for period two in time.

(ii) More general orientations of the boundary with respect of the lattice. Using a counting argument, it is possible to show that the next simple orientation will give four modes on each side of the lattice, the next one six and so on. For a general orientation (irrational with respect to the lattice) there will be an infinity of modes that disappear when the orientation becomes rational.

(iii) Nonlinear effects as a perturbation expansion around the linear solution.

(iv) Application of the method to more complicated geometries such as a wedge.

(v) Finally, these results must be applied to more complicated models such that the FCHC one. However, the complexity of the analytical computations seems prohibitive for these models and will probably require symbolic computer tools.

Acknowledgements

We have had stimulating discussions with many people over the course of this work; we would like to thank J. P. Boon, D. Frenkel, T. Ladd, P. Lallemand, Y. H. Qian, S. Zaleski, and E. Znaty to name a few. R.C. thanks Bertin et Cie and École Normale Supérieure de Lyon for their support. D.L. is grateful to the NSF for its support under the grant DMS-8914420.

References

[1] U. Frisch, B. Hasslacher and Y. Pomeau, Lattice-gas automata for the Navier–Stokes equation, Phys. Rev. Lett. 56 (1986) 1505–1508.
[2] G. Doolen, ed., Lattice Gas Methods for Partial Differential Equations (Addison–Wesley, Reading, MA, 1990).
[3] U. Frisch, D. d'Humières, B. Hasslacher, P. Lallemand, Y. Pomeau and J.P. Rivet, Lattice gas hydrodynamics in two and three dimensions, Complex Systems 1 (1987) 649–707.
[4] S. Wolfram, Cellular Automaton fluids 1: basic theory, J. Stat. Phys. 45 (1986) 471–526.
[5] D. d'Humières and P. Lallemand, Numerical simulations of hydrodynamics with lattice gas automata in two dimensions, Complex Systems 1 (1987) 599–632.
[6] A.J.C. Ladd, M.E. Colvin and D. Frenkel, Application of lattice-gas cellular automata to the Brownian motion of solids in suspension, Phys. Rev. Lett. 60 (1988) 975–978.
[7] D.H. Rothman, Cellular-automaton fluids: a model for flow in porous media, Geophysics 53 (1988) 509–518.
[8] S. Succi, E. Foti and F. Higuera, Three-dimensional flows in complex geometries with the lattice Boltzmann method, Europhys. Lett. 10 (1989) 433–438.
[9] C. Cercignani, in: The Boltzmann Equation and Its Applications (Springer, Berlin, 1988) pp. 252–260.
[10] G. McNamara and G. Zanetti, Use of the Boltzmann equation to simulate lattice-gas automata, Phys. Rev. Lett. 61 (1988) 2332–2335.
[11] F.J. Higuera, Lattice gas simulation based on the Boltzmann equation, in: Discrete Kinematic Theory, Lattice Gas Dynamics, and Foundations of Hydrodynamics, ed. R. Monaco (World Scientific, Singapore, 1989) pp. 329–342.
[12] R. Gatignol, Kinetic theory boundary conditions for discrete velocity gases, Phys. Fluids 20 (1977) 2022–2030.
[13] M. Hénon, Viscosity of a lattice gas, Complex Systems 1 (1987) 763–789.
[14] G. Zanetti, The hydrodynamics of lattice gas automata, Phys. Rev. A 40 (1989) 1539–1548.

[15] U. Frisch, Relation between the lattice Boltzmann equation and the Navier–Stokes equations, Physica D 47 (1991) 231–232; these Proceedings.
[16] P.C. Rem and J.A. Somers, Cellular automata on a transputer network, in: Discrete Kinematic Theory, Lattice Gas Dynamics, and Foundations of Hydrodynamics, ed. R. Monaco (World Scientific, Singapore, 1989) pp. 268–275.
[17] R. Cornubert, D. d'Humières and D. Levermore, in preparation.

CHAPTER 7

COMPUTER HARDWARE

Physica D 47 (1991) 263–272
North-Holland

PROGRAMMABLE MATTER: CONCEPTS AND REALIZATION

Tommaso TOFFOLI and Norman MARGOLUS
MIT Laboratory for Computer Science, Cambridge, MA 02139, USA

Received 31 January 1990

This paper is a manifesto, a brief tutorial, and a call for experiments on *programmable matter* machines.

1. Introduction

After several years of issue analysis (see refs. [19,12,26,27] and [21,5,24,23,22]), conceptual design [11,14,15,1,25,26], and hands-on experience with more modest realizations [16,20,26,28], we have finally embarked on the construction of CAM-8, a scalable cellular automata machine offering unprecedented performance in the fine-grained modeling of distributed, spatially extended systems.

Technically speaking, one could describe CAM-8 as a fine-grained multiprocessor, and elaborate on design issues [15]. Here, however, we will concentrate on the most essential feature of the CAM-8 architecture: it embodies the concept of *programmable matter.*

In programmable matter, the same cubic meter of machinery can become a wind tunnel at one moment, a polymer soup at the next; it can model a sea of fermions, a genetic pool, or an epidemiology experiment at the flick of a console key. Ten times as large a simulation will simply require ten cubic meters of machinery, instead of one. Flexibility, instant reconfigurability, variable resolution, total accessibility, and handling safety make such programmable matter worth a premium over ordinary matter.

All computers can realize programmable matter to some extent. The real question is whether the above features can be achieved today on a sufficiently large scale (in terms of spatial resolution and updating rate) and with sufficient flexibility (in terms of choice of the underlying fine-grained model) for the concept of 'programmable matter' to come to life. It is in this respect that CAM-8, a machine optimized for fine-grained material simulations, has a role to play. In brief, CAM-8 will provide *a general-purpose instrument for the systematic exploration of a previously inaccessible region of the computational spectrum.*

Our principal goal in this paper is to make the scientific community – physicists in particular – aware of the possibilities of this kind of instrument, and to invite suggestions for experiments that will make significant use of its novel capabilities.

2. Fine-grained models of matter

The properties of 'gold' as a material (for instance, specific weight, hardness, conductivity) are quite distinct from the properties of an isolated gold atom (atomic number, mass, spin, etc.). Certain statistical features emerge and become better and better characterized as we consider a larger and larger lump of gold; beyond a certain size, these bulk properties have stabilized enough that we can speak of 'gold' in a generic way, without having to specify the size or the shape of the lump.

Similarly, if we consider an indefinitely extended, uniformly interconnected network of identical personal computers, we will have defined a synthetic "material" having well-defined bulk properties. These properties remain (to a good approximation) well characterized even when we implement only a finite – but sufficiently large – portion of the network. Connecting together three personal computers does not yield a *larger* personal computer – it yields a new kind of object; but joining together three large pieces of this material does yield a larger piece of the *same* material.

If we changed the network's interconnection pattern, or if we replaced the processor at every node with a different kind of processor, we would obtain a different "material," with potentially different bulk properties.

So far we have been speaking in rather metaphorical terms – we would really like our computer networks to be able to mimic real materials. With this in mind, lets consider certain aspects of our computer-network model more carefully.

– *Fine grain.* A personal computer contains a large amount of state. Even if it had only 256 bytes of memory (instead of a million or so), it would still be able to run programs having on the order of $2^{8 \cdot 256}$ ($\approx 10^{600}$) distinct states – much longer (at one new state every microsecond) than the life of the universe. With so much potential for complexity hidden in each node, well-characterized bulk properties might emerge only on a scale spanning an astronomical number of nodes.

To ensure that bulk properties emerge at a reasonably short scale (today we can build networks with millions or billions of nodes at most) the nodes must be very simple, i.e., to each node there must be associated just a few bits' worth of state variables. This is what we mean by a *fine-grained* computing network.

Our focus on fine-grained models is motivated, of course, by conceptual as well as practical considerations. By Occam's razor, before resorting to complex building blocks one should try to synthesize complex behavior from simple building blocks.

– *Polynomial interconnection.* We have mentioned *uniform*, *indefinitely extended* networks – i.e., networks whose geometry is invariant under an infinite group of transformations – but we have not mentioned what kinds of group we have in mind. The essential distinction, illustrated in fig. 1, is between groups having *exponential* growth (the number of new nodes one can reach from a given node in n steps grows as a^n) and groups having *polynomial* growth (this number grows as n^a).

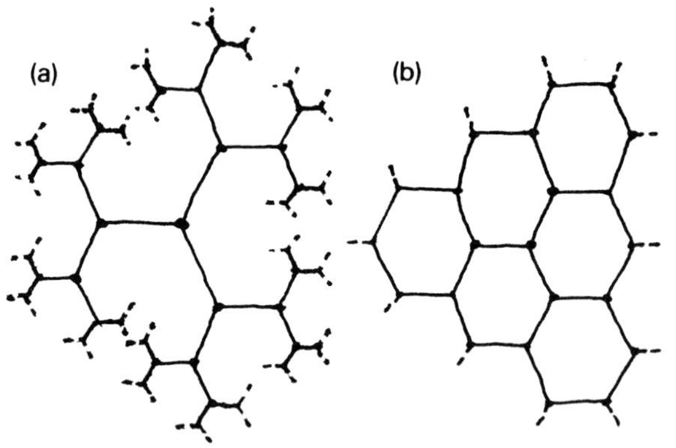

Fig. 1. In the *exponential* network at left, the number of new nodes reachable in n steps grows as 2^n; in the *polynomial* network at right, it grows as n^2.

Malthus' centuries-old argument still holds: the physical world does not support indefinite exponential growth. One can conceive of an exponential architecture – say, a binary-tree network; but one cannot concretely implement it for an indefinite number of levels using uniform technological parameters (same wire size, clock rate, etc.) at all levels. In the end, the exponential structure will have to be embedded within a polynomial architecture – at most *three-dimensional*, to be specific. Our attention, therefore, will be restricted to polynomial networks that extend indefinitely in at most three spatial dimensions.

For concreteness, let's give a simple example of the kind of synthetic "materials" we'd like to be able to program within a programmable-matter machine – namely, a *fluid* modeled by a *lattice gas*.

The idea behind lattice-gas hydrodynamics is to model a fluid by a system of particles that move in discrete directions at discrete speeds, and undergo discrete interactions. In the HPP lattice gas [9], identical particles move at unit speed on a two-dimensional orthogonal lattice, in one of the four possible directions. Isolated particles move in straight lines. When two particles coming from opposite directions meet, the pair is "annihilated" and a new pair, traveling at right angles to the original one, is "created" (fig. 2a). In all other cases, i.e., when two particles cross one another's paths at right angles (fig. 2b) or when more than two particles meet, all particles just continue

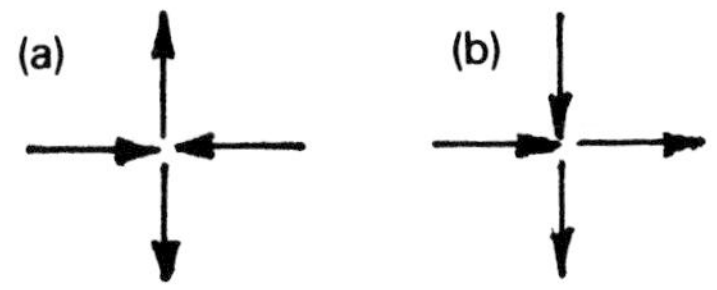

Fig. 2. In the HPP gas, particles colliding head-on are scattered at right angles (a), while particles crossing one another's paths go through unaffected (b).

straight on their paths.

As soon as the numbers involved become large enough for averages to be meaningful – say, averages over spacetime volume elements containing thousands of particles and involving thousands of collisions – a definite continuum dynamics emerges. And, in the present example, it is a rudimentary *fluid* dynamics, with quantities recognizably playing the roles of density, pressure, flow velocity, viscosity, etc.

According to Pomeau, seeing this model of his running on one of our early cellular automata machines made him realize that what had been conceived primarily as a *conceptual* model could indeed be turned, by using suitable hardware, into a *computationally accessible* model: this stimulated his interest in finding lattice-gas rules which would provide better models of fluids. A landmark was reached with the slightly more complicated FHP model [7] (it uses six rather than four particle directions), which gives, in an appropriate macroscopic limit, a fluid obeying the well-known *Navier–Stokes* equation, and thus suitable for modeling actual hydrodynamics (see ref. [10] for a tutorial). Soon after, analogous results for three-dimensional models were obtained by a number of researchers [8].

On what scale can one implement models of the above kind? Let l be the linear size of the lattice, so that a three-dimensional lattice of roughly cubic shape will consist of l^3 nodes. Typically, the duration t (number of updates of the entire network) of a significant experiment will be at least one order of magnitude greater than l (since macroscopic motion will be much slower than the maximum speed of information transmission in the computing medium). Taking $t \approx 10l$, the experiment will entail $\approx 10l^4$ individual site updates, or *events*.

With today's fastest general-purpose computers, fine-grained three-dimensional models of physically interesting phenomena (e.g., low-Reynolds-number hydrodynamics, diffusion in nonhomogeneous media, phase separation in two-component systems) can be updated at a rate of about 10^8 events/s [17,2,3]. Note that processing speed rather than memory size is the limiting factor; it is remarkable that machines with such different architectures as a Cray X-MP and a Thinking Machines CM-2 yield comparable performance in tasks of this kind.

With one day's worth of computer time, or about 10^5 s, one obtains $\approx 10^{13}$ events, corresponding to $l \approx 1000$. This allows one to get a glimpse, but just a glimpse, at bulk dynamics. In fact, to arrive at bulk properties we typically have to average over volume elements containing (at least) a thousand sites, and thus having a diameter of (at least) 10 sites. If each of these volume elements corresponds to a "pixel" in the resulting macroscopic picture, with $l \approx 1000$ we are left with a picture having a resolution of only 100 pixels per side.

If an amount of electronic hardware (chips, wires, etc.) comparable to that in an X-MP or a CM-2 is reconfigured into an appropriate architecture – such as that of CAM-8 – one obtains a flexible computer capable of running this class of models at least two (and in many cases three or four) orders of magnitude faster. This means that each of the four factors in l^3t can grow by up to one order of magnitude with respect to current computers, making new phenomena accessible to fine-grained simulation.

3. What is programmable in programmable matter?

Once you are satisfied that the above claims are valid, you may be tempted to say: "I'll grant you a gain of a few orders of magnitude over a general-purpose computer; but of course – you must understand – yours is an extremely specialized piece of hardware, designed to do just one thing over and over."

In fact, with a *really* dedicated machine – a VLSI implementation of a specific fine-grained model (think of a "silicon wind tunnel") – one could easily gain *many more* orders of magnitude in performance (cf., e.g., ref. [6]). Such a cast-in-concrete machine might eventually be able to play an important role in a production environment, but would probably be of much less use as a research tool. On the other hand, in spite of its definite specialization, CAM-8 remains a *wide-purpose* architecture.

With the benefit of hindsight, let us define *programmable matter* as a three-dimensional, uniformly textured, fine-grained computing "medium" having the following properties:

(i) It can be assembled into lumps of arbitrary size (the limits being given by economics rather than technology).

(ii) It can be dynamically reconfigured into any uniform, polynomially interconnected, fine-grained computing network,

(iii) It can be interactively driven, in the sense that its dynamical evolution can be started, interrupted, and resumed at any moment in response to the occurrence of specified internal or external events.

(iv) It is totally accessible to real-time observation, analysis, and modification.

CAM-8 substantially meets all of the above specifications. Much of its flexibility derives from a careful compromise between parallel and serial processing. Briefly, the machine consists of a three-dimensional array of *modules*, with each module responsible for a certain "volume" of simulation space consisting of millions of sites (the actual number is 2^n, where n depends on the current DRAM technology and is, in the near term, in the range 22–24, corresponding to 4–16 million sites). Site updating within a module is performed serially, with all modules operating in parallel and in lockstep. Overall, the array of modules is controlled by a *host* computer (which may be a conventional workstation).

Total simulation volume, as in point (1), can be "programmed" by simply purchasing the appropriate number of modules. As for point (2), the programmability of CAM-8 "matter" extends to the following features:

– *Number of dimensions.* The total simulation volume can be logically configured into an array having, besides the three main dimensions x, y and z, which are indefinitely extendible, a number of extra dimensions of bounded size. This number is, of course, limited by the fact that the product of the sizes along all dimensions cannot exceed the total volume of programmable matter available, and the product of the sizes of the extra dimensions cannot exceed the volume of a module.

– *Aspect ratio.* Within the above bounds, the size of each dimension is independently adjustable (by powers of 2), so that one can choose, for instance, a square array or a long-and-thin one.

– *Interconnectivity.* Direct communication between sites is not restricted to first neighbors in the array. For a given number of dimensions d and aspect ratio, there is defined a d-parallelepiped of volume slightly greater than 2^{n+d} centered on the given site, such that *any* site within that volume can be chosen as a neighbor. For example, in two dimensions, and with $n = 22$, at each step a site can have neighbors anywhere within a 4K$\times$4K square; in three dimensions, and with $n = 24$, anywhere within a $512\times512\times512$ cube. Furthermore, by slowing the simulation down by a factor k, the diameter of the neighborhood can be increased by a factor of k. This rich interconnectivity is the most significant innovation of CAM-8: earlier cellular automata machines were handcuffed by the limitations of "canned' neighborhoods (cf. refs. [4,26]). Now one can perform three-dimensional simulations, synthesize a variety of spacetime "crystallographic groups", and use widely different speeds of propagation of information within the same simulation (e.g., fast "photons" and slow "electrons").

– *Amount of states per site.* For a given overall number of bits of state, one can trade off number of sites for number of bits per site.

– *Transition function.* In the current implementation, the local transition function, that is, the rule used to update a site as a function of its neighbors, is determined by a lookup table, and thus can be assigned in an arbitrary way. This lookup table is double-buffered, so that a new table can be downloaded while an updating step is in progress, in preparation for a subsequent updating step that will use a different rule. The size of the lookup tables is currently 16×2^{16}, corresponding to 16 inputs and 16 outputs, and thus direct lookup can be used as long as the amount of input to a node does not exceed 16 bits. For a state-

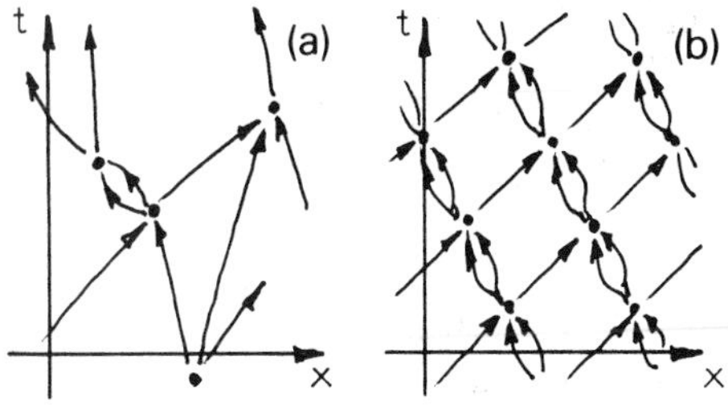

Fig. 3. (a) A spacetime diagram, consisting of signals (arcs) and events (nodes). (b) A uniform spacetime diagram.

variable set larger than that, graceful degradation occurs as one makes more and more recourse to table composition.

– *Serial/parallel ratio.* Finally, the same number of physical processors can be concentrated on relatively few sites, for maximum speed, or shared by many sites, for maximum simulation size. Intermediate choices are of course possible.

4. Programmer's model

The generality of CAM-8's programmable matter is reflected in a very simple "programmer's model" for the machine.

We are all familiar with representing the history – and thus the evolution – of a system by a spacetime diagram: a cycle-free directed graph whose arcs represent *signals* and whose nodes represent *events* in spacetime (fig. 3a).

CAM-8 supports spacetime diagrams that are uniform and fine-grained (fig. 3b). Uniform means that the layout of arcs and nodes is invariant for spacetime translations; of course, the *values* of the signals will in general change from place to place. Fine-grained means that only a small number of signals can interact at each event, and each signal has a small set of available states. (In fact, in CAM-8 all arcs carry *binary* signals; more complex signals can, of course, be synthesized by bundling together a few arcs.)

The operation of CAM-8 is simply the steady unfolding of such a uniform and fine-grained space time history.

Basically, all one has to do to program CAM-8 is:

– *Signals.* Specify the diagram's interconnection

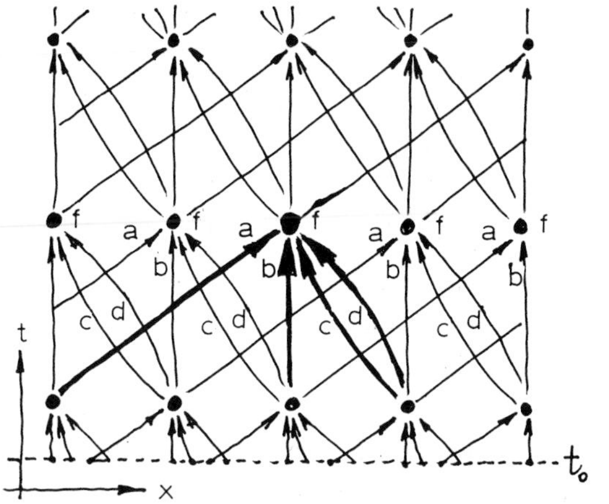

Fig. 4. A spacetime diagram with four input signals associated with each event.

pattern. Since the diagram is uniform, it is enough to specify the geometry of a "primitive unit cell". This, in turn, is simply given by a list of *neighborhood vectors*, one for each of the arcs that impinge on a node, specifying where each arc comes from.

– *Events.* Specify the event's transition function, i.e., for each possible combination of states for the signals that converge to an event, specify the desired combination of states for the signals that radiate out of it. (Note that, by construction, there is exactly one output signal for each input signal.)

– *Initial conditions.* Assign the values of the signals that cross a spacelike surface, typically of the form t_0 = const (cf. fig. 4).

For example, for the diagram of fig. 4, which represents the time evolution of a one-dimensional system, the neighborhood vectors are (-2), (0), $(+1)$, and $(+1)$, since the four signals converging on any event at time t originate from events at time $t-1$ that are offset by those amounts with respect to the destination event.

Fig. 5 illustrates the spacetime layout for the HPP gas mentioned in the previous section. The four directions of travel for the particles are labeled a, b, c, d, and the transition function is labeled f. For this system, the neighborhood vectors are

$$\begin{aligned} a&: \ (-1,-1) \\ b&: \ (+1,-1) \\ c&: \ (+1,+1) \\ d&: \ (-1,+1) \end{aligned},$$

and the transition function f is given by the table

	in *abcd*	out *abcd*		in *abcd*	out *abcd*
	0000	0000		1000	1000
	0001	0001		1001	1001
	0010	0010	*	1010	0101
	0011	0011		1011	1011
	0100	0100		1100	1100
*	0101	1010		1101	1101
	0110	0110		1110	1110
	0111	0111		1111	1111

where 1 represents the presence and 0 the absence of a particle on a particular spacetime track. Note that in only two cases (marked with an asterisk) out of sixteen does an interaction take place; in all other cases, each signal proceeds undisturbed.

Uniformity of structure at the hardware level does not mean, of course, that one cannot have properties in the modeled system that vary from place to place.

For instance, we may want to study turbulence by letting a fluid flow past a solid obstacle. Then one bit at each site will be used as a parameter (left unchanged by the transition function) specifying whether that site belongs to the obstacle or to the space occupied by the fluid. Depending on the value of that bit, different sections of the lookup table, and thus different laws, will be activated.

Similarly, by using extra state at each site, one can model systems with sources and sinks, potential wells, lenses, etc. [26,18], and even systems in which the initial conditions are specified on a spacelike surface which isn't flat [12].

In the next section we'll briefly mention other ways of "modulating" in space or time (or both) the structural parameters of the computing network.

Finally, a few words about data analysis. CAM-8 will generate new data at a steady rate of about 0.5 Giga-bits/s *per module.* Of course, one can always suspend the simulation for a moment and read out the entire state of the system ("core dump"), but this involves a lot of data transfer and puts the entire analysis burden on the host. In ordinary circumstances one would want data selection, compression, and extraction to take place continuously, in real time, and without significantly interfering with the updating itself. Therefore, flexible resources for low-level data reduction must be built-in, and must grow in direct proportion to the size of the system.

We have seen that the transition function is responsible for producing the output signals of an event as arbitrary functions of its input signals. Using the same machinery, one can produce additional binary outputs, each telling whether the input signals represent an occurrence of an arbitrarily specified condition (for example, whether there is a collision involving an even number of particles). The number of these occurrences is integrated over selected portions of space and time (for this, we take advantage of CAM-8's flexible scanning order, as explained in the next section); the corresponding counts, accumulated by dedicated hardware counters, are made available to the host computer.

For instance, for a lattice gas, one can subdivide the array into a number of volume elements each containing a few thousand sites, and (in parallel with the updating) tabulate the total number of particles moving in each direction in each step – thus constructing a macroscopic picture of the flow. Instead of many Giga-bits/s of fine-grained detail (which is, in a sense, an artifact of this type of model), the host will see only the relevant hydrodynamical data – at a rate a few thousand times slower than the raw data rate.

5. Implementation notes

To give the reader a feeling for quantitative aspects of CAM-8's modeling environment, we shall briefly discuss how some of the functional behavior described in the previous section is achieved in practice. This section may be skipped without loss of continuity; a more detailed discussion of the hardware appears in an earlier paper [15].

Consider a regular array of bits that extends indefinitely in all directions (for concreteness, you may think of a two-dimensional array – a "bit-plane"); we shall call such an array a *layer.* We will now superpose, in good registration, a number p of layers – so that at each site we have a *pile* of p bits. This entire collection of bits will be made to evolve by repeated application of the following

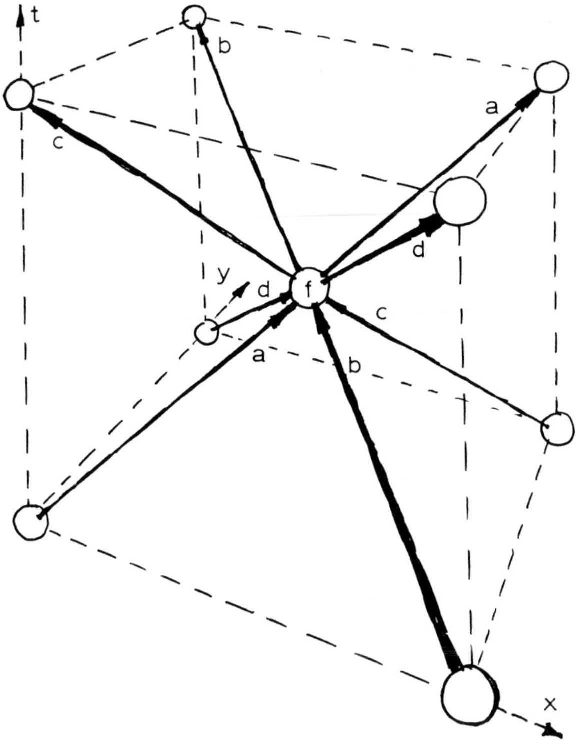

Fig. 5. Spacetime layout of the HPP gas.

procedure, called a *step*, consisting of two stages:

– *Data movement.* Each layer is independently shifted as a whole by an arbitrary number of positions in an arbitrary direction. We still end up with a pile at each site, but with a new make-up.

– *Data interaction.* We now take each pile and we send it to a p-input, p-output lookup table; the lookup table returns a new pile, which we put in place of the original one.

Note that at the interaction stage each pile is processed independently, so that the *order* in which the piles are updated is irrelevant. We could even have several copies of the lookup table, and do some (or all) of the processing in parallel. Also note that, at the data movement stage, the shift performed on each layer is a *uniform* and *data-blind* operation (each bit is moved by a fixed offset, quite independently of its position and its value); thus, it becomes possible, in a suitable implementation, to replace this operation by one that shifts the frame of reference (by incrementing a single pointer) rather than moving the data themselves.

There is an obvious one-to-one correspondence between a dynamics specified in this way and one defined by a uniform, fine-grained spacetime diagram as in the previous section. For example, in fig. 4, all the arcs labeled a make up a layer (and similarly those labeled b, c, and d); the corresponding neighborhood vector (-2) specifies the amount of shift of this layer at the data movement stage. Events represent the data interaction stage: still referring to fig. 4, the four arcs entering a node represent the make-up of a pile just before being given to the lookup table, and the four arcs leaving it, the pile as it comes out of the lookup table.

The present way of specifying a dynamics in terms of movement of layers and updating of piles closely reflects CAM-8's hardware organization, briefly described below.

It will be convenient to consider for a moment layers that wrap around on themselves along each dimension (the topology is that of a hypertorus) and so have finite size. In a CAM-8 module, there are $p = 16$ of these layers, each represented by a separate DRAM chip. To implement the *data interaction* stage, these chips are sequentially scanned in lockstep, site by site, providing a 16-bit pile (one bit per layer) at every clock cycle. The lookup table is implemented by a SRAM chip (16 inputs, 16 outputs) which accepts a pile at every clock cycle (30 ns) and returns new piles at the same rate. Each new pile is deposited back where the corresponding old one came from, again one bit per layer. One complete scan represents one updating step for the entire volume encompassed by the module.

To visualize the scanning order, think of a nested DO-loop, with one nesting level for each spatial dimension. Each of the indices of the loop deals with one component (x, y, etc.) of the site address, and drives a separate subset of the DRAM address lines. The state of the complete set of these lines represents the current site address. The "volume" represented by the DRAM chip is shaped into the desired number of dimensions and aspect ratio (cf. sect. 2) by altering the parameters of the DO-loop.

Now, even though the same site address is intended for use by all 16 DRAM chips, we route a separate copy of these lines to each individual chip. This allows us, by inserting an adder stage in the address path, to add a different *offset* to the site address for each chip. This offset determines from where – along each dimension – the scanning of the chip's internal data will actually start. As far as the stream of data that comes out of the chip during an updating step is concerned, changing the offset is equivalent to shifting the entire contents of the chip. This mechanism, then, implements the *data transport* stage.

More specifically, the current offset is kept in the layer's *offset* register, which is decremented before each updating pass by the value of that layer's neighborhood vector (cf. previous section), which is kept in the *kick* register. All these registers have the same width as the site address, and like it are subdivided into sections each corresponding to one spatial dimension. Addition is performed independently on each section, i.e., the carry is inhibited at the boundary between one section and the next. (This splitting of the site address is simply controlled by the *dimension* register, each of whose bits, when set, inhibits carry propagation at the corresponding position in the adders.)

The above discussion explains the operation of a single module, with space wrapped around within the module itself. When we lay an indefinite number of modules side-by-side in order to achieve an arbitrarily large spatial extent, we must cut open the individual pieces of space and glue them edge-to-edge. Because each module is updating only one cell at a time, only 16 wires are required to glue two adjacent modules together – one per layer.

To understand the basic technique, think of a one-dimensional chain of modules. When a given layer is shifted by less than the width of a module, all of the new data for a given module come either from what is already inside of that module, or from one of its nearest neighbors – via a connecting wire. Moreover, since *all* of the layer shifts in the same direction, each interconnection need only carry data in one direction. By updating the shifted data before storing it away, we can combine the shift and the update, just as in the single module case.

The fine-grained dynamics of CAM-8 is completely determined by the contents of the lookup table (1 Mbyte of SRAM) and the neighborhood vectors (a few bytes stored in the kick registers). Typically, this information will be the same for all modules. By rewriting tables and vectors, one gets a new dynamics. This can be done at every step, thus obtaining an arbitrarily *time-dependent* dynamics.

Since scanning order is under our control, we can even scan, for instance, all even-positioned sites first; then change the dynamics, and scan all odd-positioned sites. As long as the number of sites scanned between changes is large compared to the number of changes in the lookup table, this way of modulating the spatial aspects of the dynamics entails little or no overhead. Thus, it becomes possible to use "checkerboard" dynamics (one rule for black, one for white) and similar coarser-grained spacetime textures for the dynamics. More complicated textures can be obtained in a similar manner by using the extra dimensions discussed earlier to group bits together: using a time-dependent dynamics allows one to perform arbitrary function compositions on these groups.

Finally, a few words about efficiency. The two main resources needed by any simulation model are *memory*, for the state variables, and *processing*, for the updating of these variables. Fine-grained models offer great conceptual simplicity and the sincerity that comes from relying as far as possible on first principles; but, in exchange, they demand very large amounts of memory and processing resources.

As we have seen, CAM-8 is able to use ordinary DRAM chips for its state variables. These chips are today an inexpensive commodity, and allow one to go further, in terms of simulation size, than any other practical alternative. Similarly, the CAM-8 architecture is able to use ordinary SRAM chips for processing; these too are a relatively inexpensive commodity, and allow one to go far in terms of processing power. The "glue" that turns these raw commodities into programmable matter is a gate array of moderate complexity ($\approx$ 10K gates), 16 copies of which go into each module.

A clock period of 30 ns will use 30 ns SRAM chips at full bandwidth. While the contents of a pile, which is used as an address to the SRAM, is essentially unpredictable, the bits within the DRAM chips are accessed in a predictable *sequence*; thus, one can make use of *nibble-wide* DRAMs, for which, as long as one remains in the same memory row, four consecutive bits can be read and four written in 120 ns, i.e., in four clock cycles. Therefore, this organization allows one to use storage and processing resources at *full bandwidth* on a close-to-100% duty-cycle.

Independently of the number of modules, the updating rate for the model as a whole will range from 8 steps per second (when all of a module's DRAM capacity is actually allocated to sites) to thousands of steps per second (when, for sake of speed, each lookup table is time-shared over thousands rather than millions of sites).

Though we have in mind substantial installations consisting of hundreds or thousands of modules – with a processing power of up to a few Tera-events/s – we are also developing a minimal, two-module version (66 Mega-events/s, 8 Mega-sites), suitable for individual interactive experimentation, with a hardware complexity and cost comparable to that of an inexpensive workstation computer. Such a machine will act as its own frame buffer, able to synchronize its scan with the beam of various raster-scan monitors so that, for instance, the contents of a module scanning a 512×512 array can be viewed in real time at 60 frames/s, and 1024×1024 at 15 frames/s.

6. Conclusions

We have presented a hardware structure that is optimized for the fine-grained simulation of spatially extended physical systems. There is the attractive possibility, in the near term, of achieving systems consisting of hundreds of billions of sites, updated a few times per second. Likely applications will immediately come to mind to any scientist who has worked with lattice models; many other scientists are already aware, at least in a vague way, of the possibilities that lie behind such a tool.

The simplicity of CAM-8's functional description does not, of course, mean that physical modeling is thereby made trivial. It should, however, spare those attempting to implement fine-grained neighborhood models the significant effort needed to twist such models to fit conventional architectures efficiently. This, together with unprecedented computing power, should make such models attractive to a wide range of disciplines.

The book in ref. [26] illustrates a variety of modeling techniques suitable for the modeling environment discussed here. Much of the recent literature on cellular automata and lattice gases discusses specific fine-grained models of physics-like systems, and thus provides a wealth of examples.

The reader should be aware that our main interest is in the fundamental connections between physics and computation, and one of the reasons we developed CAM-8 is to work on certain questions in that area (cf. refs. [12,23]). As theoretical material scientists – for one – we are mere amateurs, and undoubtedly many of the most obvious applications of CAM-8 have not yet come to our attention. For this reason, we solicit proposals for imaginative and demanding experiments to be performed on this machine – on themes spanning physics, mathematics, and computer science – that will allow us to exercise its design, possibly avoid costly mistakes or omissions, and help make hardware, software, and methodology accessible to more people in a shorter time.

Acknowledgements

This research was supported in part by the Defense Advanced Research Projects Agency (N00014-89-J-1988), and in part by the National Science Foundation (8618002-IRI).

References

[1] C. Bennett, N. Margolus and T. Toffoli, Bond-energy variables for Ising spin-glass dynamics, Phys. Rev. B 37 (1988) 2254.

[2] B.H. Boghosian, W. Taylor and D. Rothman, A cellular-automaton simulation of two-phase flow on the CM-2 computer, in: Proc. Supercomputing '88, Vol. 2. Science and Applications, eds. J. Margin and S. Lundstrom, IEEE (1989) 34–44.

[3] B. Boghosian, personal communication.

[4] A. Clouqueur and D. d'Humières, RAP1, A cellular automaton machine for fluid dynamics, Complex Systems 1 (1987) 585–597.

[5] B. Chopard, A cellular automata model of large scale moving objects, J. Phys. A., to appear .

[6] A. Despain, Prospects for a lattice-gas computer, in: Lattice-Gas Methods for Partial Differential Equations, eds. G. Doolen et al. (Addison–Wesley, Reading, MA, 1990) pp. 211–218.

[7] U. Frisch, B. Hasslacher and Y. Pomeau, Lattice-gas automata for the Navier–Stokes equation, Phys. Rev. Lett. 56 (1986) 1505–1508.

[8] U. Frisch et al., Lattice gas hydrodynamics in two and three dimensions, in: Lattice-Gas Methods for Partial Differential Equations, eds. G. Doolen et al. (Addison–Wesley, Reading, MA, 1990) pp. 77–135.

[9] J. Hardy, O. de Pazzis and Y. Pomeau, Molecular dynamics of a classical lattice gas: Transport properties and time correlation functions, Phys. Rev. A 13 (1976) 1949–1960.

[10] B. Hasslacher, Discrete Fluids, Los Alamos Sci., Special Issue No. 15 (1987) 175–200, 211–217.

[11] N. Margolus, Physics-like models of computation, Physica D 10 (1984) 81–95.

[12] N. Margolus, Physics and computation, Ph. D. Thesis, Tech. Rep. MIT/LCS/TR-415, MIT Laboratory for Computer Science (1988).

[13] N. Margolus, Cellular automata machines – a new environment for modeling, in: Proceedings of the 1988 Rochester Forth Conference, ed. L. Forsley (Institute for Applied Forth Research, Rochester, NY, 1988) pp. 12–21.

[14] N. Margolus and T. Toffoli, Cellular automata machines, Complex Systems 1 (1987) 967–993.

[15] N. Margolus and T. Toffoli, Cellular automata machines, revised version of [14], in: Lattice-Gas Methods for Partial Differential Equations eds. G. Doolen et al. (Addison–Wesley, Reading, MA, 1990) pp. 219–249.

[16] N. Margolus, T. Toffoli and G. Vichniac, Cellular-automata supercomputers for fluid dynamics modeling, Phys. Rev. Lett. 56 (1986) 1694–1696.

[17] T. Shimomura, G. Doolen, B. Hasslacher and C. Fu, Calculations Using Lattice Gas Techniques, Los Alamos Science, Special Issue No. 15 (1987) 201–211.

[18] M. Smith, Representations of geometrical and topological quantities in cellular automata, Physica D 45 (1990) 271–277.

[19] T. Toffoli, Cellular automata mechanics, Tech. Rep. 208, Comp. Comm. Sci. Dept., The University of Michigan (1977) p. 255.

[20] T. Toffoli, CAM: A high-performance cellular-automaton machine, Physica D 10 (1984) 195–204.

[21] T. Toffoli, Cellular automata as an alternative to (rather than an approximation of) differential equations in modeling physics, Physica D 10 (1984) 117–127.

[22] T. Toffoli, Frontiers in computing, in: Information Processing, ed. G.X. Ritter (North-Holland, Amsterdam, 1989) p. 1.

[23] T. Toffoli, How cheap can mechanics' first principles be?, in: Complexity, Entropy, and the Physics of Information, ed. W. Zurek (Addison–Wesley, Reading, MA, 1990) pp. 301–317.

[24] T. Toffoli, Four topics in lattice gases: ergodicity; relativity; information flow; and rule compression for parallel lattice-gas machines, in: Discrete Kinetic Theory, Lattice Gas Dynamics and Foundations of Hydrodynamics, ed. R. Monaco (World Scientific, Singapore, 1989) pp. 343–354.

[25] T. Toffoli, Cellular automata machines as physics emulators, in: Impact of Digital Microelectronics and Microprocessors on Particle Physics, eds. M. Budinich et al. (World Scientific, Singapore, 1988) pp. 154–160.

[26] T. Toffoli and N. Margolus, Cellular Automata Machines – A New Environment for Modeling (MIT Press, Cambridge, MA, 1987).

[27] T. Toffoli and N. Margolus, Invertible cellular automata: a review, Physica D 45 (1990) 229–253.

[28] G. Vichniac, Simulating physics with cellular automata, Physica D 10 (1984) 96–115.

CHAPTER 8

APPLICATIONS

Physica D 47 (1991) 275–280
North-Holland

LATTICE GAS SIMULATION OF FREE-BOUNDARY FLOWS

K.A. CLIFFE, R.D. KINGDON, P. SCHOFIELD and P.J. STOPFORD
Theoretical Physics Division, AEA Technology, Harwell Laboratory, Didcot, Oxfordshire OX11 0RA, UK

Received 29 December 1989

Within a very short time, many spectacular results have been produced to demonstrate the potential of lattice gas hydrodynamics (LGH) to predict the behaviour of systems governed by the Navier–Stokes and related equations at low to moderate Reynolds numbers. However, in order for LGH to be accepted as an alternative to conventional computational fluid dynamics (CFD) methods for the modelling of flows of practical importance, it is necessary to demonstrate in some specific cases that LGH has clear advantages over CFD, either where problems remain intractable to the latter or where LGH could give savings in computational resources.

One area where CFD techniques have considerable difficulties is that of free-boundary problems, in which unstable fronts separate fluid species in the flow. The inclusion of chemical reactions (the fluid species reacting at the interfaces) renders the problem even less tractable to CFD, while presenting no difficulties in principle to LGH.

Clavin, Lallemand, Pomeau and Searby have simulated the Kelvin–Helmholtz and Rayleigh–Taylor free-boundary instabilities as well as a simple reaction front showing the Darrieus–Landau instability. Using their techniques, we have reproduced the Kelvin–Helmholtz and Rayleigh–Taylor instabilities.

One of the aims of our work is to give a practical demonstration of LGH to model the flow and reactions of vehicle exhaust emissions at street level in an urban environment.

1. Introduction

Cellular automata (CA) are discrete dynamical systems consisting of finite-state variables arranged on a uniform grid. Using appropriate grid geometries, occupation rules, collision rules and boundary conditions, all of which are straightforward to implement, it is possible to simulate a wealth of behaviour, including pattern generation, pseudo-random number generation, growth models, lattice gas models, and fluid flow.

The last application, termed lattice gas hydrodynamics (LGH), was initiated in 1986 by Frisch, Hasslacher and Pomeau (FHP) [1], who showed that, with selected collision rules, variables on a triangular lattice formed a system which could simulate the macroscopic Navier–Stokes equations in two dimensions. The LGH concept was subsequently extended to the solution of three-dimensional fluid-flow problems (such as flow in a square channel past a circular plate [2]), complex-boundary problems (such as flow in porous media [3]), and free-boundary problems (such as the Rayleigh–Taylor and Kelvin–Helmholtz instabilities [4]). The simulation of free-boundary problems is especially interesting because this is an area in which conventional computational fluid dynamics (CFD) techniques have encountered considerable problems. If one includes chemistry by considering situations where the free boundary is a reacting front (such as might occur in combusting flows or unmixed chemically reacting flows) then CFD has even more problems, while solution using LGH remains straightforward in principle. The practical difficulties of using either CFD or LGH techniques clearly need further consideration and these will be discussed in section 2.

The basis of this feasibility study is the appreciation that heterogeneous reaction–convection–diffusion problems do arise in many physical systems, including the atmosphere, and that in many of these cases previous attempts at simulation by approximating the flow to be homogeneous, or by otherwise decoupling the chemistry and flow, have failed to reproduce the observed reaction and dispersion phenomena. The applicability of LGH to atmospheric problems is discussed in section 3, with further discussion on the specific application of the simulation of nitrogen dioxide generation at street level in built-up areas, following the emission of nitrogen monoxide in vehicle exhaust. This particular phenomenon has not been successfully explained using conventional modelling techniques [5], and since it is of evident environmental importance it is worthy of consideration as a possible application of LGH.

2. Motivation for application of LGH

Two major advantages of LGH which indeed are common to all CA are purely computational in nature. The first is that the CA rules for updating site values can be expressed entirely in Boolean algebra and therefore no floating-point calculations are required. This important feature, known as "bit-democracy" (where all bits have equal weight in a calculation, rather than representing more and less-significant figures), eliminates the problem of rounding error. Accuracy is limited only by the resolution of the grid. Bit-democracy, a feature of all CA, enables one to tackle highly nonlinear problems where sensitivity to initial conditions (and consequently to numerical fluctuations) can render floating-point computations meaningless. The second major advantage of CA is that the updating rule need involve local coupling only (e.g from nearest neighbours), and so CA is ideally suited to massively parallel computing architectures and algorithms.

Apart from these intrinsic advantages of bit-democracy and parallelism, LGH is particularly well suited to many practical fluid flow problems at low to moderate Reynolds numbers because:

(a) Fluid mixtures and reactions can be modelled directly.

(b) Boundary conditions for complex geometries can be implemented more easily than in conventional CFD codes.

(c) Nonlinear behaviour does not require special treatment.

(d) Regions of high gradient (boundary layers) do not need to be resolved explicitly.

(e) Parameter spaces can be explored quickly without the necessity for extreme accuracy.

(f) Behaviour is easy to visualise.

(g) Coding is easy, relative to the effort required for CFD.

Clearly to obtain a meaningful evaluation of LGH one should compare it with conventional CFD techniques, which have benefited from continuous development over many years, creating extremely efficient commercial codes.

LGH precisely simulates small-scale behaviour and for this reason it cannot compete with CFD techniques in simulating high Reynolds number flows. Two-dimensional LGH can reach Reynolds numbers of 1000–10 000; the three-dimensional LGH simulation of Rivet, Hénon, Frisch and d'Humieres [2] reaches a Reynolds number of 190. Orszag and Yakhot [6] give dimensional scaling arguments to show that, because LGH must simulate on scales below that of the dissipation scale of the turbulent fluid, for higher Reynolds number the LGH simulation will always be bettered by CFD methods. Recently Hasslacher [7] has considered this scaling argument more carefully. By writing the Reynolds number Re in terms of the "Reynolds coefficient" Re*,

$$\mathrm{Re} = ML\,\mathrm{Re}^*,$$

where M is the Mach number and L is a global length, he identifies Re* as a crucial parameter which is conventionally assumed to be of order 1 but which in LGH is model-dependent and in three dimensions can be of order 10 or higher. This clearly leads to a more optimistic view of the applicability of LGH at high Reynolds number.

Scaling arguments aside, it remains true that conventional techniques can also be written so as to exploit highly parallel systems. However, since LGH processors do not need to perform real arithmetic, they can be made in large numbers at relatively low cost. LGH also has a considerable advantage over conventional CFD techniques where small-scale phenomena are of interest. One can argue that in cases where small-scale eddies are crucial for the co-mingling of otherwise diffuse reactants, CFD will clearly not be appropriate for predicting the observed behaviour since it smears out sub-grid-scale features. On the other hand, if the small scales are not crucial then one could equally perform a re-scaling of the LGH algorithm (through appropriate modifications to the collision rules), accessing higher Reynolds numbers. While the latter re-scaling (which amounts to "large eddy simulation") remains to be demonstrated in practice, and no doubt requires a fair amount of ingenuity, the objection to the use of CFD where flow and chemistry are inseparable is more immediate in its consequences.

One can delineate the role of LGH further by considering the CFD treatment of free-boundary problems. CFD can cope with stable interfaces, for example shock fronts, with the use of adaptive gridding techniques. However, where interface configurations are time-dependent, adaptive gridding becomes increasingly expensive in computer time in principle, and in practice has not yet been demonstrated to work. The only other option is to choose a very fine mesh covering the entire solution region, which equally leads to prohibitively high solution times. Further, if one considers a free boundary problem with chemistry, where for example the interface separates two reactants, then one has the elements of a problem which CFD has little hope of modelling faithfully but which is a simple extension of LGH.

There are two caveats to this argument. Firstly, the chemistry should ideally be nonlinear (i.e. not first-order). CFD can deal with nonlinear equations (the Navier–Stokes equations themselves are nonlinear) but the nonlinearity does introduce a sensitive dependence of chemistry upon local flow conditions, which in a free boundary problem are sufficiently inhomogeneous to cause CFD approximations to be inadequate. Secondly, in the simplest two-dimensional LGH models (e.g. "FHP-II" with two species), reaction energy is not taken into account since the particles move with constant speed and momentum is conserved. Energy-conserving models can be constructed, but one has to be careful to ensure Galilean invariance when devising the collision scheme. Furthermore one always has to consider the penalty in terms of bit-cost. FHP-II needs 7 bits per node; introducing two different particles requires 7 more bits per node to describe particle "colour"; specifying the energy of moving particles would require another 6 bits per node, for a simple two-speed model. Each additional sophistication reduces the potential size of the problem that can be considered, ultimately lowering the ceiling Reynolds number.

All things considered, it is clear that if there are practical problems which involve the solution of (probably nonlinear) chemistry in a heterogeneous free-boundary flow, then two-dimensional LGH in its present state of development has good prospects for useful application. The present ceiling Reynolds number is probably high enough for most two-dimensional applications. In the longer term, the case for modelling higher Reynolds number remains open, and with increased parallelisation and intelligent re-scaling LGH may well provide stiff competition for CFD.

3. Applicability to atmospheric problems

An understanding of some of the reaction–dispersion systems present in the atmosphere is becoming increasingly important. Many of the pollution phenomena – for example, ozone layer depletion, acid rain and street-level vehicle emissions – depend crucially upon chemistry and flow coupled together. The usual treatment of such systems has been to consider the reactants as "inert tracer" in a dispersion plume which evolves with a concentration distribution with Gaussian cross-section (a "Gaussian plume") [8]. This is clearly inadequate if the chemistry is not first-order, where there is a strong dependence of reaction rate upon local concentration. Furthermore, a Gaussian plume does not model the effects of concentration fluctuations. CFD techniques which model the flow explicitly (using, for example, algebraic stress or large eddy simulation models for the turbulence, as have been implemented in the FLOW 3D code [9]) can also take into account nonlinear chemical processes, but problems arise if there are steep concentration gradients, for example at an interface in a heterogeneous mixture. As the discussion in the previous section has indicated, there are clearly opportunities in atmospheric modelling for the application of LGH. In this section general criteria for applicability will be summarised, followed by an example application of the formation of nitrogen dioxide at street level.

With reference to the above comments and to the previous section, the criteria for applicability of LGH to modelling atmospheric systems may be stated as follows:

(a) The chemistry should be nonlinear (i.e not first-order). First-order chemistry can be decoupled from the flow and Gaussian plume modelling could well be adequate.

(b) The flow should be heterogeneous, with chemical reactions taking place across free boundaries. Flow with gentle concentration gradients can be modelled using CFD.

(c) The flow should not be highly turbulent. A Reynolds number of 1000–10000 for two-dimensional LGH is probably adequate for typical applications.

(d) The chemistry should be relatively simple. LGH is unlikely to be appropriate for modelling complex chain reactions, simply because perfectly good finite-difference codes already exist (such as FACSIMILE/CHEKMAT [10], which can solve dozens of chemical reactions together with two-dimensional diffusion) and because the addition of each new species adds further to the number of bits required to specify a LGH nodal state.

(e) The flow is incompressible and the reaction energy release can be ignored. While two-speed LGH models have been developed which could deal with compressible flow and with energy conservation, these complexities are to be avoided if possible in order to economise on bits.

(f) The flow should be effectively two-dimensional. Again, three-dimensional LGH models exist but do not reach sufficiently high Reynolds numbers to be generally applicable at present, although for a study of the initial dynamics and chemistry of a small plume they may well be adequate.

Where the above conditions are satisfied it is clear that two-dimensional LGH could make a useful contribution. The remainder of this section will be used to consider one such potential application, the formation of nitrogen dioxide (NO_2) at street level due to nitrogen monoxide (NO) emissions in vehicle exhausts. The reaction forming NO_2 from NO is second order,

$$NO + NO + O_2 \rightarrow 2NO_2.$$

Reaction rate $k = 2.0 \times 10^{-38}$ cm^6 $molecule^{-2}$ s^{-1} at 298 K [11].

The Photochemical Oxidants Review Group's Second Report [5] gives data for measurements of roadside NO_2 calculations. In some cases the NO_2/NO_x ratio was measured in the range 10–50%, compared with raw exhaust emission measurements which suggested a ratio of ≈ 5–10%. Present modelling studies suggest that $2NO/O_2$ would only give substantial amounts of NO_2 in extreme conditions. The report concludes that the elevated NO_2 concentrations at kerbsides, which exceed recommended guidelines laid down by the CEC and World Health Organisation, are not fully understood, and that it is not possible to predict in detail the impact of NO_x

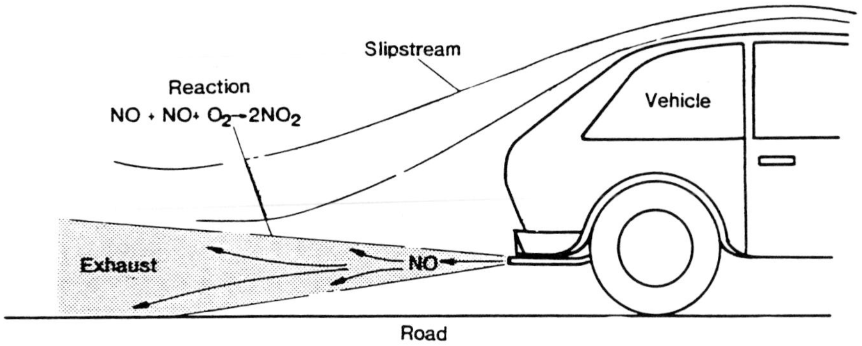

Fig. 1. The reaction–dispersion system at the exhaust pipe of a vehicle for the reaction forming nitrogen dioxide.

emission reductions from motor vehicles on peak urban NO_2 concentrations.

The reaction–dispersion system at the exhaust pipe of a vehicle is shown schematically in fig. 1. Dispersion of the NO in the exhaust gases would cause the generation of NO_2 through the $2NO/O_2$ reaction to fall off rapidly with distance from the vehicle. High local wind speeds would only serve to enhance dispersion and thereby decrease the reaction rate. Therefore a "worst-case" modelling scenario would be to model the system close to the exhaust pipe with low local wind speeds. The system would then appear as in fig. 2. This flow system is very similar to one modelled using LGH already, the Kelvin–Helmholtz jet instability [4]. This has been simulated in a LGH two-species model developed at Harwell by ourselves. One further needs to include the $2NO/O_2$ reaction. This could be modelled with just three species, NO, NO_2 and "air", with the probability of reaction in NO–air collisions taking account of oxygen fraction. The flow is incompressible and reaction energy release is probably not relevant.

One concludes that the two-dimensional LGH modelling of the system in fig. 2 is certainly possible. It is, moreover, desirable, since fig. 2 is arguably the most accurate representation of the physical system and other modelling techniques cannot model with this detail.

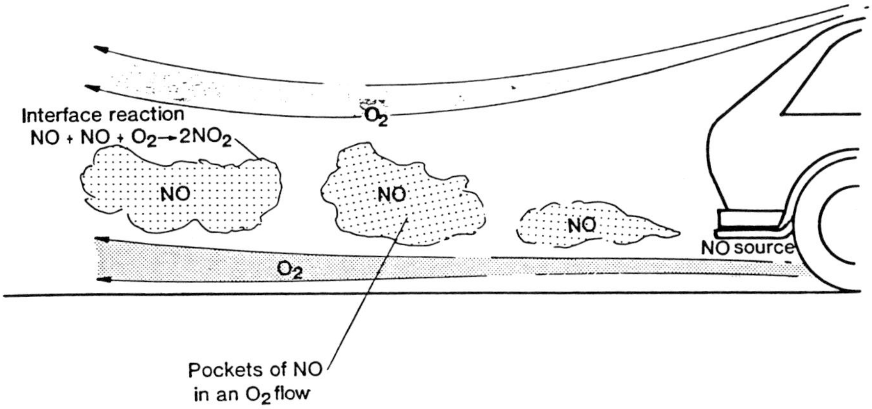

Fig. 2. Simplified "worst-case" model for the reaction–dispersion system in fig. 1.

4. Conclusions

Lattice gas thermodynamics in its present state of development is best applied to those problems where flow and chemistry are coupled, and where the chemical species are poorly mixed, giving rise to high concentration gradients and free boundaries. Such physical systems arise in many practical situations, and in the context of the atmosphere these include ozone layer depletion, acid rain and nitrogen dioxide formation at street level. The last of these possible applications has been investigated in more detail and the modelling of such a system using two-dimensional LGH is deemed possible and desirable. The simulation could provide insight into the unexpectedly high measurements of kerbside NO_2 concentrations, as well as providing a practical demonstration of the LGH technique.

A LGH model capable of simulating flows similar to that in the NO_2 problem has already been developed in Theoretical Physics Division, Harwell.

Acknowledgements

The work described in this report has been funded by the UK Department of the Environment, and by the Underlying Research Programme of AEA Technology.

References

[1] U. Frisch, B. Hasslacher and Y. Pomeau, Phys. Rev. Lett. 56 (1986) 1505–1508.
[2] J-P. Rivet, M. Hénon, U. Frisch and D. d'Humieres, Europhys. Lett. 7 (1988) 231–236.
[3] D.H. Rothman, Geophysics, 53 (1988) 509–518.
[4] P. Clavin, P. Lallemand, Y. Pomeau and G. Searby, J. Fluid Mech. 188 (1988) 437–464.
[5] R.G. Derwent, A.J. Apling, M.R. Ashmore, D.J. Ball, P.A. Clark, A.T. Cocks, R.A. Cox, D. Fowler, M.J. Gay, R.M. Harrison, P.J.A. Kay, D.P.H. Laxen, A. Martin, D. McKenna, S.A. Penkett, D.A. Warrilow, M.L. Williams and P.T. Wood, Oxides of nitrogen in the United Kingdom, A Second Report of the United Kingdom Photochemical Oxidants Review Group, preprint (1989).
[6] S.A. Orszag and V. Yakhot, Phys. Rev. Lett. 56 (1986) 1691–1693.
[7] B. Hasslacher, Discrete Fluids, Los Alamos Sci., Special Issue (1987) 175–217.
[8] F. Pasquill and F.B. Smith, Atmospheric Diffusion, 3rd Edition (Ellis Horwood, Chichester, 1983).
[9] A.D. Burns, M. Ciofalo, D.S. Clarke, S. Gavrilakis, I.R. Hawkins, I.P. Jones, J.R. Kightley and N.S. Wilkes, Progress with the HARWELL-FLOW 3D software for the prediction of laminar and turbulent flow, and heat transfer, 1987–1988, Harwell Report AERE-R13148, HMSO (1988).
[10] A.R. Curtis and W.P. Sweetenham, FACSIMILE/CHEKMAT user's manual, Harwell Report AERE-R12805, HMSO (1987).
[11] R. Atkinson, D.L. Baulch, R.A. Cox, R.F. Hampson Jr., J.A. Kerr and J. Troe, Int. J. Chem. Kinetics 21 (1989) 115–150.

Physica D 47 (1991) 281–295
North-Holland

BOUNDARY AND OBSTACLE PROCESSING IN A VECTORIZED MODEL OF LATTICE GAS HYDRODYNAMICS

Gregory RICCARDI, Charles BAUER and Hwa LIM
Department of Computer Science and Supercomputer Computations Research Institute, Florida State University, Tallahassee, FL 32306, USA

Received 21 January 1990

This paper describes a vectorized supercomputer implementation of a cellular automaton model for lattice gas hydrodynamics. A detailed description of the algorithm is given along with a careful complexity analysis and performance evaluation of it. Particular attention is paid to boundary and obstacle processing. Two applications of the program are described: an acoustic wave model and an obstructed flow through a pipe. The results of executing these models are displayed with density maps and flow diagrams.

1. Introduction

In this paper, we investigate the efficient execution of the FHP model of lattice gas hydrodynamics [2, 3, 5, 6, 8, 9] on vector supercomputers. The model uses a two-dimensional triangular lattice of nodes with particles moving between nodes. The state of the lattice is represented by one-dimensional arrays of bits, as described below. This representation allows us to exploit the vector processing capabilities of supercomputers to provide very high speed execution of fluid models. This algorithm has been developed and tested on Cyber 205 and ETA10 computers, and is equally applicable on other vector computers which support the full range of Boolean operations.

Our representation of the FHP lattice was briefly described in ref. [7]. A summary of our representation is given in section 2 along with details of the representation of boundaries and obstacles. Both deterministic and non-deterministic collision handling are provided.

The algorithm used for boundary and obstacle processing is presented in detail in section 3 using the notation of Fortran 8x. Four boundary processing styles are provided: rebounding (or simple reflection), specular reflecting, periodic and constant flow. Each individual boundary and obstacle may use any of the four styles.

This vectorized program for the FHP model has been analyzed to determine both the speed of execution on a vector supercomputer, and the number of operations which must be executed for each time step. Section 4 gives the results of that analysis. A similar analysis has been done by Hayot for execution on multi-processor scalar computers [4].

The application of this program to fluid modeling is illustrated with two fluid problems. The first shows the propagation of an acoustic wave; the second models the flow of a gas through a pipe which has an obstruction on one wall. The problems of displaying the results of the execution is discussed. A density map is used to display the distribution of particles in the lattice at various time steps. The direction of fluid flow is displayed using a flow diagram.

0167-2789/91/$03.50

2. Implementing the FHP lattice

The FHP model uses a two-dimensional lattice that tiles space into triangles. Each node is connected by arcs to its six nearest neighbors in the north, northeast, southeast, south, southwest and northwest directions. Collisions occur when the incoming particles arrive symmetrically around a node such that their net momentum is zero. These collisions cause a rotation of all particles at that node by 60°. In the "chiral" or deterministic version of the algorithm, the rotation is always in the clockwise direction. In the non-chiral version, rotation is randomly determined to be either clockwise or counterclockwise for each occurrence. The collision rules conserve both momentum and the total number of particles at a given node. On the other hand, if particles arrive at the node in an asymmetric formation, they pass straight through without collision and rotation.

The algorithm is made up of two stages, collision and propagation. In the collision stage, particles arrive at each node from a subset of the six possible directions, interact according to a set of rules, and then exit via another subset of the six directions. In the propagation stage, the particles travel from their nodes a lattice constant distance (i.e. the distance between two nearest neighbors), in the direction of their respective momenta to arrive as incoming particles at the neighboring nodes, thus preparing for the next collision stage. Since each particle at a given node must move in a different direction, only two sets of six values are required to describe the motion of the particles at that node. One set defines whether or not a particle arrives at the node along each of the six directions at the beginning of the collision stage. The other set defines whether or not a particle exits the node along each of the six directions at the beginning of the propagation stage. Clearly, these states can be coded with binary variables.

2.1. *Representing the lattice*

In order to promote vectorization, the lattice is oriented and the nodes are numbered so that the two-dimensional lattice can be represented by one-dimensional arrays. Fig. 1 illustrates an example of

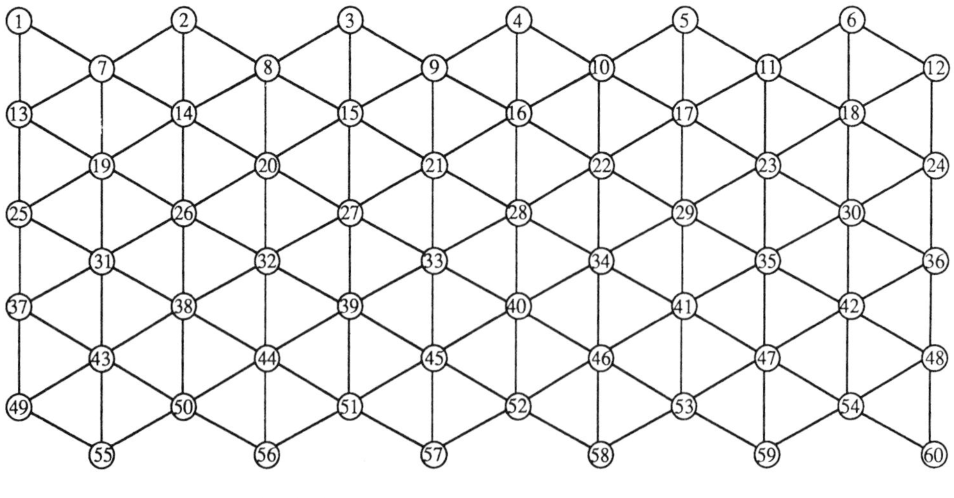

Fig. 1. Representing a 10 × 6 lattice as a vector.

this representation for a 10 by 6 lattice. A lattice of this size was used for initial test runs and validation testing of the code. Actual application runs use much larger lattices.

This is a different representation than that given by Brosa and Stauffer [1], who pack all of the particles of a node into a single word. A collision table is used to determine when to rotate the outputs. Their approach also incorporates rest particles, which are not included in our algorithm.

The vector program uses arrays of the BIT data type for the representation of the particles at the nodes. A variable of type BIT may have a value of either "1" or "0", and only a single bit of memory is needed to store its value, resulting in the equal and effective use of every bit of central memory. The program also uses a vector assignment syntax, which allows operations to be executed upon entire bit arrays.

The vector program defines twelve arrays of type BIT to represent the positions of particles in the lattice:

```
BIT  Xnorth(SIZE), Xnortheast(SIZE), Xsoutheast(SIZE)
BIT  Xsouth(SIZE), Xsouthwest(SIZE), Xnorthwest(SIZE)
BIT  Ynorth(SIZE), Ynortheast(SIZE), Ysoutheast(SIZE)
BIT  Ysouth(SIZE), Ysouthwest(SIZE), Ynorthwest(SIZE)
```

Arrays Xnorth, Xnortheast, Xsoutheast, Xsouth, Xsouthwest, and Xnorthwest represent the presence of incoming particles along each of the six directions before collision. Arrays Ynorth, Ynortheast, Ysoutheast, Ysouth, Ysouthwest, and Ynorthwest represent the presence of exiting particles along each of the six directions before propagation. Thus:

Xi(k) = 1, if a particle approaches node "k" from the "i" direction at the beginning of the collision stage;
= 0, otherwise.

Yi(k) = 1, if a particle exits node "k" along the "i" direction at the beginning of the propagation stage;
= 0, otherwise.

The interior nodes of the lattice are those that are surrounded by neighboring nodes in all six directions. Many operations are performed on the interior nodes, but not the boundary nodes which need to be handled separately. The bit array data structure can also be used as a vector mask in order to select which nodes in the lattice are affected by the given assignment statement. For example, the vector program contains a mask for the interior nodes, MASKI, which is initialized to describe the position of the interior nodes.

The boundary nodes are those nodes along the perimeter of the lattice that do not have neighbors in all six directions. This representation is different from the standard one (c.f. ref. [4]) which excludes those boundary nodes from the lattice. In the standard approach, all of these boundary nodes rebound incoming particles back to the adjacent lattice node. Fig. 2 shows the ten types of boundary nodes, which are characterized by their missing neighbors. Each of the ten types requires separate consideration in the program.

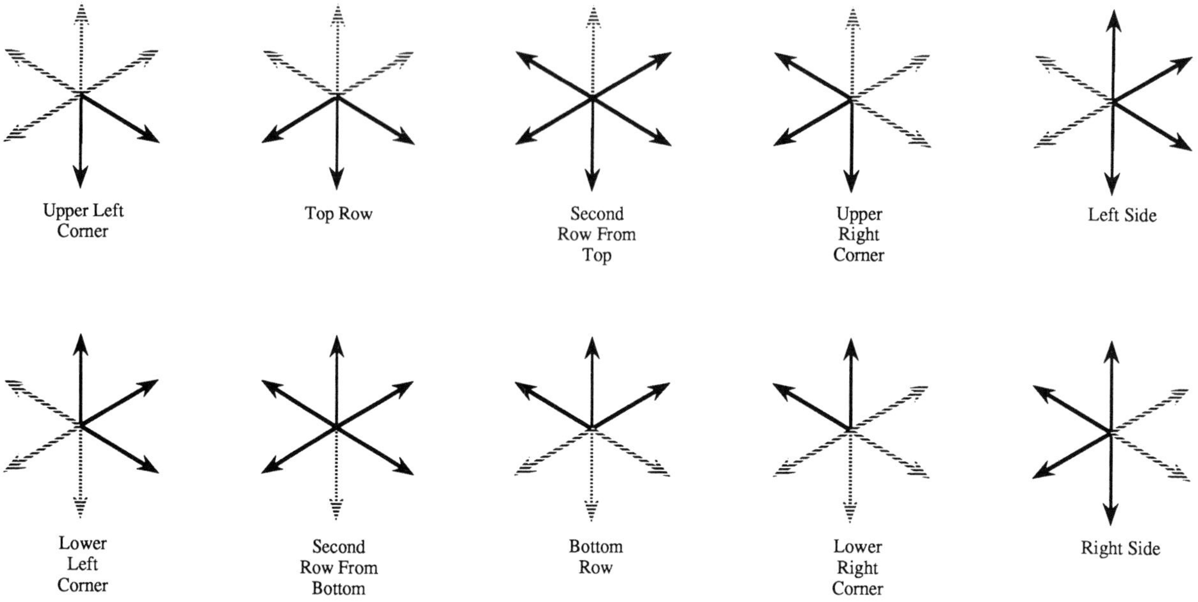

Fig. 2. Classes of boundary modes.

There are four unique corner nodes (1, 12, 49 and 60 in fig. 1) which are treated separately. There are also two rows of boundary nodes at the top of the lattice (2–6 and 7–11 in fig. 1) and two rows of boundary nodes at the bottom of the lattice (55–59 and 50–54 in fig. 1). Since nodes in a row are numbered and stored contiguously, each of these four boundary rows can be operated upon by short vector assignment statements. The nodes of the other two boundary types, the left and the right, (13, 25, and 37 for the left and 24, 36, and 48 for the right in fig. 1) are not contiguous, occurring at regular intervals of $2n$, where n is the number of nodes in a row, throughout the entire length of the lattice. The

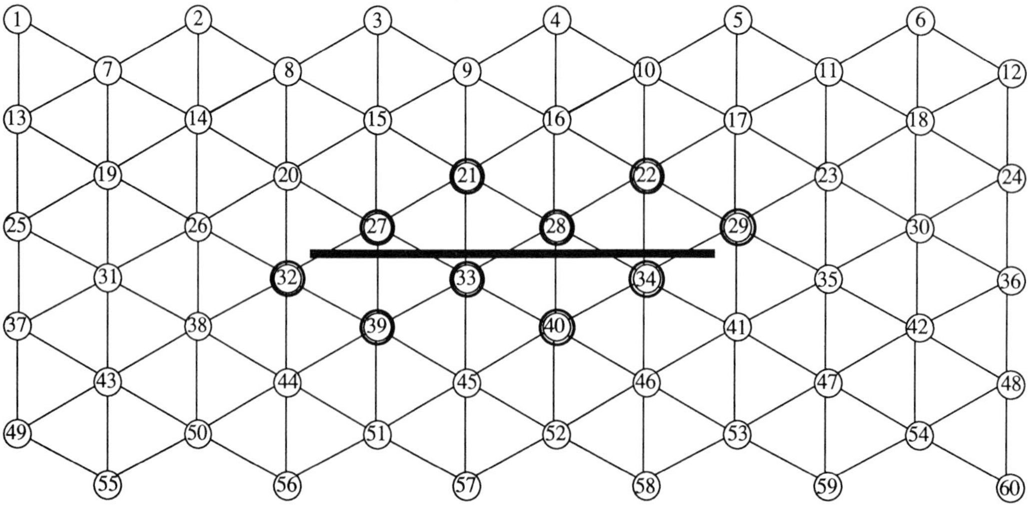

Fig. 3. An obstacle in the lattice.

program uses a vector mask to perform operations on these nodes without affecting other nodes in the lattice.

Obstacles are introduced in the lattice by extending the boundary processing to interior nodes. Fig. 3 shows the lattice of fig. 1 with a flat obstacle inside. In this case, nodes 21 and 22 are in the top row boundary type, missing only the neighbor to the south, 27 and 28 are second row, 33 and 34 are second last row, and 39 and 40 are last row nodes. This leaves nodes 29 (missing southwest) and 32 (missing northeast) to be handled as new boundary types.

The program also implements four possible boundary conditions which can exist for each of the ten boundary types. These conditions are "specular reflection", "rebounding", "total leakage", and "periodic". In specular reflection, particles entering a boundary node will "bounce away" such that the angle of incidence equals the angle of reflection. In rebounding, a particle entering a boundary node rebounds 180° back in the opposite direction from which it entered. In periodic, the boundaries of the grid "wrap around" top to bottom or left to right, so that a particle passing through a boundary node at the top of the lattice, for example, is reintroduced at the corresponding node at the bottom of the lattice.

3. Vector implementation of the FHP model

This section relates the implementation of the vector program of the FHP model, originally described in Lim et al. [7], with emphasis on recent improvements and on boundary node processing. It considers first the task of program initialization. Next it describes in detail the algorithms for the collision stage and the propagation stage.

3.1. Program initialization

The variables and data structures in the program are defined according to user-supplied parameters. The user chooses the number of rows and columns that comprise the lattice. The user also defines a boundary condition for each of the ten boundary types. The user chooses the chiral or non-chiral algorithm for collision resolution, the total number of iterations to be performed, the periodic number of iterations between output steps, the size of the sub-sections that divide the lattice for summarization purposes, and the output formal desired. In the sound propagation model, the user also provides the dimensions of the concentrated center and the percentage occupancy of particles in nodes outside the concentrated area. In the obstructed flow model, the user also specifies the location and size of the obstruction and the average initial velocity of the lattice particles.

3.2. The collision stage

In the collision stage, incoming particles converge at each node, interact, and then take exit paths according to a specified set of collision rules. In nodes that can have input from all six directions, the particles may collide and rotate. These nodes include all interior nodes and any boundary nodes with the periodic or the total leakage conditions, which allow incoming particles from the exterior of the lattice. Boundary nodes with simple reflection or specular reflection conditions do not allow any new particles to enter the lattice, so they cannot have input from all six directions and must be handled separately. The collision algorithm presented here is significantly faster than the one which we described in ref. [7].

The first step taken in the collision stage is to determine where collisions occur, that is, where two, three, or four incoming particles are arranged symmetrically around a node. The program performs two searches that test each node for the eight possible symmetrical patterns that cause collisions. The results of the two searches are combined into a vector masks COLLIDE and NOCOLL such that

COLLIDE(k) = 1 if there is a collision detected at node "k",
NOCOLL(k) = 1 if no collision is detected at node "k".

The first search selects nodes that have two or four incoming particles arranged symmetrically. The second search selects nodes that have three incoming particles arranged symmetrically. Both searches include the cases where there are either six or zero incoming particles. This is acceptable since rotation in these two cases does not alter the state of the node.

```
  COLLIDE(1:SIZE) = (XNORTH(1:SIZE) .EQV. Xsouth(1:SIZE)) .AND.
+      (Xnortheast(1:SIZE) .EQV. Xsouthwest(1:SIZE)) .AND.
+      (Xsoutheast(1:SIZE) .EQV. Xnorthwest(1:SIZE))
  COLLIDE(1:SIZE) = COLLIDE(1:SIZE) .OR.
+      ((Xnorth(1:SIZE) .EQV. Xsoutheast(1:SIZE)) .AND.
+      (Xsoutheast(1:SIZE). EQV. Xsouthwest(1:SIZE)) .AND.
+      (Xnortheast(1:SIZE) .EQV. Xsouth(1:SIZE)) .AND.
+      (Xsouth(1:SIZE) .EQV. Xnorthwest(1:SIZE)))
  NOCOLL(1:SIZE) = .NOT.COLLIDE(1:SIZE)
```

After the collisions have been located, the program determines the directions of the particles leaving each node. Interior nodes are handled first, using one of two algorithms for resolving collisions, chiral or non-chiral.

In the chiral algorithm, the COLLIDE and NOCOLL masks are used to select which value is correct for each node. For example, at every node where a particle enters from the southeast and a collision does occur, the corresponding bit in Ynorth is set, representing a 60° clockwise rotation. Also, at every node where a particle enters from the south and no collision occurs, the corresponding bit in Ynorth is set.

```
  Ynorth(1:SIZE) = (Xsoutheast(1:SIZE).AND.COLLIDE(1:SIZE)) .OR.
+      (Xsouth(1:SIZE).AND.(NOCOLL(1:SIZE)))
```

In the non-chiral algorithm, the program randomly splits COLLIDE into bit masks CLOCKW and CCLOCKW for clockwise and counterclockwise rotations. The CLOCKW, CCLOCKW, and NOCOLL masks are then used to select which value is correct for each node. For example, in the following code, at every node where a particle enters from the southeast and a clockwise collision occurs, the corresponding bit in Ynorth is set. Next, at every node where a particle enters from the southwest and a counterclockwise collision occurs, the corresponding bit in Ynorth is set. Finally, at every node where a particle enters from the south and no collision occurs, the corresponding bit in Ynorth is set.

```
  Ynorth(1:SIZE) = (Xsoutheast(1:SIZE).AND.CLOCKW(1:SIZE)) .OR.
+      (Xsouthwest(1:SIZE).AND.CCLOCKW(1:SIZE)) .OR.
+      (Xsouth(1:SIZE).AND.NOCOLL(1:SIZE))
```

Next, the output particles are calculated for the boundary nodes. Since boundary nodes with conditions of periodic or total leakage are included in the previous collision calculations, the program only considers boundaries with simple or specular reflection conditions. Recall that in a node with specular reflection, an incoming particle rebounds off the boundary in a direction such that the angle of incidence equals the angle of reflection, and that in a node with simple reflection, an incoming particle rebounds off the boundary in the opposite direction from which it approached. Also, only directions that connect the boundary nodes to interior nodes must be considered since simple reflection and specular reflection do not allow particles to leave the lattice.

Nodes in the left-side and right-side boundaries are not contiguous, but are spread throughout the entire length of the lattice. Mask vectors MASKL and MASKR are constructed at the beginning of the program and specify the positions of the left boundary nodes and right boundary nodes, respectively. Operations to be performed on the left boundary, for example, are actually performed on the entire lattice, but the MASKL mask is used to selectively save only the results for the left boundary nodes.

```
      IF (TYPEL.EQ.SPECULA) THEN
        Ynorth(1:SIZE) = Xsouth(1:SIZE).AND.MASKL(1:SIZE).OR.
     +       Ynorth(1:SIZE)
        Ynortheast(1:SIZE) = Xsoutheast(1:SIZE).AND.MASKL(1:SIZE).OR.
     +       Ynortheast(1:SIZE)
        Ysoutheast(1:SIZE) = Xnortheast(1:SIZE).AND.MASKL(1:SIZE).OR.
     +       Ysoutheast(1:SIZE)
        Ysouth(1:SIZE) = Xnorth(1:SIZE).AND.MASKL(1:SIZE).OR.
     +       Ysouth(1:SIZE)
      ELSE IF (TYPEL.EQ.SIMPLE) THEN
        Ynorth(1:SIZE) = Xnorth(1:SIZE).AND.MASKL(1:SIZE).OR.
     +       Ynorth(1:SIZE)
        Ynortheast(1:SIZE) = Xnortheast(1:SIZE).AND.MASKL(1:SIZE).OR.
     +       Ynortheast(1:SIZE)
        Ysoutheast(1:SIZE) = Xsoutheast(1:SIZE).AND.MASKL(1:SIZE).OR.
     +       Ysoutheast(1:SIZE)
        Ysouth(1:SIZE) = Xsouth(1:SIZE).AND.MASKL(1:SIZE).OR.
     +       Ysouth(1:SIZE)
      END IF
```

Calculations performed on the nodes in the top two boundaries and the bottom two boundaries have the advantage that the nodes in these boundaries are stored contiguously. Therefore, the calculations can be performed in single, short vector assignments without the need to operate on the entire lattice and without the need for masks. The following code processes the top boundary:

```
IF (TYPEtop.EQ.SPECULA) THEN
  Ysoutheast(topFIRST:topLAST) = Xsouthwest(topFIRST:topLAST)
  Ysouth(topFIRST:topLAST) = Xsouth(topFIRST:topLAST)
  Ysouthwest(topFIRST:topLAST) = Xsoutheast(topFIRST:topLAST)
ELSE IF (TYPEtop.EQ.SIMPLE) THEN
  Ysoutheast(topFIRST:topLAST) = Xsoutheast(topFIRST:topLAST)
  Ysouth(topFIRST:topLAST) = Xsouth(topFIRST:topLAST)
  Ysouthwest(topFIRST:topLAST) = Xsouthwest(topFIRST:topLAST)
END IF
```

The remaining boundaries are the four corner nodes, each of which must be handled separately. The following code, composed of individual bit assignments, processes the upper left corner:

```
IF (TYPEC1.EQ.SPECULA) THEN
  Ysoutheast(CORNER1) = Xsouth(CORNER1)
  Ysouth(CORNER1) = Xsoutheast(CORNER1)
ELSE IF (TYPEC1.EQ.SIMPLE) THEN
  Ysoutheast(CORNER1) = Xsoutheast(CORNER1)
  Ysouth(CORNER1) = Xsouth(CORNER1)
END IF
```

3.3. *The propagation stage*

In the propagation stage, exiting particles leave each node and travel a unit distance in the lattice to a neighboring node, initializing the conditions for the collision stage of the next iteration. The interior nodes by definition have neighbors in every direction, so particles can propagate to these nodes from every direction. Boundary nodes have neighboring nodes in some subset of the six directions, and thus can only receive particles from these neighbors. However, other directions on the boundary nodes lead to the exterior of the lattice and must be treated independently in the program.

The transport of particles between adjacent nodes occurs at the beginning of the propagation stage. The algorithm for two of the six directions is very straightforward. A given node k receives an incoming particle from the north if and only if there is a particle exiting node $k - 2n$ to the south, where n is the number of nodes in a row. Also, node k receives an incoming particle from the south if and only if there is a particle exiting node $k + 2n$ to the north, as illustrated by node 21 in fig. 4.

```
Xnorth(1 + 2*N:SIZE) = Ysouth(1:SIZE - 2*N).
```

Unfortunately, the algorithm for the other four directions is slightly more complex. The relative numbering of the two adjacent nodes on the previous row and the two adjacent nodes depend on

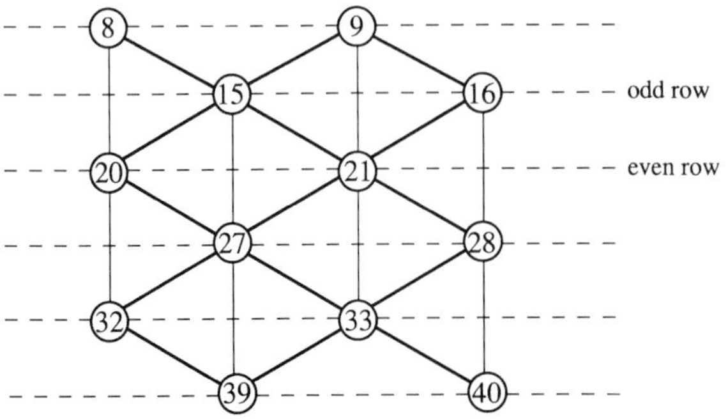

Fig. 4. Propagation by odd and even rows.

whether the node itself resides on an odd or even row. If it is on an odd row, a given node k is adjacent to node $k-n-1$ to the northwest, node $k-n$ to the northeast, node $k+n-1$ to the southwest, and node $k+n$ to the southeast. However, if it is on an even row, a given node k is adjacent to node $k-n$ to the northwest, node $k-n+1$ to the northeast, node $k+n$ to the southwest, and node $k+n+1$ to the southeast. Fig. 4 shows these relationships. Thus, the program creates mask MASKO to hold the critical information of whether or not each node is on an odd row:

```
      Xnortheast(1 + N:SIZE) = ((Ysouthwest(1:SIZE - N).AND.MASKO(1 + N:SIZE)).OR.
     +       (Ysouthwest(1 + 1:SIZE - N + 1).AND.NOT.MASKO(1 + N:SIZE)))
```

Next, the program determines the incoming particles for the boundary nodes in the directions where they have no neighbors. Any boundary with the specular reflection or simple reflection condition does not allow incoming particles from exterior directions. Thus, under these two boundary conditions, the Xi values for boundary nodes in directions without neighbors are left to be zero by default. The program handles the other two boundary conditions explicitly. Under the total leakage condition, particles may pass freely out of the lattice boundaries. Incoming particles are considered to be a constant pattern in each iteration. For example, particles may leak out of the right boundary and a constant inflow is imposed on the left boundary, like in a wind tunnel. Under the periodic condition, particles that exit the lattice from one boundary node are reintroduced to the lattice in a corresponding node in the opposite boundary. The program uses a complex set of assignments to determine the destination of particles exiting the lattice.

Nodes in the left and right boundaries occur throughout the lattice vector, so the MASKL and MASKR masks are again used to control assignments done only at these nodes. For example, if the left boundary has the total leakage condition, the incoming particles from the southwest and the northwest at the left boundary are reset to their original patterns. If the left boundary has the periodic condition, the incoming particles from the southwest and the northwest are equated to outgoing particles from the right boundary to the northeast and the southeast, respectively:

```
      IF (TYPEL.EQ.LEAKAGE) THEN
        Xsouthwest(1:SIZE) = ORIGXsouthwest(1:SIZE).AND.MASKL(1:SIZE) .OR.
     +       Xsouthwest(1:SIZE)
        Xnorthwest(1:SIZE) = ORIGXnorthwest(1:SIZE).AND.MASKL(1:SIZE) .OR.
     +       Xnorthwest(1:SIZE)
      ELSE IF (TYPEL.EQ.PERIOD) THEN
        Xsouthwest(1:SIZE) = Ynortheast(1 + 2*N - 1:SIZE + 2*N - 1).AND.
     +       MASKL(1:SIZE) .OR. Xsouthwest(1:SIZE)
        Xnorthwest(1:SIZE) = Ysoutheast(1 - 1:SIZE - 1).AND.MASKL(1:SIZE) .OR.
     +       Xnorthwest(1:SIZE)
      END IF
```

Propagation calculations performed on the two top and two bottom boundaries again have the advantage of using short, contiguous vector assignments. For example, if the top boundary has the total leakage condition, incoming particles from the northwest, the north, and the northeast are reset to their original patterns. If the top boundary is periodic, these incoming particles are equated to particles exiting the lattice to the southeast, the south, and the southwest, respectively, from the two lower boundaries:

```
IF (TYPEtop.EQ.LEAKAGE) THEN
  Xnorth(topFIRST:topLAST) = ORIGXnorth(topFIRST:topLAST)
  Xnortheast(topFIRST:topLAST) = ORIGXnortheast(topFIRST:topLAST)
  Xnorthwest(topFIRST:topLAST) = ORIGXnorthwest(topFIRST:topLAST)
ELSE IF (TYPEtop.EQ.PERIOD) THEN
  Xnorth(topFIRST:topLAST) =
+       Ysouth(topFIRST + (M - 2) * N:topLAST + (M - 2) * N)
  Xnortheast(topFIRST:topLAST) =
+       Ysouthwest(topFIRST + (M - 1) * N:topLAST + (M - 1) * N)
  Xnorthwest(topFIRST:topLAST) =
+       Ysoutheast(topFIRST + (M - 1) * N - 1:topLAST + (M - 1) * N - 1)
END IF
```

The four remaining boundaries are the four corners, each of which is a single node. Again, in the total leakage case, the inputs are reset to their original values, and in the periodic case, inputs are equated to particles leaving the lattice from other boundary nodes. All calculations are assignments of single bit values.

```
IF (TYPEC1.EQ.LEAKAGE) THEN
  Xnorth(CORNER1) = ORIGXnorth(CORNER1)
  Xnortheast(CORNER1) = ORIGXnortheast(CORNER1)
  Xsouthwest(CORNER1) = ORIGXsouthwest(CORNER1)
  Xnorthwest(CORNER1) = ORIGXnorthwest(CORNER1)
ELSE IF (TYPEC1.EQ.PERIOD) THEN
  Xnorth(CORNER1) = Ysouth((M - 2) * N + 1)
  Xnortheast(CORNER1) = Ysouthwest((M - 1) * N + 1)
  Xsouthwest(CORNER1) = Ynortheast(2 * N)
  Xnorthwest(CORNER1) = Ysoutheast(M * N)
END IF
```

4. Complexity analysis of the FHP implementation

The top section of table 1 shows the complexity analysis of the collision algorithm. The code that determines where collisions occur requires 15 vector operations upon the full length of the lattice. The code for resolving collisions in interior nodes is divided into two possible values. The chiral case requires 18 full length operations on the lattice. However, the more complex non-chiral case requires 30 full length bit vector operations plus 2 additional full length numeric vector operations (used to generate the random numbers that determine the clockwise or counterclockwise rotation at each collision). The next three lines show the operations for resolving collisions at boundary nodes. In each line, the values are 0 for boundaries with periodic or total leakage conditions, since these nodes were included in the same

Table 1
Complexity analysis[a].

	Specular			Simple			Period			Leakage		
	l	s	b	l	s	b	l	s	b	l	s	b
Collisions												
detect	15	0	0	15	0	0	15	0	0	15	0	0
interior												
chiral	18	0	0	18	0	0	18	0	0	18	0	0
nonchiral	30 + 2	0	0	30 + 2	0	0	30 + 2	0	0	30 + 2	0	0
left/right	16	0	0	16	0	0	0	0	0	0	0	0
top/bottom	0	16	0	0	16	0	0	0	0	0	0	0
corners	0	0	10	0	0	10	0	0	0	0	0	0
Propagation												
interior	18	2	0	18	2	0	18	2	0	18	2	0
left/right	0	0	0	0	0	0	8	0	0	8	0	0
top/bottom	0	0	0	0	0	0	0	8	0	0	8	0
corners	0	0	0	0	0	0	0	0	14	0	0	14
Total												
chiral	67	18	10	67	18	10	59	10	14	59	10	14
nonchiral	79 + 2	18	10	79 + 2	18	10	71 + 2	10	14	71 + 2	10	14

[a)] The values (l, s, b) are full lattice operations, short vector operations, and single bit operations.

code that resolved the interior collisions. On the other hand, boundaries with specular reflection or simple reflection conditions are resolved separately. First, the left and right boundaries together require 16 vector operations on the whole lattice, using vector masks to select only the proper boundary nodes. Second, the resolution of collisions at the top two and bottom two boundaries requires 16 short vector operations. Remember, since these boundaries are comprised of contiguous nodes, the operations affect only the boundaries themselves and not the entire lattice. Finally, the resolution of collisions at the corner nodes requires 10 single bit assignments. These three values for the boundaries conclude the analysis of the collision phase.

The second section of table 1 shows the complexity analysis for the propagation phase. The code for the interior nodes requires 18 long vector operations and 2 short vector operations. The next three lines of the chart describe the operations necessary for propagating particles that enter the lattice from the boundary nodes. Note that for specular reflections and simple reflection, particles may not enter the lattice, so these values are 0. First, the code for the left and right boundaries together requires 8 long bit vector operations. Second, the code for the four top and bottom boundaries requires 8 short vector operations. Finally, the code for the four corners requires 14 individual bit assignments. These values conclude the analysis of the propagation phase.

The full algorithm with deterministic collision handling requires only 67 whole vector operations per time step. The only scalar code in the algorithm is that to handle the boundary conditions on the four corner nodes. The non-chiral code is significantly slower because of the need to generate random numbers for collision handling. It is hoped that in the future, a random bit mask generator can be used in place of the numeric random number generator.

Next, we consider the performance of the program on a single processor of an ETA10-G supercomputer. Running under optimal conditions, with the lattice size of 508 by 127 (just under the maximum vector length of 65 535), periodic boundary conditions, chiral collision resolution, and no output summarizations calculated, the program completed 1000 iterations in 1.698 s. Dividing by the 64 516 nodes in the lattice yields a performance of 37.99 million node updates per second.

The program's data structures require 32 bit vectors that are as long as the lattice itself. (In the alternate version of the program, described below, which allows for a lattice longer than 65 535, only 26 of the data vectors need to be as long as the lattice. The other 6 are working vectors that only need to be 65 535 elements long.) As is often the case, this program offers a variety of tradeoffs. Some of the bit arrays could be removed from the algorithm, but the performance of the program would suffer. On the other hand, the speed of the program could be increased at the price of introducing more scratch arrays.

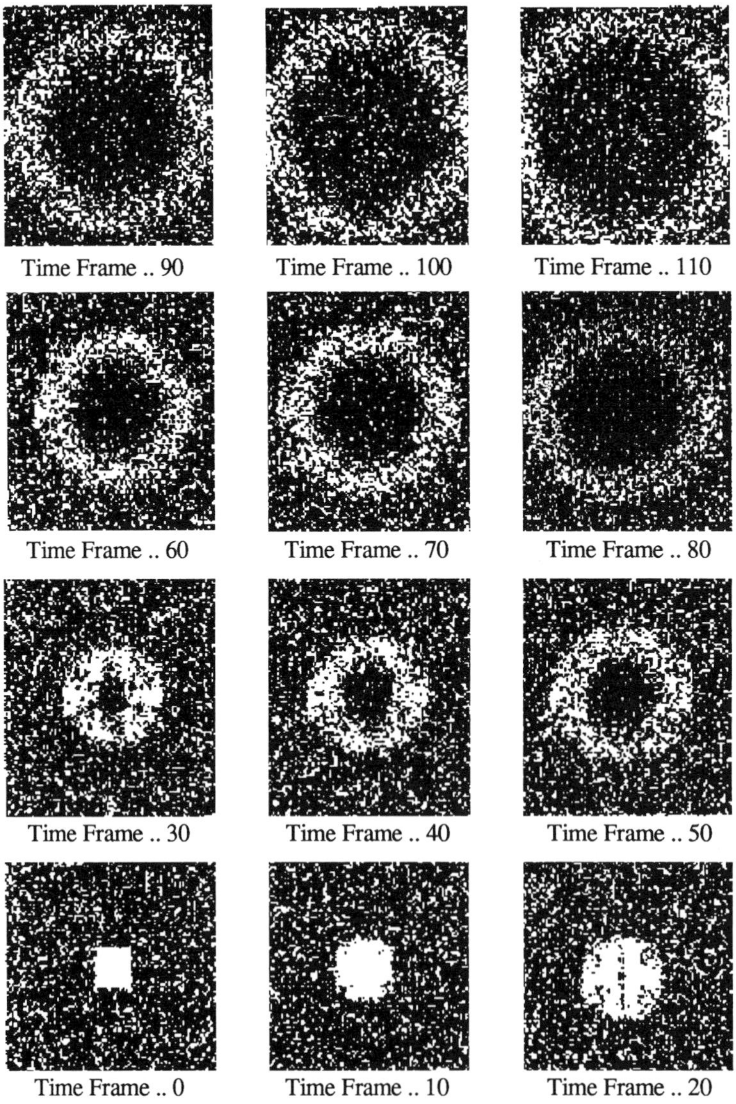

Fig. 5. Density map of sound wave propagation.

We have tried to strike a reasonable compromise between the achievement of high speed and the economy of space.

The program also requires an array of type REAL whose length equals that of the lattice to calculate random numbers in the case of non-chiral collision handling. It also uses eight numeric arrays for summarization and output. The size of these arrays depends on the granularity of the lattice summarization chosen by the user.

The maximum lattice size in the version of the program presented in this paper is limited by the maximum length of an ETA10 vector operation, 65 535. An alternate version of the program allows a longer lattice and uses a technique known as "stripmining" to perform operations on the lattice nodes in groups of 65 535. In either program, the hardware performs the vector operations most efficiently for long vector lengths, so defining the lattice length to be equal to or slightly less than a multiple of 65 535 ensures optimal performance.

5. Methods of displaying the state of the lattice

This program produces text files which represent the state of the lattice at specified time steps. The user may specify a *superlattice*, which partitions the lattice into rectangles. The size of the superlattice determines the granularity of the output data. For example, if the user desires very coarse granularity, he can define a small superlattice so that many lattice nodes fit in one superlattice rectangle. Alternately, if the user desires very fine granularity, he can arrange a large superlattice that allows a single lattice node in each superlattice rectangle. The user-defined constants specify the number of rows and columns of rectangles in the superlattice.

The lattice node numbering scheme was chosen to aid vectorization. Unfortunately, under this scheme, a lattice with an equal number of rows and columns does not have a square shape. Rather, a 4 by 1 ratio of nodes, or four rows per one column, approximates a square shape. For this reason, the lattice nodes

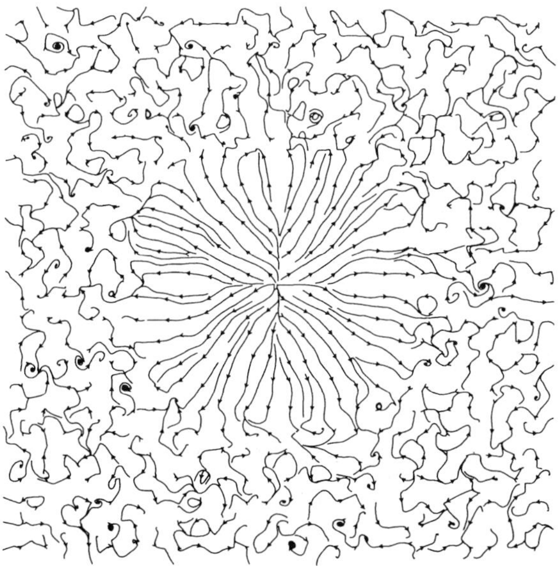
(a)

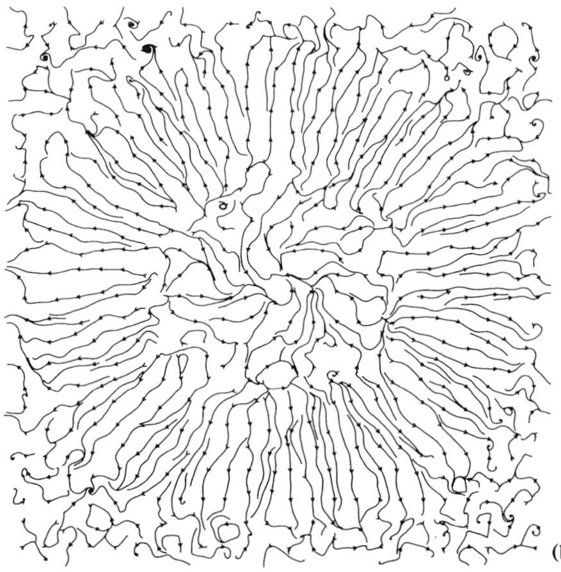
(b)

Fig. 6. Flow diagram of sound wave model. (a) Time step 25, (b) time step 75.

should be defined in a 4 by 1 ratio if the user desires a nearly square lattice space. Similarly, the user should plan a 4 by 1 ratio of lattice nodes in each superlattice rectangle if he desires these rectangles to be shaped roughly like squares.

There are two methods which are used to summarize the state of the superlattice. The first output method produces the density of the superlattice as the number of particles in each rectangle. This option is useful to observe the density distribution of the lattice particles. The second output method produces the average velocity of the particles in each superlattice rectangle. The program represents the velocity vector as a horizontal "x component" and a vertical "y component" (not related to the X and Y bit arrays).

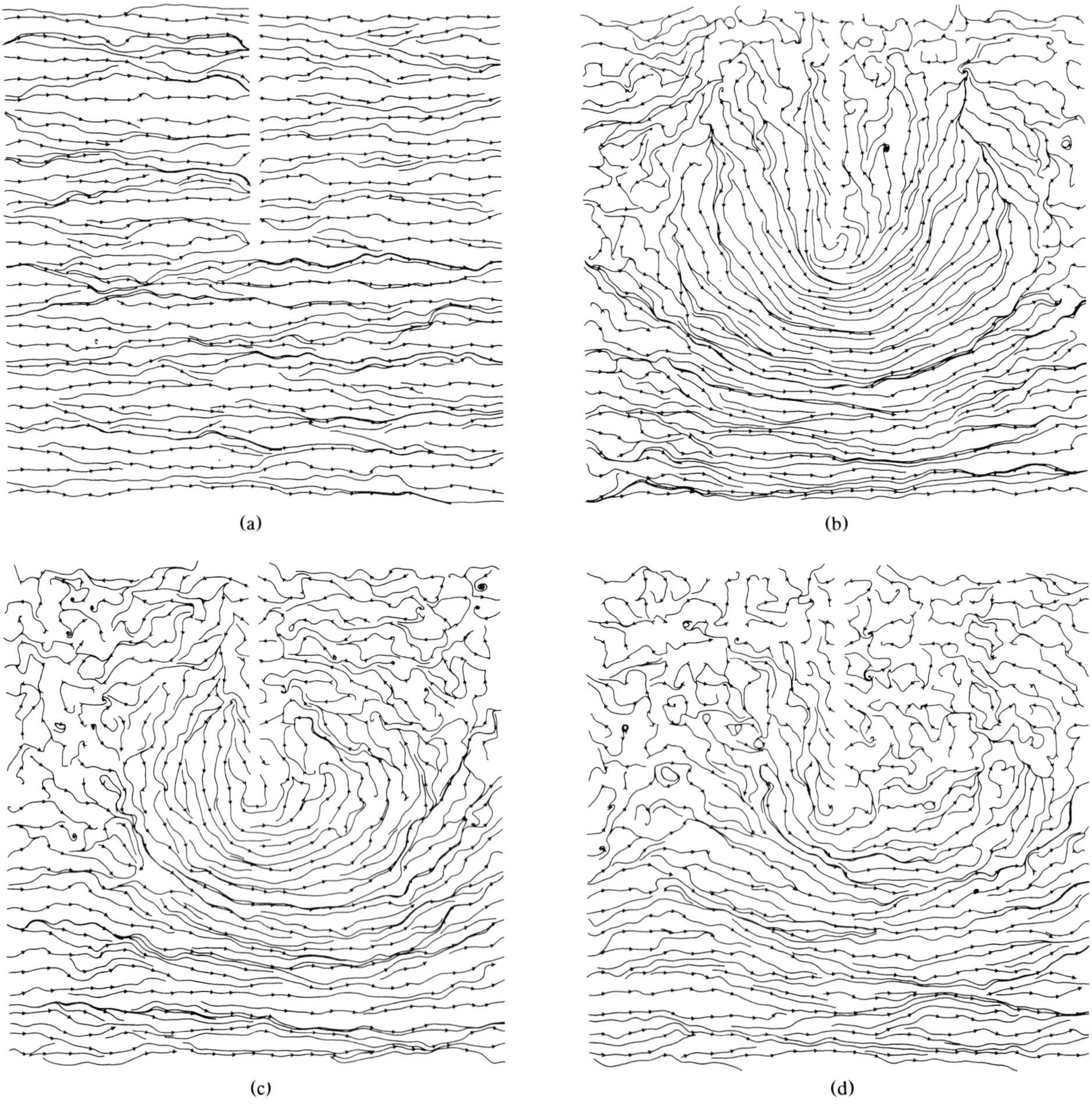

Fig. 7. Flow diagram of the obstructed flow model. (a) Time step 0, (b) time step 100, (c) time step 300, (d) time step 500.

6. Applications of the FHP model

One model implemented with the program is a sound wave propagation model. The simulation has a 412 by 103 square lattice with a concentration of particles in the middle. The central region of 100 by 25 nodes has 100% particle occupancy while the rest of the lattice is populated uniformly to 33%. The model simulates the propagation of a sound wave as the high concentration of particles in the center moves outward in a rough circle towards the boundaries of the lattice, as shown in the density map of fig. 5. The boundary conditions are all specular reflection. The superlattice is 103 by 103 rectangles, so that each rectangle contains density information about a 4 by 1 set of nodes.

Fig. 6 shows the flow diagram associated with two time steps of the sound wave model. The velocity vector information produced by the sound wave model was run through the NCAR Graphics routine STRMLN to produce the flow diagram, which was then printed on a laserprinter. The pictures show that the area of high density is flowing toward the edge of the lattice, while low density areas exhibit random flow patterns. It should be noted that STRMLN displays only the flow direction and not the speed of the flow.

Another model implemented with the program is an obstructed flow through a pipe. The simulation has a 400 by 100 square lattice with an obstacle extending from the middle of the left boundary. An initial flow is imposed such that there is a 50% occupancy of particles in the three northward directions and a 16.66% occupancy of particles in the three southward directions. The left and right boundaries and the obstacle boundaries are all specular reflection. However, the top and bottom boundaries are periodic, allowing the flow to remain northward. The superlattice is 50 by 50 rectangles, so that each rectangle contains velocity information about an 8 by 2 set of nodes.

Fig. 7 shows the flow diagram for four time steps of the obstructed flow model. By time step 500 an eddy has formed at the end of the obstacle, and the main flow is smoothly northward.

Acknowledgements

This research is supported in part by the US Department of Energy under contract number DE-FC05-85ER250000, and by the allocation of computer time by the Florida State University Computing Center.

References

[1] U. Brosa and D. Stauffer, Vectorized multisite coding for hydrodynamic cellular automata, J. Stat. Phys. 57 (1989) 399.
[2] U. Frisch, B. Hasslacher and Y. Pomeau, Phys. Rev. Lett. 56 (1986) 1505.
[3] B. Hasslacher, Discrete fluids, Los Alamos Sci. Special Issue, Vol. 15, ed. N.G. Cooper (1987) p. 175.
[4] F. Hayot, M. Mandal and P. Sadayappan, Implementation and performance of a binary lattice gas algorithm on parallel processor systems, J. Comput. Phys. 80 (1989).
[5] H. Lim, Lattice gas automata of fluid dynamics for unsteady flow, Complex Systems, Vol. 15, ed. N.G. Cooper (1988) 968.
[6] H. Lim, Cellular automaton simulations of simple boundary layer problems, Phys. Rev. A 40 (1989) 968.
[7] H. Lim, G. Riccardi, C. Bauer and S. Sharma, A vector algorithm for lattice gas hydrodynamics, Int. J. Supercomputer Appl. 3.4 (1989) 64.
[8] N. Margolus, T. Toffoli and G. Vichniac, Cellular automata supercomputers for fluid dynamics modeling, Phys. Rev. Lett. 56 (1986) 1694.
[9] S. Wolfram, Cellular automaton fluids 1: Basic theory, J. Stat. Phys. 45 (1986) 471.

APPENDIX

Physica D 47 (1991) 299–337
North-Holland

BIBLIOGRAPHY FOR NATO WORKSHOP ON LATTICE GAS METHODS FOR PDE'S: THEORY, APPLICATIONS AND HARDWARE

This bibliography is a compilation of lattice gas references, including theory and computer experiments. References are given in alphabetical order by first author, with the most recent reference first. An attempt was made to include articles published before April 1990 which refer to the April 1986 Physical Review Letters by Frisch, Hasslacher, and Pomeau. Some preprints and reports appear also. Much of this bibliography is the work of Dominique d'Humières.

Appert, C., D.H. Rothman and S. Zaleski, A liquid-gas model on a lattice, Physica D 47 (1991) 85–96; these Proceedings.

We describe a triangular lattice model able to undergo a liquid–gas transition. The model is obtained by adding an attractive force to the Frisch–Hasslacher–Pomeau gas in the form of non-local interactions. Several types of interactions are suggested and their properties are discussed. When the attractive forces are strong enough the model decomposes into a dense and a light phase. The equation of state of the model is analogous to a van der Waals equation. The theoretical prediction of the equation of state, obtained using a Boltzmann or factorization assumption, agrees well with numerical observations. The isotropy of the model is tested by a numerical computation of the two-dimensional power spectrum.

Appert, C. and S. Zaleski, A lattice gas with a liquid–gas transition, Phys. Rev. Lett. 64 (1990) 1–4.

We discuss a new momentum-conserving lattice gas model in which particles are allowed to exchange momentum between distant sites. The interactions may be tuned so that a first order transition occurs between a dense and a light phase. An equation of state may be predicted with the assumption that the lattice is in a factorized state just after the particles have propagated. This method accurately predicts the pressure of the stable and unstable states.

Balasubramanian, K., F. Hayot and W.F. Saam, Darcy's law from lattice-gas hydrodynamics, Phys. Rev. A 36 (1987) 2248–2253.

Within the hexagonal lattice-gas model, we obtain Darcy's law for flow in the presence of scatterers. The associated momentum dissipation is described by an effective damping term in the Navier–Stokes equation, which we relate to the density of scatterers. The kinematic viscosity can be obtained from the Darcy velocity profile, once the permeability is determined. We also check that in the hexagonal lattice model, after coarse graining, velocity decay and plane-parallel Poiseuille flow occur as described by the macroscopic equations.

Baudet, C., J.P. Hulin, P. Lallemand and D. d'Humieres, Lattice-gas automata: a model for the simulation of dispersion phenomena, Phys. Fluids A 1 (1989) 507–512.

The dispersion of a tracer in two-dimensional (2-D) parallel flow between two parallel plates has been studied numerically using a lattice-gas model with two different species of particles. After a stepwise change of concentration at one end of the model, the tracer distribution in the flow evolves theoretically, as predicted, toward a Gaussian profile in a time lapse in agreement with the predictions of the Taylor model. The mean square width of the front increases linearly with time after the stabilization period, as required for a diffusive spreading process. The longitudinal dispersion coefficient $D_{||}$ has been determined in a range of Peclet number values, Pe, between 4.3 and 35.4. It varies as the square of the velocity, in agreement with the Taylor–Aris model; the molecular diffusion coefficient value ($D_m = 0.62$ in lattice and simulation step units) obtained from the proportionality coefficient is in good agreement with the values obtained by other independent methods.

Benzi, R. and S. Succi, Bifurcations of a lattice gas flow under external forcing, J. Stat. Phys. 56 (1989) 69–81.

We study the behavior of a Frisch–Hasslacher–Pomeau lattice gas automaton under the effect of a spatially periodic forcing. It is shown that the lattice gas dynamics reproduces the steady-state features of the bifurcation pattern predicted by a properly truncated model of the Navier–Stokes equations. In addition, we show that the dynamical evolution of the instabilities driving the bifurcation can be modeled by supplementing the truncated Navier–Stokes equation with a random force chosen on the basis of the automaton noise.

Benzi, R. and S. Succi, Two-dimensional turbulence with the lattice Boltzmann equation, J. Phys. A 23 (1990) L1–L5.

We investigate the ability of the lattice Boltzmann equation to reproduce the basic physics of fully turbulent two-dimensional flows and present a qualitative estimate of its computational efficiency with respect to other conventional techniques.

Bernardin, D., O.E. Sero-Guillaume and C. H. Sun, Multispecies 2D lattice gas with energy levels: Diffusive properties, Physica D 47 (1991) 169–188; these Proceedings.

We consider two particular applications of a multispecies, multispeed lattice gas with energy levels. In the first one, we study the mass diffusion properties of a model where the collisions preserve the partial masses. In the second one, we are looking at heat diffusion for a model where the total mass, momentum and energy are the only conserved quantities. The diffusion equations are derived by the Chapman–Enskog method and some numerical simulations are presented.

Bernardin, D. and O.E. Sero-Guillaume, Lattice gases mixtures models for mass diffusion, Eur.J. Mech. 9 (1990) 21–46.

A general model for mixtures of gases on a lattice is proposed. Its dissipative properties are studied using a Chapman–Enskog expansion of its Boltzmann equation. For non-reactive gases, general expressions for the diffusion coefficients and the viscosity are given at low mean velocity. Two-component mixtures are more completely studied and a detailed example is proposed.

Binder, P.M. and M.H. Ernst, Lattice gas automata with time-dependent collision rules, Physica A 164 (1990) 91–104.

Particle-scatterer models with time-dependent collision rules in two-dimensional square and triangular lattices are studied. Very good agreement is seen between analytical and molecular dynamics results for the diffusion coefficient at all scatterer concentrations. The results differ significantly from the corresponding analysis for stochastic dynamics. In all square-lattice Lorentz models there is a spurious diffusion mode; its transport coefficient has the same value as the usual diffusion coefficient. The long-time tails with a high frequency modulation, $(-1)^T$, observed in computer simulations, are explained in terms of this spurious mode.

Binder, P.M., Abnormal diffusion in a wind-tree lattice gas, Complex Systems 3 (1989) 1–7.

It is found numerically that a two-dimensional left-turning particle-scatterer system does not diffuse for scatterer densities slightly above one-half, or greater. For smaller densities, the diffusion coefficient is much lower than what the Boltzmann approximation predicts; this is caused by orbiting events. An isotropic-scattering model with reflective impurities also shows deviations from the diffusion equation for various densities of isotropic and reflective scatterers, caused by retracing events.

Binder, P.M. and D. Frenkel, Direct measurement of correlation functions in lattice Lorentz gases, AMOLF preprint (1990).

Binder, P.M., Evidence for Lagrangian tails in a lattice gas, in: Cellular Automata and Modeling of Complex Physical Systems, eds. P. Manneville, N. Boccara, G.Y. Vichniac and R. Bidaux (Springer, Berlin, 1989) pp. 155–160.

An efficient method to measure particle velocity correlation functions in a lattice gas is developed. We show significant deviations from the Boltzmann-level result in the HPP gas.

Binder, P.M., Numerical experiments with lattice Lorenz gases, in: Lattice Gas Methods for Partial Differential Equations, ed. G.D. Doolen (Addison–Wesley, Reading, MA, 1989) pp. 471–480.

A new method to simulate point-scatterer systems in a lattice is described. Simulations are performed to illustrate the effect of static correlations in the diffusion coefficient and velocity correlation function, and to study the ergodicity of deterministic models.

Binder, P.M., The properties of tagged lattice fluids I: Diffusion coefficients, in: Discrete Kinetic Theory, Lattice Gas Dynamics and Foundations of Hydrodynamics, ed. R. Monaco (World Scientific, Singapore, 1989) pp. 28–37.

We study the diffusive behavior of particle-scatterer and tagged-particle models in the two-dimensional square lattice. Simulations agree with theoretical results. Several discreteness and high-density effects are observed.

Binder, P.M., D. d'Humières and L. Poujol, The properties of tagged lattice fluids II: Velocity correlation functions, in: Discrete Kinetic Theory, Lattice Gas Dynamics and Foundations of Hydrodynamics, ed. R. Monaco (World Scientific, Singapore, 1989) pp. 38–43.

We report preliminary measurements of the velocity autocorrelation function for a tagged particle in a lattice gas. These measurements agree with the Boltzmann-level theory. The Green–Kubo integration of these measurements agrees with theoretical predictions for the diffusion coefficient. To within the error bars of the simulations (0.003), we observe no long-time tails.

Binder, P.M. and D. d'Humières, Self-diffusion in a tagged-particle lattice gas, Phys. Lett. A 140 (1989) 465–468.

Collisions in lattice gas models do not conserve particle identity. Whenever one needs to follow individual particles, several choices are often consistent with a given model; these choices result in large variations of the diffusion coefficient. We illustrate this numerically and analytically.

Binder, P.M., Lattice models of the Lorentz gas physical and dynamical properties, Complex Systems 1 (1987) 559–574.

This paper examines the validity and usefulness of cellular automaton models of fluid motion by means of a simple problem in kinetic theory. We formulate three lattice models of the motion of a particle in a two-dimensional matrix of fixed, randomly placed, non-overlapping scatterers (the Lorentz gas). We measure several macroscopic and microscopic properties of this system, such as diffusion coefficients and mean-free paths. The results agree with analytical predictions, except at a high density of scatterers, where the models break down. We also study these models as discrete dynamical systems. The properties of their state-transition diagrams, which give the number of all possible trajectories of the particles and their lengths, are similar to those of chaotic and random discrete maps. This agrees with analytical predictions that this gas exhibits chaotic behavior.

Boghosian, B., W. Taylor and D.H. Rothman, A cellular automata simulation of two-phase flow on the CM-2 connection machine computer, in: Proceedings of Supercomputing '88, 2: Science and Applications, eds. J.L. Martin and S.S. Lundstrom (IEEE Computer Society Press, 1989) pp. 34–44.

Boghosian, B.M. and C.D. Levermore, A cellular automaton for Burgers' equation, Complex Systems 1 (1987) 17–30. Reprinted in: Lattice Gas Methods for Partial Differential Equations, ed. G.D. Doolen (Addison–Wesley, Reading, MA, 1989) pp. 481–496.

We study the approximation of solution to the Burgers' equation,

$$\frac{\partial n}{\partial t} + c\frac{\partial}{\partial x}\left(n - \frac{n^2}{2}\right) = \nu\frac{\partial^2 n}{\partial x^2}$$

by spatially averaging a probabilistic cellular automaton motivated by random walks on a line. The automaton consists of moving particles on a one-dimensional periodic lattice with speed one and in a random direction subject to the exclusion principle that at most one particle may move in a given direction from a given lattice site, at a given time. The exclusion principle gives rise to the nonlinearity in Eq. (1) and introduces correlations between the particles which must be estimated to obtain statistical bounds on the error. These bounds are obtained in two steps. The first is showing that the ensemble average of the automaton is a stable explicit finite differencing scheme of Eq. (1) over the lattice with a second-order convergence in the lattice spacing. The numerical diffusion of this scheme plays an important role in relating the automaton rules to Eq. (1). The next step is showing that the spatial averaging of a single evolution of the automaton converges to the spatial averaging of the ensemble as $1/\sqrt{M}$ where M is the number of lattice sites averaged. Simulations are presented and discussed.

Bonetti, M., A. Noullez and J.P. Boon, Lattice gas simulation of 2-D viscous fingering, in: Cellular Automata and Modeling of Complex Physical Systems, eds. P. Manneville, N. Boccara, G.Y. Vichniac and R. Bidaux (Springer, Berlin, 1989) pp. 239–241.

Bonetti, M., A. Noullez and J.P. Boon, Viscous fingering in a 2-D porous lattice, in: Discrete Kinetic Theory, Lattice Gas Dynamics and Foundations of Hydrodynamics, ed. R. Monaco (World Scientific, Singapore, 1989) pp. 395–399.

The lattice gas method has been applied to simulate viscous fingering in a homogeneous porous medium. A low viscosity fluid is pushed through the medium initially filled with a highly viscous fluid. The two fluids are miscible and have viscosity ratio 33. The viscosity difference is set by selecting specific collision rules for each fluid. The porous medium is simulated by a random spatial distribution of static scattering nodes. The displacement of the less viscous fluid is triggered by imposing a constant pressure gradient at the entering zone of the lattice. Cellular automata techniques are used for computation. The instability of the separation zone between the two fluids subjected to a mechanical constraint in a two-dimensional porous medium is analogous to the Saffman-Taylor instability in a Hele-Shaw cell. The simulations show that viscous penetration develops into viscous fingering with a characteristic wavelength of the order of a fraction of the transverse size of the lattice. Simulations were conducted under various physical conditions in order to ascertain the nature of the instability.

Boon, J.P., Statistical mechanics and hydrodynamics of lattice gas automata: An overview, Physica D 47 (1991) 3–8; these Proceedings.

Some of the issues raised by recent work on lattice gas automata are reviewed.

Boon, J.P., Lattice gas automata simulation of complex flows, in: Microscopic Simulations of Complex Flow, ed. M. Mareschal (Plenum Press, New York, 1990) pp. 25–46.

Boon, J.P., Lattice gas simulations: a new approach to fluid dynamics, preprint, Université Libre de Bruxelles (1989).

Boon, J.P. and A. Noullez, Lattice gas diffusion and long time correlations, in: Discrete Kinetic Theory, Lattice Gas Dynamics and Foundations of Hydrodynamics, ed. R. Monaco (World Scientific, Singapore, 1989) pp. 400–408.

Self-diffusion in lattice gases raises the problem of particle identification. Such an identification can be realized by introducing different types of particles and extending the cellular automata collision rules to include type conservation. Any particular set of collision rules induces specific diffusive behavior. Simulations are performed to track a tagged particle using an extended version of the FHP model for colored automata. Mean-square displacements are measured for systems at low and moderate densities. Corresponding velocity autocorrelation functions are computed showing negative recorrelation at moderate density. Careful analysis of the long time behavior of the velocity autocorrelation functions has been conducted to investigate the long time tail effect. We conjecture that, within the limits of presently achievable accuracy, long time tails cannot be detected by direct measurement in lattice gases.

Bowler, K.C. and R.D. Kenway, Physics on parallel computers. Part 2: Applications, Contemp. Phys. 29 (1988) 33–55.

Parallelism is an intrinsic feature of many physical systems. The design of high-performance computers is increasingly making use of the same concept by employing many processors working cooperatively to carry out a single computation. Consequently, the computer simulation of physical systems can rather naturally exploit the latest computer architectures to test theoretical models and to make measurements which are inaccessible to real experiments. In an earlier article we reviewed the design and use of parallel computers. Here we describe how they have been used to obtain insight into some fundamental problems in physics.

Bowler, K.C., A.D. Bruce, R.D. Kenway, G.S. Pawley and D.J. Wallace, Applications of parallel computing in condensed matter, Physica Scripta T19 (1987).

Computational methods permit the detailed study of microscopic properties and their macroscopic consequences in a host of problems in physics which may be inaccessible to direct experimental study and too complex for theoretical analysis. Reliable calculations from first principles, however, require enormous computing resources. In this talk we describe how parallel computing can provide a practical and cost-effective solution to this problem, illustrating the key ideas with examples from Monte Carlo and molecular dynamics simulations, electronic structure calculations and the analysis of experimental data.

Bowler, K.C., A.D. Bruce, R.D. Kenway, G.S. Pawley and D.J. Wallace, Exploiting highly concurrent computers for physics, Physics Today (Oct. 1987) 40–48.

Architectures as varied as rigid arrays of many simple processors and reconfigurable networks of transputers are being used to solve problems as diverse as lattice gauge theories and neural networks.

Brieger, L. and E. Bonomi, A stochastic cellular automaton model of nonlinear diffusion and diffusion with reaction, J. Comput. Phys. (1990), to appear.

This article presents a stochastic cellular automaton model of diffusion and diffusion with reaction. The master equations for the model are examined and we assess the difference between the implementation in which a single particle at a time moves (asynchronous dynamics) and one implementation in which all particles move simultaneously (synchronous dynamics). Biasing locally each particle's random walk, we alter the diffusion coefficients of the system. By choosing appropriately the biasing function, we can impose a desired non-linear diffusive behavior in the model. We present an application of this model, adapted to include two diffusing species, two static species and a chemical reaction in a prototypical simulation of carbonation in concrete.

Brieger, L. and E. Bonomi, A stochastic cellular automaton simulation of the non-linear diffusion equation, Physica D 47 (1991) 132–158; these Proceedings.

In this article we investigate a cellular automaton simulation of the non-linear diffusion equation. The diffusion coefficient characterizing the equation is used to locally bias the random walks of particles on a square lattice. Emphasis is placed on respecting the massively parallel nature of the automaton model, while also correctly simulating the macroscopic behavior described by the equation. The result, a highly parallel algorithm, presents an interesting possibility for parallel and dedicated machines.

Brosa, U., C. Kuettner and U. Werner, Flow through a porous membrane simulated by cellular automata and by finite elements, J. Stat. Phys. (September 1990), to appear.

Computational results concerning incompressible viscous flow through two channels connected by a porous membrane are presented. The example is extraordinary for its four different types of boundary conditions that are necessary to make the problem complete. The solution is accomplished by two methods: by cellular automata and by finite elements. The numerical means to satisfy the boundary conditions are given for both methods. Overall agreement is achieved, but significant differences show up in details.

Brosa, U., Direct simulation of a permeable membrane, J. Phys. (Paris) (1990), in press.

Cellular automata are used to compute flow through a permeable membrane. Scattering centers constitute the membrane. This is in marked contrast to the approach of classical hydrodynamics which represents a membrane by a boundary condition. With the scattering centers we obtain different, but more plausible results indicating that simple diffusion is the dominating process in a porous layer. We have thus a case where cellular automata show superiority over the classical methods of theoretical hydrodynamics.

Brosa, U. and D. Stauffer, Simulation of flow through a 2D random porous medium, preprint (1990).

The cellular automata approximation of two-dimensional hydrodynamics is used to model flow between randomly placed, partially overlapping circles. The Flow resistance is first roughly proportional to the number of circles and then increases stronger for higher numbers of such obstacles.

Brosa, U. and D. Stauffer, Vectorized multisite-coding for hydrodynamic cellular automata, J. Stat. Phys. 57 (1989) 399–403.

Simulating eight lattices for Pomeau's cellular automata simultaneously through bit-per-bit operations, a vectorized Fortran program reached 30 million updates per second and per CRAY YMP processor. We give the full innermost loops.

Burgess, D., F. Hayot and W.F. Saam, Interface fluctuations in a lattice gas, Phys. Rev. A 39 (1989) 4695–4700.

Within the framework of lattice gas hydrodynamics, we study fluctuations of an interface separating two immiscible fluids. The static fluctuations show the random walk behavior typical of systems with energies proportional to their lengths, whereas the time behavior of the interface can be described by a linear Langevin equation with noise. We discuss the thermodynamics of the interface fluctuating in the heat bath provided by the two fluids, measure and calculate the macroscopic and microscopic coefficients of surface tension, and check numerically the relevant fluctuation–dissipation relation.

Burgess, D., F. Hayot and W.F. Saam, Model for surface tension in lattice-gas hydrodynamics, Phys. Rev. A 38 (1988) 3589–3592.

We introduce in lattice-gas hydrodynamics an interface between two fluids. The interface deforms locally under the impact of fluid particles. We define and measure a surface tension from Laplace's law applied to a circular bubble. Surface tension can be varied by changing the stiffness of the interface. A Boltzmann-type calculation gives an expression of surface tension in agreement with the numerical results.

Burgess, D. and F. Hayot, Saffman–Taylor-type instability in a lattice gas, Phys. Rev. A 40 (1989) 5187–5192.

We show how a Saffman–Taylor-type fingering instability emerges in lattice-gas hydrodynamics. We discuss linear stability and the effect of noise which is intrinsic to the algorithm. Noise leads to a time-splitting instability of the evolving finger.

Burgess, C. and S. Zaleski, Buoyant mixtures of cellular automata gases, Complex Systems 1 (1987) 31–50.

The use of lattice gas (cellular automaton) models has recently been advocated as an interesting method for the simulation of fluid flow. These automata are an idealization of the real microscopic molecular dynamics. We present a model derived from the hexagonal lattice gas rules of Frisch, Hasslacher and Pomeau (FHP) that incorporates buoyant forces. We discuss its properties and derive the hydrodynamical equations in the low density limit and find the buoyant force and seepage effects characteristic of gravitating mixtures, as well as deviations from the Navier–Stokes equations in the incompressible case. An equivalent of the quasi-incompressible limit of Boussinesq exists, where the Boussinesq equations are recovered but only for steady flow. The unsteady flow equations suffer from the lack of Galilean invariance of FHP-type models. We discuss other tentative models that would overcome this difficulty. The self-diffusion coefficient is also computed from the theory, as well as the mean free path. This allows one to check some of the predictions of the Chapman-Enskog expansion for these gases. We also perform numerical simulations at a Rayleigh number of 6000, showing natural convection near a heated wall and the Rayleigh-Benard instability in a time independent regime.

Cabannes, H., On the initial-value problem in discrete kinetic theory, Eur. J. Mech. B/Fluids (1990), to be published.

We prove that for some discrete models of the Boltzmann equation, the initial value problem possesses a global solution in time, even for partially negative initial data.

Cabannes, H. and J.-P. Duruisseau-Aloyd, Construction, using MACSYMA, of exact solutions for some discrete models of the Boltzmann equation, in: International Symposium on Advanced Computers for Dynamics and Design A28 (1989) 161–166.

The purpose of this paper is to show how the use of MACSYMA, a software package of formal calculus, allows one to build exact solutions for some partial differential equation systems which appear in the kinetic theory of gases. Those equations constitute a system of semi-linear hyperbolic equations: linear with respect to the derivatives of the unknown functions and quadratic with respect to those functions. The main idea is to satisfy the equations using functions which are real parts of complex tangents. Working out the quadratic terms and the derivatives one obtains, for the parameters introduced in the initial functions, a system of algebraic equations which, in some cases, possess solutions.

Cabannes, H. and D.H. Tiem, Exact solutions for some discrete models of the Boltzmann equation, Complex Systems 1 (1987) 575–584.

For the simplest of the discrete models of the Boltzmann equations, the Broadwell model, exact solutions have been obtained by Cornille in the form of bisolitons. In the present paper, we build exact solutions for more complex models.

Cercignani, C., Discrete models in kinetic theory, in: Discrete Kinetic Theory, Lattice Gas Dynamics and Foundations of Hydrodynamics, ed. R. Monaco (World Scientific, Singapore, 1989) pp. 62–73.

In this presentation, I shall try to cover the known mathematical results concerning existence and uniqueness in discrete kinetic theory.

Chang, C.-R. and X.Y. Zhang, The dispersion patterns of acicular particles in viscous medium, IEEE Trans. Magn. 23 (1987) 2886–2888.

A lattice model of ferrofluid with time-dependent viscosity has been developed to simulate the dispersion patterns of interacting fine ferromagnetic particles during the manufacturing process of magnetic recording medium. Closure and chaining of the magnetic particles in uniform magnetic field have been obtained. The length of the chain is dependent on the packing factor. For magnetic particles possessing electric charge, the agglomeration of particles will reduce. The hysteresis corresponding to different patterns are calculated using quasi-dynamic vectorial model.

Chen, H., S. Chen, G.D. Doolen and W.H. Matthaeus, A brief description of lattice gas models for multiphase flows and magnetohydrodynamics, to be published by Santa Fe Institute (1990).

Lattice gas models for single phase fluids, multiphase fluids and for magnetohydrodynamic fluids are briefly described.

Chen, H., S. Chen, G.D. Doolen, Y.C. Lee and H.A. Rose, Multithermodynamic phase lattice-gas automata incorporating interparticle potentials, Phys. Rev. A 40 (1989) 2850–2853.

An extension of the lattice-gas automaton model for fluid equations is presented which has nonideal-gas thermodynamic properties reminiscent of first-order phase transitions. The new idea in this model is an algorithm for including interparticle potentials.

Chen, H., S. Chen and G.D. Doolen, Sound wave propagation in FHP lattice gas automata, Phys. Lett. A 140 (1989) 161–165.

Two properties of sound wave propagation are studied for the simplest lattice gas model for modeling the Navier–Stokes equation. It is shown that the propagation is anisotropic for finite flow speed. It is also shown that the density fluctuation is advected at an unphysical speed. Both of these effects are improved considerably by introducing more speeds in the model.

Chen, H., S. Chen, G.D. Doolen and Y.C. Lee, Simple lattice gas models for waves, Complex Systems 2 (1988) 259–267. Reprinted in: Lattice Gas Methods for Partial Differential Equations, ed. G.D. Doolen (Addison–Wesley, Reading, MA, 1989) pp. 497–508.

A simple lattice gas model for solving the linear wave equation is presented. In this model a photon representation is used. Energy and momentum are shown to be conserved.

Chen, H., W.H. Matthaeus and L.W. Klein, An analytic theory and formulation of a local magnetohydrodynamic lattice gas model, Phys. Fluids 31 (1988) 1439–1445.

A theoretical description of the newly developed magnetohydrodynamic lattice gas model [Phys. Rev. Lett. 58 (1987) 1845] is presented. The model is a direct extension of the lattice gas model for incompressible Navier–Stokes fluids [Phys. Rev. Lett. 56 (1986) 1505]. In the present model the magnetic force and the magnetic induction effects are formulated with local microscopic dynamical rules only, using a bidirectional random walk process. The development of the theory strongly emphasizes the symmetries connecting the microscopic and macroscopic physics. A preliminary numerical test is described.

Chen, H. and W.H. Matthaeus, Cellular automaton formulation of passive scalar dynamics, Phys. Fluids 30 (1987) 1235–1237.

Cellular automata modeling of the advection of a passive scalar in a two-dimensional flow is examined in the context of discrete lattice kinetic theory. It is shown that if the passive scalar is represented by tagging or coloring automaton particles, a passive advection–diffusion equation emerges without use of perturbation expansions. For the specific case of the hydrodynamic lattice gas model of Frisch, Hasslacher and Pomeau [Phys. Rev. Lett. 56 (1986) 1505] the diffusion coefficient is calculated by perturbation.

Chen, H. and W.H. Matthaeus, New cellular automaton model for magnetohydrodynamics, Phys. Rev. Lett. 58 (1987) 1845–1848. Reprinted in: Lattice Gas Methods for Partial Differential Equations, ed. G.D. Doolen (Addison–Wesley, Reading, MA, 1989) pp. 543–460.

A new type of two-dimensional cellular automaton method is introduced for computation of magnetohydrodynamic systems. Particle population is described by a 36-component tensor referred to a hexagonal lattice. By appropriate choice of the coefficients that control the modified streaming algorithm and the definition of the macroscopic fields, it is possible to compute both Lorentz-force and magnetic-induction effects. The method is local in the microscopic space and therefore suited to massively parallel computers.

Chen, S., H. Chen, G.D. Doolen, Y.C. Lee and H. Rose, Lattice gas models for nonideal gas fluids, Physica D 47 (1991) 97–111; these Proceedings.

This paper presents a lattice gas model with a nonideal gas equation of state. Solid–gas transitions are modeled. Applications of the model to shock waves are presented. The generalization of this model to liquid crystal flow is also discussed.

Chen, S., K. Diemer, G.D. Doolen, K.G. Eggert and B.J. Travis, Lattice gas automata for flow through porous media, Physica D 47 (1991) 72–84; these Proceedings.

Lattice gas hydrodynamic models for flows through porous media in two and three dimensions are described. The computational method easily handles arbitrary boundaries and a large range of Reynolds numbers. Darcy's law is confirmed for Poiseuille flow and for complicated boundary flows. Multiply connected pore structures similar to actual sandstone with fixed fractal dimension and porosity are generated. Permeability is calculated as a function of fractal dimension and porosity and compared with result of other methods and experiments.

Chen, S., M. Lee, K.H. Zhao and G.D. Doolen, A lattice gas model with temperature, Physica D 37 (1989) 42–59.

A temperature-dependent lattice gas model is studied. The thermohydrodynamic equations are derived using the Chapman–Enskog expansion method. The model is applied to Poiseuille flow and to heat conduction.

Chen, S., Z.-S. She, L.C. Harrison and G.D. Doolen, Optimal initial condition for lattice-gas hydrodynamics, Phys. Rev. A 39 (1989) 2725–2727.

A method for minimizing the unphysical oscillations in simple lattice-gas hydrodynamic models is presented. Numerical simulations of two types of shear flows are reported that illustrate the usefulness of this method.

Chen, S., H. Chen and G.D. Doolen, How the lattice gas model for the Navier–Stokes equation improves when a new speed is added, Complex Systems 3 (1989) 243–251.

The original lattice gas automaton model requires a density-dependent rescaling of time, viscosity and pressure in order to obtain the Navier–Stokes equation. Also, the corresponding equation-of-state contains an unphysical velocity dependence. We show that an extension of this model which includes six additional particles with a new speed overcomes both problems to a large extent. The new model considerably extends the range of allowed Reynolds number.

Chen, Sh., Y. Chen and G.D. Doolen, Lattice gas simulation of viscous flow in a cavity, Scientia Sinica (1989), in press.

Two-dimensional rectangular cavity flows are studied using a 7-bit lattice gas automaton model. The uniform velocity boundary and wall-driven boundary are applied to the upper boundary or both the upper and lower boundaries to simulate different cavity flows. The comparison of the velocity distribution of this calculation with experimental data and other numerical results is encouraging.

Cheng, Z., J.L. Lebowitz and E.R. Speer, Microscopic shock structure in model particle systems: the Boghosian–Levermore cellular automaton revisited, preprint (1990).

We carried out new computer simulations of the Boghosian–Levermore stochastic cellular automaton for the Burgers equation. The existence of an extra conservation law in the dynamics – even and odd lattice sites exchange their contents at every time step – implies that the automaton decomposes into two independent subsystems; the simulations show that the density from each subsystem exhibits a shock front which does not broaden with time. The location of the shock in a particular microscopic realization differs from that predicted by the Burgers equation by an amount which depends only on the initial microscopic density of the particle system, that is, fluctuations in the stochastic dynamics do not affect the shock profile on the time scale considered. This is in complete accord with theoretical expectations. The apparent broadening of the shock in the original Boghosian–Levermore simulations is shown to result from averaging the two subsystem densities.

Chopard, B. and M. Droz, Cellular automata approach to diffusion problems, in: Cellular Automata and Modeling of Complex Physical Systems, eds. P. Manneville, N. Boccara, G.Y. Vichniac and R. Bidaux (Springer, Berlin, 1989) pp. 130–143.

Diffusion is a general phenomenon playing an important role in many physical or chemical problems. As a result, one often observes the formation of diffusion fronts having complex properties. An analytical analysis of these problems is often not possible and thus, one uses numerical simulation. However, actual diffusion of particles is a time-consuming process. Accordingly deterministic cellular automata algorithms can be very useful. In this paper, we review several situations in which diffusion plays an important role. We then define a class of cellular automaton algorithms for diffusion. We do the explicit derivation of the macroscopic diffusion equations from the microscopic rules and discuss the possible difficulties. The algorithms are then tested in the framework of nonequilibrium diffusion and gradient percolation. It is shown that this cellular automaton approach reproduces very well the expected results both for the solutions of the diffusion equation and for the detailed properties of the diffusion front (fractal dimension). Finally, a novel class of algorithms, avoiding some difficulties previously met, is defined and the corresponding macroscopic properties are given.

Chopard, B. and M. Droz, Cellular automata model for the diffusion equation, submitted to J. Stat. Phys.

We consider a Cellular Automata rule for a synchronous random walk on a two-dimensional square lattice, subject to an exclusion principle. It is found that the macroscopic behavior of our model obeys the Telegrapher's equation with an adjustable diffusion constant. By construction, the dynamics of our model is exactly described by a linear discrete Boltzmann equation which is solved analytically for some boundary conditions. Consequently, the connection between the microscopic and the macroscopic descriptions is obtained exactly and the continuous limit studied rigorously. The typical system size for which a true diffusive behavior is observed may be deduced. It is shown that a suitable choice of the parameters of the rule allows us to consider quite small systems. In particular, our cellular automata model can simulate the Laplace equation to a precision of the order $(\lambda/L)^6$, where L is the size of the system and λ the lattice spacing.

Chopard, B., Strings: a cellular automata model of moving objects, in: Cellular Automata and Modeling of Complex Physical Systems, eds. P. Manneville, N. Boccara, G.Y. Vichniac and R. Bidaux (Springer, Berlin, 1989) pp. 246–256.

We propose a reversible and local rule for modeling large-scale moving objects called strings. These are one-dimensional objects that can be thought of as a discrete version of a chain of masses and springs. Longitudinal and transverse motions can be defined, making the string move in a three-dimensional space with adjustable momentum and energy. Basic mechanisms for collisions between strings are discussed.

Chopard, B. and M. Droz, Cellular automata approach to non-equilibrium diffusion and gradient percolation, J. Phys. A 22 (1989) 1609–1619.

We propose a deterministic approach to lattice diffusion in two dimensions. This method is implemented on a cellular automaton special purpose computer in order to study the properties of the interface of particles diffusing from a source to a sink. Fractal properties of the diffusion front of this non-equilibrium process are compared with results from percolation theory. The observed agreement indicates that diffusion fronts and gradient percolation coincide asymptotically and that the cellular automata method is a viable alternative to standard simulations for this class of problems.

Chopard, B. and M. Droz, Hierarchical equation for the correlation functions in a cellular automata model of nonequilibrium fluids, in: Discrete Kinetic Theory, Lattice Gas Dynamics and Foundations of Hydrodynamics, ed. R. Monaco (World Scientific, Singapore, 1989) pp. 74–82.

Hierarchical equations are derived for the correlation functions in a cellular automata model of a nonequilibrium fluid. It is shown that the equations obtained have a similar structure to the ones used in the framework of the fluctuating hydrodynamics. This explains the qualitative agreement between the two approaches obtained in a previous work for a fluid submitted to a temperature gradient.

Chopard, B. and M. Droz, Cellular automaton model for heat conduction in a fluid, Phys. Lett. A 126 (1988) 476–480.

A cellular automata model for a two-speed lattice gas is introduced to study heat propagation in a fluid at rest. Boltzmann and thermo-hydrodynamical equations are derived analytically. Explicit forms for the equation of state, the thermal conductivity and the Dufour coefficient are given. A good agreement is found between the theoretical predictions and numerical cellular automata simulations.

Chopard, B. and M. Droz, Cellular automata model for thermo-hydrodynamics, in: Chaos and Complexity, eds. R. Livi, S. Ruffo, S. Ciliberto and M. Buiatti (World Scientific, Singapore, 1988) pp. 302–306.

Chopard, B. and M. Droz, Cellular automata approach to nonequilibrium correlation functions in a fluid, Helv. Phys. Acta 61 (1988) 893–896.

Clavin, P., P. Lallemand, Y. Pomeau and G. Searby, Simulation of free boundaries in flow systems by lattice-gas models, J. Fluid Mech. 188 (1988) 437–464.

It has been recently proved that lattice-gas models with Boolean particles can provide a very powerful method to study viscous flows at moderate Reynolds and small Mach numbers (d'Humières, Pomeau & Lallemand 1985; Frisch, Hasslacher & Pomeau 1986; d'Humières & Lallemand 1986). We present here algorithms for an extension of these models to provide a simple and efficient way to simulate a large variety of flow problems with free boundaries. This is done by introducing two different types of particles that can react following a specific kinetic scheme based on autocatalytic reactions. In order to check the powerful character and the reliability of the method we also present preliminary results of two-dimensional computer simulations concerning problems ranging from the competition between molecular diffusion and turbulent mixing in flows presenting a Kelvin–Helmholtz instability to the spontaneous generation of turbulence in premixed flame fronts subject to the Darrieus–Landau instability. The dynamics of an interface developing a Rayleigh–Taylor instability is also considered as well as some typical problems of phase transition such as spinodal decomposition and the nucleation process.

Clavin, P., D. d'Humières, P. Lallemand and Y. Pomeau, Cellular automata for hydrodynamics with free boundaries in two and three dimensions, Compt. Rend. Acad. Sci. Paris II 303 (1986) 1169. Reprinted in translation in: Lattice Gas Methods for Partial Differential Equations, ed. G.D. Doolen (Addison–Wesley, Reading, MA, 1989) pp. 415–422.

Cellular automata are used to simulate two-dimensional hydrodynamic flows with free boundaries as found in the Rayleigh–Taylor instability. We propose an extension of these rules for three-dimensional flows with free boundaries.

Cliffe, K.A., R.D. Kingdon, P. Schofield and P.J. Stopford, Lattice gas simulation of free-boundary flows, Physica D 47 (1991) 275–280; these Proceedings.

Within a very short time, many spectacular results have been produced to demonstrate the potential of lattice gas hydrodynamics (LGH) to predict the behaviour of systems governed by the Navier–Stokes and related equations at low to moderate Reynolds numbers. However, in order for LGH to be accepted as an alternative to conventional computational fluid dynamics (CFD) methods for the modelling of flows of practical importance, it is necessary to demonstrate in some specific cases that LGH has clear advantages over CFD, either where problems remain intractable to the latter or where LGH could give savings in computational resources. One area where CFD techniques have considerable difficulties is that of free-boundary problems, in which unstable fronts separate fluid species in the flow. The inclusion of chemical reactions (the fluid species reacting at the interfaces) renders the problem even less tractable to CFD, while presenting no difficulties in principle to LGH. Clavin, Lallemand, Pomeau and Searby have simulated the Kelvin–Helmholtz and Rayleigh–Taylor free-boundary instabilities as well as a simple reaction front showing the Darrieus–Landau instability. Using their techniques, we have reproduced the Kelvin–Helmholtz and Rayleigh–Talor instabilities. One of the aims of our work is to give a practical demonstration of LGH to model the flow and reactions of vehicle exhaust emissions at street level in an urban environment.

Clouqueur, A. and D. d'Humières, R.A.P., a family of cellular automaton machines for fluid dynamics, in: Proceedings of the 12th Gwatt Workshop on Complex Systems, October 13–15, 1988, Gwatt, Switzerland.

Clouqueur, A. and D. d'Humières, RAP1, a cellular automaton machine for fluid dynamics, Complex Systems 1 (1987) 585–597. Reprinted in: Lattice Gas Methods for Partial Differential Equations, ed. G.D. Doolen (Addison–Wesley, Reading, MA, 1989) pp. 251–266.

RAP1 is a special purpose computer built to study lattice gas models. It allows the simulation of any model using less than 16 bits per node, and interactions restricted to first and second nearest neighbors on a 256×512 square lattice. The time evolution of the automaton is displayed in real time on a color monitor at a speed of 50 frames per second.

Cohen, E.G. D. and X.P. Kong, Lattice gas cellular automata, in: Lectures on Thermodynamics and Statistical Mechanics, eds. A.E. Gonzalez et al. (World Scientific, Singapore, 1989) p. 1.

A brief survey is given of the use of lattice gas cellular automata for the study of fluid flow and diffusion processes.

Colvin, M.E., A.J.C. Ladd and B.J. Alder, Maximally discretized molecular dynamics, Phys. Rev. Lett. 61 (1988) 381–384.

It is shown that the coarsest discretization of positions and velocities in molecular dynamics leads to qualitatively correct transport coefficients and quantitatively predictable, long time tails in the velocity autocorrelation function, but requires orders of magnitude less computer time than standard molecular dynamics methods.

Cornubert, R., D. d'Humieres and D. Levermore, A Knudsen layer theory for lattice gases, Physica D 47 (1991) 241–259; these Proceedings.

A Knudsen layer theory is presented for lattice gases with arbitrary boundary conditions. Analytical results are obtained for two special orientations; they exhibit anisotropic Knudsen layers provided suitable conditions are satisfied. However, the standard boundary conditions used in previous simulations are shown to be isotropic, the bulk steady state extending everywhere in the gas. This theory allows a more accurate localization of the obstacle with respect to the lattice nodes. These results are in good agreement with the numerical simulations.

Dab, D. and J.P. Boon, Cellular automata approach to reaction–diffusion systems, in: Cellular Automata and Modeling of Complex Physical Systems, eds. P. Manneville, N. Boccara, G.Y. Vichniac and R. Bidaux (Springer, Berlin, 1989) pp. 257–273.

We propose a cellular automaton model for a simple reaction–diffusion system: a three-molecular autocatalytic scheme known as the Schlögl model. Cellular automata simulations show qualitative behavior in agreement with Schlögl's phenomenological description: bistability and front propagation. Quantitatively, significant discrepancies are observed between the phenomenological predictions and the results of the simulations, which can be interpreted in terms of reactive recorrelations. These discrepancies are directly related to the ratio of reactive collisions versus elastic collisions and can be eliminated by efficient stirring. Correspondingly we observe mesoscopic effects like the formation of unsteady domains which cannot be predicted by the phenomenological theory.

Dahlburg, J.P., D. Montgomery and G.D. Doolen, Noise and compressibility in lattice-gas fluids, Phys. Rev. A 36 (1987) 2471–2474.

Computations are reported in which the hexagonal lattice gas is used to simulate two-dimensional Navier–Stokes shear flows. Limitations associated with noise in the initial loading and compressible effects associated with a velocity-dependent equation of state arise and interact with each other. A relatively narrow window in density and flow speed exhibits physical behavior.

DeMasi, A., R. Esposito, J.L. Lebowitz and E. Presutti, Hydrodynamics of stochastic cellular automata, Commun. Math. Phys. 125 (1989) 127–145.

We investigate a stochastic version of cellular automata used for simulating hydrodynamical flows, e.g. the HPP and FHP models. The extra stochasticity consists of random exchanges between neighboring cells which conserve momentum. We prove that, in suitable limits, these models satisfy the appropriate continuous Boltzmann and hydrodynamic equations, the same as those conjectured for the original models (except that there is no negative viscosity contribution). The results are obtained by proving a very strong form of propagation of chaos and by using Hilbert–Chapman–Enskog type expansions. Explicit proofs are presented for the stochastic HPP model.

DeMasi, A., R. Esposito, J.L. Lebowitz and E. Presutti, Rigorous results on some stochastic cellular automata, in: Discrete Kinetic Theory, Lattice Gas Dynamics and Foundations of Hydrodynamics, ed. R. Monaco (World Scientific, Singapore, 1989) pp. 93–101.

Computer simulations on cellular automata show that they reproduce patterns observed in real physical fluids evolving according to the solutions of the hydrodynamic equations. This is remarkable, even astonishing, since at a microscopic-particle level the dynamics seems at best a caricature of the interactions between real molecules. It shows that the macroscopic behavior of a fluid does not depend on the detailed features of the particle interactions: systems which microscopically look completely different may give rise to the same type of macroscopic equations. Scale separation is responsible for the existence of this kind of *universality phenomena.* The purpose of this paper is to explain, by a rigorous analysis, the origin of such collective behavior in a class of stochastic cellular automata; a stochastic variant of the HPP and FHP models. We shall see that these models have a very interesting and complex macroscopic structure. There are, in fact, several space–time regimes, and when the state of the microscopic system is correspondingly suitably prepared at the initial time, then, in that specific regime, it behaves according to some macroscopic equation. There is a window in space–time through which we see a kinetic behavior, described by a Boltzmann-like equation. Looking through a different lens, focused on a longer time scale, we see hydrodynamical behavior described by an Euler-like equation. Focusing on still longer times and suitably choosing the initial state we can observe the analogue of the incompressible Navier–Stokes and Euler equations.

Despain, A., C.E. Max, G.D. Doolen and B. Hasslacher, Prospects for a lattice-gas computer, in: Lattice Gas Methods for Partial Differential Equations, ed. G.D. Doolen (Addison–Wesley, Reading, MA, 1989) pp. 211–218.

A two-day workshop was held in June of 1988 to discuss the feasibility of designing and building a large computer dedicated to lattice-gas cellular automata. The primary emphasis was on applications for modeling Navier–Stokes hydrodynamics. The meeting had two goals: (1) to identify those theoretical issues which would have to be addressed before the hardware implementation of a lattice-gas machine would be possible; and (2) to begin to evaluate alternative architectures for a dedicated lattice-gas computer. This brief paper contains a summary of the main issues and conclusions discussed at the workshop.

d'Humières, D., Y.H. Qian and P. Lallemand, Finding the linear invariants of lattice gases, in: Proceedings of the Workshop on Computational Physics and Cellular Automata, Ouro Preto, Brazil, August 8–11, 1989 (1990).

Hydrodynamical phenomena can be simulated by discrete lattice gas models obeying cellular automata rules with suitable restrictions on the crystallographic symmetries of the underlying lattice. However, the derivation of the dynamical equations assumes that mass, momentum and eventually energy are the only conserved quantities during the time evolution of the automata. We shall present here an exact method to find all the conserved linear quantities of any given model, method which is equivalent to f the null space of a matrix derived from the time evolution of the automaton. Thus, the algorithmic complexity is of order N^3, where N is the number of nodes of the lattice. We shall also give a faster method when the model is invariant under the translation group of the lattice. These results will be applied to several one- and two-dimensional models. Finally, we shall note that this method gives the linear invariants of any cellular automaton.

d'Humières, D., Y.H. Qian and P. Lallemand, Invariants in lattice gas models, in: Discrete Kinetic Theory, Lattice Gas Dynamics and Foundations of Hydrodynamics, ed. R. Monaco (World Scientific, Singapore, 1989) pp. 102–113.

Hydrodynamical phenomena can be simulated by discrete lattice gas models obeying cellular automata rules with suitable restrictions on the crystallographic symmetries of the underlying lattice. After proper limits are taken, various standard fluid dynamical equations can be derived from the underlying exact microdynamical Boolean equations, including the incompressible Navier–Stokes equations in two and three dimensions. However, this derivation assumes that mass, momentum and energy are the only conserved quantities during the time evolution of the automata. Using recent results, we shall show how difficult it is to check that there are no extra or spurious conserved quantities. We shall attempt to formalize this problem more carefully and use the concept of geometrical invariants to list some generic class of them.

d'Humières, D., P. Lallemand and Y.H. Qian, Modèles monodimensionels de gaz sur réseaux, divergence de la viscosité, Compt. Rend. Acad. Sci. Paris II 308 (1989) 585–590.

We use the renormalization group method to analyze the divergence of viscosity which appears in one-dimensional lattice-gas models. Numerical simulations do confirm the theoretical prediction for the divergence.

d'Humières, D., P. Lallemand, J.P. Boon, A. Noullez and D. Dab, Fluid dynamics with lattice gases, in: Chaos and Complexity, eds. R. Livi, S. Ruffo, S. Ciliberto and M. Buiatti (World Scientific, Singapore, 1988) pp. 278–301.

In this paper we present some results concerning the use of a particularly simple method to simulate fluid flows. As initially proposed by Frisch, Hasslacher and Pomeau, the fluid is modeled as a 2-D triangular lattice gas. We first describe the initial model and give basic results concerning its thermodynamic and transport properties. We then extend the model to treat diffusion phenomena. Using various collision rules, we obtain fairly large variations in the effective diffusion coefficient for a binary mixture. We also present a theoretical analysis for lattice gas diffusion.

d'Humières, D., P. Lallemand and Y.H. Qian, Review of flow simulations using lattice gases, in: Proceedings of the International Seminar on Hyperbolic Problems, Bordeaux, France, June 13-17, 1988, Lecture Notes (Springer, Berlin), in press.

Lattice gases are first defined as a new way to perform molecular dynamics calculations in a simplified manner but at large speeds thus allowing to consider enough particles to simulate real flows. Some results of the statistical analysis of lattice gases are summarized showing that their macroscopic behavior follows closely that of real fluid flows at low Mach numbers. An example of a two-dimensional flow simulation using a simple lattice gas model is given. Finally new results are presented concerning one-dimensional shocks studied by numerical simulations for two-dimensional lattice gases, showing good agreement with theoretical analysis.

d'Humières, D., P. Lallemand and G. Searby, Dynamics of two-dimensional bubbles by the lattice gas method, in: Physicochemical Hydrodynamics Interfacial Phenomena, NATO ASI Series B174, ed. M.G. Velarde (Plenum, New York, 1988) pp. 71–86.

Lattice gases, introduced recently as a way to calculate viscous flows at moderate Mach and Reynolds number, are currently the object of intense research. Among the variants of the initial proposal of Frisch, Hasslacher and Pomeau, the case of lattice gas mixtures has proven to provide an efficient means of studying hydrodynamical flows with free boundaries. In this paper we first give some basic informations on lattice gases, then we discuss in some detail a particular lattice gas that may describe fluid mixtures. We present the results of computer experiments designed to study the basic properties of this lattice gas mixture. Finally, we show the behavior of two-dimensional bubbles as an illustration of the capabilities of lattice gases in the context of hydrodynamical flows with free boundaries. Questions raised by the current work will be addressed at the end.

d'Humières, D., A. Clouqueur and P. Lallemand, Lattice gas and parallel processors, Calcolo 25 (1987) 129–151.

This paper introduces the notion of lattice gas as a new medium to perform fluid dynamics simulations. After giving the definition of lattice gases, the results of statistical analysis of their macroscopic behavior are reviewed. It is shown that Navier–Stokes equations can be simulated. Some information is given concerning the algorithms and their possible adaptation to parallel hardware, including cellular automata. Finally the construction of a specialized machine to do lattice gas simulations will be presented.

d'Humières, D. and P. Lallemand, Hydrodynamical simulations with lattice gas, in: Supercomputing, eds. A. Lichnewsky and C. Saguez (North-Holland, Amsterdam, 1987) pp. 363–380.

Following a few comments concerning the gap between the microscopic description of fluids and their macroscopic behavior in flows, lattice gases are defined. A simple model for a two dimensional hexagonal lattice is described in detail. Results of a statistical mechanical analysis are given to show that a lattice gas follows Navier–Stokes equations at low Mach number. After a brief description of the algorithms used to simulate the dynamics of lattice gases, examples of flows are presented:
- formation of a Karman street behind a cylinder,
- flow in a channel with a step,
- Kelvin–Helmholtz instability,
- Rayleigh–Taylor instability.

Concluding remarks relate this work to cellular automata and highly parallel specialized hardware.

d'Humières, D. and P. Lallemand, Numerical solutions of hydrodynamics with lattice gas automata in two dimensions, Complex Systems 1 (1987) 599–632. Reprinted in: Lattice Gas Methods for Partial Differential Equations, ed. G.D. Doolen (Addison–Wesley, Reading, MA, 1989) pp. 297–332.

We present results of numerical simulations of the Frisch, Hasslacher, and Pomeau lattice gas model and of some of its variants. Equilibrium distributions and several linear and nonlinear hydrodynamics flows are presented. We show that interesting phenomena can be studied with this class of models, even for lattices of limited sizes.

d'Humières, D., P. Lallemand and G. Searby, Numerical experiments on lattice gases: mixtures and Galilean invariance, Complex Systems 1 (1987) 633–647. Reprinted in: Lattice Gas Methods for Partial Differential Equations, ed. G.D. Doolen (Addison–Wesley, Reading, MA, 1989) pp. 333–350.

In this paper, we first describe an extension of the standard Frisch, Hasslacher, Pomeau hexagonal lattice gas to study reaction–diffusion problems. Some numerical results are presented. We then consider the question of Galilean invariance from an experimental point of view, showing cases where the standard model is inadequate. Finally, we introduce a way to cure the Galilean disease and present some results of simulations for a few typical cases.

d'Humières, D. and P. Lallemand, Flow of a lattice gas between two parallel plates and development of the Poiseuille profile, Compt. Rend. Acad. Sci. Paris II 302 (1986) 983. Reprinted in translation in: Lattice Gas Methods for Partial Differential Equations, ed. G.D. Doolen (Addison–Wesley, Reading, MA, 1989) pp. 429–436.

After recalling briefly the use of a lattice gas to simulate aerodynamical flows, we describe an experimental situation corresponding to a 2-D flow between two parallel plates. A series of velocity distributions shows the progressive deformation from a Blasius profile for a boundary layer on a flat plate to a parabolic Poiseuille distribution. The comparison of the present results to those computed by Slichting shows the excellent quality of the predictions that can be obtained with the new simulation method proposed by Frisch, Hasslacher, and Pomeau.

d'Humières, D. and P. Lallemand, Lattice gas automata for fluid mechanics, Physica A 140 (1986) 326–335.

A lattice gas is the representation of a gas by its restriction on the nodes of a regular lattice for discrete time steps. It was recently shown by Frisch, Hasslacher and Pomeau that such very simple models lead to the incompressible Navier–Stokes equation provided the lattice has enough symmetry and the local rules for collisions between particles obey the usual conservation laws of classical mechanics. We present here recent results of numerical simulations to illustrate the power of this new approach to fluid mechanics which may give new tools for numerical studies and build a bridge between cellular automata theory and complex physical problems.

d'Humières, D. and P. Lallemand, 2-D and 3-D hydrodynamics on lattice gases, Helv. Phys. Acta 59 (1986) 1231–1234.

The 2-D and 3-D Navier–Stokes equations are obtained from lattice gas models on regular lattices using particle speeds: zero, one and $\sqrt{2}$. Preliminary investigations of the linear behavior are presented for the two dimensional case.

d'Humières, D., P. Lallemand and U. Frisch, Lattice gas models for 3D hydrodynamics, Europhys. Lett. 2 (1986) 291–297.

The 3D Navier–Stokes equations are obtained from two different lattice gas models. The first one has its sites on a cubic lattice and has particle speeds zero, one and $\sqrt{2}$. The second one is a 3D projection of a lattice gas implementation of the 4D Navier–Stokes equations, residing on a face-centered hypercube lattice.

d'Humières, D., P. Lallemand and Y. Pomeau, Simulation de l'hydrodynamique bidimensionnelle à l'aide d'un gaz sur réseau, Bull. Soc. Franç. Phys. 60 (1986) 14–15.

d'Humières, D., Y. Pomeau and P. Lallemand, Écoulement d'un gaz sur réseau dans un canal bidimensionnel développement de profil de Poiseuille, Compt. Rend. Acad. Sci. Paris II 301 (1986) 983–988.

d'Humières, D., Y. Pomeau and P. Lallemand, Two-dimensional hydrodynamics calculations with a lattice gas, in: Innovative Numerical Methods in Engineering, A Computational Mechanics Publication (Springer, Berlin, 1986) pp. 241–248.

After recalling the molecular dynamics method to calculate various macroscopic properties of a gas from interparticle potentials, a very simplified model of a gas is introduced. Point particles moving at constant speed along the links of a regular lattice are shown to behave like a real gas, and thus provide a way to give solutions of the Navier–Stokes equation. Molecular dynamics can be calculated at very high speed on a very large number of particles allowing to study macroscopic flows. Examples of unsteady two-dimensional flows at Reynolds number of order of 100 are shown. Extension of this work on specialized hardware involving a large number of automata is discussed.

d'Humières, D., Y. Pomeau and P. Lallemand, Simulation of 2-D Von Karman Streets using a lattice gas, Compt. Rend. Acad. Sci. Paris II 301 (1985) 1391. Reprinted in translation in: Lattice Gas Methods for Partial Differential Equations, ed. G.D. Doolen (Addison–Wesley, Reading, MA, 1989) pp. 423–426.

A lattice gas model is introduced in which particles travel along the links of a planar triangular lattice, with constant speed and synchronized to collide at the vertices of the lattice. The macroscopic hydrodynamic behavior of the gas is demonstrated for channel flow around a plate perpendicular to the mean flow velocity, leading to the creation of nonstationary eddies.

d'Humières, D., P. Lallemand and T. Shimomura, Cellular automata, a new tool for hydrodynamics, Los Alamos National Laboratory Report, LA-UR-85-4051 (1985).

d'Humières, D., Y. Pomeau and P. Lallemand, Une nouvelle méthode de simulation numérique en mécanique des fluides les gas sur réseau, Images Phys. 68 (1987) 89–94.

Diemer, K., K. Hunt, S. Chen, T. Shimomura and G.D. Doolen, Density and velocity dependence of Reynolds numbers for several lattice gas models, in: Lattice Gas Methods for Partial Differential Equations, ed. G.D. Doolen (Addison–Wesley, Reading, MA, 1989) pp. 137–177.

Analytic calculations of shear viscosity and a reduced Reynolds number for eleven different 2-D and six different 3-D lattice gas hydrodynamic models are compared. The mean free path is calculated for ten of the 2-D models. Only models with one non-zero speed are considered. For comparison with the analytic results, Reynolds numbers for several 2-D models are calculated, using the microscopic definition of shear viscosity, from very low velocity 2-D Couette flow simulations. A similar check, using the definition of mean free path, is done for one of the 2-D models. All analytic derivations of viscosity have ignored its velocity-dependence. Viscosity for one model is calculated using a computer simulation to demonstrate a velocity-dependence and thus give a rough idea of the accuracy of these analytic viscosity calculations. The velocity dependence of the mean free path is calculated analytically, compared with computer calculations, and shown to be strongly anisotropic.

Doolen, G.D., Lattice gas models for fluid dynamics, Physics Today 41 (1988) 39–40.

Droz, M. and B. Chopard, Cellular automata approach to physical problems, Helv. Phys. Acta 61 (1988).

The main applications of cellular automata (CA) in physical problems are briefly reviewed. After having defined what a CA is, some considerations on the motivations to use CA in physics are given. Several examples of their applications in the fields of hydrodynamics, equilibrium statistical mechanics of systems on a lattice, growth mechanisms, diffusion processes, pattern recognition, models of memory and deterministic dynamics are given. An explicit application to a model of nonequilibrium phase transition is considered in more detail. It is shown how the CA rules are constructed, how to define analytically a mean field approximation for the automaton and how the results of this mean field approximation compare with the exact numerical solution of the model. Finally some general remarks on the prospects and open problems in the field are made.

Droz, M. and B. Chopard, Non-equilibrium phase transitions and cellular automata, in: Chaos and Complexity, eds. R. Livi, S. Ruffo, S. Ciliberto and M. Buiatti (World Scientific, Singapore, 1988) pp. 307–317.

Duarte, J.A.M.S. and U. Brosa, Viscous drag by cellular automata, J. Stat. Phys. 59 (1990) 501–508.

A simple method to compute the drag coefficient of two-dimensional bodies with arbitrary shapes is presented. The procedure is based on cellular automata as an extreme idealization of the molecular dynamics of a viscous fluid. We verify the algorithm by examples and obtain results in quantitative agreement with experiments even when eddies behind obstacles are formed.

Dubrulle, B., U. Frisch, M. Hénon and J.-P. Rivet, Low-viscosity lattice gases, Physica D 47 (1991) 27–29; these Proceedings.

New three-dimensional lattice gas models with very low (and possibly negative) viscosities are studied theoretically and tested in numerical implementations.

Dubrulle, B., Method of computation of the Reynolds number for two models of lattice gas involving violation of semi-detailed balance, Complex Systems 2 (1988) 577–609.

We show how the theory of lattice gases developed by Frisch, d'Humières, Hasslacher, Lallemand, Pomeau and Rivet can be extended to cases involving violation of semi-detailed balance. This allows further reduction of the viscosity. However, since the universality of the distribution is lost, the function $g(\rho)$ becomes dependent on the collision laws and has to be evaluated by a suitable generalization of the work of Hénon on viscosities. Cases with and without rest particles are considered. The lattice Boltzmann approximation is used.

Dufty, J., Time correlation functions and hydrodynamic modes for lattice gas cellular automata, in: Microscopic Simulations of Complex Flow, ed. M. Mareschal (Plenum, New York, 1990) pp. 259–268.

Dufty, J.W. and M.H. Ernst, Hydrodynamics modes and Green–Kubo relations for lattice gas cellular automata, preprint, Utrecht University (1989).

Dufty, J.W. and M.H. Ernst, Green–Kubo relations for lattice gas cellular automata, J. Phys. Chem. 93 (1989) 7015–7019.

The correlation function description of transport in simple fluids is applied to discrete lattice gas cellular automata. The linearized hydrodynamic equations are derived to Navier–Stokes order and transport coefficients are identified in terms of Green–Kubo time correlation function expressions.

Elton, B.H., A numerical analysis of lattice gas and lattice Boltzmann methods in the computation of solutions to nonlinear advective–diffusive systems, Ph.D. Thesis, University of California, Davis (September 1990)

This dissertation introduces a numerical theory for the massively parallel lattice gas and lattice Boltzmann methods for computing solutions of nonlinear advective–diffusive systems. The analysis covers convergence of the methods for periodic domains in two spatial dimensions. The convergence theory includes the discrete Chapman–Enskog expansion in obtaining consistency, and conditions of monotonicity in establishing stability. Convergence of some lattice methods is studied, including two for some two-dimensional nonlinear diffusion equations, one for the one-dimensional lattice method [B. Boghosian and C.D. Levermore, Complex Systems 1 (1987)] for the one-dimensional Burgers equation and one for a nonlinear two-dimensional advection–diffusion equation. Convergence is formally proven for the first three methods, revealing that they are second order accurate, conservative, conditionally monotone finite difference methods. Computational results for all the lattice methods is presented that supports the theoretical results. In addition, a domain decomposition method using mesh refinement is presented for lattice gas and lattice Boltzmann methods. Computational evidence for lattice gas methods is reported, as the domain decomposition strategy is applied to a lattice gas for the one-dimensional viscous Burgers equation.

Elton, B.H., C.D. Levermore and G. H. Rodrigue, Lattice Boltzmann methods for some 2-D nonlinear diffusion equations: Computational results, in: Proceedings of the Workshop on Asymptotic Analysis and Numerical Solution of Partial Differential Equations, held at Argonne National Laboratory, Feb. 26–28, 1990 (Marcel Dekker, New York, 1990).

In this paper we examine two lattice Boltzmann methods (that are a derivative of lattice gas methods) for computing solutions to two two-dimensional nonlinear diffusion equations of the form

$$\frac{\partial u}{\partial t} = \nu \left(\frac{\partial D(u)}{\partial x} \frac{\partial u}{\partial x} + \frac{\partial D(u)}{\partial y} \frac{\partial u}{\partial y} \right),$$

where $u = u(\boldsymbol{x}, t)$, $\boldsymbol{x} \in R^2$, ν is a constant, and $D(u)$ is a nonlinear term that arises from a Chapman–Enskog asymptotic expansion. In particular, we provide computational evidence supporting recent results [B. Elton, A numerical analysis of lattice gas and lattice Boltzmann methods in the computation of solutions to nonlinear advective–diffusive systems, Ph.D. dissertation, University of California, Davis, September 1990.] showing that the methods are second order convergent (in the L_1-norm), conservative, conditionally monotone finite difference methods. Solutions computed via the lattice Boltzmann methods are compared with those computed by other explicit, second order, conservative, monotone finite difference methods. Results are reported for both the L_1- and L_∞-norms.

Elton, B.H. and G.H. Rodrigue, Sub-structuring for lattice gases, in: Third International Symposium on Domain Decomposition Methods for Partial Differential Equations (SIAM, Philadelphia, PA, 1990) pp. 451–461.

In this paper we apply domain decomposition techniques to lattice gases. We demonstrate how a lattice gas can be modeled by a non-linear partial differential equation using the same classical techniques as were used by Chapman and Enskog in approximating the kinetics of a real gas. As an example, we show how a one-dimensional viscous Burgers' equation models a specific lattice gas. We then use the domain decomposition ideas for solving this differential equation to develop a sub-structuring technique for this exemplary lattice gas.

Ernst, M.H., Statistical mechanics of cellular automata fluids, in: Liquids, Freezing and the Glass Transition, Les Houches Summer School Proceedings, 51, eds. J.-P. Hansen and D. Levesque (Elsevier Science Publishers, Amsterdam), in press.

Contents: Introduction, Continuous Fluids Out of Equilibrium, Statistical Mechanics of CA-fluids, Green–Kubo Relations, and Applications.

Ernst, M.H., Linear response theory for cellular automata fluids, in: Fundamental Problems in Statistical Mechanics, Vol. 7, ed. H. van Beijeren (North-Holland, Amsterdam, 1990).

Lecture Notes. In these lectures we have developed nonequilibrium statistical mechanics of cellular automata fluids, in close parallel with linear response theory for continuous fluids. The main message conveyed in these notes is that lattice gas cellular automata with b velocity states may be considered as bona fide, although extremely simplified, statistical mechanical models of nonequilibrium systems with many degrees of freedom, that are able to model many aspects of hydrodynamics and transport properties of fluids.

Ernst, M.H., Mode-coupling theory and tails in CA fluids, Physica D 47 (1991) 198–211; these Proceedings.

After a summary of important effects of mode-coupling theory in continuous fluids, the theory is extended to CA fluids and the long-time tails calculated for the velocity correlation function and for the stress–stress correlation function, including the spurious contributions from the staggered momentum densities. At finite densities the Fermi statistics strongly suppresses the long-time singularities arising from two fluid modes, but not those from tagged particle diffusion.

Ernst, M.H., G.A. van Velzen and P.M. Binder, Breakdown of the Boltzmann equation for lattice gas cellular automata, Phys. Rev. A 39 (1989) 4327–4329.

In lattice gases the Boltzmann equation is not valid at low densities, if the collision rules admit reflections of unlike particles, because of long-lived correlated ring-type collisions. This is shown for a simple mixture, a Lorentz gas, by comparing theoretical and molecular dynamics results for the diffusion coefficient.

Ernst, M.H. and P.M. Binder, Lorentz lattice gases: Basic theory, J. Stat. Phys. 51 (1988) 981–990.

We present several ballistic models of the Lorentz gas in two-dimensional lattices with deterministic and stochastic deflection rules and their corresponding Liouville equations. Boltzmann-level-equation results are obtained for the diffusion coefficient and velocity autocorrelation function for models with stochastic deflection rules. The long-time behavior of the mean square displacement is briefly discussed and the possibility of abnormal diffusion indicated. Even if the diffusion coefficient exists, its low density limit may not be given correctly by the Boltzmann equation.

Ernst, M.H. and J.W. Dufty, Hydrodynamics and time correlation functions for cellular automata, J. Stat. Phys. 58 (1990) 57–86.

Hydrodynamic excitations in lattice gas cellular automata are described in terms of equilibrium time correlation functions for the local conserved variables. For large space and time scales the linearized hydrodynamic equations are obtained to Navier–Stokes order. Exact expressions for the associated susceptibilities and transport coefficients are identified in terms of correlation functions. The general form of the time correlation functions for conserved densities in the hydrodynamic limit is given and illustrated by some examples suitable for comparison with computer simulation. The transport coefficients are related to time correlation functions for the conserved fluxes in a way analogous to the Green–Kubo expressions for continuous fluids. The general results are applied for a one-component fluid and several types of binary diffusion. Also discussed are the effects of unphysical slow modes such as staggered particle or momentum densities.

Ernst, M.H. and J.W. Dufty, Green–Kubo relations for lattice gas cellular automata, Phys. Lett. A 138 (1989) 391–395.

Green–Kubo relations are derived for linear transport coefficients (viscosities, diffusion) in lattice gases using the correlation function description of transport in fluids. The only ingredients are the local microscopic conservation laws without any further specifications of the microdynamical laws.

Eykholt, R., A.R. Bishop, P.S. Lomdahl and E. Domany, A nonequilibrium-to-equilibrium mapping and its application to the perturbed sine-Gordon equation, Physica D 23 (1986) 102–111.

Given a partial differential equation (PDE) in one time and D spatial dimensions which is driven by noise $\zeta(\boldsymbol{x};t)$ with a known distribution $P[\zeta]$, one may find the distribution $P[\psi]$ of the solution $\psi(\boldsymbol{x};t)$. Furthermore, the distribution may be written as $P[\psi] \sim e^{-\beta H[\psi]}$ with $H[\psi]$ an effective Hamiltonian in $D+1$ dimensions (time having become an extra spatial dimension). Then, the most probable solution of the PDE is that function which minimizes $H[\psi]$. Here, we describe this method and illustrate it for the damped, driven sine-Gordon equation in one spatial dimension ($D = 1$).

Falk, H., Dynamical spin system: Exact solution and mean recurrence time, Physica D 31 (1988) 389–396.

A model involving a chain of $N \geq 2$ spins $s_i = \pm 1$, $i = 1, \ldots, N$, evolving synchronously in discrete time t via a nonlinear, autonomous transformation $s_i(t+1) = s_i(t)s_{i+1}(t)$, $i = 1, \ldots, N-1$; $s_N(t+1) = s_N(t)$, is presented. The transformation equations are solved explicitly and the detailed decomposition of state space into ergodic sets is Foundations On the assumption of equally likely initial states, the mean recurrence time is calculated and its variance is discussed. The model displays a strikingly sensitive dependence on the number of spins, and this is reflected in the "staircase" behavior of the mean recurrence time. Remarks are made regarding the connection between the behavior of the model and the ground states of a related two-dimensional Ising model.

Frenkel, D. and M. van der Hoef, A test of mode-coupling theory, in: Microscopic Simulations of Complex Flow, ed. M. Mareschal (Plenum, New York, 1990) pp. 281–296.

Frenkel, D., Long-time decay of velocity autocorrelation function of two-dimensional lattice gas cellular automata, in: Cellular Automata and Modeling of Complex Physical Systems, eds. P. Manneville, N. Boccara, G.Y. Vichniac and R. Bidaux (Springer, Berlin, 1989) pp. 144–154.

A method is introduced to compute the velocity autocorrelation function (VACF) of a tagged particle in a lattice gas. This method yields at least a million-fold improvement over the conventional method to measure such correlation functions in lattice-gas cellular automata (LGCA). For lattice Lorentz gases, the gain is even several orders of magnitude larger. Using this method, a t^{-1} algebraic tail in the VACF of a tagged particle in a two-dimensional LGCA is clearly observed.

Frenkel, D. and M.H. Ernst, Simulation of diffusion in a two-dimensional lattice gas cellular automaton: A test of mode-coupling theory, Phys. Rev. Lett. 63 (1989) 2165–2168.

We compute the velocity autocorrelation function of a tagged particle in a two-dimensional lattice-gas cellular automaton using a method that is about a million times more efficient than existing techniques. A t^{-1}-algebraic tail in the tagged-particle is found to agree quantitatively with the predictions of mode-coupling theory. However, the magnitude of logarithmic corrections to the t^{-1}-tail is much smaller than expected.

Frisch, U., Relation between the lattice Boltzmann equation and the Navier–Stokes equations, Physica D 47 (1991) 231–232; these Proceedings.

It is shown that the lattice gas Boltzmann equation may be rewritten in a form which brings out its close relation with the Navier–Stokes equations. Various consequences are pointed out.

Frisch, U., D. d'Humières, B. Hasslacher, P. Lallemand, Y. Pomeau and J.-P. Rivet, Lattice gas hydrodynamics in two and three dimensions, Complex Systems 1 (1987) 649–707. Reprinted in: Lattice Gas Methods for Partial Differential Equations, ed. G.D. Doolen (Addison–Wesley, Reading, MA, 1989) pp. 75–136.

Hydrodynamical phenomena can be simulated by discrete lattice gas models obeying cellular automata rules [Frisch et al., Phys. Rev. Lett. 56 (1986) 1505; D. d'Humières et al., Europhys. Lett. 2 (1986) 219]. It is here shown for a class of D-dimensional lattice gas models how the macrodynamical (large-scale) equations for the densities of microscopically conserved quantities can be systematically derived from the underlying exact microdynamical Boolean equations. With suitable restrictions on the crystallographic symmetries of the lattice and after proper limits are taken, various standard fluid dynamical equations are obtained, including the incompressible Navier–Stokes equations in two and three dimensions. The transport coefficients appearing in the macrodynamical equations are obtained using variants of the fluctuation–dissipation theorem and Boltzmann formalisms adapted to fully discrete situations.

Frisch, U., B. Hasslacher and Y. Pomeau, Lattice-gas automata for the Navier–Stokes equation, Phys. Rev. Lett. 56 (1986) 1505–1508. Reprinted in: Lattice Gas Methods for Partial Differential Equations, ed. G.D. Doolen (Addison–Wesley, Reading, MA, 1989) pp. 11–18.

We show that a class of deterministic lattice gases with discrete Boolean elements simulates the Navier–Stokes equation, and can be used to design simple, massively parallel computing machines.

Frisch, U. and J.-P. Rivet, Gaz sur réseau pour l'hydrodynamique formule de Green–Kubo, Compt. Rend. Acad. Sci. Paris II 303 (1986) 1065–1068. Reprinted in translation in: Lattice Gas Methods for Partial Differential Equations, ed. G.D. Doolen (Addison–Wesley, Reading, MA, 1989) pp. 437–442.

The Green–Kubo formalism, allowing the evaluation of transport coefficients of macroscopic systems in terms of equilibrium correlation functions, is extended to lattice gas hydrodynamics.

Fritz, J., On the hydrodynamic limit of a one-dimensional Ginzburg–Landau lattice model. The a priori bounds, J. Stat. Phys. 47 (1987) 551–571.

The simplest Ginzburg–Landau model with conservation law is investigated. The initial state is specified by an inhomogeneous profile of the chemical potential associated with the conserved quantity, that is, the mean spin. It is shown that the mean spin satisfies a nonlinear diffusion equation in the hydrodynamic limit. The proof is based on the nice, parabolic structure of the model. A standard perturbation technique is used.

Fritz, J., Review of the work of V.I. Oseledec on the stationary states of HPP-FHP-type cellular automata, unpublished.

In a series of fairly long and nontrivial papers (CMP 1976–84) B.M. Gurevich and Yu.M. Suhov investigated the set of stationary states of infinite Hamiltonian systems. They have shown that every stationary state satisfying some regularity conditions is an equilibrium state. HPP-FHP automata mimic classical dynamics, and Oseledec managed to extend this method to some classes of cellular automata including FHP models with randomized collisions. In fact, he proves that every stationary Gibbs state with an additional mixing property is a Bernoulli measure. Here we sketch his nice proof and basic results for the simplest HPP-FHP models. Partly we follow a recent, still unpublished paper by Gurevich on asymptotically additive integrals. Oseledec's beautiful treatment of stochastic automata (FHP models) is also included.

Gabetta, E., From stochastic mechanics to the discrete Boltzmann equation, Mathl. Comput. Modelling (1990), to be published.

Gatinol, R. and F. Coulouvrat, Constitutive laws for discrete velocity models of gas, in: Discrete Kinetic Theory, Lattice Gas Dynamics and Foundations of Hydrodynamics, ed. R. Monaco (World Scientific, Singapore, 1989) pp. 121–145.

For a model of gas described by the discrete Boltzmann equations, we study the effect of multiple collisions. By using the Chapman–Enskog method, we obtain the constitutive laws. The Navier–Stokes and Burnett equations associated with the model are explicitly given in the case of a gas near a homogeneous state. An analogy with classical hydrodynamics is shown.

Goles, E. and G.Y. Vichniac, Invariants in automata networks, J. Phys. A 19 (1986) L961–L965.

We give two extensions of Pomeau's additive invariant for reversible cellular automata and networks.

Gunstensen, A.K. and D.H. Rothman, A lattice-gas model for three immiscible fluids, Physica D 47 (1991) 47–52; these Proceedings.

Lattice-gas methods have recently proven very useful for the study of immiscible mixtures of two fluids, with applications ranging from two-phase flow in porous media to spinodal decomposition of binary fluids. Whereas the original one-phase lattice gas models the fluid as a collection of identical particles, in the immiscible two-phase lattice gas the particles are colored red or blue and the collisions between particles are chosen to achieve surface tension. We introduce a new lattice-gas model which extends the two-phase immiscible lattice gas to the simulation of a mixture of three immiscible fluids, i.e. red, green and blue. This extension achieves more than the obvious generalization: immiscible mixtures of three fluids yield phenomena that can be qualitatively different from analogous phenomena observed with two fluids. To demonstrate this point, we show simulations of phase separation of three immiscible fluids and three-phase flow in porous media.

Gunstensen, A.K. and D.H. Rothman, A Galilean-invariant immiscible lattice gas, Physica D 47 (1991) 53–63; these Proceedings.

Recently, lattice-gas methods have been introduced as a technique for the simulation of one- and two-phase fluid flow. These methods model the fluid as a collection of particles which propagate on a regular lattice and undergo collisions at the nodes of the lattice. In an asymptotic limit, lattice gases simulate the Navier–Stokes equations. However, these models suffer from a lack of Galilean invariance. An important physical manifestation of the lack of invariance is that the fluid vorticity advects with a velocity different from the velocity of the fluid. We introduce a new, Galilean-invariant, model for simulating immiscible two-phase flow. Unlike previous Galilean-invariant models, the collisions in this new model satisfy semi-detailed balance, which is achieved by the inclusion of a large number of rest particles with zero velocity. Since adding many rest particles is not computationally tractable, the presence of a large number of such particles is simulated by weighting the outcome of the collisions by a factor related to the frequency with which the collisions would have occurred if the rest particles had been explicitly included in the model. We demonstrate that, in the new model, the vorticity advects at the same velocity as the fluid. We also show that the model obeys Laplace's formula for surface tension and demonstrate an application of the new model to the Rayleigh–Taylor instability. Growth rates as a function of wavenumber computed in the early stages of the instability compare well to theoretical predictions.

Gunstensen, A.K., A fast implementation of the FHP lattice gas, MIT Porous Flow Project, Report No. 1 (1988).

Hardy, J., O. de Pazzis and Y. Pomeau, Molecular dynamics of a classical lattice gas transport properties and time correlation functions, Phys. Rev. A 13 (1976) 1949–1961.

Hardy, J., Y. Pomeau and O. de Pazzis, Time evolution of two-dimensional model system I: invariant states and time correlation functions, J. Math. Phys. 14 (1973) 1746–1759.

Hardy, J. and Y. Pomeau, Thermodynamics and hydrodynamics for a modeled fluid, J. Math. Phys. 13 (1972) 1042–1051.

Hasslacher, B. and D.A. Meyer, Knot invariants and cellular automata, Physica D 45 (1990) 328–344.

There is a deep connection between the theory of invariants of knots in three dimensions and certain classes of solvable models in statistical mechanics. We study the consequences of this for a class of cellular automaton systems. Knot invariants furnish both a non-perturbative framework for the classification and analysis of such models and a primary tool with which to study the emergence of thermodynamic behavior in extended systems.

Hasslacher, B., Spontaneous curvature in a class of lattice gas field theories, Physica D 47 (1991) 19–23; these Proceedings.

We describe a class of cellular automata having a natural lattice gas interpretation which also develop non-perturbative curvature singularities. Variations of these models could be useful in describing curvature transisions in crystals, membranes and superconducting materials.

Hasslacher, B., Discrete fluids, Los Alamos Sci. 15 (1988) 175–200, 211–217.

The lattice gas automaton is an approach to computing fluid dynamics that is still in its infancy. In this three-part article, one of the inventors of the model presents its theoretical foundations and its promise as a general approach to solving partial differential equations and to parallel computing.

Hatori, T. and D. Montgomery, Transport coefficients for magnetohydrodynamic cellular automata, Complex Systems 1 (1987) 735–752. Reprinted in: Lattice Gas Methods for Partial Differential Equations, ed. G.D. Doolen (Addison–Wesley, Reading, MA, 1989) pp. 351–370.

A Chapman–Enskog development has been used to infer theoretical expressions for coefficients of kinematic viscosity and magnetic diffusivity for a two-dimensional magnetohydrodynamic cellular automaton.

Hayot, F., Fingering instability in a lattice gas, Physica D 47 (1991) 64–71; these Proceedings.

I describe work done over the last two years concerning a Saffman–Taylor-type instability in lattice gas hydrodynamics. The emergence of typical macroscopic laws, such as Darcy's and Laplace's, from a microscopic lattice gas is shown. The successful modeling of a fingering instability between two immiscible fluids hinges on the development of an appropriate interface algorithm.

Hayot, F. and M.R. Lakshmi, Cylinder wake in lattice gas hydrodynamics, Physica D 40 (1989) 415–420.

We measure, using a lattice gas algorithm, the time development of a wake behind a cylinder at Reynolds numbers equal to 30 and 60. We measure also drag at these values and determine Re = 90 to be the Reynolds number where vortex shedding starts. Agreement with experimental results is good. We also measure vortex shedding frequency at Re = 108 from transverse velocity and at Re = 180 from lift on the cylinder. The corresponding Strouhal numbers are obtained and discussed.

Hayot, F., M. Mandal and P. Sadayappan, Implementation and performance of a binary lattice gas algorithm on parallel processor systems, J. Compl. Phys. 80 (1989) 277–287.

We study the performance of a lattice gas binary algorithm on a real arithmetic machine, a 32 processor INTEL iPSC hypercube. The implementation is based on so-called multi-spin coding techniques. From the measured performance we extrapolate to larger and more powerful parallel systems. Comparisons are made with 'bit' machines, such as the parallel Connection Machine.

Hayot, F., The effect of Galilean non-invariance in lattice gas automaton one-dimensional flow, Complex Systems 1 (1987) 753–761. Reprinted in: Lattice Gas Methods for Partial Differential Equations, ed. G.D. Doolen (Addison–Wesley, Reading, MA, 1989) pp. 371–382.

In the simple case of one-dimensional flow between plates, we show the effect of Galilean non-invariance of the usual hexagonal lattice gas mode. This effect leads to a distorted velocity profile when the velocity exceeds a value of 0.4. Higher-order corrections to the Navier–Stokes equations are considered in a discussion of the numerical importance of the distortion.

Hayot, F., Unsteady, one-dimensional flow in lattice-gas automata, Phys. Rev. A 35 (1987) 1774–1777.

We study numerically unsteady, one-dimensional flow between parallel plates in the hexagonal lattice-gas automaton. An initial tangential velocity instability is created at one plate and its propagation into the system is investigated. This propagation depends on the boundary conditions at the opposite plate. From the observed fluid behavior, viscosity is estimated.

Hayot, F., Viscosity in lattice gas automata, Physica D 28 (1987) 210–214.

Using the Green–Kubo formalism, we derive an expression, valid at any density, for the viscosity for hexagonal lattice gas automata. Estimates of an effective viscosity, valid for time scales relevant to numerical experiments, are given for the two-dimensional hexagonal lattice gas.

Hénon, M., On the relation between lattice gases and cellular automata, in: Discrete Kinetic Theory, Lattice Gas Dynamics and Foundations of Hydrodynamics, ed. R. Monaco (World Scientific, Singapore, 1989) pp. 160–161.

Hénon, M., Optimization of collision rules in the FCHC lattice gas and addition of rest particles, in: Discrete Kinetic Theory, Lattice Gas Dynamics and Foundations of Hydrodynamics, ed. R. Monaco (World Scientific, Singapore, 1989) pp. 146–159.

Various collision rules for the FCHC 24-velocity lattice are discussed. Global rules give a Reynolds coefficient $R_x^{\max} \simeq 2$. Detailed rules, obtained by a fine-tuning optimization, give $R_x^{\max} = 7.57$. Formulas are given for the FCHC lattice with the addition of rest particles, with arbitrary probabilities for each value of the number of particles present at a node. With a maximum of 1, 2, and 3 rest particles, the best values obtained for $R_x^{\max}$ are 8.46, 10.22 and 10.71.

Hénon, M., Isometric collision rules for the four-dimensional FCHC lattice gas, Complex Systems 1 (1987) 475–494.

Collision rules are presented for the four-dimensional face-centered hypercubic-lattice (FCHC). The velocity set after collision is deduced from the velocity set before collision by an isometry, chosen so as to preserve the momentum and minimize the viscosity. A detailed implementation recipe is given. The shear viscosity is computed; the result shows that essentially all memory of the previous velocities is lost at each collision. Another set of collision rules, based on a random choice of the output state, has similar properties. The isometric principle can also be applied to the two-dimensional square (HPP) and triangular (FHP) lattices: one recovers the usual rules with minor differences.

Hénon, M., Viscosity of a lattice gas, Complex Systems 1 (1987) 763–789. Reprinted in: Lattice Gas Methods for Partial Differential Equations, ed. G.D. Doolen (Addison–Wesley, Reading, MA, 1989) pp. 179–208.

The shear viscosity of a lattice gas can be derived in the Boltzmann approximation from a straightforward analysis of the numerical algorithm. This computation is presented first in the case of the Frisch–Hasslacher–Pomeau two-dimensional triangular lattice. It is then generalized to a regular lattice of arbitrary dimension, shape, and collision rules with appropriate symmetries. The viscosity is shown to be positive. A practical recipe is given for choosing collision rules so as to minimize the viscosity.

Herrmann, H.J., Special purpose computers in statistical physics, Physica A 140 (1986) 421–427.

A new trend in physics, namely the building of special purpose computers (SPC), is reviewed. Special emphasis is given to the following questions: Why does one build SPC's? When is it worthwhile to build an SPC? How does one proceed if one wants to build an SPC? Finally the most important results that have been obtained for statistical physics through SPC's up to now will be sketched.

Higuera, F.J., S. Succi and R. Benzi, CFD with the lattice Boltzmann equation, in: Microscopic Simulations of Complex Flow, ed. M. Mareschal (Plenum, New York, 1990) pp. 71–84.

Higuera, F.J., Lattice gas simulation based on the Boltzmann equation, in: Discrete Kinetic Theory, Lattice Gas Dynamics and Foundations of Hydrodynamics, ed. R. Monaco (World Scientific, Singapore, 1989) pp. 162–177.

The Boltzmann equation for lattice gases is applied as a numerical method for incompressible flows. The lattice Boltzmann equation is free from fluctuations and is more adequate than the direct lattice gas simulation for implementing general non-deterministic collision rules and for handling the possible symmetries of the distribution functions. Optimal collision rules are discussed and applied to the simulation of two- and three-dimensional flows. A new kind of Bose-like lattice gas is defined by dropping out the exclusion principle. It is shown that the viscosity of these gases can be arbitrarily small.

Higuera, F.J., S. Succi and R. Benzi, Lattice gas dynamics with enhanced collisions, Europhys. Lett. 9 (1989) 345–349.

An efficient strategy is developed for building suitable collision operators to be used in a simplified version of the lattice gas Boltzmann equation. The resulting numerical scheme is shown to be linearly stable. The method is applied to the computation of the flow in a channel containing a periodic array of obstacles.

Higuera, F.J. and J. Jimenez, Boltzmann approach to lattice gas simulations, Europhys. Lett. 9 (1989) 663–668.

An alternative simulation procedure is proposed for lattice hydrodynamics, based on the lattice Boltzmann equation instead of on the microdynamical evolution. The averaging step, used by the latter method to derive macroscopic quantities, is suppressed, as well as the associated fluctuations. The collision operator is expressed in terms of its linearized part, and condensed into a few parameters, which can be selected, independently of a particular collision rule, to decrease viscosity as much as desired.

Higuera, F.J. and S. Succi, Simulating the flow around a circular cylinder with a lattice Boltzmann equation, Europhys. Lett. 8 (1989) 517–521.

It is shown that the lattice Boltzmann equation deriving from the Frisch–Hasslacher–Pomeau cellular automaton, being free from microscopic fluctuations, provides a new appealing tool to simulate realistic incompressible hydrodynamics. Numerical results pertaining to a two-dimensional flow past a cylinder are reported and compared with numerical and experimental data available in the literature.

Hogeweg, P., Cellular automata as a paradigm for ecological modeling, Appl. Math. Comp. 27 (1988) 81–100.

We review cellular automata as a modeling formalism and discuss how it can be used for modeling (spatial) ecological processes. The implications of this modeling paradigm for ecological observation are stressed. Finally we discuss some shortcomings of the cellular-automaton formalism and mention some extensions and generalizations which may remedy these shortcomings.

Huang, J.-I., Y.-H. Chu and C.-S. Yin, Lattice-gas automata for modeling acoustic wave propagation in inhomogeneous media, Geophys. Res. Lett. 15 (1988) 1239–1241.

We report recent developments in using lattice-gas automata to simulate acoustic wave propagation in inhomogeneous media and give experimental results that include transmitted and reflected waves. Two sets of rules are adopted to govern the evolution of particles in the lattice; one applicable to lattice sites on the interface, the other to interior sites. Both sets of rules are simple and deterministic, thereby preserving the computational advantages of lattice-gas automata. Further investigation bears long-term promise of yielding a novel and efficient forward modeling tool for geologically realistic models.

d'Humières, D., [listed under D instead of H]

Kadanoff, L., G. McNamara and G. Zanetti, From automata to fluid flow comparisons of simulation and theory, Phys. Rev. A 40 (1989) 4527–4541.

Lattice-gas automata have been proposed as a new way of doing numerical calculations for hydrodynamic systems. Here, a lattice-gas simulation is run to see whether its behavior really does correspond, as proposed, to that of the Navier–Stokes equation. The geometry used is the two-dimensional version of laminar pipe flow. Three checks on the existing theory are performed. The parabolic profile of momentum density arising from the dynamics is quantitatively verified. So is the equation of state, which arises from the statistical mechanics of the system. Finally, the well-known logarithmic divergence in the viscosity is observed in the automaton and is shown to disagree with the earliest theoretical predictions in this system. Proper agreement is achieved, however, when the theory is extended to include three extra (recently discovered) conserved quantities. In this way, checks of both linear and nonlinear parts of the hydrodynamic description of lattice-gas automata have been achieved.

Kadanoff, L.P., G.R. McNamara and G. Zanetti, A Poiseuille viscometer for lattice gas automata, Complex Systems 1 (1987) 791–803. Reprinted in: Lattice Gas Methods for Partial Differential Equations, ed. G.D. Doolen (Addison–Wesley, Reading, MA, 1989) pp. 383–398.

Lattice gas automata have been recently proposed as a new technique for the numerical integration of the two-dimensional Navier–Stokes equation. We have accurately tested a straightforward invariant of the original model, due to Frisch, Hasslacher, and Pomeau, in a simple geometry equivalent to two-dimensional Poiseuille (Channel) flow driven by a uniform body force. The momentum density profile produced by this simulation agrees well with the parabolic profile predicted by the macroscopic description of the gas given by Frisch et al.. We have used the simulated flow to compute the shear viscosity of the lattice gas and have found agreement with the results obtained by d'Humières et al.. [LANL reprint] using shear wave relaxation measurements, and, in the low density limit, with theoretical predictions obtained from the Boltzmann description of the gas [J.-P. Rivet and U. Frisch, Compt. Rend. Acad. Sci. Paris II 302 (1986) 732].

Kadanoff, L.P., On two levels, Physics Today 39 (1986) 7–9.

Kadanoff, L.P. and J. Swift, Transport coefficients near the critical point. A master equation approach, Phys. Rev. 165 (1968) 310–322.

Kawashima, S., A. Watanabe, M. Maeji and Y. Shizuta, Publication of the Research Instute for Mathematical Sciences, Kyoto University, No. 22 (1986).

In preceding papers, we studied the general theory of the discrete Boltzmann equation and formulated several conditions under which the solutions in the large for the Cauchy problem exist and approach the Maxwellian corresponding to the initial data as $t \to \infty$. We continue in this paper the study of concrete discrete models. Our aim is to verify the stability condition for the 32-velocity model introduced by Cabannes. Since the size of the model is relatively large, the computation needed becomes necessarily lengthy.

Kong, X.P. and E.G.D. Cohen, Anomalous diffusion in a lattice-gas wind-tree model, Phys. Rev. B 40 (1989) 4838.

Two new strictly deterministic lattice-gas automata derived from Ehrenfest's wind-tree model are studied. While in one model normal diffusion occurs, the other model exhibits abnormal diffusion in that the distribution function of the displacements of the wind particle is non-Gaussian, but its second moment, the mean-square displacement, is proportional to the time, so that a diffusion coefficient can be defined. A connection with the percolation problem and a self-avoiding random walk for the case in which the lattice is completely covered with trees is discussed.

Kong, X.P. and E.G.D. Cohen, Lorentz lattice gases, abnormal diffusion and polymer statistics, J. Stat. Phys. (1990).

Diffusive behavior in various Lorentz lattice gases, especially wind-tree like models, is discussed. Comparisons between lattice and continuum models as well as deterministic and probabilistic models are made. In one deterministic model, where the scatterers behave like double-sided mirrors, a new kind of abnormal diffusion is found, viz., the mean square displacement is proportional to the time, but the probability density distribution function is non-Gaussian. The connections of this mirror model with the percolation problem and the statistics of polymer chains on a lattice are also discussed.

Kong, X.P. and E.G.D. Cohen, A kinetic theorist's look at lattice gas cellular automata, Physica D 47 (1991) 9–18; these Proceedings.

The diffusion behavior of a number of Lorentz lattice gases is studied in its dependence on collision rules and its similarity to corresponding Lorentz gases that are continuous in space, but have the same discrete velocity space. The difference in behavior resulting from probabilistic or deterministic collision rules is discussed.

Kugelmass, S.D. and K. Steiglitz, Design and construction of LGM-1: a lattice gas machine with linear speedup, in: Proceedings of the 22nd Annual Conference on Information Sciences and Systems, Princeton University, March 16–18, 1988.

Kugelmass, S.D., R. Squier and K. Steiglitz, Performance of VLSI engines for lattice gas computations, Complex Systems 1 (1987) 939–965.

We address the problem of designing and building efficient custom VLSI-based processors to do computations on large multi-dimensional lattices. The design tradeoffs for two architectures which provide practical engines for lattice updates are derived and analyzed. We find that I/O constitutes the principal bottleneck of processors designed for lattice computations, and we derive upper bounds on throughput for lattice updates based on Hong and Kung's graph-pebbling argument that models I/O. In particular, we show that $R = O(BS^1/d)$, where R is the site update rate, B is the main memory bandwidth, S is the processor storage, and d is the dimension of the lattice.

Ladd, A.J.C., Hydrodynamic interactions and transport coefficients in a suspension of spherical particles, in: Microscopic Simulations of Complex Flow, ed. M. Mareschal (Plenum, New York, 1990) pp. 129–140.

Ladd, A.J.C. and D. Frenkel, Dynamics of colloidal dispersions via lattice-gas models of an incompressible fluid, in: Cellular Automata and Modeling of Complex Physical Systems, eds. P. Manneville, N. Boccara, G.Y. Vichniac and R. Bidaux (Springer, Berlin, 1989) pp. 242–245.

Progress in applying lattice-gas models to the simulation of colloidal dispersions is described. A new set of collision rules for the four-dimensional face-centered-hypercubic lattice, which lead to an exactly isotropic viscosity and which can be implemented with a small table lookup, have been developed. However, they are only suitable for low Reynolds number flows. A new implementation of a constant-velocity boundary condition for lattice-gas models is described.

Ladd, A.J.C., M.E. Colvin and D. Frenkel, Application of lattice-gas cellular automata to the Brownian motion of solids in suspension, Phys. Rev. Lett. 60 (1988) 975–978.

An adaptation of lattice-gas cellular automata to the simulation of solid–fluid suspensions is described. The method incorporates both dissipative hydrodynamic forces and thermal fluctuations. At low solid densities, theoretical results for the drag force on a single disk and the viscosity of a suspension of disks are reproduced. The zero-shear-rate viscosity has been obtained over a range of packing fractions and results indicate that simulations of three-dimensional suspensions are feasible.

Lavallée, P., J.P. Boon and A. Noullez, Boundaries in lattice gas flows, Physica D 47 (1991) 233–240; these Proceedings.

A one-dimensional lattice gas model is used to study the interaction of fluid flows with solid boundaries. Various interaction mechanisms are examined. Lattice Boltzmann simulations show that bounce-back reflection is not the only interaction that yields "no-slip" boundary conditions (zero velocity at a fixed wall) and that Knudsen-type interaction is also appropriate.

Lavallée, P., J.P. Boon and A. Noullez, Lattice Boltzmann equation for laminar boundary flow, Complex Systems 3 (1989) 317–330.

A simple method based on the lattice Boltzmann equation is presented for the evaluation of the velocity profile of fluid flows near walls or in the vicinity of the interface between two fluids. The method is applied to fluid flow near a wall, to channel flow, and to the transition zone between two fluids flowing parallel to each other in opposite directions. The results show good agreement with microdynamical lattice gas simulations and with classical fluid dynamics.

Lavallée, P., J.P. Boon and A. Noullez, Boundary interactions in a lattice gas, in: Discrete Kinetic Theory, Lattice Gas Dynamics and Foundations of Hydrodynamics, ed. R. Monaco (World Scientific, Singapore, 1989) pp. 206–214.

A method using the cellular automata approach is presented to evaluate the velocity profile in a fluid flow near solid boundaries. The usual method is to simulate the flow via the hexagonal lattice gas microdynamical equations from which the velocity profile is determined. The present approach is to solve the lattice Boltzmann equations using average, continuous values for the link populations. The velocity profile and the time evolution of the profile obtained in this way for a flow parallel to a flat plate are compared with actual simulations and with the boundary layer theory.

Lawniczak, A., D. Dab, R. Kapral and J.P. Boon, Reactive lattice gas automata, Physica D 47 (1991) 132–158; these Proceedings.

A probabilistic lattice gas cellular automaton model of a chemically reacting system is constructed. Microdynamical equations for the evolution of the system are given; the continuous and discrete Boltzmann equations are developed and their reduction to a generalized reaction–diffusion equation is discussed. The microscopic reactive dynamics is consistent with any polynomial rate law up to the fourth order in the average particle density. It is shown how several microscopic CA rules are consistent with a given rate law. As most CA systems, the present one has spurious properties whose effects are shown to be unimportant under appropriate conditions. As an explicit example of the general formalism, a CA dynamics is constructed for an autocatalytic reactive scheme known as the Schlögl model. Simulations show that in spite of the simplicity of the underlying discrete dynamics, the model exhibits phase separation and wave propagation phenomena expected for this system. Because of the microscopic nature of the dynamics, the role of internal fluctuations on the evolution process can be investigated.

Lebowitz, J.L., E. Orlandi and E. Presutti, Convergence of stochastic cellular automaton to Burgers' equation fluctuations and stability, Physica D 33 (1988) 165–188.

We prove that, for almost all realizations of the Boghosian–Levermore stochastic cellular automaton model, the density profile converges, in the scaling limit, to the solution of Burgers' equation. The proof goes via the propagation of chaos and yields tight bounds on the fluctuations. These estimates also yield stability properties of the (smooth) shock front: at long times it remains well-defined on a microscopic scale, but its location fluctuates.

Lebowitz, J.L., E. Presutti and H. Spohn, Microscopic models of hydrodynamic behavior, J. Stat. Phys. 51 (1988) 841–862.

We review recent developments in the rigorous derivation of hydrodynamic-type macroscopic equations from simple microscopic models: continuous-time stochastic cellular automata. The deterministic evolution of hydrodynamic variables emerges as the law of large numbers, which holds with probability one in the limit in which the ratio of the microscopic to the macroscopic spatial and temporal scales go to zero. We also study fluctuations in the microscopic system about the solution of the macroscopic equations. These can lead, in cases where the latter exhibit instabilities, to complete divergence in behavior between the two at long macroscopic times. Examples include Burgers' equation with shocks and diffusion–reaction equations with traveling fronts.

Lebowitz, J.L., Microscopic origin of hydrodynamic equations derivation and consequences, Physica A 140 (1986) 232–239.

We describe some recent progress in deriving autonomous hydrodynamic-type equations for macroscopic variables from model stochastic microscopic dynamics of particles on a lattice. The derivations also yield the microscopic fluctuations about the deterministic macroscopic evolution. These grow, with time, to become infinite when the deterministic solution is unstable. A form of microscopic pattern selection is also Foundations

Leko, T.D., Comment on lattice-gas automata for the solution of partial differential equations, Math. Mech. 68 (1988) T462–T463.

Levermore, C.D. and B.M. Boghosian, Deterministic cellular automata with diffusive behavior, in: Cellular Automata and Modeling of Complex Physical Systems, eds. P. Manneville, N. Boccara, G.Y. Vichniac and R. Bidaux (Springer, Berlin, 1989) pp. 118–129. Also in: Discrete Kinetic Theory, Lattice Gas Dynamics and Foundations of Hydrodynamics, ed. R. Monaco (World Scientific, Singapore, 1989) pp. 44–61.

It is a classical result that an ensemble of independent unbiased random walks on the one-dimensional lattice, Z, and moving at discrete times, Z+, has a continuum limit given by a diffusion equation. More recently, systems of randomly walking particles interacting via an exclusion principle have been studied. Another interesting problem is that of using deterministic dynamical systems for the same purpose. Of course, to the extent that the underlying microscopic dynamics of atoms in real diffusing media are deterministic, we know that this should be possible. In this work, we describe two completely deterministic cellular automata that exhibit diffusive behavior in one dimension, possibly with spatial inhomogeneity. We analyze these automata both theoretically and experimentally to investigate their continuum limits. In the first of these, we experimentally find significant deviations from the Chapman–Enskog theory; these deviations are due to a buildup of correlations that invalidates the Boltzmann molecular chaos assumption. In the second, the relevant correlations have been suppressed, and good agreement with the Chapman–Enskog theory is obtained.

Lim, H.A., Cellular automaton simulations of simple boundary-layer problems, Phys. Rev. A 40 (1989) 968–980.

A lattice-gas automaton is a variant of a cellular automaton. Its cellular universe is a regular triangular lattice and particles reside on the lattice nodes. The time evolution of this discrete dynamical system of particles proceeds in two alternating phases: collision and propagation. Such a model, though very simple and deterministic, is capable of producing very complex behaviors. A boundary layer develops whenever a real, viscous fluid flows along a solid boundary. We simulate boundary-layer and related problems in the incompressible limit of fluid dynamics using a lattice-gas automaton. Our lattice-gas automaton simulations show that viscosity effects on Couette flows (flows between parallel plates), Stokes flows, and Blasius flows (flows across a plate) give results as predicted by the Navier–Stokes equations. By considering different geometries and by carefully varying the gas properties, we obtain in particular the time-dependent velocity profiles, which are in good agreement with theoretical predictions. These inferences may be viewed as further support for the internal consistency of the lattice-gas approach, and they also substantiate the belief that the lattice-gas automaton can be a useful, viable tool for simulating fluid dynamics.

Lim, H.A., Lattice gas automata of fluid dynamics for unsteady flow, Complex Systems 2 (1988) 45–58.

We study lattice gas automata of fluid dynamics in the incompressible flow limit. It is shown that the viscosity effect on the transition layer from the steady uniform velocity in one stream to the steady uniform velocity in another, adjacent stream produces the correct profile. We further study the intrinsic damping or smoothing action of viscous diffusion and show that the results agree with those obtained from the Navier–Stokes equation. In both cases, we obtain a kinetic viscosity $\nu \sim 0.65$, consistent with the prediction of the Boltzmann approximation.

Lindgren, K., Entropy and correlations in dynamical lattice systems, in: Cellular Automata and Modeling of Complex Physical Systems, eds. P. Manneville, N. Boccara, G.Y. Vichniac and R. Bidaux (Springer, Berlin, 1989) pp. 27–40.

Information theory provides concepts for analyzing correlations and randomness in lattice systems of any dimension. Quantities expressing the information in correlations of different lengths are reviewed. A method for calculating the measure entropy, the average entropy per lattice site, is presented. It takes successively into account correlations of larger and larger distance, and the series converges faster than the expression given by the definition of the average entropy. The formalism is applied to deterministic and probabilistic cellular automata. The behavior of the spatial measure entropy in time is analyzed and its relevance to statistical mechanics is discussed. For a general class of lattice gas models it is proven that the measure entropy and the thermodynamic entropy closely relates this theorem to the second law of thermodynamics.

Liu, F. and N. Goldenfeld, Deterministic lattice model for diffusion-controlled crystal growth, Physica D 47 (1991) 124–131; these Proceedings.

An efficient lattice model is developed to study the late stages of diffusion-controlled crystal growth. We establish the existence of a dense branching morphology and its relation to diffusion-limited aggregation. We find a clear morphological transition from kinetic-effect-dominated growth to surface-tension-dominated growth, marked by a difference in the way growth velocity scales with undercooling. We also study the evolution of interfacial instability and find a scaling behavior for the interface power spectra, indicating the non-linear selection of a unique length scale.

Long, L.N., R.M. Coopersmith and B.G. McLachlan, Cellular automatons applied to gas dynamic problems, Proceedings of the AIAA 19th Fluid Dynamics, Plasma Dynamics and Lasers Conference, Honolulu (1987).

This paper compares the results of a relatively new computational fluid dynamics method, cellular automatons, with experimental data and analytical results. This technique has been shown to qualitatively predict fluid-like behavior; however, there have been few published comparisons with experiment or other theories. Comparisons are made for a one-dimensional supersonic piston problem, Stokes First Problem, and the flow past a normal flat plate. These comparisons are used to assess the ability of the method to accurately model fluid dynamic behavior and to point out its limitations. Reasonable results were obtained for all three test cases. While this is encouraging, the fundamental limitations of cellular automatons are numerous. In addition, it may be misleading, at this time, to say that cellular automatons are a computationally efficient technique. Other methods, based on continuum or kinetic theory, would also be very efficient if as little of the physics were included.

Maddox, J., Mechanizing cellular automata, Nature 321 (1986) 107.

Mareschal, M. and E. Kestermont, Order and fluctuations in nonequilibrium molecular dynamics simulations of two-dimensional fluids, J. Stat. Phys. 48 (1987) 1187–1201.

Finite systems of hard disks placed in a temperature gradient and in an external constant field have been studied, simulating a fluid heated from below. We used the methods of nonequilibrium molecular dynamics. The goal was to observe the onset of convection in the fluid. Systems of more than 5000 particles have been considered and the choice of parameters has been made in order to have a Rayleigh number larger than the critical one calculated from the hydrodynamic equations. The appearance of rolls and the large fluctuations in the velocity field are the main features of these simulations.

Margolus, N. and T. Toffoli, Cellular automata machines, in: Lattice Gas Methods for Partial Differential Equations, ed. G.D. Doolen (Addison–Wesley, Reading, MA, 1989) pp. 219–250.

The advantages of an architecture optimized for cellular automata simulations are so great that, for large-scale CA experiments, it becomes absurd to use any other kind of computer.

Margolus, N., T. Toffoli and G. Vichniac, Cellular-automata supercomputers for fluid-dynamics modeling, Phys. Rev. Lett. 56 (1986) 1694–1697.

We report recent developments in the modeling of fluid dynamics and give experimental results (including dynamical exponents) obtained with cellular automata machines. Because of their locality and uniformity, cellular automata lend themselves to an extremely efficient physical realization; with a suitable architecture, an amount of hardware resources comparable to that of a home computer can achieve (in the simulation of cellular automata) the performance of a conventional supercomputer.

McCauley, J.L., Chaotic dynamical systems as automata, Z. Naturforsch. 42a (1987) 547–555.

We discuss the replacement of discrete maps by automata, algorithms for the transformation of finite-length digit strings into other finite-length digit strings, and then discuss what is required in order to replace chaotic phase flows that are generated by ordinary differential equations by automata without introducing unknown and uncontrollable errors. That question arises naturally in the discretization of chaotic differential equations for the purpose of computation. We discuss as examples an autonomous and a periodically driven system, and a possible connection with cellular automata is also discussed. Qualitatively, our considerations are equivalent to asking when can the solution of a chaotic set of equations be regarded as a machine, or a model of a machine.

McNamara, G.R. and G. Zanetti, Use of the Boltzmann equation to simulate lattice-gas automata, Phys. Rev. Lett. 61 (1988) 2332–2335. Reprinted in: Lattice Gas Methods for Partial Differential Equations, ed. G.D. Doolen (Addison–Wesley, Reading, MA, 1989) pp. 289–296.

We discuss an alternative technique to the lattice-gas automata for the study of hydrodynamic properties; namely, we propose to model the lattice gas with a Boltzmann equation. This approach completely eliminates the statistical noise that plagues the usual lattice-gas simulations that demand much less computer time. It is estimated to be more efficient than the lattice-gas automata for intermediate-to-low Reynolds number, $Re < 100$.

Molvig, K., P. Donis, R. Miller, J. Myczkowski and G. Vichniac, Multi-species lattice gas automata for realistic fluid dynamics, in: Cellular Automata and Modeling of Complex Physical Systems, eds. P. Manneville, N. Boccara, G.Y. Vichniac and R. Bidaux (Springer, Berlin, 1989) pp. 206–231.

We present lattice-gas automata which yield in the macroscopic limit genuine 3D Euler equations that are free of the discreteness artifacts (lack of Galilean invariance and anomalous pressure) present in previous models. The automata involve particles with different speeds. Particles move on FCHC sublattices corresponding to their speeds, and the different sublattices interact via an energy exchange collision. These features are necessary for the physical requirements of isotropy and energy transport. Assigning different rates to the direct and inverse energy exchange collisions allows one to force the g-function to unity, yielding macroscopic Galilean invariance. The treatment of the energy degree of freedom also achieves the elimination of the spurious term in u^2 in the pressure. Kinetic, hydrodynamical and thermodynamical properties for this class of multi-species automata are derived formally and are also investigated with numerical simulations.

Molvig, K., P. Donis, J. Myczkowski and G. Vichniac, Multi-species lattice gas hydrodynamics, unpublished.

A new class of lattice gas automata are developed, based on multiple species of particles with energy exchanging interactions. These automata are endowed with a genuine energy that is distinct from mass and can be constructed so as to be free of all the artifacts of discretization that characterized previous lattice gas models. The model possesses true Galilean invariance with allowance for compressibility, proper scalar pressure exhibiting equipartition, and an energy transport equation. Its thermal behavior is explored and a practical method to achieve Galilean invariance is presented.

Molvig, K., P. Donis, J. Myczkowski and G. Vichniac, Continuum fluid dynamics from a lattice gas, preprint, MIT (1988).

A lattice gas automata theory is presented which removes the discreteness artifacts that plague the current models. The automata are endowed with a genuine energy that is distinct from mass. The macroscopic equations possess Galilean invariance with allowance for compressibility and temperature variations, proper scalar pressure exhibiting equipartition (without the Mach number dependent anomaly), and an energy transport equation, all fully three dimensional.

Molvig, K., P. Donis, J. Myczkowski and G. Vichniac, Removing the discreteness artifacts in 3D lattice-gas fluids, in: Discrete Kinetic Theory, Lattice Gas Dynamics and Foundations of Hydrodynamics, ed. R. Monaco (World Scientific, Singapore, 1989) pp. 409–418.

A lattice gas automaton theory is presented which yields Galilean invariant macroscopic equations with allowance for compressibility and temperature variations, proper scalar pressure exhibiting equipartition (without the Mach number dependent anomaly), and an energy transport equation, all fully three-dimensional.

Montgomery, D. and G.D. Doolen, Magnetohydrodynamic cellular automata, Phys. Lett. A 120 (1987) 229–231.

A generalization of the hexagonal lattice gas model of Frisch, Hasslacher and Pomeau is shown to lead to two-dimensional magnetohydrodynamics. The method relies on the ideal point-wise conservation law for the vector potential.

Montgomery, D. and G.D. Doolen, Two cellular automata for plasma computations, Complex Systems 1 (1987) 831–838. Reprinted in: Lattice Gas Methods for Partial Differential Equations, ed. G.D. Doolen (Addison–Wesley, Reading, MA, 1989) pp. 461–470.

Plasma applications of computational techniques based on cellular automata are inhibited by the long-range nature of electromagnetic forces. One of the most promising features of cellular automata methods has been the parallelism that becomes possible because of the local nature of the interactions, leading (for example) to the absence of Poisson equations to be solved in fluid simulations. Because it is in the nature of a plasma that volume forces originate with distant charges and currents, finding plasma cellular automata becomes largely a search for tricks to circumvent this nonlocality of the forces. We describe automata for two situations where this appears possible: two-dimensional magnetohydrodynamics (2D MHD) and the one-dimensional electrostatic Vlasov–Poisson system. Insufficient computational experience has accumulated for either system to argue that it is a serious alternative to existing methods.

Nadiga, B.T., J.E. Broadwell and B. Sturtevant, Study of multispeed cellular automaton, in: Rarefied Gas Dynamics: Theoretical and Computational Techniques, eds. E.P. Muntz, D.P. Weaver and D.H. Campbell, Progress in Astronautics and Aeronautics, Vol. 118 (AIAA, Washington, DC, 1989) pp. 155–170.

Most cellular automata intended to describe fluid motion simulate single-speed particles moving on square or hexagonal lattices. In the latter case, two-dimensional low-Mach-number flows have been shown by Frisch et al. (Lattice gas automata for the Navier–Stokes equation, Phys. Rev. Lett. 56 (1986) 1505–1508) to obey the Navier–Stokes equations for incompressible flow. These authors also discuss the various difficulties associated with the models, in particular, the restriction to low speeds. Furthermore, it is clear that with only one allowed speed, temperature or energy cannot be specified independently of the velocity. d'Humières et al. (Lattice gas models for 3d hydrodynamics, Europhysics Letters 2 (1986) 291–297) describe what appears to be the simplest multispeed model for flows in both two and three dimensions. The present paper describes the results of an exploratory investigation of heat conduction and shock-wave formation with the two-dimensional model. The irreversible macroscopic behavior of this microscopically reversible system is also examined.

Nasilowski, R., An arbitrary-dimensional cellular-automaton fluid model with simple rules, Proceedings of the Interdisciplinary Seminar on Dissipative Structures in Transport Processes and Combustion Bielefeld 1989 (Springer, Berlin, 1990).

A novel cellular automaton model of fluid dynamics is considered, similar to the lattice gas automata recently introduced by Frisch, Hasslacher and Pomeau, and others. The new model combines in it many desirable technical features that are absent or not satisfactory in the classical models: arbitrary space dimensionality, simple orthogonal lattice structure, as well as deterministic, easily implementable transition rules which involve no more than two cells at a time. Nevertheless, our model, at least theoretically, should be able to simulate nontrivial fluid dynamical behavior, since the nonlinear convection term takes the desired isotropic form in the incompressible limit at a particular mean mass density.

Nickel, G.H., Cellular automaton rules for solving the Milne problem, Phys. Lett. A 133 (1988) 219–224.

The methodology for deriving cellular automaton rules to solve radiation transport problems is developed. Calculations using several of these rules to solve the Milne problem in Cartesian geometry show good agreement with the exact solution.

Noullez, A. and J.P. Boon, Long-time correlations in a 2D lattice gas, Physica D 47 (1991) 212–215; these Proceedings.

We present the results of CA simulations of a moderately dense lattice gas using an extended version of the FHP model for colored automata. By tracking a tagged particle, we construct its velocity autocorrelation function and we show that the corresponding power spectrum exhibits a low-frequency contribution characteristic of the long-time power law behavior ($\approx t^{-1}$, in 2D).

Oono, Y. and S. Puri, Computationally efficient modeling of ordering of quenched phases, Phys. Rev. Lett. 58 (1987) 836–839.

Computationally efficient discrete space–time models of phase-ordering dynamics of thermodynamically unstable systems (e.g., spinodal decomposition) are proposed. Two-dimensional lattice (100×100) simulations were preformed to obtain scaled form factors.

Oono, Y. and S. Puri, Study of phase-separation dynamics by use of cell dynamical systems. I. Modeling, Phys. Rev. A 38 (1987) 434–453.

We present a computationally efficient scheme of modeling the phase-ordering dynamics of thermodynamically unstable phases. The scheme utilizes space–time discrete dynamical systems, viz., cell dynamical systems (CDS). Our proposal is tantamount to proposing new *Ansätze* for the kinetic-level description of the dynamics. Our present exposition consists of two parts: part I (this paper) deals mainly with methodology and part II [S. Puri and Y. Oono, Phys. Rev. Lett. 38 1542 (1988)] gives detailed demonstrations. In this paper we provide a detailed exposition of model construction, structural stability of constructed models (i.e., insensitivity to details), stability of the scheme, etc. We also consider the relationship between the CDS modeling and the conventional description in terms of partial differential equations. This leads to a new discretization scheme for semi-linear parabolic equations and suggests the necessity of a branch of applied mathematics which could be called qualitative numerical analysis,

Oono, Y. and C. Yeung, A cell dynamical system model of chemical turbulence, J. Stat. Phys. 48 (1987) 593–644.

A cellular-automaton-like caricature of chemical turbulence on an infinite one-dimensional lattice is studied. The model exhibits apparently turbulent space–time patterns. To make this statement precise, the following problems or point are discussed: (1) The infinite-system-size limit of such cell-dynamical systems and its observability is defined. (2) It is proved that the invariant state in the large-system-size limit of the turbulent phase exhibits spatial patterns governed by a Gibbs random field. (3) Potential characteristics of turbulent space–time patterns are critically surveyed and a working definition of (weak) turbulence is proposed. (4) It is proved that the invariant state of the turbulent phase is actually (weak) turbulent. Furthermore, we conjecture that the turbulent phase of our model is an example of a K system that is not Bernoulli.

Orszag, S.A. and V. Yakhot, Reynolds number scaling of cellular-automaton hydrodynamics, Phys. Rev. Lett. 56 (1986) 1691–1693. Reprinted in: Lattice Gas Methods for Partial Differential Equations, ed. G.D. Doolen (Addison–Wesley, Reading, MA, 1989) pp. 269–274.

We argue that the computational requirements for presently envisaged cellular-automaton simulations of continuum fluid dynamics are much more severe than for solution of the continuum equations.

Papatzacos, P., Cellular automaton model for fluid flow in porous media, Complex Systems 3 (1989) 383–405.

A cellular automaton model for the simulation of fluid flow in porous media is presented. A lattice and a set of rules are introduced, such that the flow equations in the continuum limit are formally the same as the equations for one-phase liquid flow in porous media. The model is valid in two as well as three dimensions. Numerical calculations of some simple problems are presented and compared with known analytical results. Agreement is within estimated errors.

Platkowski, T., Shock wave profiles for discrete velocity models of mixtures, in: Discrete Kinetic Theory, Lattice Gas Dynamics and Foundations of Hydrodynamics, ed. R. Monaco (World Scientific, Singapore, 1989) pp. 248–257.

Exact shock wave profiles are studied for a discrete velocity model of gas mixtures. Conditions on the existence of the shock wave solutions and the entropy overshoots in the interior of the shock are investigated for the considered model. Preliminary results for other possible models of mixtures with exact shock solutions are reported. Corresponding profiles for discrete velocity models of single gases are also briefly discussed.

Protopopescu, V., Global existence for symmetric discrete velocity models, in: Discrete Kinetic Theory, Lattice Gas Dynamics and Foundations of Hydrodynamics, ed. R. Monaco (World Scientific, Singapore, 1989) pp. 258–267.

For symmetric discrete velocity models, some generalizations of previous global existence proofs are given to include bounded geometries, continuous models, and stationary states with no momentum conservation. Additional results and simpler proofs are obtained for completely symmetric models.

Puri, S. and Y. Oono, Study of phase-separation dynamics by use of cell dynamical systems. II. Two-dimensional demonstrations, Phys. Rev. A 38 (1988) 1542–1565.

We present detailed results on the form factors of two-dimensional systems undergoing phase-ordering processes, using both deterministic and stochastic cell dynamical systems. We show the robustness of the asymptotic form factors against quench depth, noise amplitude, etc. The effect of noise is essentially to delay the number of steps needed to reach the asymptotic behavior. In the case with a nonconserved order parameter, we demonstrate that the form factor obtained by T. Ohta, D. Jasnow and K. Kawasaki [Phys. Rev. Lett. **49**, 1223 (1982)] is asymptotically very accurate. We also present preliminary results for off-critical quenches.

Qian, Y.H., D. d'Humières and P. Lallemand, A short note on Green-Kubo formula of viscosity in one-dimensional lattice gas models of single particle mass, preprint, ENS (1989).

Qian, Y.H., D. d'Humières and P. Lallemand, Diffusion simulation with a deterministic one-dimensional lattice-gas model, preprint, ENS (1989).

A one-dimensional lattice-gas model is proposed and used to simulate diffusion processes in one dimension. Explicit forms of transport coefficients are given as a function of density and kinetic energy within the Boltzmann approximation. Without definitions of temperature and pressure, a steady non-trivial solution is given analytically in the non-convective case when the kinetic energy is kept constant.

Rapaport, D.C., Microscale hydrodynamics: discrete-particle simulation of evolving flow patterns, Phys. Rev. A 36 (1987) 3288–3299.

The technique of molecular-dynamics simulation – in which the equations of motion of a system of interacting particles are solved numerically to yield the temporal evolution of the system – is used in a study of the flow of a two-dimensional fluid past a circular obstacle. The flow is observed to develop with time, passing through a series of well-defined patterns that bear a striking similarity with flow patterns observed experimentally in liquid and gas flow; the patterns include stationary eddies, periodic shedding of vortices, and a vortex street characterized by a Strouhal number close to the experimental value. Very large systems – by current molecular-dynamics standards – need to be used in order to accommodate the obstacle and the region occupied by the structured wake, and the present work includes the largest such simulations carried out to date. Though more extensive work is called for, the results suggest that continuum hydrodynamics is applicable down to much shorter length scales than hitherto believed, and that the molecular-dynamics approach can thus be used to study certain kinds of hydrodynamic instabilities.

Rapaport, D.C. and E. Clementi, Eddy formation in obstructed fluid flow: a molecular-dynamics study, Phys. Rev. Lett. 57 (1986) 695–698.

Two-dimensional fluid flow past a circular obstacle has been simulated at the microscopic level by means of a molecular-dynamics approach. At sufficiently large Reynolds number the flow field is observed to exhibit characteristics common to real fluids, namely the appearance of eddies, periodic eddy separation, and an oscillatory wake. Very large systems – typically 160 000 particles – are required in order to provide adequate space for these flow patterns to develop.

Rem, P.C. and J.A. Somers, Cellular automata on a transputer network, in: Discrete Kinetic Theory, Lattice Gas Dynamics and Foundations of Hydrodynamics, ed. R. Monaco (World Scientific, Singapore, 1989) pp. 268–275.

We present a variant of the three-dimensional FCHC lattice gas model by d'Humières et al., including boundary conditions that are consistent with the Navier–Stokes equation. The model has been designed to run on a transputer network and has been optimized for storage requirements. We also present a two-phase extension of the model, based on local interaction of particles.

Riccardi, G., C. Bauer and H. Lim, Boundary and obstacle processing in a vectorized model of lattice gas hydrodynamics, Physica D 47 (1991) 281–295; these Proceedings.

This paper describes a vectorized supercomputer implementation of a cellular automaton model for lattice gas hydrodynamics. A detailed description of the algorithm is given along with a careful complexity analysis and performance evaluation of it. Particular attention is paid to boundary and obstacle processing. Two applications of the program are described: an acoustic wave model and an obstructed flow through a pipe. The results of executing these models are displayed with density maps and flow diagrams.

Rieger, M. and P. Vogel, New lattice gas method for semiconductor transport simulations, Proceedings of the 6th International Conference on Hot Carriers, Phoenix, Solid State Electronics 32 (1989) 1399–1403.

We present a new lattice gas simulation technique for semiclassical semiconductor transport, which is comparable in accuracy to Monte Carlo simulations, but is capable of dealing with arbitrarily complex spatial inhomogeneities, nonlinear dynamical effects and carrier instabilities with only a minimal additional cost in computer time, and runs orders of magnitude faster on parallel hardware.

Rivet, J.-P., M. Hénon, U. Frisch and D. d'Humières, Simulating fully three-dimensional external flow by lattice gas methods, in: Discrete Kinetic Theory, Lattice Gas Dynamics and Foundations of Hydrodynamics, ed. R. Monaco (World Scientific, Singapore, 1989) pp. 276–285.

We have built a three-dimensional 24-bit lattice gas algorithm with improved collision rules. Collisions are defined by a look-up table with 2^{24} entries, fine-tuned to maximize the Reynolds number. External flow past a circular plate at Reynolds number around 190 has been simulated. The flow is found to evolve from axisymmetric to fully 3D. Such simulations take a few minutes of CRAY-2 per circulation time (based on plate diameter and upstream velocity).

Rivet, J.-P., Hydrodynamique par la méthode des gas sur réseaux, Ph.D. Thesis, Université de Nice (1988).

Rivet, J.-P., M. Hénon, U. Frisch and D. d'Humières, Simulating fully three-dimensional external flow by lattice gas methods, Europhys. Lett. 7 (1988) 231–236.

We have built a three-dimensional 24-bit lattice gas algorithm with improved collision rules. Collisions are defined by a look-up table with 2^{24} entries, fine-tuned to maximize the Reynolds number. External flow past a circular plate at Reynolds number around 190 has been simulated. The flow is found to evolve from axi-symmetric to fully 3D. Such simulations take a few minutes of CRAY-2 per circulation time (based on plate diameter and upstream velocity).

Rivet, J.-P., Green–Kubo formalism for lattice gas hydrodynamics and Monte-Carlo evaluation of shear viscosities, Complex Systems 1 (1987) 839–851. Reprinted in: Lattice Gas Methods for Partial Differential Equations, ed. G.D. Doolen (Addison–Wesley, Reading, MA, 1989) pp. 399–414.

A Green–Kubo formula, relating the shear viscosity to discrete time correlation functions, is derived via a Liouville equation formalism for a class of non-deterministic lattice gas models. This allows a Monte-Carlo calculation of the viscosity. Preliminary results are presented for the Frisch–Hasslacher–Pomeau two-dimensional lattice gas model.

Rivet, J.-P., Simulation d'écoulements tridimensionnels par la méthode des gaz sur réseau premiers résultats, Compt. Rend. Acad. Sci. Paris II 305 (1987) 751–756.

The pseudo-4-D lattice gas model of d'Humières, Lallemand and Frisch with the collision rules of Hénon is implemented. Viscosity measurements are presented and the Taylor–Green vortex is simulated at a Reynolds number close to one hundred.

Rivet, J.-P. and U. Frisch, Lattice gas automata in the Boltzmann approximation, Compt. Rend. Acad. Sci. Paris II 302 (1986) 267–272. Reprinted in translation in: Lattice Gas Methods for Partial Differential Equations, ed. G.D. Doolen (Addison–Wesley, Reading, MA, 1989) pp. 443–450.

Shear and bulk viscosities are determined for two lattice gas automata simulating the two-dimensional Navier–Stokes equations.

Rothman, D.H., Macroscopic laws for immiscible two-phase flow in porous media: results from numerical experiments, J. Geophys. Res. (1990), in press.

Flow through porous media may be described at either of two length scales. At the scale of a single pore, fluids flow according to the Navier–Stokes equations and the appropriate boundary conditions. At a larger, volume-averaged scale, the flow is usually thought to obey a linear Darcy law relating flow rates to pressure gradients and body forces via phenomenological permeability coefficients. Aside from the value of the permeability coefficient, the slow flow of a single fluid in a porous medium is well-understood within this framework. The situation is considerably different, however, for the simultaneous flow of two or more fluids not only are the phenomenological coefficients poorly understood, but the form of the macroscopic laws themselves is subject to question. I describe a numerical study of immiscible two-phase flow in an idealized two-dimensional porous medium constructed at the pore scale. Results show that the macroscopic flow is a nonlinear function of the applied forces for sufficiently low levels of forcing, but linear thereafter. The crossover, which is not predicted by conventional models, occurs when viscous forces begin to dominate capillary forces; i.e., at a sufficiently high capillary number. In the linear regime, the flow may be described by the linear phenomenological law $u_i = \sum_j L_{ij} f_j$, where the flow rate u_i of the ith fluid is related to the force f_j applied to the jth fluid by the matrix of phenomenological coefficients L_{ij}, which depends on the relative concentrations of the two fluids. The diagonal terms are proportional to quantities commonly referred to as relative permeabilities. The cross terms represent viscous coupling between the two fluids; they are conventionally assumed to be negligible and require special experimental procedures to observe in a laboratory. In contrast, in this numerical study the cross terms are straightforward to measure and are found to be of significant size. The cross terms are additionally observed to be approximately equal, which is the behavior predicted by Onsager's reciprocity theorem. However, persistent transient effects can render the reciprocity unobservable. The numerical study is performed with a discrete numerical model of the molecular dynamics of immiscible mixtures called the immiscible lattice gas. The immiscible lattice gas models both the Navier–Stokes equations and surface tension. Numerical tests presented here additionally provide quantitative validation of the method's ability to simulate wetting phenomena and the effects of capillary pressure. Whereas the numerical study of the linear phenomenological laws utilizes a highly simplified porous medium with one pore and two throats, numerical examples of wetting and nonwetting invasion experiments in a geometrically complex 2-D porous medium are also provided.

Rothman, D.H., Immiscible lattice gases: New results, new models, in: Cellular Automata and Modeling of Complex Physical Systems, eds. P. Manneville, N. Boccara, G.Y. Vichniac and R. Bidaux (Springer, Berlin, 1989) pp. 232–238.

Recent results obtained with immiscible lattice-gas models are reviewed. Particular attention is devoted to a description of a Galilean-invariant model and a study of spinodal decomposition. A variation of the basic model that simulates a fluid with a negative viscosity is also reviewed.

Rothman, D.H., Lattice gas automata for immiscible two-phase flow, in: Discrete Kinetic Theory, Lattice Gas Dynamics and Foundations of Hydrodynamics, ed. R. Monaco (World Scientific, Singapore, 1989) pp. 286–299.

I demonstrate preliminary applications of the immiscible lattice gas model to problems in two-phase flow. Two problems are emphasized: flow in porous media and the Rayleigh–Taylor instability. Numerical experiments show the existence of a critical capillary pressure in simulations of two-phase flow in a channel. Predictions of linear stability theory are observed in a simulation of the Rayleigh–Taylor instability.

Rothman, D.H., Negative-viscosity lattice gases, J. Stat. Phys. 56 (1989) 1119–1127.

A new irreversible collision rule is introduced for lattice-gas automata. The rule maximizes the flux of momentum in the direction of the local momentum gradient, yielding a negative shear viscosity. Numerical results in 2-D show that the negative viscosity leads to the spontaneous ordering of the velocity field, with vorticity resolvable down to one lattice-link length. The new rule may be used in conjunction with previously proposed collision rules to yield a positive shear viscosity lower than the previous rules provide. In particular, Poiseuille flow tests demonstrate a decrease in viscosity by more than a factor of 2.

Rothman, D.H., Cellular-automaton fluids: a model for flow in porous media, Geophysics 53 (1988) 509–518.

Numerical models of fluid flow through porous media can be developed from either microscopic or macroscopic properties. The large-scale viewpoint is perhaps the most prevalent. Darcy's law relates the chief macroscopic parameters of interest – flow rate, permeability, viscosity, and pressure gradient – and may be invoked to solve for any of these parameters when the others are known. In practical situations, however, this solution may not be possible. Attention is then typically focused on the estimation of permeability, and numerous numerical methods based on knowledge of the microscopic pore–space geometry have been proposed. Because the intrinsic inhomogeneity of porous media makes the application of proper boundary conditions difficult, microscopic flow calculations have typically been achieved with idealized arrays of geometrically simple pores, throats, and cracks. I propose here an attractive alternative which can freely and accurately model fluid flow in grossly irregular geometries. This new method solves the Navier–Stokes equations numerically using the cellular-automaton fluid models introduced by Frisch, Hasslacher and Pomeau. The cellular-automaton fluid is extraordinarily simple – particles of unit mass traveling with unit velocity reside on a triangular lattice and obey elementary collision rules – but is capable of modeling much of the rich complexity of real fluid flow. Cellular-automaton fluids are applicable to the study of porous media. In particular, numerical methods can be used to apply the appropriate boundary conditions, create a pressure gradient, and measure the permeability. Scale of the cellular-automaton lattice is an important issue: the linear dimension of a void region must be approximately twice the mean free path of a lattice gas particle. Finally, an example of flow in a 2-D porous medium demonstrates not only the numerical solution of the Navier–Stokes equations in a highly irregular geometry, but also numerical estimation of permeability and a verification of Darcy's law.

Rothman, D.H. and J.M. Keller, Immiscible cellular-automaton fluids, J. Stat. Phys. 52 (1988) 1119–1127. Reprinted in: Lattice Gas Methods for Partial Differential Equations, ed. G.D. Doolen (Addison–Wesley, Reading, MA, 1989) pp. 275–282.

We introduce a new deterministic collision rule for lattice-gas (cellular-automaton) hydrodynamics that yields immiscible two-phase flow. The rule is based on a minimization principle and the conservation of mass, momentum, and particle type. A numerical example demonstrates the spontaneous separation of two phases in two dimensions. Numerical studies show that the surface tension coefficient obeys Laplace's formula.

Rothman, D.H. and S. Zaleski, Spinodal decomposition in a lattice-gas automaton, J. Phys. (Paris) 50 (1989) 2161–2174.

We analyze a lattice-gas model of spinodal decomposition recently introduced by Rothman and Keller. This immiscible lattice gas (ILG) is of special interest because it not only conserves momentum and particle number, but is also capable of hydrodynamic simulations of interfaces. Here we perform a jointly theoretical and empirical study of the statistical behavior of the ILG. We first obtain a theoretical prediction of the diffusion coefficient by solving a discrete Boltzmann equation. We then confirm, by numerical simulations of diffusion and domain growth kinetics, that the spinodal curve of the ILG is approximately given by the line in the space of density and concentration where the theoretical diffusion coefficient vanishes. The ILG is also an interesting example of a system with irreversible microdynamics.

Rothman, D.H., Modeling seismic *P*-waves with cellular automata, Geophys. Res. Lett. 14 (1987) 17–20.

Cellular automata are arrays of discrete variables that follow local interaction rules and are capable of modeling many physical systems. A successful recent application has been in fluid dynamics, in which the Navier–Stokes equations are solved by creating a model in which space, time, and the velocity of particles are all discrete. Acoustic waves can be obtained from these fluid models when perturbations of the idealized fluid are small. Because seismic *P*-waves can be approximated by the acoustic wave equation, cellular automata can be adapted for seismic wave computations. This study shows how to model *P*-waves in two dimensions by using a modified form of the cellular-automaton rules for fluids. Propagation, reflection, and the computation of synthetic seismograms are demonstrated. Because no arithmetic calculations are needed and each lattice site can be updated simultaneously, this method is well suited for implementation of massively parallel computers. Among the many potential advantages are unconditional stability, no round-off errors, and the possibility for devising novel approaches for modeling waves in inhomogeneous media.

Ruijgrok, Th.W. and E.G.D. Cohen, Deterministic lattice gas models, Phys. Lett. A 133 (1988) 415

Two new deterministic lattice versions of the wind-tree model, with fixed or moving mirrors placed randomly on a square lattice are considered. In both cases, the diffusion coefficient appears to be given by the Boltzmann expression for all densities of the mirrors.

Rujàn, P., Cellular automata and statistical mechanical models, J. Stat. Phys. 49 (1987) 139–142.

We elaborate on the analogy between the transfer matrix of usual lattice models and the master equation describing the time development of cellular automata. Transient and stationary properties of probabilistic automata are linked to surface and bulk properties, respectively, of restricted statistical mechanical systems. It is demonstrated that methods of statistical physics can be successfully used to describe the dynamic and the stationary behavior of such automata. Some exact results are derived, including duality transformations, exact mappings, disorder and linear solutions. Many examples are worked out in detail to demonstrate how to use statistical physics in order to construct cellular automata with desired properties. This approach is considered to be a first step toward the design of fully parallel, probabilistic systems whose computational abilities rely on the cooperative behavior of their components.

Sakaguchi, H., Phase transitions in coupled Bernoulli maps, Prog. Theor. Phys. Progr. Lett. 80 (1988) 7–12.

A coupled map system is proposed which is deterministic and shows phase transitions. Our system is composed of a large number of the Bernoulli maps. Chaotic behaviors of the individual maps make the coupled system ergodic. By using two different models, equilibrium and nonequilibrium phase transitions are studied analytically and numerically.

Salem, J. and S. Wolfram, Thermodynamics and hydrodynamics with cellular automata, in: Theory and Applications of Cellular Automata, ed. S. Wolfram (World Scientific, Singapore, 1986) pp. 362–366.

Searby, G., V. Zahnlé and B. Denet, Lattice-gas mixtures and reactive flows, in: Discrete Kinetic Theory, Lattice Gas Dynamics and Foundations of Hydrodynamics, ed. R. Monaco (World Scientific, Singapore, 1989) pp. 300–314.

It is shown why lattice gas algorithms for lattice gas mixtures must be invariant to Galilean transformations. We describe an extension of the FHP 2-D lattice gas that is 'pseudo' Galilean invariant and which has up to three chemical species. We show an example of the effect of the dynamics of coherent structures on the global reaction rate of a reactive mixing layer.

Sente, B., M. Dumont and P. Dufour, Simulation of surface reactions in heterogeneous catalysis: Sequential and parallel aspects, in: Cellular Automata and Modeling of Complex Physical Systems, eds. P. Manneville, N. Boccara, G.Y. Vichniac and R. Bidaux (Springer, Berlin, 1989) pp. 274–279.

In order to investigate the steady state properties of the bimolecular surface reaction $A + \frac{1}{2}B_2 \Longrightarrow AB$ on catalytic surfaces, we derive kinetic equations of the mean field approximation (MFA) type and we compare resulting kinetic phase transition points, bistability and hysteresis characteristics with those obtained by Monte Carlo simulations performed on a sequential computer or on a parallel one (which requires the construction of a probabilistic CA machine).

Sero-Guillaume, O.E. and D. Bernardin, A lattice gases model for heat transfer and chemical reaction, Eur. J. Mech. 9 (1989) 177–196.

A model of a multi-species, multi-speed lattice gas with energy levels is considered. Each particle can be composed of atoms, and the collisional invariants are the number of atoms of each type, the total momentum and the total energy. Using the Gibbs universal distribution, the thermodynamic relations at equilibrium are studied. The evolution equations of the macroscopic quantities associated with invariants are derived by the Chapman–Enskog method. The ability of this kind of model to simulate thermal processes and chemical reactions, at low hydrodynamic velocities is shown.

Shimomura, T., G.D. Doolen, B. Hasslacher and C. Fu, Calculations using lattice gas techniques, Los Alamos Science Special Issue (1987) 201–210. Reprinted in: Lattice Gas Methods for Partial Differential Equations, ed. G.D. Doolen (Addison–Wesley, Reading, MA, 1989) pp. 3–10.

An overview of advantages and disadvantages of lattice gas methods is presented along with color snapshots of two- and three-dimensional flow fields.

Somers, J.A. and P.C. Rem, Analysis of surface tension in two-phase lattice gases, Physica D 47 (1991) 39–46; these Proceedings.

One of the promising fields of application for lattice gas methods is two-phase flow simulation. An important issue in the design of two-phase lattice gas models is managing complexity. Continuing the basic principles of single-phase lattice gases, the two-phase collision operator should be kept as simple as possible, for the sake of both easy implementation and theoretical validation. At the same time it is required that parameters such as surface tension, diffusion and viscosity obtain physically relevant values. In this paper we present a two-dimensional two-phase model, based on a strictly local 16-bit collision operator, which incorporates a respectable surface tension relatively independent of the density.

Somers, J.A. and P.C. Rem, The construction of efficient collision tables for fluid flow computations with cellular automata, in: Cellular Automata and Modeling of Complex Physical Systems, eds. P. Manneville, N. Boccara, G.Y. Vichniac and R. Bidaux (Springer, Berlin, 1989) pp. 161–177.

A well-known problem with the implementation of the three-dimensional FCHC lattice gas is the size of the collision table. Especially when stationary particles are added in order to decrease the kinematic viscosity, a full collision table would occupy more than 100 megabytes of memory. In this paper, we discuss an optimization strategy for the FCHC lattice gas with three stationary particles, which produces a collision table that fits into 64 kilobytes of memory.

Spiga, G., G. Dukek and V.C. Boffi, Scattering kernel formulation of the discrete velocity model of the extended Boltzmann system, in: Discrete Kinetic Theory, Lattice Gas Dynamics and Foundations of Hydrodynamics, ed. R. Monaco (World Scientific, Singapore, 1989) pp. 315–328.

In the frame of the scattering kernel formulation, a three-dimensional six-velocity model is proposed for modeling the extended kinetic theory of gas mixtures. The effects of removals and of a background medium on a gas of test particles are considered in some detail. Solution techniques for the determination of exact analytical solutions are devised and discussed for a particular case.

Succi, S., E. Foti and M. Gramignani, Flow through geometrically irregular media with the lattice gas automata, MECCANICA (1990), to be published.

We present a numerical verification of Darcy's law and the determination of of the permeability of a three-dimensional porous medium starting by the knowledge of its microstructure. The flow is here simulated by using a particular class of cellular automata within the Boltzmann approximation.

Succi, S., A. Cancelliere, C. Chang, E. Foti, M. Gramignani and D. Rothman, A direct computation of the permeability of three-dimensional porous media, International Conference on Numerical Methods in Groundwater Resources, Venice, 1990, to be published in Comp. Mech. Institute.

We present a series of high-resolution numerical simulations of flows in porous media with the Lattice Boltzmann method. Quantitative evaluations of the medium's permeability as a function of the porosity are presented.

Succi, S., R. Benzi and F. Higuera, The lattice Boltzmann equation: A new tool for computational fluid dynamics, Physica D 47 (1991) 219–230; these Proceedings.

We present a series of applications which demonstrate that the lattice Boltzmann equation is an adequate computational tool to address problems spanning a wide spectrum of fluid regimes, ranging from laminar to fully turbulent flows in two and three dimensions.

Succi, S., E. Foti and F. Higuera, Three-dimensional flows in complex geometries with the lattice Boltzmann method, Europhys. Lett. 10 (1989) 433–438.

Succi, S., R. Benzi, E. Foti, F. Higuera and F. Szelényi, Lattice Boltzmann computing on the IBM 3090 vector multiprocessor, in: Cellular Automata and Modeling of Complex Physical Systems, eds. P. Manneville, N. Boccara, G.Y. Vichniac and R. Bidaux (Springer, Berlin, 1989) pp. 178–185.

We discuss the basic theory of the Lattice Boltzmann Equation (LBE) and present some applications to two- and three-dimensional hydrodynamics. In particular, we report on the study of the flow in a plane channel containing a periodic array of identical plates promoting flow instabilities at Reynolds numbers well below the critical value for the classical Poiseuille flow. With concern to three-dimensional applications, we present some preliminary studies aimed to assess the parameter range in which the LBE can be used to investigate low-Reynolds flows in porous media. The ultimate goal of this project is to compute the permeability of a three-dimensional medium taking into account its microgeometrical configuration. Performance data pertaining to the implementation on a six-headed IBM 3090 vector multiprocessor are also offered.

Succi, S., F. Higuera and F. Szelényi, Three-dimensional flows with the lattice Boltzmann equation on the IBM 3090/VF, ACM Proceedings of the International Conference on Supercomputing, Kreta (1989) p. 128

We illustrate the basic features of the Lattice Boltzmann Equation (LBE), a new finite-difference scheme that arises from the microdynamics of the Frisch–Hasslacher–Pomeau cellular automaton once, instead of tracking the individual history of each particle, one only solves for the mean values of the particle populations living in the lattice. Details on the actual coding of the LBE scheme are presented and discussed along with performance data pertaining to the vector and parallel implementation on the IBM 3090/600 vector multiprocessor. Finally, two applications in the field of two- and three-dimensional hydrodynamics are briefly illustrated.

Succi, S., R. Benzi and F. Higuera, Lattice gas and Boltzmann simulations of homogeneous and inhomogeneous hydrodynamics, in: Discrete Kinetic Theory, Lattice Gas Dynamics and Foundations of Hydrodynamics, ed. R. Monaco (World Scientific, Singapore, 1989) pp. 329–342.

We describe the application of lattice gas techniques to two distinct problems arising in the context of fluctuating and non-fluctuating fluid dynamics respectively. In the first case we study the bifurcations of the Frisch–Hasslacher–Pomeau automaton under external forcing and in the second we present a successful application of the lattice Boltzmann equation to the study of a flow past a circular obstacle.

Succi, S., D. d'Humières and F. Szelényi, Lattice gas hydrodynamics on IBM 3090/VF, IBM J. Res. Develop. 33 (1989) 136–148.

Succi, S., Triangular versus square lattice gas automata for the analysis of two-dimensional vortex fields, J. Phys. A 21 (1988) L43–L49.

The consequences of the lack of isotropy of the momentum flux tensor of the Hardy–Pomeau–De Pazzis (HPP) fluid are discussed. It is shown that this lack of isotropy is tantamount to introducing a force which is incompatible with a correct evolution of two-dimensional vortex configurations. In addition, a qualitative discussion is presented on the physical reasons why this problem can be cured by moving to the six-link lattice introduced by Frisch, Hasslacher and Pomeau (FHP).

Succi, S., R. Benzi and P. Santangelo, An investigation of fractal dimensions in two-dimensional lattice gas turbulence, J. Phys. A 21 (1988) 1771–776.

We investigate the dissipation mechanisms acting in a two-dimensional lattice gas automaton by inspecting the structure functions of the turbulent velocity field associated with the Boolean configuration of the automaton. In particular we investigate whether the Boolean noise produced by the automaton can promote fractal structures within the flow. We show that this is not the case and the presence of the noise only results in a non-analyticity of the flow field which can be progressively eliminated upon averaging the Boolean field on coarser and coarser grids. As a result, we find that the non-fractal nature of homogeneous two-dimensional turbulence is not affected by the presence of the microscopic noise.

Succi, S., P. Santangelo and R. Benzi, High-resolution lattice-gas simulation of two-dimensional turbulence, Phys. Rev. Lett. 60 (1988) 2738–2741.

The mechanisms of two-dimensional turbulence are investigated by means of a very high-resolution lattice-gas simulation. The results from this simulation are quantitatively compared with the direct integration of the Navier–Stokes equation. The dissipation of the flow simulated by the lattice gas is estimated by a simple scaling argument on the microscopic noisy field of the automaton.

Succi, S., Cellular automata modeling on IBM 3090/VF, Comput. Phys. Commun. 47 (1988) 173–180.

The basic features concerning the implementation of the Hardy–Pomeau–De Pazzis (HPP) cellular automaton on a general purpose computer are illustrated, with explicit reference to the IBM 3090 vector processor. Special attention is paid to the choice of a proper data organization allowing it to take full advantage of the high processing rates offered by a high speed vector processor like the IBM 3090.

Takesue, S., Fourier's law and the Green–Kubo formula in a cellular-automaton model, Phys. Rev. Lett. 64 (1990) 252–255.

The properties of energy transport are numerically investigated with the use of a one-dimensional cellular-automaton model called $26R$. The validity of Fourier's law and the Green–Kubo formula for thermal conductivity κ is demonstrated for this model in the limit of large systems. Nonlinear correction to Fourier's law and the recovery of left–right symmetry are also discussed.

Takesue, S., Reversible cellular automata and statistical mechanics, Phys. Rev. Lett. 59 (1987) 2499–2502.

Reversible cellular automata are used to investigate the thermodynamic behavior of large systems. Additive conserved quantities are regarded as the energy of these models. By the consideration of a large system as the sum of a subsystem and a heat bath, it is numerically shown that a canonical distribution is realized under certain conditions concerning the conserved quantities.

Tarnowski, D., Les supercalculateurs bientôt démodés?, La Recherche 174 (1986) 272–273.

Toffoli, T. and N. Margolus, Programmable matter: concepts and realization, Physica D 47 (1991) 263–272; these Proceedings.

This paper is a manifesto, a brief tutorial, and a call for experiments on programmable matter machines.

Toffoli, T., Four topics in lattice gases: ergodicity; relativity; information flow; and rule compression for parallel lattice-gas machines, in: Discrete Kinetic Theory, Lattice Gas Dynamics and Foundations of Hydrodynamics, ed. R. Monaco (World Scientific, Singapore, 1989) pp. 343–354.

We briefly present four topics that have recently arisen in the study of lattice gases. (a) Specific ergodicity asks, for an invertible cellular automaton, what fraction of the total information needed to identify an individual state is devoted to specifying the position of this state on its orbit. We give empirical evidence that this question has a definite answer. (b)

We prove that a wide class of lattice gases are Lorentz invariant in the limit as the lattice spacing goes to zero. (c) We present a nontrivial situation where information flows as a locally additive quantity, much as energy and momentum, and thus takes on a strikingly tangible aspect. (d) In the light of recent progress in identifying usable lattice gas rules for 3-D hydrodynamics, we discuss rule compression trade-offs for parallel lattice gas machines.

Toffoli, T., Information transport obeying the continuity equation, IBM J. Res. Develop. 32 (1988) 29–36.

We analyze nontrivial dynamical systems in which information flows as an additive conserved quantity – and thus takes on a strikingly tangible aspect. To arrive at this result, we first give an explicit characterization of equilibria for a family of lattice gases.

Toscani, G., Recent developments on the existence theory for the discrete velocity models, in: Discrete Kinetic Theory, Lattice Gas Dynamics and Foundations of Hydrodynamics, ed. R. Monaco (World Scientific, Singapore, 1989) pp. 355–370.

Some recent results on the existence and uniqueness of the solution to the initial value problem for the discrete velocity models of the Boltzmann equation are reviewed. A comparison between the one-dimensional and three-dimensional in space state-of-the-art is discussed.

Van der Hoef, M.A. and D. Frenkel, Long-time tails of the velocity autocorrelation function in 2D and 3D lattice gas cellular automata: a test of mode-coupling theory, Phys. Rev. A (1990), to be published.

Van der Hoef, M.A. and D. Frenkel, Tagged particle diffusion in 3D lattice gas cellular automata, Physica D 47 (1991) 191–197; these Proceedings.

We report simulations of tagged particle diffusion in three-dimensional lattice gas cellular automata (LGCA). In particular, we looked at the decay of the velocity autocorrelation function (VACF), using a new technique that is about a million times more efficient than the conventional techniques. For longer times the simulations clearly show the algebraic $t^{-3/2}$ tail of the VACF. We compare the observed long-time tail with the predictions of mode-coupling theory. In three dimensions, the amplitude of this tail is found to agree within the (small) statistical error with these predictions.

Van Velzen, G.A. and M.H. Ernst, Breakdown of the approximation for a lattice Lorentz gas, in: Discrete Kinetic Theory, Lattice Gas Dynamics and Foundations of Hydrodynamics, ed. R. Monaco (World Scientific, Singapore, 1989) pp. 371–383.

We study lattice gas models with collision rules that allow or forbid reflections between unlike particles for a simple mixture (fixed scatterers and independent moving particles). In the reflecting models, the diffusion coefficient, calculated from the Boltzmann equation, differs strongly from the value obtained from an effective medium approximation, even at the lowest densities of scatterers. The reasons for the non-validity of the Boltzmann equation are explained. The effective medium approximation results are confirmed by molecular dynamics simulations.

Vichniac, G., Cellular-automata fluids, in: Instabilities and Nonequilibrium Structures, eds. E. Tirapegui and D. Villasael (Reidel, Dordrecht, 1989).

Vives, E. and A. Planes, Lattice-gas model of orientable molecules: Application to liquid crystals, Phys. Rev. A 38 (1988) 5391–5400.

We have analyzed a two-dimensional lattice-gas model of cylindrical molecules which can exhibit four possible orientations. The Hamiltonian of the model contains positional and orientational energy interaction terms. The ground state of the model has been investigated on the basis of Karl's theorem. Monte Carlo simulation results have confirmed the predicted ground state. The model is able to reproduce, with appropriate values of the Hamiltonian parameters, both, a smectic–nematic-like transition and a nematic–isotropic-like transition. We have also analyzed the phase diagram of the system by mean-field techniques and Monte Carlo simulations. Mean-field calculations agree well qualitatively with Monte Carlo results but overestimate transition temperatures.

Wallace, D.J., Scientific computation on SIMD and MIMD machines, Phil. Trans. R. Soc. London Ser. A 326 (1988) 481–498.

The ICL Distributed Array Processor and Meiko Computing Surface have been successfully applied to a wide range of scientific problems. I give an overview of selected applications from experimental data analysis, molecular dynamics and

Monte Carlo simulation, cellular automata for fluid flow, neural network models, protein sequencing and NMR imaging. I expose the problems and advantages of implementations on the two architectures, and discuss the general conclusions which one can draw from experience so far.

Wayner, P., Modeling chaos, BYTE (May 1988) 253–258.

A parallel CPU architecture can take you where shorter clock ticks, smarter instructions and more on-chip memory can't go.

Wells, J.T., D.R. Janecky and B.J. Travis, A lattice gas automata model for heterogeneous chemical reactions at mineral surfaces and in pore networks, Physica D 47 (1991) 115–123; these Proceedings.

A lattice gas automata (LGA) model is described which couples solute transport with chemical reactions at mineral surfaces and in pore networks. Chemical reactions and transport are integrated into a FHP-I LGA code as a module so that the approach is readily transportable to other codes. Diffusion calculations are compared to finite element Fickian diffusion results and provide an approach to quantifying space–time ratios of the models. Chemical reactions at solid surfaces, including precipitation/dissolution, sorption, and catalytic reaction, can be examined with the model because solute diffusion and mineral surface processes are all treated explicitly. The simplicity and flexibility of the LGA approach provide the ability to study the interrelationship between fluid flow and chemical reactions in porous materials, at a level of complexity that has not previously been computationally possible.

Wolfram, S., Cellular automaton fluids 1: Basic theory, J. Stat. Phys. 45 (1986) 471–526. Reprinted in: Lattice Gas Methods for Partial Differential Equations, ed. G.D. Doolen (Addison–Wesley, Reading, MA, 1989) pp. 19–74.

Continuum equations are derived for the large-scale behavior of a class of cellular automaton models for fluids. The cellular automata are discrete analogues of molecular dynamics, in which particles with discrete velocities populate the links of a fixed array of sites. Kinetic equations for microscopic particle distributions are constructed. Hydrodynamic equations are then derived using the Chapman–Enskog expansion. Slightly modified Navier–Stokes equations are obtained in two and three dimensions with certain lattices. Viscosities and other transport coefficients are calculated using the Boltzmann transport equation approximation. Some corrections to the equations of motion for cellular automaton fluids beyond the Navier–Stokes order are given.

Yakhot, V., B.J. Bayly and S.A. Orszag, Analogy between hyperscale transport and cellular automaton fluid dynamics, Phys. Fluids 29 (1986) 2025–2027. Reprinted in: Lattice Gas Methods for Partial Differential Equations, ed. G.D. Doolen (Addison–Wesley, Reading, MA, 1989) pp. 283–288.

It is argued that the dynamics of a very large-scale (hyperscale) flow superposed on the stationary small-scale flow maintained by a force $\mathbf{f}(\mathbf{x})$ is analogous to the cellular automaton hydrodynamics on a lattice having the same spatial symmetry as the force $\mathbf{f}$.

Zaleski, S., Weakly compressible fluid simulations at high Reynolds numbers, in: Discrete Kinetic Theory, Lattice Gas Dynamics and Foundations of Hydrodynamics, ed. R. Monaco (World Scientific, Singapore, 1989) pp. 384–393.

Lattice gases have recently been introduced by Frisch, Hasslacher and Pomeau as an attractive method for the simulation of incompressible flow. These gases obey on the large scale a set of hydrodynamical equations with an artificial kind of compressibility. In this paper lattice gas simulations are compared with finite difference solutions of the same artificially compressible equations. In both methods the Mach number must be kept small to approximate incompressible flow, as in Chorin's classical scheme. It is found that the lattice gas has an excessively fine grid in certain classes of problems. In typical shear flow situations the lattice gas has a grid size of order $\lambda Mc/\nu$ where λ is the boundary layer size, ν is the viscosity of the gas, M is the Mach number and c is the speed of sound. Thus it is reasonably efficient if the Mach number is not too small and the boundary layer needs to be finely resolved.

Zanetti, G., Counting hydrodynamic modes in lattice gas automata models, Physica D 47 (1991) 30–35; these Proceedings.

A Monte Carlo scheme for the search of extensive conserved quantities in lattice gas automata models is described. It is based on an approximation to the microscopic dynamics and it amounts to estimating the dimension of the eigenspace with eigenvalue 1 of a linear operator related to the lattice gas automata model evolution operator linearized around equilibrium distributions. The applicability of this technique is limited to models with collision rules satisfying semi-detailed balance.

Zanetti, G., Lattice gas automata comparison of simulation and theory, in: Microscopic Simulations of Complex Flow, ed. M. Mareschal (Plenum, New York, 1990) pp. 47–56.

Zanetti, G., The hydrodynamics of lattice gas automata, Phys. Rev. A 40 (1989) 1539–1548.

The linear and nonlinear hydrodynamics of the two-dimensional lattice gas automata (LGA) are discussed. The physics of the LGA is found to be richer than previously expected. Together with sound and shear waves (characteristic of simple fluids) there are three new hydrodynamic modes. The conserved quantities corresponding to the latter arise from a feature of the microscopic definition of the LGA; i.e., the particles of the microscopic gas occupy the sites of a regular lattice and can only hop from one site to its nearest neighbors. The presence of these new conserved quantities has unexpected results on the macroscopic behavior of the fluid. In fact, there is a nonlinear coupling between the two classes of modes, and while the new conserved densities are merely convected by the momentum density, the current of the latter contains terms that depend only on the new modes. Thus the presence of a finite amount of the new conserved densities produces flow patterns that are not solutions of the Navier–Stokes equation. Although only the two-dimensional hexagonal lattice gas is discussed, the arguments described here apply with equal force to currently proposed three-dimensional models.

Zehnlé, V. and G. Searby, Lattice gas experiments on a non-exothermic diffusion flame in a vortex field, J. Phys. (Paris) 50 (1989) 1083–1097.

It is a known shortcoming of lattice gas models for fluid flow that they do not posses Galilean invariance. In the case of a single component incompressible flow, this problem can be compensated by a suitable rescaling of time, viscosity and pressure. However, this procedure cannot be applied to a flow containing more than one species. We describe here an extension of the Frisch–Hasslacher–Pomeau collision model which restores a pseudo Galilean invariance. We then present a simulation of a 2-D reactive shear layer in the configuration of a diffusion flame subjected to the Kelvin–Helmholtz instability.

Physica D 47 (1991) 338
North-Holland

LIST OF CONTRIBUTORS

Physica D 47 (1991) 339
North-Holland

ANALYTIC SUBJECT INDEX